M
ar
the Complex
Surety Case

edited by **Philip L. Bruner**

TORT AND INSURANCE PRACTICE SECTION

AMERICAN BAR ASSOCIATION

Cover design by Jim Colao.

The materials contained herein represent the opinions of the authors and editors and should not be construed to be the action of either the American Bar Association or the Tort and Insurance Practice Section unless adopted pursuant to the bylaws of the Association.

Nothing contained in this book is to be considered as the rendering of legal advice for specific cases, and readers are responsible for obtaining such advice from their own legal counsel. This book and any forms and agreements herein are intended for educational and informational purposes only.

Printed in the United States of America.

02 01 00 99 98 5 4 3 2 1

Kauffman, Caryn A., 1972–
Complex construction surety case manual / Caryn A. Kauffman. Philip L. Bruner.
p. cm.
ISBN 1-57073-567-0
1. Contractors--Bonding--United States--Trial practice.
2. Insurance. Surety and fidelity--United States--Trial practice.
3. Suretyship and guaranty--United States--Trial practice.
4. Complex litigation--United States. I. Bruner, Philip L., 1939–
II. Title.
KF1228.K38 1998
346.73'08684--dc21 98–17202
CIP

Discounts are available for books ordered in bulk. Special consideration is given to state bars, CLE programs, and other bar-related organizations. Inquire at ABA Publishing, Book Publishing, American Bar Association, 750 North Lake Shore Drive, Chicago, Illinois 60611.

CONTENTS

CONTRIBUTORS

ROBERT J. BERENS • Partner in the Phoenix, Arizona law firm of Dyer, Mann & Berens. Mr. Berens is certified as a bankruptcy specialist by the Arizona State Bar and the American Bankruptcy Board of Certification. He received his B.S. degree from the University of Arizona in 1983, his M.B.A. degree from Arizona State University in 1985, and his J.D. degree from the University of San Diego in 1988.

KENNETH C. BORDEN • Member of the Miami, Florida law firm of Kimbrell & Hamann, P.A. Mr. Borden is a graduate of Emory University where he received a B.A. in Accounting in 1989. He received his J.D. degree from the University of Miami School of Law in 1994.

ESPERANZA DIAZ BRISCOE • Member of the Miami, Florida law firm of Kimbrell & Hamann, P.A. Ms. Briscoe received her undergraduate degree from Florida International University in Miami, Florida in 1986 and her J.D. degree from Tulane Law School in New Orleans, Louisiana in 1991.

PHILIP L. BRUNER • Head of the Construction Law Group and a partner in the Minneapolis, Minnesota law firm of Faegre & Benson. Mr. Bruner is a Founding Fellow of the American College of Construction Lawyers, a Fellow of the American Bar Foundation, and a Fellow of the National Contract Management Association. He is a past Chair of the ABA/TIPS Fidelity and Surety Law Committee and a past Chair of the International Construction Division of the American Bar Association Forum Committee on the Construction Industry. Mr. Bruner received his undergraduate degree from Princeton University in 1961, his J.D. degree from the University of Michigan Law School in 1964, and his M.B.A. degree from Syracuse University in 1967.

JAMES F. CROWDER, JR. • Partner in the Miami, Florida law firm of Kimbrell & Hamann, P.A. Mr. Crowder is a past Chair of the ABA/TIPS Fidelity and Surety Law Committee. He recently served as an advisor to the American Law Institute on the Restatement (Third) of Suretyship and Guaranty and is a member of the Long Range Planning Committee of the ABA Tort and Insurance Practice Section. Mr. Crowder received his undergraduate degree from Emory University in Atlanta, Georgia and his law degree from the University of Florida.

JOHN W. DRESTE • Partner in the law firm of Ernstrom & Dreste. Mr. Dreste is a member of the ABA/TIPS Fidelity and Surety Law Committee. He received his B.S. degree from the Rochester Institute of Technology in 1981 and his law degree from State University of New York at Buffalo in 1984.

J. WILLIAM ERNSTROM • A founding partner of the law firm of Ernstrom & Dreste. He is Vice-Chair of the Contract Documents Committee for the Associated General Contractors of America. Mr. Ernstrom also serves as a member of the Board of Directors of the AGC of America and the Board of the National Associate Members Council. For ten years, he served as Chair of the New York State Construction Dispute Resolution Council. Mr. Ernstrom received his undergraduate degree from Hamilton College and his law degree from Cornell Law School.

VINCENT G. FASANO • Senior Vice President and Chief Claims Officer for the Reliance Surety Company. Mr. Fasano is a past Vice-Chair of the ABA/TIPS Fidelity and Surety Law Committee. He has also served as Chairman of the Board of Directors of the National Bond Claims Association. Mr. Fasano received his undergraduate degree from Parsons College and his law degree from Suffolk University Law School.

DONALD G. GAVIN • Partner in the Washington, D.C. and Vienna, Virginia law firm of Wickwire Gavin, P.C. He is a past Chair of the ABA Public Contract Law Section and is a Vice-Chair of the ABA/TIPS Fidelity & Surety Law Committee. He received his B.S. degree from the Wharton School of the University of Pennsylvania and his J.D. degree from the University of Pennsylvania. Mr. Gavin also has a Masters of

Laws degree in Government Procurement Law from the National Law Center of George Washington University.

PAUL GOLDEAN • Associate with the Dallas, Texas firm of Strasburger & Price. Mr. Goldean received his J.D. degree from Southern Methodist University in 1997. He received his B.B.A. degree from the University of Texas at El Paso in 1994.

DEBORAH S. GRIFFIN • Partner in the Boston, Massachusetts law firm of Peabody & Arnold. Ms. Griffin is a Vice-Chair of the ABA/TIPS Fidelity and Surety Law Committee, and a charter member of the American College of Construction Lawyers. She received her undergraduate degree from Bryn Mawr College in 1971 and her J.D. degree from Boston University School of Law in 1974.

GUY W. HARRISON • Member of the Miami, Florida law firm of Kimbrell & Hamann, P.A. Mr. Harrison received his B.A. degree from Duke University and his J.D. from the University of Florida College of Law. Mr. Harrison is a member of the ABA/TIPS Fidelity and Surety Law Committee.

JOHN H. HINDERAKER • Partner in the Minneapolis, Minnesota law firm of Faegre & Benson where he has tried over one hundred jury cases to verdict. Mr. Hinderaker is a member of the ABA/TIPS Fidelity and Surety Law Committee. He received his undergraduate degree from Dartmouth College in 1971 and his law degree from Harvard Law School in 1974.

JULIAN F. HOFFAR, III • Partner in the law firm of Watt, Tieder & Hoffar, L.L.P. Mr. Hoffar is a Fellow of the American College of Construction Lawyers. He received his B.A. degree from Wittenberg University in 1969 and his J.D. degree from George Washington University in 1972.

ALICIA E. KIRSHENBAUM • Associate with the Florida law firm of Carlton Fields in its Miami, Florida office. Ms. Kirshenbaum obtained her undergraduate degree from the University of Arizona in 1993. She received her J.D. degree from the University of Miami Law School in 1997.

JAY M. MANN • Partner in the Phoenix, Arizona law firm of Dyer, Mann & Berens. Mr. Mann is a past Vice-Chair of the ABA/TIPS Fidelity and Surety Law Committee. He received his undergraduate degree from the University of Illinois in 1972 and his J.D. degree from Loyola University School of Law (Chicago) in 1975.

ROBERT A. MCCALL • Partner in the Boston, Massachusetts law firm of Peabody & Arnold. Mr. McCall received his undergraduate degree from Harvard University in 1983 and his J.D. degree from Northeastern University School of Law in 1988.

PATRICK J. O'CONNOR, JR. • Partner in the Minneapolis, Minnesota law firm of Faegre & Benson. He is a Vice-Chair of the ABA/TIPS Fidelity and Surety Law Committee. Mr. O'Connor served as an advisor to the American Law Institute on the Restatement (Third) of Suretyship and Guaranty. He is also a member of Steering Committee 7 (Bonds, Liens, and Insurance) of the ABA Forum Committee on the Construction Industry.

JEFFREY S. OSGOOD • Partner in the Dallas, Texas firm of Strasburger & Price. Mr. Osgood is a member of the ABA/TIPS Fidelity and Surety Law Committee and a member of the ABA Forum on the Construction Industry. Mr. Osgood received his B.A. degree from Colorado College in 1971 and his law degree from the University of Houston in 1974.

JOHN J. PETRO • Partner in the Columbus, Ohio law firm of McNamara and McNamara. Mr. Petro is a founding Fellow of the American College of Construction Lawyers and a past Chair of the ABA/TIPS Fidelity and Surety Law Committee. He is also Subcommittee Chair of the Editorial Subcommittee for *Fidelity and Surety News*, a publication of the Fidelity and Surety Law Committee. Mr. Petro received his B.A. degree from Ohio State University in 1959 and his law degree from the Ohio State University College of Law in 1962.

DOUGLAS M. REIMER • Partner in the Chicago, Illinois office of the McDermott, Will & Emery law firm. He is a past Chair of the ABA/TIPS Fidelity and Surety Law Committee and past Chair of the ABA Forum Committee on the Construction Industry. Mr. Reimer received his B.A. degree in 1955 from Lawrence College and his J.D. degree in 1958 from Northwestern University School of Law.

HUGH E. REYNOLDS, JR. • Partner in the Indianapolis, Indiana law firm of Locke, Reynolds, Boyd & Weisell. He is a past Chair of the Tort and Insurance Practice Section of the American Bar Association and a past Chair of its Fidelity and Surety Law and Appellate Advocacy Committees. He is also a past Chair of the Governing Committee of the ABA Forum on the Construction Industry. Mr. Reynolds is a Founding Fellow of the American College of Construction Lawyers and served as an advisor for the Restatement (Third) of Suretyship and Guaranty. Mr. Reynolds received his B.S. degree from the University of Notre Dame in 1950 and his J.D. degree from the University of Michigan Law School in 1953.

PATRICIA H. THOMPSON • Shareholder in the Florida law firm of Carlton Fields, resident in the firm's Miami office. Ms. Thompson is Chair-elect of the ABA/TIPS Fidelity and Surety Law Committee. She received her B.A. degree from Saint Olaf College in 1973 and her J.D. degree from Vanderbilt University in 1976.

JOYCE TSONGAS • Founder of Tsongas Associates, a trial consulting firm based in Portland, Oregon. Ms. Tsongas was a founding member and the second President of the American Society of Trial Consultants. Ms. Tsongas has done consulting work on numerous large and complex cases, including working extensively with the plaintiffs in the Exxon Valdez Oil Spill Litigation and with Dow Corning in the successful breast implant coverage litigation. Her special interests include witness preparation and case strategy. Ms. Tsongas' academic background is in speech communication and psychology. She obtained her graduate degree in speech communication from Purdue University.

PETER A. "TONY" WARNER • President of Warner Construction Consulting, Inc. Mr. Warner provides consulting services to the construction industry in the areas of scheduling, impact and delay analysis. Mr. Warner has served as a special master and an expert in many litigated disputes. He is also an instructor in the Graduate School Construction Management Program at Catholic University.

ROBERT G. WATT • Managing Partner of the law firm of Watt, Tieder & Hoffar, L.L.P. Mr. Watt is a member of the ABA/TIPS Fidelity and Surety Law Committee. He received his B.A. degree from Grinnell College in 1967 and his J.D. degree from George Washington University in 1971.

JON M. WICKWIRE • Founding partner of the Washington, D.C. and Vienna, Virginia law firm of Wickwire Gavin, P.C. Mr. Wickwire is a Fellow of the American College of Construction Lawyers. He received his B.A. degree from the University of Maryland in 1965 and his J.D. degree from Georgetown University in 1968.

RUSSELL A. YAGEL • Member of the Miami, Florida law firm of Kimbrell & Hamann, P.A. Mr. Yagel received his undergraduate degree from Stetson University and his J.D. degree from Capital University. Mr. Yagel is a licensed CPA (inactive status).

INTRODUCTION

This manual on managing and litigating the complex construction surety case is the first effort by the American Bar Association's Fidelity & Surety Law Committee to produce a truly advanced publication that addresses many of the most sophisticated issues arising in the preparation for and trial of the "mega" construction surety case—a case that routinely involves multiple parties, multiple project locations, multiple claimants, multiple sources for "salvage," multiple forums for dispute resolution, and a host of technical engineering, accounting, and construction issues. When called upon by Duncan Clore and Paul Devin, the 1996-97 and 1997-98 Chairs respectively of the ABA/TIPS Fidelity and Surety Law Committee, to serve as program chair and as editor of this manual, I was given free rein to assemble a faculty of talented senior trial lawyers, surety officers, and trial consultants with broad experience in the representation of sureties in complex construction cases. Their cumulative experience exceeds several hundred years. This manual is a testament to the collective expertise of those lawyers.

In the preparation and editing of this manual, I was ably assisted by and wish to express special appreciation and gratitude to Caryn A. Kauffman, a member of my law firm of Faegre & Benson here in Minneapolis. Her contributions were most valuable and insightful.

This manual, like the other excellent manuals published by the ABA/TIPS Fidelity and Surety Law Committee on *The Law of Suretyship*, *Bond Default*, and *The Law of Payment Bonds*, will find a permanent place in the extensive literature on suretyship. It should be studied carefully by all surety claims officers and surety trial lawyers.

Philip L. Bruner
Minneapolis, Minnesota
Editor
Past Chair, Fidelity and Surety Law Committee

CHAPTER 1

STRATEGIC "GENERALSHIP" OF THE COMPLEX CONSTRUCTION SURETY CASE

Philip L. Bruner

I. Introduction

Construction suretyship is among the most complex of all fields of law. Suretyship's extraordinary complexity is due to the cumulative effect of: (1) the absence of "construction suretyship law" as a defined legal subject, thereby requiring practitioners to search for applicable legal principles among established legal subjects of contract, tort, equity, and statute law and their subsidiary subjects of suretyship, creditors' rights, insurance, professional liability, and damages in order to address disputes arising from the issuance of contractor bonds and the performance of construction contracts; (2) the surety's multiple independent legal obligations to act in good faith with its bond obligee,[1]

1 With respect to the surety's obligation to its obligee, *see, e.g., Cates Constr. Co. Inc. v. Talbot Partners*, 62 Cal. Rptr. 2d 548 (Cal. Ct. App. 1997) (holding surety liable for punitive damages of $15 million for relying on its principal's position without conducting an adequate investigation); *Transamerica Premier Ins. Co. v. Brighton Sch. Dist.*, 940 P.2d 348 (Colo. 1997) (holding that a surety was liable for "bad faith breach of the performance bond" for failing to respond promptly to the obligee's notice of default and termination).

its bond principal and indemnitors,[2] and payment bond claimants,[3] whose positions frequently conflict and routinely place the surety in the middle of a savage legal "cross-fire;" (3) the highly technical nature of factual issues addressing design adequacy and construction conformance, which issues ordinarily require detailed expert investigation to identify and address them correctly;[4] (4) the surety's interest in mitigating loss by minimizing cost in the completion of defaulted contracts, by collecting all proper contract funds and by pursuing third parties (such as subcontractors, suppliers, design professionals, accountants, lenders, and other sureties) who may have caused or contributed to the loss or otherwise may be responsible for the loss;[5] (5) the multiplicity of forums and venues (arbitration versus litigation; federal court versus state court; interpleader and consolidation; trial by jury, court, or a special master) for resolution of

---

2 With respect to the surety's obligation to its principal and indemnitors, *see, e.g., St. Paul Fire & Marine Ins. Co. v. Pepsico, Inc.*, 160 F.R.D. 464 (S.D.N.Y. 1995) (holding that a surety that blindly paid claims without investigation or defense was denied indemnity); *Arntz Contracting Co. v. St. Paul Fire & Marine Ins. Co.*, 54 Cal. Rptr. 2d 888 (Cal. Ct. App. 1996) (upholding a trial court ruling disallowing the surety's recovery of completion costs from the contractor and its indemnitors on the ground that the surety's removal and replacement of certain stucco, sheet metal, and roofing work upon demand of the owner was "unnecessary and unwarranted" and in breach of an implied covenant of good faith). It is not uncommon for principals and indemnitors to accuse the surety of acting too hastily in accepting the obligee's declaration of default. *See Windowmaster Corp. v. Morse-Diesel, Inc.*, 722 F. Supp. 1532 (N.D. Ill. 1988).

3 With respect to the surety's duties to payment bond claimants, *see, e.g., Farmer's Union Cen. Exchange, Inc. v. Reliance Ins. Co.*, 675 F. Supp. 1534 (D.N.D. 1987); *Loyal Order of Moose, Lodge 1392 v. International Fidelity Ins. Co.*, 797 P.2d 622 (Alaska 1990); *K-W Indus. v. National Sur. Corp.*, 754 P.2d 502 (Mont. 1988); *Szarkowski v. Reliance Ins. Co.*, 404 N.W.2d 502 (N.D. 1987). Each case holds that a payment bond surety may be liable for bad faith claim handling to payment bond claimants. *Farmer's Union* was decided in favor of the surety and provides the best analysis of this obligation. *See also* Robert C. Niesley, *Payment Bond Claims Handling and the Law of Bad Faith, in* THE LAW OF PAYMENT BONDS (Lybeck & Shreves eds., 1998).

4 *See infra*, Chapter 9 for insight into technical construction issues.

5 *See infra*, Chapter 5 for theories under which third parties are pursued.

disputes, which require careful consideration;[6] (6) the frequency with which a single project may spawn multiple suits involving a large number of parties; (7) the difficulty of judges and lay juries in reaching a fair determination in any construction case, even in the quiet of the courtroom, of "precisely when the disorder and constant readjustment, which is to be expected by any [contractor] on the job site, became so extreme, so debilitating and so unreasonable as to constitute a breach of contract;"[7] (8) the ever present pressures generated by the limited time within which a surety frequently must act, the limited facts usually available for consideration prior to having to act, the frequently conflicting perspectives and information provided by the obligee and principal, the high stakes for a wrong decision, and the real risk of "second guessing" in future litigation; and (9) the recognition that in the complex environment of construction suretyship, rarely can a wrong decision be reversed;[8] and (10) the stark reality that "major construction

---

6 *See infra* Section V on dealing with multiple claims, multiple suits, multiple claimants, and multiple forums.

7 *Blake Constr. Co. v. C.J. Coakley Co. Inc.*, 431 A.2d 569 (D.C. App. 1981). In that case the United States Court of Appeals for the District of Columbia described the construction environment and its own frustration in determining a breach of a construction contract as follows:

> [E]xcept in the middle of a battlefield, nowhere must men coordinate the movement of other men and all materials in the midst of such chaos and with such limited certainty of present facts and future occurrences as in a huge construction project such as the building of this $100 million hospital. Even the most painstaking planning frequently turns out to be mere conjecture and accommodation to changes must necessarily be of the rough, quick, and *ad hoc* sort, analogous to ever-changing commands on the battlefield. Further, it is a difficult task for a court to be able to examine testimony and evidence in the quiet of a courtroom several years later concerning such confusion and then extract from them a determination of precisely when the disorder and constant readjustment, which is to be expected by any subcontractor on a job site, become so extreme, so debilitating and so unreasonable as to constitute a breach of contract between a contractor and a subcontractor. This was the formidable undertaking faced by the trial judge in the instant case . . .

Id. at 575.

projects generate major construction litigation" and the "management of either is perilous."[9]

Oliver Wendell Holmes, when in the twilight of his career on the United States Supreme Court, harked back to his years as a Union officer in the Civil War and observed: "A new and valid idea is worth more than a regiment and fewer [people] can furnish the former than can command the latter." Successful resolution of any complex construction suretyship case requires "new and valid" ideas and "generalship."[10] Sound and imaginative judgment must distinguish and select the truly creative ideas from the myriad "garden variety" thoughts. As Holmes infers, a "new and valid idea" is more likely to come from those individuals with good judgment who can think strategically rather than from those who simply "follow the book." Individuals who are good strategical thinkers ordinarily possess most or all of these qualities: (1) the "big picture" VISION to see the entire field of conflict and its myriad skirmishes, of which a particular dispute may be recognized as only a part; (2) the WILL to pursue (and settle) the dispute rather than "pass the buck" to others; (3) the EXPERIENCE to understand and evaluate the complex facts underlying a construction suretyship dispute against the background of construction and surety industry practices and customs; (4) the MANAGEMENT ABILITY to attract and marshal economically the best professional help (advocates, experts, and consultants), to oversee a competent investigation, and to identify the best fact witnesses to support the legal presentation; (5) the

---

8 *See generally* Philip L. Bruner, *et al.*, *The Surety's Response to the Obligee's Declaration of Default and Termination: "To Perform or Not to Perform—That is the Question,"* 17 CONSTR. LAW. 3 (Jan. 1997). Among other things, this article discusses the consequences of an owner's wrong decisions to terminate or not to terminate a bonded contract, and the surety's wrong decisions to perform or not to perform under its performance bond.

9 *Morse/Diesel, Inc. v. Trinity Indus., Inc.*, 67 F.3d 435, 437 (2d Cir. 1995).

10 One authority on "generalship," United States General George S. Patton, offered this observation:

> One does not plan and then try to make circumstances fit those plans. One tries to make plans fit the circumstances. I think the difference between success and failure in high commands depends upon the ability, or lack of it, to do just that.

KNOWLEDGE of legal weapons available in the arsenal of construction and surety claims and defenses; (6) the SKILL to select those legal weapons, in combination with successful dispute resolution forums, most likely to result in the "successful" resolution of the dispute; (7) the OBJECTIVITY to "stay above the fight" and maintain a broad perspective above those laboring in the factual and legal "trenches;" (8) the WISDOM and FLEXIBILITY to resolve the dispute, where possible, by negotiation or other suitable means; (9) the CHARACTER at all times to deal ethically, to avoid personal rancor or incivility, and to always "take the high road;" and (10) the MAGNANIMITY to remain unegotistical and professional in "success."

The twenty-first century will belong to strategic thinkers (those "big picture" people who can create and oversee to successful conclusion a strategic road map to success) among surety claims officers and outside counsel. Strategic thinkers are nurtured through the careful study of "vicarious experiences" of others described in reported legal decisions that explain why someone "won" and someone "lost" in the on-going legal wars of construction suretyship.

II.
The Strategic Road Map: Tailoring Legal Theories To Desired Litigation Results

It is the function of "generalship" to create the strategic "road map" of creative ideas to address the multitude of complexities. The first point of reference on the road map is the appreciation of the significance that different legal theories play in adjudication of the multitude of claims asserted in a major contractor default. In large default cases, it is not unusual for owners, contractors, subcontractors, designers, material suppliers, sureties, and insurers all to be named as parties in one or more lawsuits and all to assert claims, cross-claims, third-party claims, and counterclaims based on different liability theories, each encrusted with its own time-honored distinctions and methods for measurement of recovery. Where claims are asserted against one or more parties under the same set of facts but on multiple liability theories, the initial issue is selecting the proper legal theories upon which the various claims and defenses should be asserted.

"Construction suretyship law," like America itself, is a "melting pot" of legal theories developed for broad application in earlier and simpler times. The panoply of claim "weapons" and "shields" available in the arsenal of the construction surety originates from four sources of legal principles developed over the centuries by English common law, subsequently carried into and refined by American jurisprudence and most recently distilled to in the *Restatements*.[11] These sources are:

> *Contract Law*, which includes a wide range of concepts applicable to the creation, enforcement, and discharge of express or implied contracts and to the contract measure of recovery of foreseeable "expectancy" damages attributable to the breach of such agreements;
>
> *Equity Law*, which is based on basic concepts of fairness and good faith and includes theories such as subrogation, restitution, indemnity, unjust enrichment, waiver, estoppel, and mistake, for which equitable relief of rescission, reformation, or injunction may be available to enforce rights and for which the measure of recovery for conferred benefit is quantum meruit (meaning "reasonable value");
>
> *Tort Law*, which imposes liability for violation of legal duties and includes theories such as negligence, misrepresentation, fraud, intentional interference, nuisance, intentional harm (the prima facie tort), fiduciary and other special relationships, defamation, product liability, and other theories for which the measure of recovery is any damages "proximately caused" and for which punitive damages may be awarded in egregious cases; and
>
> *Statutory Law*, which includes: (a) rights and remedies created by statute (*e.g.*, the public contractor's bond statute, the public contract bidding statute, the Uniform Arbitration Act, the Uniform Commercial Code, contractor licensing statutes, mechanic's lien statutes, etc.); and (b) statutory constraints which limit, modify, or supersede common law legal rights created under contract, tort, or equity (*e.g.*, statutes of limitations, statutes of frauds, etc.).

These legal principles of contract, equity, tort, and statute offer their own distinct rights and remedies, each of which must be clearly

---

11 *See, e.g.*, RESTATEMENT (THIRD) OF SURETYSHIP & GUARANTY (1995); RESTATEMENT (SECOND) OF CONTRACTS (1979); RESTATEMENT (SECOND) OF TORTS (1965); RESTATEMENT (SECOND) OF RESTITUTION (1988); RESTATEMENT OF SECURITY (1941); RESTATEMENT (SECOND) OF PROPERTY (1976).

understood when disputes arise.[12] Each claim and defense must be analyzed and evaluated by counsel for strengths and weaknesses based on the result intended. This "sorting out" process requires review of contract agreements, common law principles, and legal duties imposed by statute.

In cases in which an express contract or bond exists, courts traditionally have accepted them as the "bench mark" by which rights and duties are measured and have looked first to the terms of the contract or bond to verify the agreed rights and remedies of the parties. The law presumes that sophisticated business people will protect themselves by charging a fair price for the risks to be assumed and work to be performed under a written contract.[13] This common sense presumption also underlies the measure of damages for breach of the contract, which limits recovery solely to "expectancy" damages that were reasonably foreseeable at the time the contract was entered into as the likely result of a breach, and which seeks to place the non-breaching party in the same position that it would have occupied if the contract had been fully performed. Damages waived by express contract provision are not recoverable.[14] Consequential damages not within the contemplation of the parties at the time the contract was entered into also are not recoverable.[15] As a practical matter, whether any particular

---

12 *See* Philip L. Bruner, *Construction Claim Recovery Measures: Untying the Gordian Knot,* XX THE FORUM 278 (Winter 1985); DOBBS, REMEDIES.

13 *See Wells Bros. Co. of N.Y. v. United States*, 254 U.S. 83, 86 (1920), in which the court enforced a "no damage for delay" clause as follows: "Men who take $1 million contracts for government buildings are neither unsophisticated nor careless. Inexperience and inattention are more likely to be found in other parties to such contracts than the contractors, and the presumption is obviously strong that the men signing such a contract . . . protected themselves against such delays [and other assumed or foreseeable risks] as are complained of by the higher price extracted for the work."

14 *See McNally Wellman Co. v. New York State Elec. & Gas Corp.*, 63 F.3d 1188 (2d Cir. 1995) (upholding clause that protected the contractor from liability for any "special, incidental, indirect or consequential damages . . . claimed in contract, equity, tort (including negligence), or otherwise"); *Roy A. Elam Masonry, Inc. v. Fru-Con Constr. Corp.*, 922 S.W.2d 783 (Mo. Ct. App. 1996) (upholding "no damages for delay" clause).

15 *See Hadley v. Baxendale*, 156 Eng. Rep. 145 (1854) (barring recovery of consequential lost profits); *American Home Assurance Co. v. Larkin Gen.*

element of damage arising out of a construction project is "consequential" or not is a factual issue contested in virtually every case.[16] In order to minimize such disputes, parties to negotiated contracts frequently specify damages within their contemplation or include contractual "liquidated damages" provisions to fix a reasonable amount of damages payable for certain types of breaches.[17] Contracts also frequently are written to include clauses which define the relief available to a non-breaching party in the event of identified types of breaches.[18]

In contrast with contract theories, tort theories generally provide greater advantages for recovery of unforeseeable consequential damages (*e.g.*, "mental anguish," "emotional distress," and other subjective consequences personal to the non-breaching party) that nevertheless can be proven to have been "proximately caused" by the defendant's

---

Hosp. Ltd., 593 So. 2d 195 (Fla. 1992) (holding surety not liable for consequential delay damages arising from the principal's default); *see also* Article 4.3.10 of AIA A201-1997 General Conditions of Contract, under which the owner and contractor each waive consequential damages against the other.

16 *Compare Downey Inc. v. Bradley Ctr. Corp.*, 524 N.W.2d 915 (Wis. Ct. App. 1994) (lost profits on other contracts recoverable) *with Topco, Inc. v. Montana Dep't of Highways*, 912 P.2d 805 (Mont. 1996) (lost profits unrecoverable as "speculative" and "uncertain").

17 *See Kingston Constructors, Inc. v. Washington Metro. Area Transit Auth.*, 930 F. Supp. 651 (D.D.C. 1996) (liquidated damages unreasonable and an illegal penalty); *Utley-James of La., Inc. v. State*, 671 So. 2d 473 (La. Ct. App. 1995) (liquidated damages assessed even though owner incurred no actual delay damages).

18 *See G.M. Shupe v. United States*, 5 Cl. Ct. 662, 677 (1984) ("plaintiff also seeks recovery on the theory of breach of contract. However, it seems clear that plaintiff's claims, and the claims of its subcontractor, are addressable under various provisions of the contract. Accordingly, plaintiff is limited to the relief provided by the contract, *i.e.* equitable adjustments, and is not entitled to any further or different relief"); *see also Mellon Stewart Constr. Inc. v. Metropolitan Water & Reclamation Dist.*, 94-C-1915, 1995 WL 124133 (N.D. Ill. 1995), in which the court ruled that contract payment provisions for extra work constituted an acknowledgment that the possibilities of delay and increased costs were contemplated at the outset of the contract and thus barred the contractor from terminating the contract for such alleged breaches.

wrongful conduct[19] and for recovery of punitive damages in egregious cases.[20] Such advantages have resulted in "more or less inevitable efforts of lawyers to turn every breach of contract into a tort."[21] This "collision between tort and contract"[22] frequently results in mere breaches of contract being dressed up by lawyers as tort claims for the obvious purpose of pursuing expanded compensatory damages and consequential or exemplary damages otherwise unrecoverable under contract theories. Whereas tort claims may support the award of exemplary or "punitive" damages, even intentional breaches of contract between businessmen will not.[23] To sort out appropriate from inappropriate tort claims, courts look first to the contract as the "benchmark" by which to evaluate and measure the rights and liabilities of the parties to the contract. Only if a party's conduct in performing or failing to perform a contract is so outrageous or otherwise violative of special legally protected relationships[24] as to the rise to the level of an

---

19 *See Bonded Concrete, Inc. v. T. J. Madden Constr. Co.*, 655 N.Y.S.2d 153 (N.Y. App. Div. 1997) (holding "mental anguish" and "emotional distress" may not be recovered for mere breach of contract).

20 *See Trustees of Columbia Univ. v. Gwathmey Siegel Assocs. Architects*, 601 N.Y.S.2d 116 (N.Y. App. Div. 1993) (punitive damages upheld agianst architect for lack of care); *Tacon Mechanical Contractors, Inc. v. Grant Sheet Metal Inc.*, 889 S.W.2d 666 (Tex. Ct. App. 1994) (punitive damages upheld against contractor who frequently withheld funds from sub without justification).

21 *See* PROSSER, HANDBOOK OF THE LAW OF TORTS § 92 (4th ed. 1971).

22 Wright & Nichols, *The Collision of Tort & Contract in the Construction Industry*, 21 U. RICH. L. REV. 457 (1987).

23 *Compare Great American Ins. Co. v. General Builders, Inc.*, 934 P.2d 257 (Nev. 1997) *and Charles Shaid of Pa., Inc. v. George Hyman Constr. Co.*, 947 F. Supp. 844 (E.D. Pa. 1996) *with Transamerica Premier Ins. Co. v. Brighton Sch. Dist.*, 940 P.2d 348 (Colo. 1997); *see also Morrow v. L. A. Goldschmidt Assoc.*, 492 N.E.2d 181 (Ill. 1986); *Barr/Nelson, Inc. v. Tonto's Inc.*, 336 N.W.2d 46 (Minn. 1983); 22 AM. JUR. 2D *Damages* § 245.

24 Special legally protected relationships between contractual parties include: (1) a fiduciary relationship between partners or joint venturers; (2) an employment relationship between employer and employee; (3) an insurance relationship between insurer and insured; (4) a professional relationship between doctor and patient or lawyer and client; (5) a contractual relationship involving the safety of third parties; and (6) a contractual

"independent tort" will a court invoke a tort measure of recovery.[25] The only significant disadvantage of tort theory over contract theory is that a tort claimant may not recover breach of contract "benefit of the bargain" damages.

In comparison with both contract and tort, equity theories permit contract rescission or reformation, injunctive relief to enforce rights, and recovery in quantum meruit for the value of services rendered. In some jurisdictions a non-breaching party to a contract either may sue for its contract "expectancy" damages or may reject the express contract and

---

relationship affected by adhesion in which disparate knowledge and bargaining power are involved. *See Iron Mountain Sec. Storage Corp. v. American Specialty Foods, Inc.*, 457 F. Supp. 1158 (E.D. Pa. 1978).

25 *See Great Am. Ins. Co. v. General Builders, Inc.*, 934 P.2d 257 (Nev. 1997) (refusing to allow a surety to be sued for the tort of bad faith, even though the surety had revoked its performance and payment bonds and thereby caused the principal to forfeit the contract, because the court found that a surety contract involved no special element of reliance or fiduciary duty necessary to permit a bad faith tort claim); *see also Blake Constr. Co. v. C.J. Coakley Co. Inc.*, 431 A.2d 569 (D.C. App. 1981), in which a subcontractor sued the prime alleging, among other things, separate claims for breach of contract and tortious duress arising out of the prime's withholding of payments to the sub, and the District of Columbia ruled as follows:

When two parties herein dispute over performance under a contract, the recourse lies in the legal suit under the contract and not in an action in tort unless there are exceptional circumstances of a violation of duty imposed by statute or law beyond the contract's terms. Certainly in the present case, where two experienced businesses are involved in a large and complex construction project, something more must be presented to support an action in tort than was shown on this record.

Id. at 577 n.5; *see also Iron Mountain Sec. Storage Corp. v. American Specialty Foods, Inc.*, 457 F. Supp. 1158 (E.D. Pa. 1978); *Sanders v. Zeagler*, 686 So. 2d 819 (La. 1997); *Gennari v. Weichert Co. Realtors*, 691 A.2d 350 (N.J. 1997); *Bonded Concrete, Inc. v. T.J. Madden Constr. Co.*, 655 N.Y.S.2d 153 (N.Y. App. Div. 1997); *Brock Bridge Ltd. Partnership, Inc. v. Development Facilitators, Inc.*, 689 A.2d 622 (Md. App. 1997); *Fisher Sand & Gravel Co. v. South Dakota Dep't of Transp.*, 558 N.W.2d 864 (S.D. 1997); *Blagrove v. JB Mechanical, Inc.*, 934 P.2d 1273 (Wyo. 1997).

sue under implied contract for the value of services rendered. An election to rescind an express contract limits the claimant's right of recovery to quantum meruit and not "expectancy" damages.[26] In other jurisdictions, the existence of an express contract bars recovery under a theory of implied contract.[27] Quantum meruit recovery, when available, will not be limited by the contract measure of damages, and full recovery of costs may be had even though performance under the contract would have resulted in a loss without any anticipated profits.[28]

III.
The Strategic Road Map: Evaluating The Surety's Performance Bond Exposure[29]

Most complex construction surety cases start with competing contentions by the principal and obligee that the other party has materially breached the contract. The obligee usually declares the

---

26 *See Harris v. Desisto*, 932 S.W.2d 435 (Mo. Ct. App. 1996), in which a developer brought suit against a contractor both to rescind a construction agreement and to recover damages for its breach. The trial court granted rescission of the agreement and awarded the developer $53,000 for breach of contract damages. On appeal, Missouri Court of Appeals ruled that the two remedies were inconsistent in that the equitable remedy for rescission of the contract seeks *disaffirmance* of the contract whereas the legal remedy for money damages proceeds upon *affirmance* of the contract. Under the circumstances, the court set aside the breach of contract award and allowed only the equitable remedy for rescission to stand. Where a legal remedy for breach of contract is available, courts ordinarily do not allow recovery on other theories in the absence of special circumstances.

27 *See American Towers Owners Ass'n, Inc. v. CCI Mechanical, Inc.*, 930 P.2d 1182 (Utah 1996), in which the Utah Supreme Court refused to allow an owner to measure its damages on the basis of the equitable remedy of quantum meruit for unjust enrichment where the subject of the suit was covered by an express contract between the owner and contractor.

28 *United States* v. *Algernon Blair Inc.*, 479 F.2d 638 (4th Cir. 1973).

29 Portions of this section are drawn and updated from Philip L. Bruner, *et al.*, *The Surety's Response to the Obligee's Declaration of Default and Termination: "To Perform or Not to Perform—That is the Question,"* 17 CONST. LAW 3 (Jan. 1997), and will appear in the Practitioners Hornbook Treatise on Construction Law being co-authored by Mr. Bruner and Mr. O'Connor under engagement with West Publishing Company.

principal in default, sometimes terminates the contract, and then calls upon the surety to perform under the terms of its performance bond. The obligee also may withhold funds from the contractor that result in unpaid subcontractors and material suppliers filing claims against the principal and surety under the payment bond. Frequently the principal and its indemnitors contest vigorously the obligee's allegations of breach and their subcontractors' entitlement to payment. The surety's most critical strategic task is to timely and critically view the entire "battlefield" for the purpose of determining whether or not the surety should respond to the demands on its bonds. The initial inquiry is whether the principal or obligee has materially breached the bonded contract; whether third parties must provide indemnity or contribution;[30] whether off-sets or defenses exist against bond claims; and whether the bond affords coverage. The threshold determination of "material breach" is made not only by the surety but by every obligee that decides to terminate its contractor for default and call on the surety under its performance bond to respond to declaration of default and termination. The surety's liability under its performance bond ordinarily is invoked only if a "material breach" has been committed by the principal which was not induced or proceeded by the obligee's own supervening material breach of contract, such as nonpayment, maladministration of the construction process, refusal to grant proper time extensions or other relief required by the contract, and the obligee thereafter properly terminated the contract and properly called upon the surety to perform under a valid performance bond. The surety's first step on the strategic road map is a detailed technical evaluation of the construction issues involved in order to determine the validity of the obligee's allegation of material breach.

A. What Constitutes A "Material Breach"

A bond "default" that triggers a surety's obligation to perform under its performance bond must rise to the level of a material breach of the bonded contract warranting termination by the owner. Ordinary items of uncompleted or nonconforming work, endemic to every construction project and routinely completed or corrected later in the construction

---

30 *See infra*, Chapter 5 for a detailed analysis of the surety's recourse against third parties.

process or as "punch list" or "warranty" work, rarely constitute material breaches. The same is true for delays in completion where the contract did not designate "time is of the essence" or where the parties contracted for exclusive remedies for delay. The existence of a material breach ordinarily is a question of fact for the jury to decide. In determining whether any failure of performance is a material breach, the jury usually is instructed to look at the following circumstances: (1) the specific basis for termination established by the terms of the bonded contract;[31]

---

31 Specific grounds and procedures for the obligee's termination ordinarily are set forth in the termination clause of the bonded contract. A standard termination clause used in many construction industry contracts is Article 14.2 of the American Institute of Architects General Conditions for Construction A201, 1997 edition, which reads:

14.2 TERMINATION BY THE OWNER FOR CAUSE

14.2.1 The Owner may terminate the Contract if the Contractor:

.1 persistently or repeatedly refuses or fails to supply enough properly skilled workers or proper materials;

.2 fails to make payment to Subcontractors for materials or labor in accordance with the respective agreements between the Contractor and the Subcontractors;

.3 persistently disregards laws, ordinances, or rules, regulations or orders of a public authority having jurisdiction; or

.4 otherwise is guilty of substantial breach of a provision of the Contract Documents.

14.2.2 When any of the above reasons exist, the *Owner, upon certification by the Architect that sufficient cause exists to justify such action*, may without prejudice to any other rights or remedies of the Owner and *after giving the Contractor and the Contractor's surety, if any, seven days' written notice*, terminate employment of the Contractor and may, subject to any prior rights of the surety:

.1 take possession of the site and of all materials, equipment, tools, and construction equipment and machinery thereon owned by the Contractor;

.2 accept assignment of subcontracts pursuant to Paragraph 5.4; and

.3 finish the Work by whatever reasonable method the Owner may deem expedient. Upon request of the Contractor, the Owner shall furnish to the Contractor a detailed accounting of the costs incurred by the Owner in finishing the Work.

(2) the extent to which the injured party will be deprived of the benefit which he reasonably expected; (3) the extent to which the injured party can be adequately compensated for the part of that benefit of which he will be deprived; (4) the extent to which the party failing to perform or offer to perform will suffer forfeiture; (5) the likelihood that the party failing to perform or to offer to perform will cure his failure, taking account of all circumstances, including any reasonable assurances; and (6) the extent to which the behavior of the party failing to perform or to offer to perform comports with standards of good faith and fair dealing.[32] The burden of proof of default is upon the obligee, even if designated a defendant in litigation.[33]

B. What Constitutes A "Wrongful" Termination

The obligee's wrongful termination of the bonded contract is itself a breach of contract that relieves the surety of liability under its performance bond.[34] The obligee's termination of the principal under the termination

---

14.2.3 When the Owner terminates the Contract for one of the reasons stated in Subparagraph 14.2.1, the Contractor shall not be entitled to receive further payment until the Work is finished.

14.2.4 If the unpaid balance of the Contract Sum exceeds costs of finishing the Work, including compensation for the Architect's services and expenses made necessary thereby, and other damages incurred by the Owner and not expressly waived, such excess shall be paid to the Contractor. If such costs and damages exceed the unpaid balance, the Contractor shall pay the difference to the Owner. The amount to be paid to the Contractor or Owner, as the case may be, shall be certified by the Architect, upon application, and this obligation for payment shall survive termination of the Contract.

(emphasis added).

32 RESTATEMENT (SECOND) OF CONTRACTS § 241 (1979); 4 ARTHUR L. CORBIN, CORBIN ON CONTRACTS, § 946 (1951).

33 *See Lisbon Contractors, Inc. v. United States*, 828 F.2d 759 (Fed. Cir. 1987).

34 *Miller v. City of Broken Arrow*, 660 F.2d 450 (10th Cir. 1981); *United States Home Corp. v. Suncoast Util.*, 454 So. 2d 601 (Fla. 1984).

clause of a bonded contract can be upheld only if the obligee sustains its burden of proof that (1) the principal materially breached its contract; (2) the principal's breach was not induced or preceded by the obligee's own supervening material breach of contract, such as nonpayment, maladministration of the construction process, or refusal to grant proper time extensions or other recognized "contract defenses;" (3) the obligee's termination was not improperly motivated or conceived in bad faith and was made independently and with the exercise of discretion by its representative having authority to terminate the contract; (4) the principal and surety were given ample notice of deficiencies so as to understand what needed to be "cured" to avoid termination; (5) the obligee properly followed its contractually-specifiedtermination procedure; (6) the contract was not "substantially performed;" and (7) the obligee did not "waive" any contract completion dates or requirements upon which it relied to support its termination. A detailed analysis of the termination procedure is as follows:

1. The Importance Of Good-Faith Motive

An obligee's default termination of a bonded contract may be invalidated by the owner's wrongful motive. A termination for default based on motives other than legitimate and well-documented concerns over the principal's inability to perform may cause the termination action to be deemed to be wrongful.[35]

---

35 Illustrative of this point is *Paul Hardeman, Inc. v. Arkansas Power & Light Co.*, 380 F. Supp. 298 (E.D. Ark. 1974). In *Hardeman*, the owner's "subjective, emotional, and visceral" response to a contractor's claim for additional costs was a termination for default to get the contractor off the job. Although the court found that there had been "some technical breaches of contract by both parties," and that "neither party was totally without blame," the court concluded:

> [The owner] was not justified in dealing with the problem in this draconic manner. Its action in doing so was wrongful and constituted an overriding breach of the agreement which not only prevents it from obtaining judgment for the costs of the completion of the work, but, to the contrary, forms a legal basis for [the contractor] to obtain monetary relief against [the owner].

Termination clauses frequently provide that the decision to terminate will be made by the "owner's representative" designated under the contract after issuance of a certificate by the design professional. To encourage the exercise of the right of termination for default only for fair and proper reasons, courts have rigorously enforced the procedural requirements of the termination clause. Issuance of the specified architect's certificate stating that sufficient cause exists to terminate the contract is a condition precedent to termination.[36] The decision to terminate for default must be reached through the exercise of independent discretion by the person designated under the contract. If that person follows the direction of a superior or others to terminate the contractor for default or simply fails to make an independent inquiry and decision, the termination may be set aside.[37]

Id. at 64; *see also Board of County Comm'rs v. Umbehr*, 116 S. Ct. 2342 (1996) (holding that independent contractors have First Amendment protection against contract termination in retaliation for contractor's criticism of public body); *McDonnell Douglas Corp. v. United States*, 35 Fed. Cl. 358 (1996) (termination based on no more than a "technical or bear default" is improper); *Darwin Constr. Co. Inc. v. United States*, 811 F.2d 593, 596 (Fed. Cir. 1987) ("[T]he contractor's status of technical default served only "as a useful pretext for taking the action found necessary on other grounds unrelated to the [contractor's] performance or to the propriety of the extension of time"); *Green Island Contracting Corp. v. State*, 458 N.Y.S.2d 828 (1983), *aff'd*, 473 N.Y.S.2d 55 (N.Y. App. Div. 1984); *Ballenger Corp. v. City of Columbia*, 331 S.E.2d 365 (S.C. Ct. App. 1985); *Torncello v. United States*, 681 F.2d 756 (Ct. Cl. 1982).

36 *Peter Kiewit Sons' Co. v. Summit Constr. Co.*, 422 F.2d 242 (8th Cir. 1969); *American Continental Life Ins. Co. v. Ranier Constr. Co.*, 607 P.2d 372 (Ariz. 1980); *Oden Constr. Co. v. Helton*, 65 So. 2d 442 (Miss. 1953).

37 Illustrative of this point is *Schlesinger v. United States*, 390 F.2d 702 (Ct. Cl. 1968). In *Schlesinger*, a default termination was set aside because no one in the government contracting office exercised independent discretion in the termination decision. The court set aside the termination with this admonition:

The record affirmatively shows that nobody in the Navy, neither the Contracting Officer nor his superiors, exercised the discretion they possessed under the [termination for default clause]. The [contractor's] status of technical default served only as a useful pretext for the taking of action felt to be necessary on other grounds unrelated to the [contractor's] performance or the propriety of an extension of time . . . Such abdication of responsibility we have always refused to sanction where there is

2. The Importance Of A Proper "Cure" Notice

Virtually all termination for default clauses require the owner to give the contractor "notice of default." The purpose of this notice is to give the contractor an opportunity to "cure" the defaults and thereby preclude termination. Even where the contract does not contain an express provision to furnish the contractor with a "cure notice," the contract may be deemed to include an implied obligation on the part of the owner to give a contractor fair notice and a reasonable opportunity to correct deficiencies or nonconformances prior to termination of the contract.[38] Termination of a contract for default without giving the required "cure

---

> administrative discretion under the contract . . . . This protective rule should have special application for default-termination which has the drastic consequence of leaving the contractor without any further compensation.

Id. at 709; *see also McDonnell Douglas Corp. v. United* States, 35 Fed. Cl. 358 (1996) (holding that "a termination imposed without the benefit of reasoned discretion could be upheld only if contractor performance were so deficient that termination for default was the only possible result"); *John A. Johnson Constr. Corp. v. United States*, 132 F. Supp. 698 (Ct. Cl. 1955) (holding that "ulterior motives" involving possible litigation do not constitute a valid ground for termination in the exercise of independent discretion).

38 *See McClain v. Kimbrough Constr. Co., Inc.*, 806 S.W.2d 194 (Tenn. Ct. App. 1990) in which the Tennessee Court of Appeals ruled that the owner's implied obligation to give the contractor a cure notice would be read into every construction contract:

> Requiring notice is a sound rule designed to allow the defaulting party to repair the defective work, to reduce the damages, to avoid additional defective performance, and to promote the informal settlement of disputes [case citations omitted]. Thus, even when the parties have not included a "take over" clause in their contract, courts have imposed on contractors the duty to give subcontractors notice and an opportunity to cure before terminating the contract for faulty performance.

Id. at 198; *see also Pollard v. Saxe & Yolles Dev. Co.*, 525 P.2d 88, 92 (Cal. 1974) (holding that "[t]he requirement of notice of breach is based on a sound commercial rule designed to allow the defendant opportunity for repairing the defective item, reducing damages, avoiding defective products in the future, and negotiating settlements").

notice" is a material breach of contract,[39] which overrides the contractor's uncured breaches for which no cure notice was provided.[40] Termination of a bonded contract for default and retention of a completing contractor without notice to the surety is a material breach that renders the bond null and void.[41]

The owner's "cure notice" must fairly apprise the contractor of the specific defaults upon which the owner intends to rely in terminating the contract if the defaults are not cured.[42] The notice should "describe the

---

39 Illustrative of this point is *Cuddy & Mountain Concrete v. Citadel Constr., Inc.* 824 P.2d 151 (Idaho Ct. App. 1992). In *Cuddy*, an owner became frustrated with the contractor's poor production due to bad weather, made arrangements to hire a successor contractor, and then terminated the contractor for default without providing the contractually-required seven-day written notice. The owner also had his project manager "supplement" previously written daily project logs to provide a "more specific record of [the contractor's] work in case something happened after the termination." The jury not only found the contract to have been breached by the owner's failure to give a "cure" notice, but awarded punitive damages as compensation for the owner's heavy-handed conduct. *Id.*; *see also Burras v. Canal Constr. & Design Co.*, 470 N.E.2d 1362 (Ind. Ct. App. 1984); *Bruning Seeding Co. v. McArdle Grading Co.,* 439 N.W.2d 789 (Neb. 1989); *United States ex rel. Cortolano & Barone v. Morano Constr.*, 724 F. Supp. 88 (S.D.N.Y. 1989).

40 *See Carter v. Krueger*, 916 S.W.2d 932 (Tenn. Ct. App. 1995) (holding that the contractor "is not barred from recovering damages even though he may have been the first to breach the contract . . . because the [owner] failed to give notice of claimed defects and failed to give the [contractor] an opportunity to cure the defects"); *see also* RESTATEMENT (SECOND) OF CONTRACTS § 237 cmt. b, § 242 cmts. a & b (1979).

41 *See Dragon Constr., Inc. v. Parkway Bank & Trust*, 678 N.E.2d 55 (Ill. Ct. App. 1997) (holding that the obligee's failure to give the surety timely notice of default as required by Article 14.2 of AIA A201 and timely opportunity to respond under its bond prior to hiring its own completion contractor constituted a material breach that voided the bond); *see also Insurance Co. of N. Am. v. Metropolitan Dade County*, No. 97-1050, 1997 WL 667601 (Fla. Dist. Ct. App. 1997) (following *Dragon*).

42 *See Blaine Econ. Dev. Auth. v. Royal Elec. Co., Inc.*, 520 N.W.2d 473 (Minn. Ct. App. 1994). In *Blaine*, an owner was found to have wrongfully terminated a construction contract without having first given a proper "cure notice." Although the owner claimed that project correspondence and agendas constituted adequate notice of performance deficiencies, the Court

inadequate performance and must fairly advise [the contractor] that [the owner] considers the inadequate performance serious enough that, without prompt correction, the contract will be terminated."[43] The owner may not rely upon reasons different than those set forth in the "cure notice" to justify termination unless those reasons were then "non-curable."[44]

A fundamental question relating to the adequacy of any "cure notice" is whether the "cure" of a material condition of performance is ascertainable by a contractor of ordinary competence, or must be analyzed and defined by the owner's design professional of record. The necessity and reasonableness of any "cure" involving construction defects of a material nature necessarily must take into consideration the intent of the plans and specifications as evidenced by the underlying design

---

of Appeals of Minnesota ruled that those documents neither informed the contractor that the problems were jeopardizing the contract nor directed the contractor to do anything about them within a specific time. *See also Hannon Elec. Co. v. United States*, 31 Fed. Cl. 135, 148 (1994), *aff'd*, 52 F.3d 343 (Fed. Cir. 1995) (holding that a "cure notice" that "thoroughly defined the parameters of the problem" was adequate); *Miree Painting Co. v. Woodward Constr. & Design, Inc.*, 627 So. 2d 389 (Ala. Ct. App. 1992), *rev. on other grounds*, 627 So. 2d 393 (Ala. 1993) (holding that a "punch list" is neither a cure notice nor a termination notice, and that in the absence of a proper "cure notice," the termination was wrongful).

43 *Id.* at 477.

44 *See McDonnell Douglas Corp. v. United States*, 35 Fed. Cl. 358 (Ct. Cl. 1996), in which the United States Court of Federal Claims concluded:

> A cure notice places the contractor on notice of performance deficiencies that the Government believes will endanger the successful completion of the contract. The cure notice allows the contractor to argue that the deficiencies are excused, or to provide reasonable assurances that the problems can be overcome. Thus, a termination for default is improper if the reasons stated for termination are different from the reasons given in the cure notice. *See generally Composite Laminates, Inc. v. United States*, 27 Fed. Cl. 310 (1992) (the contractor must be on adequate notice of the reasons for default termination for the termination to be proper). The Government may not propose reasons for default at trial that the contractor might have cured had it been on notice of the problem during contract performance. The "post hoc" justification exception is limited to those circumstances in which the cure notice could not have been a factor, *i.e.*, where the reason asserted at trial was non-curable.

calculations and criteria, applicable building codes and standards of workmanship,[45] and the legal doctrines of substantial performance and economic waste.

The owner's insistence upon a construction defect "cure" in strict compliance with contract plans and specifications, without regard to design criteria, economic waste, or substantial performance, is unreasonable and may constitute a breach of contract. Whether a construction defect "cure" must strictly comply with the contract plans and specifications, and even the necessity for any "cure" at all, are determinations that the design professional of record properly should make. The design professional also is the proper party to advise the contractor of the specific "fix" necessary to "cure" the nonconformance so as to meet the intent of the plans and specifications. A contractor justifiably may look to the owner and design professional of record for direction as to just what repair or replacement work is necessary. Consideration of design safety factors, usually known only by the design professional, is fundamental to both the materiality of a construction defect and the need for and extent of the "cure." For this and other reasons, most standard termination clauses provide that the architect or engineer of record must certify that "sufficient cause exists" to justify termination.[46]

3. *The Owner's Waiver Of The Right To Terminate For Default*

A material breach by the contractor may be waived by the owner by failure to take timely action to terminate the contract for default and by owner encouragement to the contractor to continue performance.[47]

---

45 *Id.* (footnote omitted) *see Carter v. Krueger*, 916 S.W.2d 932 (Tenn. Ct. App. 1995) (opining that "in the absence of express plans and specifications, the standard of workmanship prevailing in the area coupled with conformity to the applicable codes . . . is the standard by which the [contractor's] performance is tested").

46 *See, e.g.*, Article 14.2.2, AIA Form A201 (1997 ed).

47 *See DeVito v. United States*, 413 F.2d 1147 (Ct Cl. 1969). In *Devito*, the government's failure to exercise its right of termination for forty-eight days, together with its knowledge of the contractor's continued performance, constituted a waiver of the contractor's breach:

However, waiver will not be inferred by a failure of the owner to terminate while the owner takes a reasonable time to investigate conditions under a reservation of rights.[48]

Waiver of the specifications also may arise as a result of the owner's failure to take timely action to terminate the contract for default and by owner "course of dealing."[49] A default termination based upon

---

> Time is of the essence in any contract containing fixed dates for performance. When a due date is passed and a contract has not been terminated for default within a reasonable time, the inference is created that time is no longer of the essence so long as the constructive election not to terminate continues and the contractor proceeds with performance. The proper way thereafter for the time to again become of the essence is for the government to issue a notice under the default clause setting a reasonable but specific time for performance upon pain of default termination. The election to waive performance remains in force until the time specified in the notice, and thereupon time is reinstated as being of the essence. The notice must set a new time for performance that is both reasonable and specific from the standpoint of the performance capabilities of the contractor at the time that the notice is given.

Id. at 1154; *see also Structural Sys., Inc. v. Borg-Warner Health Prods.*, 654 S.W.2d 300 (Mo. 1983); *Martin J. Simko Constr. Inc. v. United States*, 11 Cl. Ct. 257 (1986), *vacated in part on other grounds*, 852 F.2d 540 (Fed. Cir. 1988); *Sun Cal Inc. v. United States*, 21 Cl. Ct. 31 (1990); *Gamm Constr. Co. v. Townsend*, 336 N.E.2d 592 (Ill. Ct. App. 1975); *compare Florida Dep't of Ins. v. United States*, 81 F.3d 1093 (Fed. Cir. 1996).

48 *See Indemnity Ins. Co. of N. Am. v. United States*, 14 Cl. Ct. 219 (1988).

49 *See Hannon Elec. Co. v. United States*, 31 Fed. Cl. 135 (1994), *aff'd*, 52 F.2d 343 (Fed. Cir. 1995), in which the court addressed this issue as follows:

> The government's unprotesting observation or acceptance of performance that it later contends was incorrect can have one of two legal consequences. [Citation omitted.] First, if the interpretation of ambiguous specifications is in dispute, such actions serve to show that the contractor's interpretation is reasonable and that the government shared the contractor's interpretation of the specifications before the dispute arose. [Citation omitted.] In this context, the contractor need prove little more than the fact that such conduct occurred. It need not show that it relied on the government's conduct, or that the actions were those of a government official with the authority to waive contract requirements. . . . A contractor also can use the

noncompliance with waived contract terms or specifications constitutes a breach of contract.

C. Has the Obligee Materially Breached The Contract

The surety, upon termination, steps into the shoes of its principal under the bonded contract and is entitled to assert against the obligee its principal's claims for additional compensation and defenses against termination and other liability arising out of or related to the contract.[50] The surety therefore must analyze the terms of the bonded contract and the facts surrounding the contract to determine whether the owner and contractor each have performed their respective material obligations. If the obligee is found to have materially breached the bonded contract prior to any breach by the principal, or to have induced the principal's breach, the obligee's termination of the bonded contract is itself a material breach that may exonerate both the principal and the surety from liability, and will allow the contractor to recover damages either under the contract or in quantum meruit.[51]

It is a fundamental principle of law that an owner's material breach of contract relieves a contractor of its obligation to perform and precludes a

---

government's unprotesting observation or acceptance of noncontractual performance to demonstrate that the government has waived a contract requirement. Professor John Cibinic has named this argument the "constructive waiver of specifications" [citation omitted]. A constructive waiver of specifications occurs not where the contract is ambiguous, but rather, where the government "has administered an initially unambiguous contract in such a way as to give a reasonably intelligent and alert opposite party the impression that a contract requirement has been suspended or waived."

Id. at 146 (citing *Gresham & Co. v. United States*, 470 F.2d 542, 555 (Ct. Cl. 1972)); *see also McDonnell Douglas Corp. v. United States*, 35 Fed. Cl. 358 (1995) (opining that "the Navy's less harsh position regarding contract specifications . . . during contract performance confirms that the termination for default [for failure to comply with the specifications] was improper"); *Miller Elevator Co. v. United States*, 30 Fed. Cl. 662 (1994).

50 *See* RESTATEMENT (THIRD) OF SURETYSHIP & GUARANTY §§ 34-36 (1995).

51 *See, e.g., Carter v. Krueger*, 916 S.W.2d 932 (Tenn. Ct. App. 1995).

termination by the owner for default.[52] Some of the events in the construction process that may constitute either an excuse for contractor nonperformance or a basis for claim for an outright material breach against the owner are described below:

1. Substantial Performance

The owner may not terminate for default a construction contract that has been substantially performed.[53] Substantial performance of a construction contract is typically defined as that degree of performance which provides the owner with construction suitable for the purpose for which it is intended.[54] Failure of the contractor to complete all work under a substantially performed contract allows the owner merely to utilize any

---

52 See *Marathon Oil Co. v. Hollis*, 305 S.E.2d 864 (Ga. Ct. App. 1983). In *Marathon*, the court allowed a contractor to recover, notwithstanding arguments that the contractor failed to complete on time. Among other things, the court concluded that: (1) the contractor "did perform the majority of the work properly pursuant to the contract and that it stood ready to correct any defects," (2) the contractor was not able to complete its work according to the original schedule because of owner-caused interruptions or other delays for which the contractor was not responsible, and (3) the owner waived the contract completion date. *See also Horton Indus., Inc. v. Village of Moweaqua*, 492 N.E.2d 220 (Ill. Ct. App. 1986); *D.E.W., Inc. v. Depco Forms, Inc.*, 827 S.W.2d 379 (Tex. Ct. App. 1992).

53 The term "substantial performance" is used in many cases interchangeably with the term "substantial completion." *See, e.g., Worthington Corp. v. Consolidated Aluminum Corp.*, 544 F.2d 227, 230-31 (5th Cir. 1976) (using substantial completion and substantial performance interchangeably); *Husar Indus. Inc. v. A.G. Huber & Sons, Inc.*, 675 S.W.2d 565, 572-73 (Mo. Ct. App. 1985) (same); *see also* RESTATEMENT (SECOND) OF CONTRACTS § 237 cmt. d, § 241 cmt. b (1979).

54 *See, e.g., O & M Constr., Inc. v. State, Div. of Admin.*, 576 So. 2d 1030, 1035 (La. Ct. App.) ("substantial performance . . . means that the construction is fit for the purpose intended . . ."), *Husar Indus., Inc. v. A.L. Huber & Son, Inc.*, 674 S.W.2d 565 (Mo. Ct. App. 1984) ("a building is substantially complete when it has reached the state in its construction so that it can be put to the use for which it was intended"); *Dittmer v. Nokleberg*, 219 N.W.2d 201, 206 (N.D. 1974) ("there is substantial performance . . . when all the essentials necessary to the full accomplishment of the [building's intended] purposes . . . are performed").

retained contract funds to complete the remaining work if the contractor fails to do so.[55]

In essence, substantial performance means that the contractor has completed its work to such an extent that it cannot be said to have *materially* breached the contract.[56] Absent a material breach of contract, the contractor may not be terminated for default and thereby deprived of its contact price as adjusted to compensate the owner for any uncompleted work. In view of judicial dislike for the forfeiture aspects of default

---

55 *See Prudential Ins. Co. of Am. v. Stratton*, 685 S.W.2d 818 (Ark. Ct. App. 1985), *aff'd*, 685 S.W.2d 818 (Ark. Ct. App.1985) ("[A] contractor who has substantially performed is entitled to recover the contract price, less the difference in value between the work as done and as contracted to be done, or less the cost of correcting defective work where this can be done without great expense or material injury to the structure as a whole"); *O & M Constr., Inc. v. State*, 576 So. 2d at 1035 ("a building contractor is entitled to recover the contract price even though defects or omissions are present when he has substantially performed the building contract"); *Dixon v. Nelson*, 107 N.W.2d 505, 507 (S.D. 1961) (holding that a contractor who has rendered substantial performance is entitled to the contract price with a deduction for minor defects and nonperformance).

56 *See J.M. Beeson Co. v. Sartori*, 553 So. 2d 180 (Fla. Dist. Ct. App. 1989), *aff'd*, 584 So. 2d 572 (Fla. Ct. App. 1991). In *Beeson*, a contractor who accepted a contract for the construction of a shopping center, which included several anchor tenants and out-parcels, completed all but one anchor tenant facility within the contract time. The owner alleged that the contractor was in default and sought to assess liquidated damages. The court, however, ruled in favor of the contractor as follows:

> The doctrine of "substantial performance" is held by this court in *Ocean Ridge Development Corp. v. Quality Plastering, Inc.*, 247 So. 2d 72, 75 (Fla. Ct. App. 1971): "Substantial performance is that performance of a contract which, while not full performance, is so nearly equivalent to what was bargained for that it would be unreasonable to deny the promisee the full contract price subject to the promisor's right to recover whatever damages have been occasioned him by the promisee's failure to render full performance. *See* 3A *Corbin on Contracts*, § 702 et seq." To say that substantial performance is performance which is nearly equivalent to what is bargained for, as the case law defines the term, in essence means that the owner can use the property for the use for which it is intended . . . .

Id. at 182.

terminations, courts frequently have rejected owner allegations of material breaches in the face of substantial performance.[57]

Since the owner may not terminate the contract for default after it has been substantially performed, the owner necessarily is not permitted to pursue the surety under the performance bond. In recognition of this rule, owners sometimes require contractors at the time of substantial completion to provide separate "maintenance" or "warranty" bonds to ensure completion of punch list work or subsequently discovered latent defects, or may state expressly in the performance bond that the surety's obligations will extend to incomplete or defective work remaining after substantial completion. Whether a surety ultimately will be held liable under its performance bond for latent defects discovered after substantial performance depends upon a variety of factors including: (1) whether the precise language of the bond affords coverage; (2) if the bond affords

---

57 *See United States ex rel. N. Maltese & Sons, Inc. v. Juno Constr.*, 759 F.2d 253 (2d Cir. 1985) ("Maltese had complied sufficiently with American Institute of Steel Construction Standards insofar as tolerances were concerned and the root openings in the majority of the connections between the columns and beams fell within acceptance mill tolerances"); *Radiation Tech., Inc. v. United States*, 366 F.2d 1003 (Ct. Cl. 1966) ("It is our view that even where time is of the essence, *i.e.*, where performance must occur by a given date, this factor must not demand that performance be measured in terms of strict conformity. It does require that performance be timely, but assuming this, there would thereafter remain for inquiry the question as to whether performance was substantial in other respects."); *Wilson v. Kapetan, Inc.*, 595 A.2d 369 (Conn. Ct. App. 1976) ("Under Louisiana law, substantial performance is a question of fact and a primary factor is whether the plant can be put to the use for which it was originally intended."); *Campagna v. Smallwood*, 428 So. 2d 1343 (La. Ct. App. 1983) ("substantial performance by a contractor is readily found, despite the existence of a large number of defects in both material and workmanship, unless the structure is totally unfit for the purpose for which it was originally intended"); *Husar Indus., Inc. v. A.L. Huber & Son, Inc.*, 674 S.W.2d 565 (Mo. Ct. App. 1984) ("A building is substantially complete when it has reached the state in its construction so that it can be put to the use for which it was intended even though comparatively minor items remain to be furnished or performed . . ."); *Patco Homes, Inc. v. Rochetti*, 522 N.Y.S.2d 903 (N.Y. App. Div. 1987); *M&W Masonry Constr., Inc. v. Head*, 562 P.2d 957 (Okla. Ct. App. 1976) ("We hold under the evidence plaintiff substantially performed its agreement and therefore had a right to quit for lack of an installment progress payment.").

coverage, the amount of time which has passed between the date of substantial completion and the date of discovery of the latent defects; (3) the extent and nature of the defects; and (4) the case law of the relevant jurisdiction.[58]

2. *The Owner's Duties Incident To Project Design*

It has long been a fundamental principle of law that the party in control of preparing and furnishing the detailed design documents is responsible for design inadequacies. Thus, where the owner issues detailed design documents found to be defective, fails to disclose critical information to the contractor, or maintains control over the contractor's construction methods, the contractor's nonperformance will be excused.

a. The Owner's Implied Warranty Of Design Adequacy

The owner's implied warranty of the adequacy of its detailed design documents furnished to the contractor is a firmly established principle of construction law,[59] which has been adopted in virtually all American

---

58 *See, e.g., Regents of the Univ. of Cal. v. Hartford Accident & Indem. Co.*, 131 Cal. Rptr. 112 (Cal. Ct. App. 1976) *vacated on other grounds*, 21 Cal. 3d 624 (1978); *School Bd. of Pirellas County v. St. Paul Fire & Marine Ins. Co.*, 449 So. 2d 872 (Fla. Dist. Ct. App. 1984); *Florida Bd. of Regents v. Fidelity & Deposit Co.*, 416 So. 2d 30 (Fla. Ct. App. 1982); *Hunters Pointe Partners v. United States Fidelity & Guar. Co.*, 486 N.W.2d 136 (Mich. Ct. App. 1992); *Vitenas v. Centanni*, 381 So. 2d 531 (La. Ct. App. 1980).

59 *See United States v. Spearin*, 248 U.S. 132 (1918). In *Spearin*, a construction site was flooded by a break in a sewer line constructed in conformance with government design documents. The government took the position that the contractor was responsible for the site clean-up because the project had not been completed and accepted and the contract required the contractor to assume responsibility for the site and examine the site and the plans. The contractor showed that it had complied fully with the government's plans and specifications in constructing the broken sewer line. In a lucid opinion written by Justice Brandeis, the Court held:

> [I]f the contractor is bound to build according to plans and specifications prepared by the owner, the contractor will not be responsible for the consequences of defects in the plans and specifications. . . . [T]he insertion of the articles prescribing the character, dimensions, and location of the

jurisdictions.[60] This obligation has been held to extend from a general contractor to his subcontractors, even though the general contractor simply passed on the design information received from the owner,[61] but not from a surety who issues the owner's design information to its completing contractor.[62] The owner's approval of shop drawings and submittals has been held to incorporate their details into the owner's design for implied warranty purposes.[63] The implied warranty may be invoked only insofar as the contractor reasonably relies upon the owner's detailed design

---

> sewer imported a warranty that if the specifications were complied with, the sewer would be adequate. This implied warranty is not overcome by the general clauses requiring the contractor to examine the site, to check the plans, and to assume responsibility for the work until completion and acceptance. The obligation to examine the site did not impose him the duty of making a diligent inquiry into the history of the locality, with a view to determining, at his peril, whether the sewer specifically prescribed by the government would prove adequate. The duty to check the plans did not impose the obligation to pass upon their adequacy to accomplish the purpose in view.

Id. at 135.

60 *See* Annotation, *Construction Contractor's Liability to Contractee for Defects or Insufficiency of Work Attributable to the Latter's Plans and Specifications,* 6 A.L.R.3D 1394 (citing authorities from various jurisdictions which have adopted the implied warranty of design specifications).

61 *See APAC Carolina, Inc. v. Town of Allendale*, 41 F.3d 157 (4th Cir. 1994).

62 *See D.H. Blattner & Sons, Inc. v. Firemen's Ins. Co.*, 535 N.W.2d 671 (Minn. Ct. App. 1995).

63 *See Northeastern Plate Glass Corp. v. Murray Walter, Inc.*, 537 N.Y.S.2d 657 (N.Y. App. Div. 1989). In *Northeastern*, the owner was held responsible for a defective window wall panel anchorage system that resulted in four panels falling off a state university dormitory building within two years of their installation. The defective anchoring details were suggested and furnished by the contractor's wall panel supplier. In imposing responsibility upon the owner, the court concluded that improper design rather than improper materials or installation was the cause of the failure and ruled that the owner retained responsibility for the loss because "the owner and its architects retained all ultimate control over and responsibility for" the construction plans and specifications. *See also Toombs & Co. v. United States*, 4 Cl. Ct. 535 (1984), *aff'd*, 770 F.2d 183 (Fed. Cir. 1985).

information. Absent reasonable reliance upon the owner's defective design, the contractor's non-performance is not excused.[64]

b. The Owner's Implied Warranty Of Commercial Availability Of Specified Construction Materials

A corollary to the owner's implied warranty of design adequacy is that, when the owner specifies a single-source product, the owner also impliedly warrants the commercial availability of the product.[65] If the product is not commercially available, the contractor and surety cannot be held responsible for delays or damages caused by the product's unavailability.

c. The Government Contractor Defense

A variant to the owner's implied warranty of the adequacy of its design is the "government contractor" defense. Under this defense, a contractor who performs work or furnishes materials pursuant to the government's detailed design documents is exonerated from liability to both the government and injured third persons.[66]

---

64 *Al Johnson Constr. Co. v. United States*, 854 F.2d 467 (Fed. Cir. 1988); *Tyger Constr. Co. Inc. v. United States*, 31 Fed. Cl. 177 (1994).

65 *See Edward M. Crough, Inc. v. Department of Gen. Serv.*, 572 A.2d 457 (D.C. App. 1990). In this case, the school district issued a re-roofing specification that required the contractor to install roofing materials manufactured by a single company. When the manufacturer later refused to sell the materials except at an exorbitant price, the owner allowed the contractor to substitute different materials after an extensive delay. Upon completion, the contractor sued the owner for its extra costs of performance based upon the owner's alleged breach of its implied warranty of commercial availability. The court upheld the implied warranty of commercial availability but concluded, however, that the warranty did not extend to specified materials which were in fact available but at an exorbitant price.

66 *See Board of Ed. v. W.R. Grace Corp.*, 609 A.2d 92 (N.J. Super. 1992). In this case, the school district's claim against the contractor for installation of construction materials containing asbestos was denied because the materials had been specified in detail by the owner and its project architect, and the contractor had merely complied with the architect's plans and specifications. In reaching this decision, the court followed long-standing precedent. *See*

d. The Owner's Implied Duty Of Disclosure

Implied in every contract is a duty on the part of the owner to disclose information material to the contractor's performance that is not available from other sources.[67] Failure by the owner to disclose material information is a breach of contract which can result in both the contractor and the surety being exonerated from liability for damages resulting from the breach.

e. Blended Design And Performance Specifications

Owners and designers frequently prepare specifications containing both design details and performance requirements to be satisfied by the contractor. Sometimes these performance requirements take the form of performance acceptance tests. When design details and performance acceptance tests requirements are incompatible, the issue of liability must be carefully analyzed.[68]

---

also Boyle v. United Tech., 487 U.S. 500 (1988); *Yearsley v. W. A. Ross Constr. Co.*, 309 U.S. 18 (1940); *Ryan v. Feeny & Sheehan Bldg. Co.*, 145 N.E. 321 (N.Y. 1924).

67 *See City of Indianapolis v. Twin Lakes Enters., Inc.*, 568 N.E.2d 1073 (Ind. Ct. App. 1991). In this case, the city was found to have breached its implied duty of disclosure by failing to inform a dredging contractor about large obstructions previously dumped by the city into the reservoir to be dredged:

> [T]he City did not inform Twin Lakes about the extent of the dumping but insisted that Twin Lakes continue to dredge the site, which it did. The contract and the circumstances in this case were of such a nature that Twin Lakes could only partly perform its obligations. In fact, Twin Lakes continued to perform the dredging after the money provided under the original contract and change orders had been expended until it was expelled from the site. The failure to inform Twin Lakes of the nature and extent of the obstructions, in the face of assumed obligations to the contrary, went to the very heart of the agreement to dredge the site. As such, the breach of that obligation constituted a material breach of contract.

Id. at 1080.

68 Illustrative of this point is *W.H. Lyman Constr. Co. v. Village of Gurnee*, 403 N.E.2d 1325 (Ill. Ct. App. 1980). In this case, the Village's contract

f. The Owner's Approval Of Contractor Plant And Equipment Or "Work Plan"

Contract documents frequently include provisions that require the owner to approve the contractor's plant and equipment or "work plan" prior to the owner's issuance of a notice to proceed. By approving the contractor's work plan, the owner implicitly accepts as reasonable the assumptions which underlie the plan (such as assumptions relating to expected site conditions) and may later be precluded from arguing that the plan was inadequate to complete the work.[69]

---

documents for construction of a sewer (1) contained both detailed plans and specifications describing the work, and (2) imposed an infiltration test which the completed sewer was required to pass before the Village would accept the work and make final payment to the contractor. The contract also contained a disclaimer of responsibility by the owner and engineer regarding the contractor's inability to meet the infiltration test. After the sewer was competed in conformance with the contract documents, the contractor was unable to pass the infiltration test and the owner refused to pay. The court resolved the subsequent suit in favor of the contractor as follows:

We construe this provision [the acceptance test and disclaimer] as an impermissible attempt on the part of the Village to shift the responsibility for the sufficiency and adequacy of the plans to the contractor without providing the contractor the corresponding benefit of having something to say about the plans he is strictly bound to follow. The contractor's duty is to perform his part of the contract in a workmanlike manner, not to evaluate the suitability of the specifications or, in the language of *Spearin*, "to pass upon their adequacy to accomplish the purpose in view."

Id. at 1332; *see also Fruin-Colnon Corp. v. Niagara Frontier Transp. Auth.*, 585 N.Y.S.2d 248 (N.Y. App. Div. 1992); *Chantilly Constr. Corp. v. Dept. of Highways*, 369 S.E.2d 438 (Va. Ct. App. 1988).

69 *See United States v. Atlantic Dredging Co.*, 253 U.S. 1 (1919). In this case, a dredging contractor encountered compacted sand and gravel with cobbles and complained that these materials were different from conditions represented in the plans and specifications to be "mainly mud or mud with an admixture of sand" (*i.e.*, non-compacted). The contract documents had reserved to the owner the right to inspect and approve the contractor's plant and equipment prior to allowing the contractor to proceed with the work.

Extensive owner "field control" over products furnished by the contractor also may excuse the contractor's non-performance.[70] The same is true with respect to construction methods specified by the owner that prove unsuitable due to unanticipated conditions.[71] Similarly, a "methods"

The owner had exercised this right and given approval. The significance of the owner's approval was explained by the court as follows:

> [T]he government's care of its interests extended to the inspection of the instrumentalities of the contractor, and required the character and capacity of the plant which was to be used, to be submitted for inspection and approval. In fulfillment of the requirement, the company submitted its plan. It was only efficient for dredging material of the character mentioned in the specifications and described on the map, and it was so approved. The significance of the submissions and approval is manifest. The character and capacity of the plant conveyed to the [government] the fact that the company was accepting as true the representation of the specifications and the map of the materials to be dredged; and reciprocally the approval of the plant by the [government] was an assurance to the company of the truth of the representation, and the justification of reliance upon it.

Id. at 5.

70 *See C.J. Langenfelder & Son, Inc. v. Pennsylvania Dep't of Transp.*, 404 A.2d 745 (Pa. 1979). In this case, the owner was held to have impliedly warranted that the contractor's concrete was adequate for the owner's intended purposes, even though the concrete mix design was designed and furnished by the concrete supplier. The owner had exercised detailed inspection and control in approving the concrete mix design, inspecting the ingredients prior to mixing and testing the concrete prior to placement. The court recognized the owner's extensive field control over the concrete in reaching this conclusion: "When the [DOT] seeks to exert such *close control* over the source, the formulation, and the production of the material to be used in the construction project, it cannot disclaim responsibility for delays caused by defects in that material." *Id.* at 751 (emphasis added); *compare Stabler Constr. Inc. v. Pennsylvania Dep't of Transp.*, 692 A.2d 1150 (Pa. 1997).

71 *See McCree & Co. v. State*, 91 N.W.2d 713 (Minn. 1958). In *McCree*, a highway embankment specification required that embankment fill be compacted to the specified density. The contractor, using compaction methods dictated by the owner's design, failed to achieve the desired density. The owner argued that contractor workmanship was the cause of the problem. The Supreme Court of Minnesota, however, sided with the contractor as follows:

specification also may shift liability for delayed completion from the contractor to the owner.[72]

g. The Owner's Implied Warranty Of Design Versus The Contractor's Warranty Of Materials

Contract documents usually contain clauses under which the contractor warrants its construction materials against defects. Owners frequently argue that the contractor's express warranty against material defects overrides the owner's implied warranty of design where the product fails as a result of design considerations.[73] The cases, however,

---

> [T]he owner's actions in furnishing detailed plans and specifications control not only the particular result to be accomplished, but also the particular construction methods to be followed and used and supports an implied warranty in keeping with the intention and expectation of the parties that the plans, specifications, and soils conditions were such as would permit successful conclusion of the work. The State was the party in control who dictated the entire contract and retained control from start to finish.

Id. at 725.

72 *See Midwest Dredging Co. v. McAninch Corp.*, 424 N.W.2d 216 (Iowa 1988). In this case, the state specified that the contractor should use a twenty-two inch transfer pipe to hydraulically transfer highway embankment fill material from a borrow site to the embankment site, but then sought to hold the contractor responsible for failure to complete the contract when the contractor encountered a significant number of boulders in the borrow site material larger than the transfer pipe diameter. The Iowa Supreme Court held as follows:

> DOT drew up plans based on its tests and required that a specific dredging and piping technique be employed. By doing so, DOT impliedly represented that the material from Borrow C could be hydraulically dredged and piped in accordance with its plans and [specifications]. DOT went beyond mere presentation of boring results and an implied warranty consequently arose.

Id. at 222.

73 *See Rhone Poulenc Rorer Pharmaceuticals, Inc. v. Newman Glassworks*, 112 F.3d 695 (3d Cir. 1997), in which a contractor's warranty of workmanship was held (in a two to one decision) to supersede the owner's

suggest that both warranties are equally effective and the determination of which warranty has been breached is a question of fact for the jury.[74]

h. The Owner's Responsibility For Latent Ambiguities In Its Design

Arguments over the intent of the owner's design documents are frequent in construction. Contractor non-performance due to latent ambiguities or conflicts in design remain the responsibility of the owner. Courts will apply various rules of construction in an effort to resolve ambiguity.[75] These rules include consideration of "reasonable logical interpretation," the "whole agreement," the "principal apparent purpose," "trade custom and usage," "concurrent interpretation and practice," "course of dealing" under prior contracts, "silent acquiescence" to the other party's interpretation, and, as a last resort, "construing against the drafter."[76]

---

implied warranty of design adequacy with respect to delaminated opacifiers on curtainwall spandral glass.

74 *See Trustees of Indiana Univ. v. Ætna Cas. & Sur. Co.*, 920 F.2d 429 (7th Cir. 1990). In this case, the owner and architect had approved the use of "courthouse blend" bricks as the face material for four buildings being constructed at Indiana University's Southeast Regional Campus in Albany, Indiana. Under the contract, the contractor warranted its work to be free of defects. When the bricks began to "spall" and deteriorate two years after installation, the bricks were tested and found to be excessively porous and with a high rate of water absorption. At trial, the owner claimed that the porous condition was due to the bricks being "under-fired during the manufacturing process," and thus defective. The contractor argued, however, that the "courthouse blend" bricks, as properly manufactured, were not "suited to the environmental conditions [severe freeze/thaw conditions] in which they were used." A jury decided in favor of the contractor and its surety and against the owner. On appeal, the United States Court of Appeals for the Seventh Circuit affirmed that the jury could find that the owner's design specification requiring installation of courthouse blend bricks was defective rather than the bricks themselves and was not "trumped" by the contractor's express warranty of its materials.

75 *See* RESTATEMENT (SECOND) OF CONTRACTS §§ 212-223 (1979); 17 AM. JUR. 2D *Contracts* § 272-281.

76 *See Darwin Constr. Co. v. United States*, 31 Fed. Cl. (1994). In *Darwin*, a specialized trade meaning was used to interpret the contractor's obligations under a landscaping contract. The question was whether a "plant list" and a "planting by lots" provision were intended to be synonymous or separate.

3. *The Owner's Implied Duty Of Cooperation*

American jurisprudence applies in all contracts the obligation to cooperate in the performance of the contract and not to delay, hinder, or interfere with the performance of other parties.[77] Over the years, owners have been found on numerous occasions to have breached their implied duty of cooperation in the context of construction contracts. Such instances have included failure to provide timely site access,[78] failure to complete other work necessary to allow the contractor to proceed,[79] and failure to reasonably schedule and coordinate the work.[80]

Owners frequently seek to hold contractors responsible for non-performance over which contractors have no control. Unless contractual language expressly provides that contractors will be responsible to complete on time without regard to cost, contractors ordinarily are entitled to extensions of time for delays beyond their control and may even be entitled to compensation for delay within the owner's control. "Excusable" causes of delay have included unusually severe weather, strikes and labor disturbances, acts of God, acts of government, and other

---

The court ruled that the ambiguity was to be construed in accordance with specialized trade usages which indicated that the "plantings by lots" list was intended only to designate plant locations, rather than additional plants to be furnished.

77 *See Gulf Mobile & Ohio Railway Co. v. Illinois Cent. Railway Co.*, 128 F. Supp. 311 (N.D. Ala. 1954), *aff'd*, 225 F.2d 816 (5th Cir. 1955), in which this doctrine was articulated as follows:

> A contracting party impliedly obligates himself to cooperate in the performance of his contract and the law will not permit him to take advantage of an obstacle to performance which he has created or which lies within his power to remove.

Id. at 324.

78 *See Capitol City Drywall Corp. v. C.G. Smith Constr. Co.*, 270 N.W.2d 608 (Iowa 1978); *Douglas Northwest Inc. v. Bill O'Brien & Sons Constr. Inc.*, 828 P.2d 565 (Wash. Ct. App. 1992); *R.G. Pope Constr. Co., Inc. v. Guard Rail of Roanoke, Inc.*, 244 S.E.2d 774 (Va. 1978).

79 *See J.J. Brown Co. v. J.L. Simmons Co.*, 118 N.E.2d 781 (Ill. App. Ct. 1954).

80 *See Tribble & Stephens Co. v. Consolidated Serv.*, 744 S.W.2d 945 (Tex. Ct. App. 1988).

events beyond the control of both the owner and contractor. "Compensable" delays, which excuse contractor non-performance and entitle the contractor to be compensated for the delay, include the entire gamut of delays within the exclusive control of the owner under typical construction contracts: providing contract "direction," making timely and proper inspections, providing timely review and approval of shop drawings, furnishing timely owner materials and equipment, providing adequate design documents, furnishing proper coordination of other contractors on site, making changes in the work, and a host of other similar responsibilities.

Owner allegations that the contractor is behind schedule must be analyzed carefully to determine whether the contractor in fact is responsible for the delays or is entitled to more time (and perhaps money) because the delays were caused either by the owner or by supervening events beyond the control of both the owner and contractor. An owner's failure to grant proper time extensions may preclude the owner for terminating the contractor for default based upon delay.[81] An owner's inclusion in the contract of exclusive remedies for contractor delay, such as a "claim submission" clause, "liquidated damages" clause, or special payment clause, also may preclude termination for delay.[82]

4. The Owner's Responsibility For Differing Site Conditions

Unless expressly disclaimed in the contract, the owner may be responsible for non-performance due to conditions at the site that differ materially from those represented in the contract documents or otherwise ordinarily encountered. Most contracts today contain an explicit "differing site conditions" clause under which the owner assumes the risk of such conditions. The purpose of such a clause is to make it unnecessary for contractors to include large contingencies in their bids to cover the risk of encountering unanticipated adverse subsurface conditions or concealed

---

81 *See J.D. Hedin Constr. Co. v. United States*, 408 F.2d 424 (Ct. Cl. 1969); *W.F. Magann Corp. v. Diamond Mfg. Co., Inc.*, 580 F. Supp. 1299 (D.S.C. 1984), *aff'd in part, rev'd in part*, 775 F.2d 1202 (4th Cir. 1985).

82 *See Vermont Marble Co. v. Baltimore Contractors, Inc.*, 520 F. Supp. 922 (D.D.C. 1981); *Mellon Stewart Constr. Inc. v. Metropolitan Water Reclamation Dist.*, No. 94C1915, 1995 WL 124133 (N.D. Ill. 1995); *Construction Contracting & Mgmt. v. McConnell*, 815 P.2d 1161 (N.M. 1991).

conditions in existing structures.[83] Even in the absence of an express clause, however, owners have been held liable for differing site conditions under various theories including (1) breach of implied warranty;[84] (2) owner misrepresentation of conditions in the contract documents;[85] (3) owner failure to disclose material information;[86] and (4) mutual mistake.[87]

5. *Owner Failure To Properly Administer The Contract*

Contractors have been excused for non-performance caused by owner maladministration of the contract in the following situations:

a. Change Orders

Virtually all construction contracts contain a clause allowing the owner unilaterally to issue written change orders during contract performance to change contract requirements without being in breach of contract. The clause enables the owner to order desired changes within the general scope of the contract, to take advantage of new or improved construction methods and techniques, to correct errors and plans, to amend the construction schedule to deal with unforeseen circumstances or field conditions, or to address changed owner needs. Upon issuance of a change order, the owner has an obligation to pay the contractor for the extra work and to extend extra time needed for performance.[88] Verbal

---

83 *See Foster Constr. v. United States*, 435 F.2d 873 (Ct. Cl. 1970); *Metropolitan Sewer Comm'n v. R.W. Constr.*, 241 N.W.2d 371 (Wis. 1976), *aff'd*, 255 N.W.2d 293 (Wis. 1977).

84 *See Big Chief Drilling Co. v. United States*, 26 Cl. Ct. 1276 (1992).

85 *See Hollerbach v. United States*, 233 U.S. 165 (1914).

86 *See PT&L Constr. Co. v. New Jersey Dep't of Transp.*, 531 A.2d 1330 (N.J. 1987).

87 *See John Burns Constr. Co. v. Interlake, Inc.*, 433 N.E.2d 1126 (Ill. App. Ct. 1982); *S.J. Groves & Sons & Co. v. State*, 273 S.E.2d 465 (N.C. Ct. App. 1980).

88 *See Bagwell Coatings, Inc. v. Middle South Energy, Inc.*, 797 F.2d 1298 (5th Cir. 1986) (holding that a contractor was entitled to a "total cost" recovery for the owner's breach of contract, where the owner's construction manager, after giving a contractor "assurances that it would work out its cost problems" related to change orders, continued to "haggle over minutia" and refused to authorize additional compensation for changes because of "alleged documentation inadequacies").

instructions that change the contract may permit recovery under the doctrine of "constructive change."[89]

b. Owner Nonpayment

Owner nonpayment has served as the basis for setting aside default terminations in numerous cases.[90] The obligation to pay will be considered in the context of surrounding circumstances, even where contract language makes it discretionary.[91]

---

89 *See A.S. McGaughan Co. Inc. v. Barram*, 113 F.3d 1256 (Fed. Cir. 1997); *Julian Speer Co. v. Ohio State Univ.*, 680 N.E.2d 254 (Ohio Ct. Cl. 1997).

90 *See, e.g., Cundy Asphalt Paving Constr., Inc. v. Angelo Materials Co.*, 915 P.2d 1181 (Wyo. 1996) (holding that contractor's wrongful nonpayment due to dispute over quality of asphalt constituted a breach of the subcontract and justified the subcontractor in abandoning performance); *Darrell J. Didericksen & Sons v. Magna Water*, 613 P.2d 1116 (Utah 1980) ("It was only when Magna refused to provide acceptable change orders that the work stopped, as the contractor refused to proceed further without written direction or authorization as required."); *White River Dev. Co. v. Meco Sys., Inc.*, 806 S.W.2d 735 (Mo. Ct. App. 1991) ("When a progress payment due is not made, a builder is entitled to suspend performance and await either other assurances satisfactory to it or actual payment."); *Blake Constr. Co., Inc. v. C. Coakley Co., Inc.*, 431 A.2d 569 (D.C. App. 1981) ("When Blake failed to assure Coakley that it would be compensated for the additional expenses it incurred beyond those contemplated in the subcontract, Blake also breached the implicit terms of the subcontract and Coakley acted properly in discontinuing performance under the subcontract."); *John W. Johnson, Inc. v. Basic Constr. Co.*, 429 F.2d 764 (D.C. Cir. 1970) ("We accordingly decide that Basic was required to give Johnson a commitment for payment for the extra painting even though it received none from [the owner], that it breached the contract when it refused to do so, that Johnson's abandonment was fully justified and that it was entitled to be compensated for the work it performed."); *Tennessee Asphalt Co. v. Purcell Enters.*, 631 S.W.2d 439 (Tenn. Ct. App. 1982) (substantial delay in payment constituted a material breach); *United States ex rel Endicott Enters. v. Star Brite Constr. Co.*, 848 F. Supp. 1161 (D. Del. 1994).

91 *United States v. Lennox Metal Mfg. Co.*, 225 F.2d 302 (2d Cir. 1955) (surrounding circumstances dictated that the government was required to make progress payments even though the contract language made them discretionary).

c. "Cardinal" Changes

A truly fundamental change in the nature of the bonded contract on the part of the owner is a breach of contract which discharges the contractor from its obligation to further perform.[92] This type of change is usually referred to in the case law as a "cardinal" change. Although the case law provides some guidance in determining whether a cardinal change has occurred, courts routinely caution that "each case must be analyzed on its own facts and in light of its own circumstances."[93]

There are two distinct tests for cardinal changes. First, a cardinal change is said to occur when the obligee affects an alteration of the work so drastic that it effectively requires the contractor to perform duties materially different from those originally bargained for.[94] This expression of the cardinal change doctrine is sometimes referred to as the "scope of the contract" test.[95] Second, in the context of projects awarded through a competitive bidding process, a cardinal change can occur when the contract is modified in such a way so as to materially change the field of competition.[96] This expression of the doctrine is sometimes referred to as

---

92 *In re Liquidation of Union Indem. Ins. Co.*, 632 N.Y.S.2d 788 (N.Y. App. Div. 1995) (holding that a material alteration that discharged the surety of its performance bond obligation was created by two change orders that added work specifically excluded under the base contract and nearly doubled the original contract price); *see also Hancock Elec. Corp. v. Washington Metro. Area Transit Auth.*, 81 F.3d 451 (4th Cir. 1996); *Airprep Tech., Inc. v. United States*, 30 Fed. Cl. 488 (1994) (holding that the contractor's alleged default was excused because that the contractor's failure of performance arose from the government's insistence upon additional requirements that constituted a cardinal change to the contract); *C. Norman Peterson Co. v. Container Corp. of Am.*, 218 Cal. Rptr. 592 (Cal. Ct. App. 1985).

93 *Edward R. Marden Corp. v. United States*, 442 F.2d 364, 369 (Ct. Cl. 1971); *Wunderlich Contracting Co. v. United States*, 351 F.2d 956, 966 (Ct. Cl. 1965).

94 *See General Dynamics Corp. v. United States*, 585 F.2d 457, 462 (Ct. Cl. 1978).

95 *See Aragona Constr. Co. v. United States*, 165 Ct. Cl. 382, 391 (1964).

96 *See Cray Research, Inc. v. Dept. of Navy*, 556 F. Supp. 201, 203 (D.D.C. 1982).

the "scope of the competition" test, and requires an analysis of the substance of the change rather than its magnitude.[97]

The surety should check for cardinal changes by applying the scope of the contract test or the scope of the competition test in the following situations:

(1) contract increases substantially, *i.e.*, by more than one hundred where the dollar value of the work to be performed under the percent; [98]

(2) where the method of performing the contract changes, so that the risks of performance are significantly altered;[99]

(3) in the competitive bidding context, where the method of performance changes in a way that potential bidders could not reasonably have anticipated;[100] or

(4) where the method of payment to or financing of the contractor changes in such a way so as to expose the surety to greater risk of loss.[101]

---

97 *See, e.g., Webcraft Packaging.*, Comp. Gen. Dec. B-194087, 79-2 CPD 120 (holding that post-bid change in requisition requirement, permitting use of standard paper rather than specialized paper, constituted a cardinal change, because a large number of potential suppliers of standard paper were prevented from bidding).

98 *See, e.g., Edward R. Marden Corp.*, 442 F.2d at 370 (cardinal change found where cost of completing construction more than doubled due to structural design errors); *Peter Kiewit Sons' Co. v. Summit Constr. Co.*, 422 F.2d 242 (8th Cir. 1969) (cardinal change found where cost of backfilling operation increased from $600,000 to approximately $2 million); *Employers Ins. of Wausau v. Construction Mgmt. Eng'rs, Inc.*, 377 S.E.2d 119 (S.C. Ct. App. 1989) (cardinal change occurred where contract value increased from $2.3 million to $6.2 million).

99 *See, e.g., Memorex Corp.*, 61 Comp. Gen. 42, B-200722, 81-2 CPD ¶ 334 (change in procurement method from outright purchase to a lease-to-ownership plan shifted substantial risks from the government to contractor and, therefore, constituted a cardinal change).

100 *See, e.g., American Air Filter Co.*, Comp. Gen. B-188408, 78-1 CPD ¶ 36; *W.H. Mullins*, Comp. Gen. B-207200, 83-1 CPD ¶ 158.

The availability of the cardinal change defense will depend heavily upon the facts and circumstances of the particular case. However, when a cardinal change in the bonded contract is found to occur, both the contractor and the surety will be discharged from further performance obligations.

d. Failing To Give Direction

The owner also breaches the contract if it fails to act reasonably in carrying out its contract obligations, such as by refusing to instruct the contractor on how to proceed in the face of adverse conditions,[102] or failing to exercise reasonable discretion to stop the work in the face of adverse conditions.[103]

6. *Impossibility/Impracticability Of Performance*

Construction contracts are sometimes simply impossible or impracticable to complete in strict conformance with their requirements. The doctrines of impossibility and impracticability of performance constitute legal excuses for non-performance. Impossibility excuses contractual non-performance found to be impossible by supervening causes beyond the control and not foreseeable by either party, such as weather or acts of the government. Impracticability, also called "practical impossibility," excuses non-performance of contracts made impossible "as a practical matter" because they only can be performed at an excessive or unreasonable cost. "Practical impossibility" thus excuses non-

---

101 *See, e.g., United States ex rel. Army Athletic Ass'n v. Reliance Ins. Co.*, 799 F.2d 1382 (9th Cir. 1986) (cardinal change occurred where contract was modified so that bond principal was no longer able to obtain adequate security for the surety's obligation); *Reliance Ins. Co. v. Colbert*, 365 F.2d 530, 534 (D.C. Cir. 1966) (cardinal change occurred where obligee revised payment schedule to provide significantly accelerated payments to contractor). Changes in the method of payment or financing must be distinguished from overpayments of the principal by the obligee. The former may constitute a cardinal change which fully discharges the surety. The latter is a defense which may discharge the surety to the extent of the overpayment. *See* Part III.C.5, *supra* (discussing cardinal changes).

102 *See Miller v. City of Broken Arrow*, 660 F.2d 450 (10th Cir. 1981).

103 *See Bignold v. King County*, 399 P.2d 611 (Wash. 1965).

performance of the contracts actually possible but commercially impracticable within the basic contract objectives contemplated by the parties.[104]

7. ***The Owner's Implied Waiver Of Contract Requirements***

The owner may impliedly waive requirements of the contract and thereafter may be estopped to compel a contractor to perform in conformance with those requirements. Waiver due to knowledge of and acquiescence in deviations or due to an express course of dealing have long been recognized.[105] Waiver of the contract completion date by taking no action to terminate the contractor's performance in encouraging the contractor to perform also has long been recognized.[106] The owner also may waive work requirements and specifications through a course of dealing of accepting nonconforming work.[107]

---

104 *See Transatlantic Financing Corp. v. United States*, 363 F.2d 312, 315 (D.C. Cir. 1966):

> The doctrine of [practical impossibility] ultimately represents the evershifting line, drawn by courts hopefully responsive to commercial practices and mores, at which the community's interest in having contracts enforced according to their terms is outweighed by the commercial senselessness of requiring. . . .

See generally RESTATEMENT (SECOND) OF CONTRACTS §§ 261-272 (1979).

105 *See Continental Ins. Co. v. City of Va. Beach*, 908 F. Supp. 341 (E.D. Va. 1995), in which a performance bond surety was discharged of all liability to a city obligee where the city materially deviated from its obligations under the bonded contract by completing payment to the contractor without properly inspecting and testing the work as required under the contract. *See also Allen & Hara, Inc. v. Barrett Wrecking, Inc.*, 898 F.2d 512 (7th Cir. 1990); *Transpower Constructors v. Grand River Dam Auth.*, 905 F.2d 1413 (10th Cir. 1990); *Lasker-Goldman Corp. v. City of New York*, 633 N.Y.S.2d 771 (N.Y. App. Div. 1995).

106 *See DeVito v. United States*, 413 F.2d 1147 (Ct. Cl. 1969).

107 *See Miller Elevator Co. v. United States*, 30 Fed. Cl. 662 (1994); *Hannon Elec. Co. v. United States*, 31 Fed. Cl. 135 (1994), *aff'd*, 52 F.2d 343 (Fed. Cir. 1995) (expressly adopting the doctrine of constructive waiver of specifications in situations where the government administered an initially unambiguous contract in such a way as to give the contractor the

8. *The Owner's Insistence Upon Strict Compliance In The Face Of Economic Waste*

Owners are entitled to require contractors to comply with contract specifications.[108] The owner's right to insist upon strict compliance is not unlimited, however, and is circumscribed sharply by the doctrine of economic waste.[109] An owner is not permitted to require the replacement of nonconforming work where the cost of correction is economically wasteful, and the work is otherwise adequate for its intended purpose.[110]

9. *Hypertechnical Inspection*

Owner insistence upon contractor conformance to standards more rigorous than those set forth in the plans and specifications is improper.

---

impression that the contract requirements had been waived); *see also McDonnell Douglas Corp. v. United States*, 35 Fed. Cl. 358 (1996) (opining that "the Navy's less harsh position regarding contract specifications. . . during contract performance confirms that the termination for default [for failure to comply with the specifications] was improper").

108 *See Elastomeric Roofing Assocs., Inc. v. United States*, 26 Cl. Ct. 1106 (1992); *Farwell Co. v. United States*, 148 F. Supp. 947 (Ct. Cl. 1957).

109 *See Grossman Holdings, Ltd. v. Hourihan*, 414 So. 2d 1037 (Fla. 1982); *Witty v. C. Casey Homes, Inc.*, 430 N.E.2d 191 (Ill. App. Ct. 1981); *Young & Jacobs v. Kent*, 129 N.E. 889 (N.Y. 1921).

110 Illustrative of this point is *Granite Constr. Co. v. United States*, 962 F.2d 998 (Fed. Cir. 1992). In this case, the government insisted that the contractor remove and replace a nonconforming water stop already embedded in the vertical "cold joints" between monolithic concrete slabs poured in place to serve as the walls of a lock and dam on the Mississippi River. The contractor protested that the water stop, although not strictly conforming to the specification, was nevertheless adequate for intended purposes. The government, invoking strict compliance, insisted that the water stop be removed and replaced at a cost to the contractor of more than $3.8 million. At trial, the contractor successfully proved that the water stop exceeded by more than twenty times the project safety margin and could withstand water pressure more than forty times that which would ever exist. Under the circumstances, the United States Court of Appeals for the Federal Circuit ruled that the government's insistence upon strict compliance constituted economic waste and that the contractor was entitled to recover the $3.8 million cost incurred to replace the water stop.

Contractor failure to comply with such excessive standards cannot serve as a basis for termination of the contract for default. The issue of excessive standards frequently arises in situations where the owners' field inspectors insist that the work meet new standards or tests not clearly spelled out in the contract.[111]

10. Fraud Or Duress

Fraud or duress on the part of the owner that induces the contractor to enter into a contract or waive contract rights is a defense that can be relied upon by the surety.[112]

11. Release Or Settlement Of Claims

The owner's release or compromise of claims against the contractor will also release the surety. Many construction contracts provide that the owner's final payment will release owner claims except those expressly reserved.[113] During the course of contract performance, disputes also are settled and compromised. The surety should look carefully into the issues

---

111 *See State v. Buckner Constr. Co.*, 704 S.W.2d 837 (Tex. Ct. App. 1985). In this case, the state hypertechnically inspected work under a contract for the painting of structural steel on twenty-seven highway bridges. Although the specifications required that loose paint be removed by sandblasting, the owner's field inspectors insisted upon complete removal of all paint, to the bare steel, and utilized non-specified tests to determine compliance. Such non-specified tests included taping adhesive tape to the bridge steel, pulling the tape off, and then insisting that the contractor keep blasting if any flecks of paint were seen on the underside of the tape. At trial, the jury concluded that the owner's hypertechnical inspection constituted a breach of contract and awarded the contractor a "total cost" recovery exceeding by almost three times the original contract price.

112 *See Taylor & Jennings, Inc. v. Bellino Bros. Constr. Co.*, 393 N.Y.S.2d 203 (N.Y. App. Div. 1977) (holding that a surety for a subcontractor could assert the defense of fraud in the inducement against the contractor's claim on the performance bond).

113 *See, e.g.*, AIA Document A201, General Conditions of Contract for Construction ¶ 9.10.4 (1997); *Renown, Inc. v. Hensel Phelps Constr. Co.*, 201 Cal. Rptr. 242 (Cal. Ct. App. 1984); *cf. Continental Ins. Co. v. City of Virginia Beach*, 908 F. Supp. 341 (E.D. Va. 1995).

of release and settlement.[114] For example, have the owner and contractor entered into a new agreement during the course of construction as a result of certain conditions encountered. A surety may be released as of a result of a novation.[115] Conversely, duress by the owner can serve as a basis for setting aside a contractor's waiver and release of claim.[116]

D. Has The Contractor (Or Its Subcontractor) Materially Breached The Contract

The surety, upon termination, also must verify that its principal and subcontractor have performed all material obligations. Some events in the construction process to be verified carefully are described below:

---

114 *See Uhle v. Tarlton Corp.*, 938 S.W.2d 594 (Mo. Ct. App. 1997); *Vulcan Painters, Inc. v. MCI Constructors, Inc.* 41 F.3d 1457 (11th Cir. 1995); *Citadel Corp. v. Sun Chemical Corp.*, 443 S.E.2d 489 (Ga. Ct. App. 1994); *Nix v. Henry C. Beck Co.*, 572 So. 2d 1214 (Ala. 1990). In *Nix*, the Supreme Court of Alabama upheld the release of an owner's $1.5 million claim against the contractor who installed asbestos fireproofing material during the construction of a shopping mall. The owner's claim was barred by a broad general release executed by the owner as part of an earlier $2,000 settlement of certain sprinkler system problems. The court ruled that the broad general release was sufficient to release all claims related to the project. *See also George Hyman Constr. Co. v. United States*, 30 Fed. Cl. 170 (1993), *aff'd*, 39 F.3d 1197 (Fed. Cir. 1994) (holding that the *Severin* Doctrine was applicable to bar a prime contractor's claim on behalf of a released subcontractor).

115 *See MacKenzie Lab, Inc. v. Lawrence*, 80 F. Supp. 710 (D. Md. 1948); *Fidelity Deposit Co. of Md. v. Olney Assocs., Inc.*, 530 A.2d 1 (Md. Ct. App. 1987); *Zuni Constr. Co., Inc. v. Great Am. Ins. Co.*, 468 P.2d 980 (Nev. 1970).

116 *See Willms Trucking Co., Inc. v. J.W. Constr. Co., Inc.*, 442 S.E.2d 197 (S.C. Ct. App. 1994). In this case, the South Carolina Court of Appeals set aside for duress a contractor's waiver and release of lien because the owner took advantage of the contractor's extreme financial pressure to pay its subcontractors and suppliers by paying the contractor much less than it would have been entitled to for certain extra work.

1. The Contractor's Implied Duty To Seek Clarification Of Patently Ambiguous Design Documents

Courts have long refused to interpret an ambiguity in design documents against an owner where the contractor was or should have been aware of the ambiguity prior to bidding, was not misled, and had an opportunity to obtain pre-bid clarification of the ambiguity from the owner. Under this rule, an ambiguity will be construed against the contractor where the ambiguity is a *major patent discrepancy, obvious omission, or drastic conflict* in provisions (as opposed to being subtle and misleading) of which the contractor using due care was or should have been aware prior to entering into the contract.[117] The "patent ambiguity" doctrine even has been applied to prevent a contractor from benefiting

---

117 *Darwin Constr. Co. v. United States*, 31 Fed. Cl. 453 (1994); *H.B. Zachary Co. v. United States*, 28 Fed. Cl. 77 (1993), *aff'd*, 17 F.3d 1443 (Fed. Cir. 1994) (holding ambiguity regarding the painting of a steel roof deck was sufficiently obvious to have required the contractor to have made inquiry prior to bidding); *see also Community Heating & Plumbing Co. v. Kelso*, 987 F.2d 1575 (Fed. Cir. 1993); *Fortec Constructors v. United States*, 760 F.2d 1288 (Fed. Cir. 1985); *Seville Constr. Inc. v. United States*, 35 Fed. Cl. 242 (1996). *aff'd*, 108 F.3d 1395 (Fed. Cir. 1997); *Emerald Isle Elec. Inc. v. United States*, 28 Fed. Cl. 71 (1993); *Delcon Constr. Corp. v. United States*, 27 Fed. Cl. 634 (1993); *Gresham, Smith & Partners v. United States*, 24 Cl. Ct. 796 (1991); *Blinderman Constr. Co. v. United States*, 17 Cl. Ct. 860 (1989); *Monarch Painting Corp. v. United States*, 16 Cl. Ct. 280 (1989); *Bromley Contracting Co., Inc. v. United States*, 14 Cl. Ct. 69 (1987), *aff'd*, 861 F.2d 729 (Fed. Cir. 1988); *Brutoco Eng'g & Constr., Inc. v. United States*, 12 Cl. Ct. 104 (1987), *aff'd*, 833 F.2d 1023 (Fed. Cir. 1987); *Baltimore Contractors, Inc. v. United States*, 12 Cl. Ct. 328 (1987); *G.M. Shupe, Inc. v. United States*, 5 Cl. Ct. 662 (1984); *Enrico Roman, Inc. v. United States*, 2 Cl. Ct. 104 (1983); *Blount Bros. Constr. Co. v. United States*, 346 F.2d 962 (Ct. Cl. 1965); *Electronics Group, Inc. v. Central Roofing Co., Inc.*, 518 N.E.2d 369 (Ill. Ct. App. 1987); *D'Annunizo Bros., Inc. v. New Jersey Transit Corp.*, 586 A.2d 301 (N.J. Super. 1991); *Gardner-Zemke Co. v. State*, 790 P.2d 1010 (N.M. 1990); *Pennsylvania Dep't of Transp. v. Anjo Constr. Co.*, 487 A.2d 455 (Pa. 1985); *City of San Antonio v. Forgy*, 769 S.W.2d 293 (Tex. Ct. App. 1989).

from an unbalanced bid.[118] The contractor's error detection duty may rest on express[119] as well as implied contract obligations.

2. *The Contractor's Implied Warranty Of Workmanship*

In undertaking to perform work and furnish materials, the contractor impliedly warrants that the work will be constructed in a "workmanlike manner" free of defects.[120] This implied warranty is cumulative with and in addition to any express contract warranty that does not disclaim implied warranties.[121]

---

118 *See Green Island Constr. Co. v. County of Chenango*, 622 N.Y.S.2d 132 (N.Y. App. Div. 1995), in which the Appellate Division of the New York Supreme Court applied the patent discrepancy doctrine to prevent a contractor from reaping the benefit of its unbalanced bid submitted after the contractor discovered during bid preparation a gross understatement in certain estimated quantities. During construction, when the contractor asked to be paid at the unbalanced unit price for quantities excavated in excess of the contract estimated quantities, the county argued that the contractor was required under the contract to report any contract document error of which it had "actual knowledge" and, having failed to do so, was not entitled to its unbalanced unit price for the extra quantities. The court concluded that, by unbalancing its bid and failing to give the county notice of anticipated quantity overruns in excess of the erroneous quantity estimates, the contractor assumed the risk that the actual work performed would exceed the estimated quantities set forth in the specifications and was entitled to no extra compensation for the quantity overrun.

119 *See Bethesda Lutheran Church v. Twin City Constr. Co.*, 356 N.W.2d 344 (Minn. Ct. App. 1984); *Jos. P. Jansen Co., Inc. v. Milwaukee Area Dist. Bd. of Voc., Tech. & Adult Ed.*, 312 N.W.2d 813 (Wis. 1981).

120 *See Kellogg Bridge Co. v. Hamilton*, 110 U.S. 108 (1884); *Leisure Resorts, Inc. v. Frank J. Rooney*, 654 So. 2d 911 (Fla. 1995); *Zielinski v. Miller*, 660 N.E.2d 1289 (Ill. App. Ct. 1995); *Paine v. Spottiswoode*, 612 A.2d 235 (Me. 1992); *Seely v. Loyd H. Johnson Constr. Co.*, 470 S.E.2d 283 (Ga. Ct. App. 1996).

121 *See Tonkin v. Bob Eldridge Constr. Co., Inc.*, 808 S.W.2d 849 (Mo. Ct. App. 1991), a contractor who recaulked leaky curtainwall joints on two apartment buildings during a one-year express warranty period thereafter refused to do further repair work and argued that the one-year warranty had expired. The owner then replaced the curtainwall and sued the contractor for the cost of replacement due to the contractor's breach of its implied warranty of workmanship. The court ruled in favor of the owner:

3. The Contractor's Implied Duty Of Cooperation

The implied duty of cooperation is equally applicable to the contractor as it is to the owner. Neither party may hinder or delay the other. This duty obviously is not applicable to delays or hindrances caused by others whom a party does not direct or control.[122] Delays that are foreseeable at the time of contract award or otherwise within the control of the contractor will provide no relief to the contractor. The owner may recover damages for the contractor's unexcused late completion of the contract.[123] The recourse of the prime for such unexcused delay may be passed on to the

---

> [A]n express warranty against defects for a limited period of time does not act as a limitation upon the obligation to perform the contract in a workmanlike manner [case citation omitted]. Thus, whatever the effect of the warranty would be in the contract the contractor had a duty to perform in a proper, workmanlike manner.

122 *Id.* at 853. *See Triangle Sheet Metal Works, Inc. v. James H. Merritt & Co.*, 588 N.E.2d 69, 70 (N.Y. 1991), *aff'g*, 566 N.Y.S.2d 644 (N.Y. App. Div. 1991), in which a sub's delay claim against a prime for delays caused by others over whom the prime had no control was denied.

> This case falls squarely within the general rule that, absent a contractual commitment to the contrary, a prime contractor is not responsible for delays that its subcontractor may incur unless those delays are caused by some agency or circumstance under the prime contractor's direction or control.

The same result was reached in *Phoenix Elec. Inc. v. Lehr Constr. Corp.*, 631 N.Y.S.2d 146 (N.Y. App. Div. 1995), and *Port Chester Electrical Constr. Corp. v. HBE Corp.*, 878 F.2d 820 (2d Cir. 1992), *aff'd*, 89 F.3d 826 (2d Cir. 1995).

123 *See P.T. & L. Constr. Co. v. State*, 578 N.Y.S.2d 921 (N.Y. App. Div. 1992), in which the New York Appellate Division found that delays in completing the highway construction job were the fault of the contractor: (1) the contractor failed to properly mobilize on the project because it had too many other projects in progress; (2) the contractor assumed the risk that excessive rain during the latter half of the construction could delay the completion of the project; (3) the contractor assumed the risk of delays caused by the Arab oil embargo; and (4) the contractor's delays were due to its own financial difficulties. Unexcused delays of subcontractors or suppliers constitute unexcused delays of the prime contractor. *See also J.J. Brown Co. v. J.L. Simmons Co.*, 118 N.E.2d 781 (Ill. 1954).

culpable sub and its surety.[124] Similarly, if the prime's delay is inexcusable, the sub may recover delay damages from the prime.[125]

4. *The Implied Product Warranties Of Merchantability And Fitness Of Purpose*

Under the Uniform Commercial Code (UCC), manufacturers and contractors, as sellers of construction products and materials, impliedly warrant their "merchantability"[126] and "fitness for a particular purpose."[127] Prior to the UCC, the law implied no such warranties and even under the UCC such warranties may be disclaimed.[128] Although the UCC applies only to the sale of "goods" and not "services," most jurisdictions distinguish between contract for "goods" and "services" by applying a "predominance" test, and although most courts have concluded that a general construction contract is a contract for "services," a broad extension

---

124 *See Spang Indus., Inc. v. Aetna Cas. & Sur. Co.*, 512 F.2d 365 (2d Cir. 1975); *American Fidelity Fire Ins. Co. v. Woody's Elec. Serv., Inc.*, 407 So. 2d 947 (Fla. Dist. Ct. App. 1981).

125 *See United States Indus. v. Blake Constr. Co.*, 671 F.2d 539 (D.C. Cir. 1982); *United States ex rel. Gray-Bar Elec. Co. v. Copeland & Sons, Inc.*, 568 F.2d 1159 (5th Cir. 1978).

126 "Merchantability" means, *inter alia*, that goods will (a) pass without objection in the trade under the contract description, (b) be fit for ordinary purposes for which the goods are used, (c) run, within variations, of even kind, quality, and quantity, and (d) meet other requirements established by a course of dealing or usage of trade. UCC § 2-314.

127 "Fitness for a particular purpose" means: Where the seller at the time of contracting has reason to know any particular purpose for which the goods are required and that the buyer is relying on the seller's skill or judgment to select or furnish suitable goods, there is unless excluded or modified under the next section an implied warranty that the goods shall be fit for such purpose. UCC § 2-315. The implied warranty of fitness has been applied to hold contractors and manufacturers to their assurances about product performance. *See Certain-Teed Products Corp. v. Goslee Roofing & Sheet Metal Inc.*, 339 A.2d 302 (Md. Ct. Spec. App. 1975); *Duall Bldg. Restoration v. 1143 East Jersey Ave., Assocs., Inc.*, 652 A.2d 1225 (N.J. Super. App. 1995).

128 *See* Marshall, *The Applicability of the Uniform Commercial Code to Construction Contracts*, 28 EMORY L. J. 335 (1979).

of the UCC warranties into construction nevertheless has resulted.[129] More expansively, one court has deemed the sale of completed commercial buildings to be the sale of a "product" covered by implied warranties.[130]

5. *The Implied Duty To Provide Technical Product Information*

Contractors routinely perform specified work or furnish specified materials in reliance on explicit assurances of performance set forth in product literature. Suppliers have an implied obligation to furnish contractors with sufficient product information to assure proper and satisfactory use of the product.[131]

6. *The Implied Duty To Warn*

Construction products sometimes are found years later to be defective. Product manufacturers, after having a basis for reasonably believing products to be generally defective, are required by law to take steps to modify product literature so as not to misrepresent the products. Failure to do so may result in substantial compensatory and punitive damages.[132]

---

129 *O'Laughlin v. Minnesota Natural Gas Co.*, 253 N.W.2d 826 (Minn. 1977) (holding that the UCC warranties applied to a furnace installed by a contractor because the price of the furnace predominated over the price of installation services).

130 *Pollard v. Saxe & Yolles Dev. Co.*, 115 Cal. Rptr. 648 (Cal. 1974).

131 *See American Rock Mechanics, Inc. v. Thermex Energy Corp.,* 608 N.E.2d 830 (Ohio Ct. App. 1992), a supplier furnished explosives to an excavation contractor for use in drilling and blasting subsurface rock. The supplier recommended its T-600 explosive, but did not explain that the T-600 explosive required significant "warm-up time" prior to use. After the contractor experienced difficulties using the T-600 explosives, the contractor learned that the problem lay in the explosive's temperature. The supplier, however, had failed to furnish any information about the need to warm up the explosive. The Ohio Court of Appeals upheld a jury verdict in favor of the contractor, and concluded that the supplier had breached its implied contractual duty to provide technical data to the contractor regarding the characteristics and requirements for use of the T-600 explosive.

132 *See State ex rel. Stephan v. GAF Corp.,* 747 P.2d 1326 (Kan. 1987), in which the manufacturer was held liable for puntive damages for selling a

7. *The Contractor's Site Investigation Duty*

Construction contracts that call for a site investigation prior to bidding impose accountability upon the contractor for site conditions that reasonably could have been observed or anticipated.[133] Thus, contractors have been denied relief for changed conditions where they failed to conduct site inspections and such inspections would have disclosed sufficient indications of the conditions complained of.[134]

---

product known to be defective but misrepresented as suitable in sales literature. GAF had published for years in *Sweet's Catalog* a specification for a roof system called the 212N. The published specification represented that the roof would last ten years. In reliance upon the published specification, an architect specified in GAF 212N roof system. When the roof system failed well within the ten-year period, the owner sued GAF. Evidence at trial showed that GAF had prior knowledge of a number of roof failures involving the 212N design, but failed to withdraw its published specification or otherwise warn designers and owners of the system deficiencies. The jury awarded the owner $1 million punitive damages and the Kansas Supreme Court affirmed. The Court said:

A manufacturer such as GAF not only has available to it the benefit of its own research and that of others, but it has the experience gained from having roofs constructed out of its products according to its specifications in many different areas. Where, as here, the manufacturer has actual knowledge of recurring defects and problems when roofs are built according to its specifications, the law imposes upon it the duty to either change the specifications, withdraw the specifications, or warn those likely to follow those recommendations.

Id. at 1330.

133 *See Clark v. United States,* 5 Cl. Ct. 447 (1984), *aff'd,* 770 F.2d 180 (Fed. Cir. 1985); *Mojave Enters. v. United States,* 3 C1. Ct. 353 (1983).

134 *See Gene Hock Excavating, Inc. v. Town of Hamburg,* 643 N.Y.S.2d 268 (N.Y. App. Div. 1996); *Green Constr. Co. v. Dep't of Transp.,* 643 A.2d 1129 (Pa. 1994); *Bilotta Constr. Corp. v. Village of Mamaroneck,* 604 N.Y.S.2d 966 (N.Y. App. Div. 1993) (existing grade elevations inaccurate and laid to be only "approximate"); *Green Constr. Co. v. Kansas Power & Light Co.,* 1 F.3d 1005 (10th Cir. 1993) (high moisture content in soil was observable); *McCormick Constr. Co. v. United States,* 18 Cl. Ct. 259 (1989), *aff'd,* 907 F.2d 159 (Fed. Cir. 1990) (boulders to be expected by well driller on site situated on an "alluvial fan"); *Cook v. Oklahoma Bd. of Public Affairs,* 736 P.2d 140 (Okla. 1987) (wet soil conditions); *Weeks*

Similarly, where the contractor failed to check available records showing underground utilities or revealing rock conditions, no claim was allowed.[135] The contractor, however, is *not* accountable for conditions that a reasonable site inspection would have disclosed if the owner prevented or interfered with the completion of such an inspection, *e.g.*, denied the contractor access to the site; afforded inadequate time to inspect the site prior to bidding;[136] or made express representations in the plans about site conditions upon which the contractor reasonably relied.[137] Unless the owner informs the contractor in the bidding documents to conduct a more detailed site investigation, such as taking its own test borings, the contractor's inspection need only be "reasonable under the circumstances."[138] A contractor, in conducting a "reasonable" inspection, ordinarily is not required to make separate test borings in addition to those furnished by the owner,[139] is not bound to make "scientifically educated and skeptical analysis" of the general

---

Dredging & Contracting, Inc. v. United States, 13 Cl. Ct. 193 (1987) (gravel conditions encountered in dredging operation); *Le Gare, Inc. v. Brookhaven Residential Sales, Inc.,* 487 A.2d 360 (Pa. Super. 1985); *National Concrete Foundation Co. v. United States,* 170 Ct. Cl. 470 (1965) (outcroppings of rock suggested subsurface rock).

135 *See Umpqua River Navigation Co. v. Crescent City Harbor Dist.,* 618 F.2d 588 (9th Cir. 1980); *Tri-County Excavating Inc. v. Borough of Kingston,* 407 A.2d 462 (Pa. 1979).

136 *See Frederick Snare Corp. v. Maine-New Hampshire Interstate Bridge Auth.,* 41 F. Supp. 638 (D.N.H. 1941); *Alley Constr. Co. v. State,* 219 N.W.2d 922 (Minn. 1974).

137 *See Fattore Co. v. Metropolitan Sewerage Comm'n,* 454 F.2d 537 (7th Cir. 1971); *Jack B. Parson Constr. Co. v. Utah Dep't of Transp.,* 725 P.2d 614 (Utah 1986); *Coatesville Contractors & Eng'rs, Inc. v. Borough of Ridley Park,* 506 A.2d 862 (Pa. 1986).

138 *See, e.g., Sergent Mech. Sys., Inc. v. United States,* 34 Fed. Cl. 505 (1995), in which an Air Force contractor encountered environmentally protected vegetation and other obstacles during excavation and removal of underground storage tanks. The United States Court of Federal Claims concurred with the contractor that a reasonable site inspection did not require inspection of the vegetation by a botanist and would not have disclosed any of the conditions encountered.

139 *See Town of Longboat Key v. Carl E. Widell & Son,* 362 So. 2d 719 (Fla. Dist. Ct. App. 1978).

situation and is not obliged to consult scholarly works.[140] Further, the contractor is not necessarily on notice of "common knowledge" in the area.[141] The contractor nevertheless may be obliged to assume sole responsibility for soil conditions if the owner clearly and specifically disclaims responsibility for subsurface conditions and for the accuracy of available soils information.[142]

8. *The Contractor's Obligation To Give Timely Notice Of Claim*

Virtually all construction contracts require the contractor to give the owner timely notice of conditions for which the contractor intends to pursue a claim for additional compensation.[143] The purpose of requiring such notice is to permit the owner to know the extent of its liabilities before they are incurred and to control its liabilities.[144] Such notice

---

140 *See Metropolitan Sewerage Comm'n v. R.W. Constr., Inc.*, 241 N.W.2d 371 (Wis. 1976).

141 *See North Harris County Junior College Dist. v. Fleetwood Constr. Co.*, 604 S.W.2d 247 (Tex. Civ. App. 1980).

142 *See McDevitt & Street Co. v. Marriott Corp.*, 713 F. Supp. 906 (E.D. Va. 1989), *aff'd in part, rev'd in part*, 948 F.2d 1281 (4th Cir. 1991); *L. J. McNulty, Inc. v. Village of Newport*, 187 N.W.2d 616 (Minn. 1971); *S&M Constructors v. City of Columbus*, 434 N.E.2d 1349 (Ohio 1982).

143 *See, e.g.,* Article 4.3 of the AIA A201-1997 General Conditions of Contract, which requires notice of claim to be given within twenty-one days after discovery of the condition giving rise to the claim or after occurrence, whichever is later.

144 *See Watson Lumber Co. v. Guennewig*, 226 N.E.2d 270 (Ill. Ct. App. 1967), in which the court observed:

> In a building and construction situation, both the owner and the contractor have interests that must be kept in mind and protected. The contractor should not be required to furnish items that were clearly beyond and outside of what the parties originally agreed that he would furnish. The owner has a right to full and good faith performance of the contractor's promise, but has no right to expand the nature and extent of contractor's obligation. On the other hand, *the owner has a right to know the nature and extent of his promise, and a right to know the extent of his liabilities before they are incurred. Thus he has a right to be protected against the contractor voluntarily going ahead with extra work at this own expense. He also has a right to control his liabilities.* Therefore, the law required his consent be evidenced before he can be charged for an extra.

requirements are routinely enforced,[145] unless waived by conduct or agreement of the parties[146] or unless the owner was fully aware of the condition and had an adequate opportunity to investigate.[147]

E. The Strategic "Road Map": Identifying The Best Bond Defenses Available To The Surety

In addition to its "contractor defenses" under the bonded contract, the surety is entitled to the protection of its own "bond defenses applicable specifically to its surety bond obligations."[148]

1. Alteration Of The Bonded Contract

Historically, sureties were favored by the law and were entitled to have the terms of their bonds strictly enforced. Courts routinely held that any alteration to the bonded contract, whether material or not, served to discharge the surety's obligation.[149] The more recent authorities hold that only material changes to the underlying contract which prejudice the surety will operate to discharge the surety to the extent of the prejudice.[150] This is the result of provisions in most modern performance bonds which require the surety to waive its right to object to changes in or extensions of

---

Id. at 276 (emphasis added).

145 *See Neal & Co. v. City of Dillingham*, 923 P.2d 89 (Alaska 1996); *Sutton Corp. v. Metro. Dist. Comm'n*, 667 N.E.2d 838 (Mass. 1996).

146 *See Metropolitan Sewerage Comm'n v. R.W. Constr. Inc.*, 241 N.W.2d 371 (Wis. 1976).

147 *See Brinderson Corp. v. Hampton Rds. Sanitation Dist.*, 825 F.2d 41 (4th Cir. 1987).

148 *See generally* RESTATEMENT (THIRD) OF SURETYSHIP & GUARANTY §§ 37-49 (1995).

149 *See, e.g., Continental Ins. Co. v. City of Va. Beach*, 908 F. Supp. 341 (E.D. Va. 1995); *In re Liquidation of Union Indemnity Ins. Co.*, 632 N.Y.S.2d 788 (N.Y. App. Div. 1995); *Congregation Ohavei Shalom, Inc. v. Comyns Bros., Inc.*, 507 N.Y.S.2d 28, 29 (N.Y. App. Div. 1986).

150 *See, e.g., United States Fidelity & Guaranty Co. v. United States*, 298 F. 365 (D.C. Cir. 1924) (surety released only to the extent of the cost of the contract modifications); *Equitable Fire & Marine Ins. Co. v. Tiernan Bldg. Corp.*, 190 So. 2d 197 (Fla. Dist. Ct. App. 1966) (same); *Zuni Constr. Co. v. Great American Ins. Co.*, 468 P.2d 980 (Nev. 1970) (surety discharged only to the extent that its risk was materially increased).

the bonded contract.[151] Even if a particular bond contains the standard waiver language, the surety can argue that it is entitled to be discharged to the extent that its rights are prejudiced by "cardinal" changes in the underlying contract.[152]

In summary, mere alteration of the bonded contract is usually insufficient to discharge the obligations of a compensated corporate surety. If the surety has not waived its right to object to alterations and can prove that the alteration is both material and prejudicial, it may be able to obtain a discharge to the extent it is prejudiced. Only in circumstances where a cardinal change as occurred can the surety successfully argue for complete discharge.[153]

2. *Changes In The Obligee Or The Principal*

Modern corporate sureties have a right to be concerned about a prejudicial increase in risk when the change involves the contracting parties. Unless the bond contains language specifically prohibiting assignment, a change in the obligee by virtue of assignment will not discharge the surety.[154] Similarly, a mere change in the form of the bond principal (*i.e.*, from sole proprietorship to partnership or from partnership to corporation) normally will not serve to discharge the surety. Sureties should remain alert to circumstances in which the bond principal undertakes, with the consent of the obligee, to complete the bonded contract through a joint venture with an unbonded entity. This situation could expose the surety to increased risk of loss, especially if the surety has no knowledge of or confidence in the quality of the work performed by the unbonded entity.

---

151 *See, e.g.*, AIA A-311 (1970) ("The surety hereby waives notice of any alteration or extension of time made by the owner."). Historically, an extension of time for performance of the bonded contract could operate to discharge the surety. However, sureties now find it extremely difficult to demonstrate prejudice caused by an extension of time. Moreover, since most bonds require sureties to waive their objections to time extensions, this potential defense is no longer significant.

152 *See In re Liquidation of Union Indem. Ins. Co.*, 632 N.Y.S.2d 788 (N.Y. App. Div. 1995).

153 *See supra* Section III.C.5.c (discussing cardinal changes).

154 *Hunters Pointe Partners Ltd. v. United States Fidelity & Guaranty Co.*, 442 N.W.2d 778 (Mich. Ct. App. 1989).

On the other hand, if the obligee allows the principal to substitute another contractor and then releases the principal from further liability, the surety is likewise discharged. Similarly, if the principal attempts to assign its contractual obligations under the bonded contract while remaining secondarily liable as a guarantor, the surety may be entitled to discharge if it can demonstrate that performance by the assignee will materially increase the risk of its bonded obligation.[155]

3. *Improper Payment Of Contract Funds By The Obligee*

Improper payment of contract funds by the obligee is a defense which rarely allows the surety to obtain a full discharge but which often allows the surety to reduce *pro tanto* the cost of its performance obligation.[156] Under most types of bonds, the surety's obligation to perform is conditioned upon the obligee having fully performed its obligations under the bonded contract. One of the obligee's primary obligations is to release payments to the principal in accordance with the terms of the contract.[157]

Most construction contracts provide for progress payments to be made only upon certification by the contractor and the architect that the work has actually been completed and require the obligee to withhold ten percent of each payment as retainage until all the work is completed. Many contracts also require the contractor to obtain consent of the surety prior to the obligee releasing final payment. The obligee's failure to adhere to these contractual payment requirements, by making progress

---

155 *See Leika Hosp. & Health Ctr. v. Xonics Med. Sys, Inc.*, 948 F.2d 271 (6th Cir. 1991); *Chicago College of Osteopathic Medicine v. George A. Fuller Co.*, 719 F.2d 1335 (7th Cir. 1983), *aff'd*, 776 F.2d 198 (7th Cir. 1985).

156 *See, e.g., National Sur. Corp. v. United States*, 118 F.3d 1542 (Fed. Cir. 1997); *National Union Indemnity Co. v. G. E. Bass & Co.*, 369 F.2d 75, 77 (5th Cir. 1966) (stating that the "modern rule" allows a surety to be discharged from its performance bond obligations to the extent that it suffers injury as a result of the obligee's overpayments). *But see Southwood Bldrs., Inc. v. Peerless Ins. Co.*, 366 S.E.2d 104, 107 (Va. 1988) (holding that surety was entitled to complete discharge as a result of obligee's overpayments).

157 *See, e.g., Southwood Bldrs.*, 366 S.E.2d at 107; *Airtrol Eng'g Co. v. United States Fidelity & Guaranty Co.*, 345 So. 2d 1271, 1273 (La. Ct. App. 1977) ("The bonding company is entitled to expect that payments will be made in accordance with the contract").

payments before the work is completed, by paying for work which the obligee knew or should have known was defective, by prematurely releasing the contract retainage, or by making final payment without the required consent of surety will serve to discharge the surety's performance obligation to the extent that the surety is injured as a result.[158]

For the surety to prevail on an overpayment defense, it must demonstrate that it has suffered an injury as a result of the overpayment. Thus, the surety may not be entitled to a partial discharge if the obligee can demonstrate that prematurely-released funds were used to purchase labor or materials which contributed to the completion of the project.[159] Finally, the obligee may be able to defend against a claim of improper payment by claiming that it relied in good faith upon the certification of progress approved by the architect or engineer.[160] In cases where the obligee claims to have relied upon the reasonable, albeit incorrect, certification of the architect or engineer, the surety may wish to consider pursuing a claim against the architect or engineer for negligence.[161]

In summary, to the extent that the obligee fails to fulfill its contractual obligations by improperly releasing payments to the principal, the surety is entitled to be partially discharged. The extent of the surety's discharge

---

158 *See, e.g., Continental Ins. Co. v. City of Virginia Beach*, 908 F. Supp. 341 (E.D. Va. 1995) (holding that a surety was discharged by the city's release of final payment without inspecting and testing the work as required by the bonded contract); *Ohio Cas. Ins. Co. v. United States.*, 12 Cl. Ct. 590, 596 (1987) (holding that surety was discharged to the extent that government had properly failed to terminate contractor and improperly released contract funds to contractor); *Merchants Bonding Co. v. Pima County*, 860 P.2d 510 (Ariz. Ct. App. 1993) (holding that surety was entitled to recover as damages the amount of final payment where the obligee had released final payment to the principal without the surety's consent and over the surety's protest); *Central Towers Apartments, Inc. v. Martin*, 453 S.E.2d 789, 797 (Tenn. Ct. App. 1969) (holding that surety was discharged to the extent that owner prematurely released contract retainage to contractor).

159 *See Transamerica Ins. Co. v. City of Kennewick*, 785 F.2d 660, 661 (9th Cir. 1986).

160 *See Argonaut Ins. Co. v. Town of Cloverdale*, 699 F.2d 417, 419-20 (7th Cir. 1983).

161 *See City of Houma v. Municipal & Indus. Pipe Serv.*, 884 F.2d 886 (5th Cir. 1989) (holding engineering firm liable to surety for negligent certification of progress payments).

will depend upon the surety's ability to demonstrate an actual injury as a result of the obligee's overpayment.

4. Statutory Defenses

A number of states have enacted suretyship codes which gives sureties specific rights vis-a-vis the principal and the obligee. These statutes typically allow the surety to demand that the obligee commence suit against the principal before pursuing the surety[162] or specify circumstances under which the surety will be exonerated from its performance bond obligations.[163] These statutory rights can be extremely valuable to the surety, particularly in situations where the principal has filed for bankruptcy and may not be easily amenable to suit. Thus, the surety

---

162 *See, e.g.,* CAL. CIVIL CODE § 2845 (1993) ("A surety may require the creditor, . . . to proceed against the principal, or to pursue any other remedy in the creditor's power which the surety cannot pursue, and which would lighten the surety's burden . . . ."); MO. REV. STAT. § 433.010 (1992) ("Any person bound as surety for another in any bond . . . may, at any time after an action has accrued thereon, require, in writing, a person having such right of action forthwith to commence suit against the principal debtor and other parties liable"); OKLA. STAT. tit. 15, § 379 (1993) ("A surety may require his creditor to proceed against the principal, or to pursue any other remedy in his power which the surety cannot himself pursue, and which would lighten his burden . . ."); *see also A. J. Kellos Constr. Co. v. Balboa Ins. Co.*, 661 F.2d 402 (5th Cir. 1981) (holding that obligee's failure to comply with a similar Georgia statute, requiring the obligee to initiate legal action against the principal prior to pursuing the surety, entitled surety to be discharged from its performance bond obligations).

163 For example, Section 377 of the Oklahoma statutes provides:

A surety is exonerated:

1. In like manner with a guarantor;
2. To the extent to which he is prejudiced by any act of the creditor which would naturally prove injurious to the remedies of the surety or inconsistent with his rights, or which lessens his security; or,
3. To the extent which he is prejudiced by an omission of the creditors to do anything, when required by the surety, which it is his duty to do.

OKLA. STAT. tit. 15, § 377 (1993).

should always be aware of the law applicable to its bonded obligation and should investigate the scope and availability of these statutory rights.[164]

5. Notice

Many private construction bonds contain language requiring the obligee to notify either or both the surety and principal of its claim. The issue of whether the obligee has given adequate notice generally arises in the context of payment bond claims.[165] Nevertheless, a performance bond surety may very well be entitled to a defense based upon untimely or insufficient notice, depending upon the language of the bond or governing statutory or common law.[166] This issue may arise in the context of whether the obligee has met the requirement of declaring the principal in default. Correspondence or other contacts from an obligee expressing

---

164 Determining what law is applicable to the surety's obligation is not always a simple matter and is a topic far too broad for this chapter. Nevertheless, the surety should always look to the language of the bond and the underlying construction contract (which is usually incorporated into the bond) to see if either contains a choice of law provision.

165 *See* M. I. Less, *Notice Requirements for Bond Claims, in* THE LAW OF PAYMENT BONDS (Lybeck & Shreves, eds., ABA 1998); *Allgood Elec. Co. Inc. v. Martin K. Eby Constr. Co.*, 959 F. Supp. 1573 (M.D. Ga. 1997); *W.T. Andrew Co. v. Mid-State Sur. Corp.*, 562 N.W.2d 206 (Mich. Ct. App. 1997).

166 *See Insurance Co. of N. Am. v. Metropolitan Dade County*, 1997 WL 667601 (Fla. Dist. Ct. App. 1997); *Dragon Constr., Inc. v. Parkway Bank & Trust*, 678 N.E. 2d 55 (Ill. Ct. App. 1997) (holding that a surety was relieved of liability under its performance bond because the obligee failed to give timely notice of default and termination and failed to give the surety an opportunity to perform under its bond prior to the obligee's retention of its own completing contractor); *see also* AIA Document A312 which states that the surety's obligation arises only after, among other things:

> The owner has notified the contractor and the surety . . . that the owner is considering declaring a contractor default and has requested and attempted to arrange a conference with the contractor and the surety to be held not later than 15 days after receipt of such notice to discuss methods of performing the construction contract. . . .

AIA Document A312 at ¶ 3.1.

dissatisfaction with the principal's performance may not be sufficient to trigger the surety's obligation as no declaration of default has occurred.[167]

If the bond does not contain language specifically providing that the obligee must notify the surety of a default by the principal, the general rule of law is the that surety takes notice of its principal's default.[168] Still, obligees must be concerned about giving the surety adequate notice of their principal's default, even in the absence of specific language requiring notice. Sureties have successfully argued that an implied condition of a performance bond is the obligee's obligation to notify the surety of the principal's defaults and give it the opportunity to remedy them.[169]

6. Contractual And Statutory Limitations

Most performance bonds contain language specifying the period of time within which an obligee must commence suit against the surety. The AIA Document A311 requires suit to be instituted before the expiration of two years from the date on which final payment under the contract falls due. AIA Document A312 requires suit to be instituted within two years after contractor default, or within two years after the contractor ceased working, or within two years after the surety refuses or fails to perform its obligations under the bond, whichever occurs first.

A number of states have enacted statutes which nullify a private party's attempts at shortening contractual limitations.[170] As a consequence, in certain jurisdictions a surety may find that it is unable to rely upon its contractual limitation provision.[171] In the absence of a statute

---

167 *See L&A Contracting Co. v. Southern Concrete Serv., Inc.*, 17 F.3d 106 (5th Cir. 1994) (surety's obligation never arose as obligee never claimed principal to be in default although it requested surety fulfill the contract so as to prevent further delays).

168 *See Continental Bank & Trust Co. v. American Bonding Co.*, 605 F.2d 1049 (8th Cir. 1979); *United States v. Ohio Cas. Ins. Co.*, 399 F.2d 387 (6th Cir. 1968); *Insta/Com, Inc. v. Ætna Cas. & Sur. Co.*, 589 S.W.2d 494 (Tex. Civ. App. 1979).

169 *See Blackhawk Heating & Plumbing Co. v. Seaboard Sur. Co.*, 534 F. Supp. 309 (N.D. Ill. 1982).

170 *See, e.g.*, FLA. STAT. ANN. § 95.03 (1982); MISS. CODE ANN. § 15-1-5 (1972).

171 *See Sheehan v. Morris Irrigation*, 410 N.W.2d 569 (S.D. 1987) (South Dakota's law was subsequently changed to reflect the fact that subjecting

specifically invalidating a contractual limitation provision, the clause will be upheld if the period of time provided is reasonable.[172] As a practical matter, most case law supports the view that a limitation period of twelve months from the accrual of the cause of action on a bond is reasonable.[173] Even limitations periods as short as six months have been upheld.[174]

Performance bonds issued in connection with a public work are governed by the applicable state's public works statute. Most jurisdictions will incorporate into statutory bonds the terms and conditions mandated for such bonds by the applicable public works act.[175] When the terms of the bond conflict with the requirements of the public works statute, the statutory terms will control if the project is determined to be a public work.[176] An exception to this rule occurs where the terms of the bond provide greater protection than the statute. In these cases, the bond generally will be interpreted in accordance with the more generous terms.[177] Sometimes it is difficult to determine whether the project in

---

sureties to long statutes of limitations periods was counterproductive as less bonding credit became available for contractors due to the long exposure times).

172 *See, e.g., United Commercial Travelers of Am. v. Wolfe*, 331 U.S. 586, 608 (1947).

173 *See, e.g., Burlew v. Fidelity & Cas. Co.*, 64 F.2d 976 (6th Cir. 1933); *In re 1616 Reminic Ltd. Partnership v. Atchison & Keller Co.*, 14 B.R. 484 (Bankr. E.D. Va. 1981); *Meyer v. Building & Realty Serv. Co.*, 196 N.E. 250 (Ind. 1935); *Camelot Excavating Co. v. St. Paul Fire & Marine Ins. Co.*, 301 N.W.2d 275 (Mich. 1981).

174 *Fitger Brewing Co. v. American Bonding Co.*, 149 N.W. 539 (Minn. 1914); *Ilse v. Ætna Indemnity Co.*, 125 P. 780 (Wash. 1912).

175 *See, e.g., Farm Bureau Mutual Ins. Co. v. Wright*, 686 S.W.2d 778 (Ark. 1985); *C.P.S. Distributors, Inc. v. Federal Ins. Co.*, 685 P.2d 783 (Colo. Ct. App. 1984); *American Druggists Ins. v. Thompson Lumber Co.*, 349 N.W.2d 569 (Minn. Ct. App. 1984); *Felix Contracting Corp. v. Federal Ins. Co.*, 468 N.Y.S.2d 39 (N.Y. App. Div. 1983).

176 *See, e.g., Wichita Sheet Metal Supply, Inc. v. Dahlstrom & Ferrell Constr. Co.*, 792 P.2d 1043 (Kan. 1990); *Sheldon Pollack Corp. v. Pioneer Concrete of Texas, Inc.*, 765 S.W.2d 843 (Tex. Civ. App. 1989); *Grimes v. Bosque County*, 240 S.W.2d 511 (Tex. Civ. App. 1951).

177 *See, e.g., Nelson Roofing & Contracting, Inc. v. C.W. Moore Co.*, 245 N.W.2d 866 (Minn. 1976); *American Cas. Co. v. Irvin*, 426 F.2d 647 (5th Cir. 1970); *Reliance Ins. Co. v. Trane Co.*, 184 S.E.2d 817 (Va. 1971); *but*

question is private or public.[178] It can be particularly difficult to determine the nature of a project if there is both private and public participation in the undertaking.

If the bond does not contain a contractual limitation provision, or if that provision is deemed unreasonable under state law and the bonded project is not public, then the state's statute of limitations provision applicable to contract actions will most likely apply. The bond's contractual limitation provision/statutory limitation will often have the effect of cutting off any argument that the surety is liable for latent defects or extended warranty obligations.[179]

There may be times when the statute of limitations applicable to the surety is longer than that for its principal. In these cases, the surety quite naturally would seek to be relieved of its obligation on the grounds that its liability is no greater than that of its principal and because its principal cannot be found liable to the obligee nor can the surety.[180] Sureties generally have been able to successfully assert their principal's statute of limitations defenses.[181] Still, sureties have been subject to suit where the statute of limitations has run with respect to their principal in instances where there is a separate statute of limitations governing suretyship arrangements.[182]

---

see Transamerica Ins. Co. v. Housing Auth., 669 S.W.2d 818 (Tex. App. 1984).

178 *See* O'Connor, *Statutory Bonds or Common-Law Bonds: The Public-Private Dilemma*, 29 TORT & INS. L. J. 77 (1993).

179 *See Kiva Constr. & Eng'g, Inc. v. International Fidelity Ins. Co.*, 749 F. Supp. 753 (W.D. La. 1990), *aff'd*, 961 F.2d 213 (5th Cir. 1992); *District School Bd. of DeSoto County v. Safeco Ins. Co.*, 434 So. 2d 38 (Fla. Dist. Ct. App. 1983).

180 *See State v. Bi-States Constr. Co., Inc.*, 269 N.W.2d 455 (Iowa 1978) (statute of limitations ran against the principal, therefore, surety also not liable).

181 *See County of Hudson v. Terminal Constr. Corp.*, 381 A.2d 355 (N.J. Super. 1977); *see also* Thomas & Leo, *Application of Statutes of Limitation Governing Construction Activity to Construction Bond Sureties*, 10 CONSTR. LAW (Jan. 1990).

182 *See Regents of the Univ. of Cal. v. Hartford Accident & Indem. Co.*, 581 P.2d 197 (Cal. 1978); *Bellevue School District v. Brazier Constr. Co.*, 691 P.2d 178 (Wash. 1984).

F. Fraud

Fraud or misrepresentation by the obligee which induces the surety to issue the bond will generally result in the surety's obligation being voidable.[183] If the misrepresentation is by someone other than the obligee, the obligee must know or have reason to know of the misrepresentation.[184] For example, if the obligee knows that the principal is in deep financial difficulty or is already in default on other obligations and fails to tell the surety, this failure to disclose can give rise to a discharge of the surety's obligations.[185] If the principal and surety fail to disclose all material facts to the surety, thereby inducing it to issue the bond, the surety is entitled to be relieved of its obligations. For example, if the obligee and principal know that the surety would not issue a bond covering the full contract price due to the principal's limited bond credit and work out an arrangement whereby only a portion of the work is bonded without telling the surety of the true scope of the entire undertaking, the surety is entitled to discharge.[186]

IV.
The Strategic "Road Map": Analyzing Recourse Against Third Parties For Indemnification Or Contribution

In most complex construction suretyship cases, a host of third parties may have potential contractual or common law obligations to indemnify the surety and its principal in whole or in part for any loss. Such parties may include: (1) the principal's comprehensive general liability insurer, who typically affords coverage under its broad form property endorsement for defective work or materials furnished by the principal's subcontractors or suppliers, and under its contractual liability

---

183 *See* RESTATEMENT (SECOND) OF CONTRACTS § 164 (1979); *Marine Bank, N.A. v. Meat Counter, Inc.*, 826 F.2d 1577 (7th Cir. 1987).

184 *See* RESTATEMENT (SECOND) OF CONTRACTS § 164 cmt. e (1979); RESTATEMENT OF SECURITY § 119 cmt. a (1941); *G&S Foods, Inc. v. Vavaroutsos*, 438 F. Supp. 122 (N.D. Ill. 1977).

185 *See St. Paul Fire & Marine Ins. Co. v. Commodity Credit Corp.*, 646 F.2d 1064 (5th Cir. 1981).

186 *See Employers of Wausau v. Construction Mgrs. of Fla.*, 377 S.E.2d 119 (S.C. Ct. App. 1989); *compare Sanders v. Zeagler*, 670 So. 2d 748 (La. Ct. App. 1996), *rev'd in part*, 686 So. 2d 819 (La. 1997).

coverage for express indemnity obligations; (2) subcontractors' and suppliers' insurers who provided the principal with certificates of insurance confirming coverage for their insureds' contractual liability to the principal or who otherwise named the principal as an "additional insured" assumed in the bonded contract; and (3) the builder's risk insurer who may afford coverage for damage to the work caused by insured perils.[187] Other parties who may have liability for indemnity or contribution include: (1) design professionals whose defective design or negligence in administration resulted in damage not protected by the "economic loss" rule; (2) accountants who negligently prepare financial information upon which the surety relied in extending bond credit also may have direct exposure to the surety; and (3) lenders who negligently disburse funds to the contractor that should have been withheld and available to protect the surety against loss.[188]

V.
The Strategic "Road Map": Dealing With Multiple Claims, Multiple Suits, Multiple Claimants, And Multiple Forums

Construction surety litigation frequently involves multiple claims asserted by multiple claimants arising out of the same project. The truly complex construction surety dispute generally arises out of the principal's default on multiple bonded contracts being performed in different states with multiple obligees, claims, claimants, and suits attributable to each defaulted contract and pursued in multiple forums of arbitration and state and federal court. Controlling the litigation to avoid both adverse and inconsistent results is a critical focus. Issues for consideration include:

A. The Surety's Litigation "Posture"

No surety who desires to settle all claims can ignore the strategic development of its litigation posture. The best offensive and defensive positions must be "staked out" and asserted. The entire field of conflict

---

187 *See infra* Chapter 6 for a detailed discussion of the surety's rights against insurers.

188 *See infra* Chapter 5 for a detailed discussion of the surety's rights against third parties.

must be viewed to confirm which battles are most important, *i.e.* most likely to establish favorable or unfavorable technical or legal precedents, and to decide how the conflicts may be consolidated and controlled. Clear goals and objectives should be established promptly. This task must be addressed during the surety's initial assessment of technical issues. This assessment necessarily will involve detailed discussions with the principal and forensic experts to analyze significant technical and legal issues and thus to minimize the "fog of the battlefield." Such discussions may dictate the early establishment of a joint defense agreement between the surety and principal as to common issues. The joint defense agreement will permit a cooperative sharing of information but assumes that the litigation positions of the surety and principal will be consistent and non-adversarial.

The next critical threshold issue for early determination by the surety is identification of potential allies and obvious adversaries. This follows naturally from the analysis of technical and legal issues. Alignment as parties in litigation is usually of little significance. For instance, subcontractors who sue the surety on its payment bond frequently seek recovery for contract claims (*e.g.*, defective design, changes, differing site conditions, etc.) for which the owner ordinarily is liable under the bonded contract and as to whom the subcontractors, principal and surety may well be allies. Conversely, owner claims against a performance bond surety for defective work or delay may well be attributable to actions of various subcontractors or suppliers as to whom the owner and surety may have allied interests. An early strategic action for a surety in any complex construction case is to determine the likely potential alignment of the participants in a construction project on the issues presented to litigation and then to become knowledgeable about litigation tactics of adversaries' commonality of interests can lead to joint prosecution and defense arrangements beneficial to the surety.

A third strategic decision for the surety is whether to pursue arbitration, when provided for in the bonded contract, in the name of the principal as to some or all claims by or against the principal under the bonded contract. Arbitration under the Construction Industry Arbitration Rules of the American Arbitration Association has been mandated for decades by the standard general conditions forms promulgated by the American Institute of Architects and the engineering profession.[189] The advantages of arbitration include: (1) resolution by arbitrators selected by the parties with extensive experience with the construction process; (2) proceedings more expedited than court litigation; (3) private hearings; and (4) limited scope of judicial review of the arbitration award. Perceived disadvantages include: (1) expense (the arbitration tribunal and arbitrators are paid fees whereas the court clerk and judge are not); (2) limited pre-hearing discovery rights; (3) limitations on joining other parties and claims in a common proceeding; (4) conduct of the arbitration hearing without traditional due process and evidence admissibility protections inherent in the judicial process; and (5) unless otherwise limited by the arbitration agreement, broad and extensive powers of the arbitrators.[190] Notwithstanding these disadvantages, there remains a widely held perception that technically complex issues should not be submitted for decision to local courts or juries. Arbitration of disputes between owner and contractor under the bonded contract is particularly advantageous to the surety where the disputes otherwise would be decided by a local jury likely to be sympathetic to the owner's defective work or termination claims.[191]

---

189 *See* Article 4.6, AIA A201-1997 General Conditions of Contract.

190 *See David Co. v. Jim W. Miller Constr. Inc.*, 444 N.W.2d 836 (Minn. 1989) (upholding an arbitration award that required the contractor to buy the defectively built project from the owner at the full resale price that the developer would have received if the project had been properly constructed); M. Egan, *Dispute Resolution*, an unpublished paper presented to the 1995 Mid-Winter Meeting of the ABA/TIPS Fidelity & Surety Law Committee (Jan. 27, 1995).

191 *See Osceola County Rural Water Sys., Inc. v. Subsurfco, Inc.*, 914 F.2d 1072 (8th Cir. 1990). In this case, a nonprofit owner that operated a rural water system commenced suit against the contractor and its surety in a rural Iowa court seeking $7.2 million to reconstruct a 680-mile rural water distribution system that was losing approximately forty percent of the water pumped into the system due to leaks. The owner demanded trial by jury in a county served by the troubled water system. When the contractor demanded arbitration, the court litigation was stayed. In the arbitration, three arbitrators with extensive construction experience concluded that the contractor had installed the pipeline in accordance with the owner's design specifications, and that the leakage was attributable to deficiencies in the specifications. The arbitrators therefore denied the owner's entire claim and awarded the contractor its contract retainage. The surety ultimately closed its file without any loss.

Because the surety's performance bond invariably references and incorporates the bonded contract, courts routinely allow sureties also to arbitrate performance bond disputes with owners and grant sureties stays of judicial proceedings pending arbitration of disputes between contractors and owners under bonded contracts.[192] Arbitrations commenced and controlled by the principal should bear as much attention of the surety as if commenced and controlled by the surety. In most jurisdictions, the surety will be bound by an arbitration award rendered against the principal in a proceeding of which the surety had notice,[193] subject only to any "surety" defenses as distinguished from "contract" defenses.[194] Since an arbitration award is deemed in "accord and satisfaction" of all the compulsory claims that were or could have been submitted to arbitration by or against the principal, any award that becomes binding on the surety also may be converted to a judgment against the surety and will collaterally estop the surety from seeking separate or additional recoveries in its own name in another forum against the party that arbitrated with the principal.

A fourth strategic decision is when and how to posture various claims or issues for settlement. The most dangerous claims of opposing parties should be resolved promptly, if possible, certainly prior to trial.

---

192 *See Matson, Inc. v. Lamb & Assoc. Packaging Inc.*, 947 S.W.2d 324 (Ark. 1997); *In re Fidelity Ins. Co. (Saratoga Springs Public Library)*, 653 N.Y.S.2d 729 (N.Y. App. Div. 1997); *Hoffman v. Fidelity & Deposit Co. of Md.*, 734 F. Supp. 192 (D.N.J. 1990); *Commercial Union Ins. Co. v. Gilbane Bldg. Co.*, 992 F.2d 386 (1st Cir. 1993); *Thomas O'Connor & Co. v. Insurance Co. of N. Am.*, 697 F. Supp. 563 (D. Mass. 1988); *compare Hartford Accident & Indem. Co. v. Scarlett Harbor Assoc. Ltd. Partnership*, 695 A.2d 153 (Md. Ct. App. 1997) (adopts minority position that incorporation of contract by reference in bond does not give the surety the contractor's right to arbitrate granted in the bonded contract); *see also* J.D. Ferrucci, *Effect of an Arbitration Provision in the Principal's Contract With a Claimant, in The Law of Payment Bonds* (Lybeck & Shreves, eds., 1998).

193 *See, e.g., Sheffield Assembly of God, Inc. v. American Ins. Co.*, 870 S.W.2d 926 (Mo. Ct. App. 1994); *Hanover Ins. Co. v. Scruggs Co.*, 292 S.E.2d 493 (Ga. Ct. App. 1982); *see also* J. D. Ferrucci, *Preclusive Effect on the Surety of Prior Judgment or Arbitration Award Against the Principal, in* THE LAW OF PAYMENT BONDS (Lybeck & Shreves eds., 1998).

194 *See Transamerica Premier Ins. Co. v. Collins & Co. Gen. Contractors, Inc.*, 735 F. Supp. 1050 (N.D. Ga. 1990); *Rashid v. Schenck Constr. Co.*, 438 S.E.2d 543 (W. Va. 1993).

Such claims, when presented by the opposition ahead of poor claims can enhance the jury's acceptance of the poor claims by association. Piecemeal settlement of opponents' claims may be advantageous in "culling out the wheat from the chaff." Mediation of disputes can be effective if all mediating parties are serious about settling their disputes—which usually occurs only after some preliminary discovery and motions have been completed.[195]

A fifth strategic issue for the litigating surety is how to best "humanize" the litigation for greatest impact at the settlement table and in the courtroom.[196] Compensated surety companies, many of which still have the word "insurance" in their corporate names, evoke little sympathy in the hearts of most jurors and many judges. The surety's best chance of humanizing the litigation is to utilize highly credible witnesses and focus on "real people" (*e.g.*, the project owners or the bond principal's owners and indemnitors) as the people with a personal interest in the outcome. The best witnesses and the "human" parties ordinarily should take the lead in pursuing meritorious affirmative claims and asserting valid defenses against claims of others. The surety strategically may then stand strategically in the shadows of its principal and merely echo "me too" to the principal's affirmative claims and defenses. In cases in which the principal or its indemnitors are not parties, the surety's assertion of third party indemnity claims against them as part of a "humanization" strategy will assure a better understanding by jurors and judges that "real people" are in fact the real parties in interest with much personal loss (both monetary and reputation) at stake.

B. Controlling Adjudication Of Common Issues Affecting Claims Venued In Different Forums

A major strategic initiative of the surety in any multi-forum dispute is to assure consistent rulings on important issues and to minimize the expense of pretrial proceedings. Early strategic consideration must therefore be given to various procedural approaches available for controlling litigation. The sophisticated procedural alternatives available in the federal court system dictate that cases commenced in

---

195 *See infra* Chapter 2 for a discussion on the mediation process.

196 *See infra* Chapter 13.

state court that have a basis for federal jurisdiction should be removed to federal court.[197] These federal procedural alternatives are as follows:

1. *Related Cases Venued In The Same Court*

Cases with common issues or related to the same subject matter and pending in the same court ordinarily should be assigned to a single judge who may determine either to consolidate the cases for pretrial proceedings or for trial of common issues of law or fact, or at least may coordinate pretrial proceedings in all cases.[198] Cases pending in different divisions of the court may be transferred upon request.[199]

2. *Related Cases Venued In Different Federal Courts*

Although the plaintiffs' choice of forum is entitled to substantial deference, related cases involving common issues of law or fact and pending in different federal courts may be consolidated in a single district by transfer of venue for the "convenience of parties and witnesses" or "in the interest of justice."[200] Transfer, however, is available only to those districts where the case "might have been brought," which authorizes transfer only if personal jurisdiction and venue will lie in the transferee court. Otherwise, the transfer is improper even if the plaintiffs consent.[201] Even without transfer, cases may be controlled by coordination between courts.[202]

---

197 Even if not all cases may be removed, a stay of state court actions should be sought pending resolution of common issues of fact or law in federal court actions.

198 *See* FED. R. CIV. P. 42(a); MANUAL FOR COMPLEX LITIGATION § 31.11 (3rd ed. 1995).

199 28 U.S.C. § 1404(b); *see also* MANUAL FOR COMPLEX LITIGATION §§ 20.123, 31.12 (3rd ed. 1995).

200 28 U.S.C. §§ 1404(a), 1406.

201 *Hoffman v. Blaski*, 363 U.S. 335 (1960).

202 *See* MANUAL FOR COMPLEX LITIGATION § 31.14 (3rd ed. 1995).

3. *Multi-District Transfers By The Judicial Panel On Multi-District Litigation*

The judicial panel on multi-district litigation may transfer civil actions pending in more than one district involving one or more common questions of fact for coordinated or consolidated *pretrial* proceedings.[203] This procedure seeks to promote efficient and economical pretrial proceedings through elimination of duplication of discovery, reduce litigation costs, and avoid conflicting pretrial rulings and schedules.

4. *Class Action*

Although class actions are rare in complex construction litigation, issues involving claims for defective construction products sold to or installed by a class of contractors might well serve as a basis for certification of a class under Rule 23 of the Federal Rules of Civil Procedure. In the typical class action, the plaintiffs seek to be recognized as representatives of the plaintiff class and seek a decision by the federal court to certify a class which, following certification, will be bound by determination on common questions of law and fact and by the common pretrial proceeding.[204]

5. *Special Master*

The rules of the federal court[205] and many state courts authorize the appointment of special masters in both jury and non-jury cases where the issues are complicated, or involve difficult computation of damages or other "exceptional conditions."[206] Such exceptional conditions do not include general complexity of litigation, projected length of trial, or congestion of the court's calendar.[207] Appointment of a special master is

---

203 28 U.S.C. § 1407.

204 *See* MANUAL FOR COMPLEX LITIGATION, ch. 30 (3rd ed. 1995).

205 FED. R. CIV. P. 53.

206 *See generally* E. L. Neel, *Interpleader and Special Masters: Procedural Options for the Beleaguered Surety*, unpublished paper presented to the 1995 annual meeting of the ABA Fidelity & Surety Law Committee (Aug. 1995).

207 *La Buy v. Howes Leather Co.*, 352 U.S. 249 (1957).

to be "the exception and not the rule," and matters referred to the special master ordinarily will not involve "fundamental issues of liability" because the special master does not sit as a confirmed judge pursuant to Article III of the United States Constitution.[208] As a practical matter, special masters frequently are appointed for their expertise in particular fields such as construction, finance and accounting, and the distinction between special masters under Rule 53 of the Federal Rules of Civil Procedure and court appointed experts under Rule 706 of the Federal Rules of Evidence is becoming blurred.[209] Unlike the expert witness, however, the special master may not be cross-examined and there is no provision for discovery of the special master's findings other than as set forth in the special master's report to the court. In multiple cases involving common issues of law and fact venued in federal and state courts, one effective method of controlling uniformity of factual findings is to have the same special master appointed in all cases.

6. *Magistrate Judge*

Like a special master, a magistrate judge may be referred issues upon which to make factual determinations based on evidence presented at adversarial hearings.[210] Factual findings ordinarily are presented to the district court in the form of proposed findings of fact served upon all parties. Magistrate judges also routinely control pretrial proceedings, and their rulings on nondispositive matters may be set aside only if "clearly erroneous or contrary to law and only if objected to within ten days of service."[211]

7. *Intervention*

Under both federal and state court rules, a surety having an interest in issues as to which an adverse decision may affect it, may intervene to protect its interest.[212] Intervention may be appropriate where necessary to lay claim to contract funds, present contract defenses, avoid collateral

---

208 *See Stauble v. Warrob Inc.*, 977 F.2d 690 (1st Cir. 1992).
209 *See* MANUAL FOR COMPLEX LITIGATION § 21.52 (3rd ed. 1995).
210 *See id.* § 21.53.
211 *See* FED. R. CIV. P. 72A.
212 *See, e.g.*, FED. R. CIV. P. 24.

estoppel, pursue recourse against third parties, and protect special surety interests.

8. *Federal Interpleader*

An effective but little used procedural device for sureties facing potential multi-district litigation is federal statutory interpleader.[213] Although Rule 22 of the Federal Rules of Civil Procedure also provides for "rule interpleader," statutory interpleader is preferred because of a number of significant advantages: [214] (1) only "minimal" federal diversity is required between any two of any number of claimants;[215] (2) nationwide service of process is available; (3) nationwide stay on other actions is available; (4) the surety, after depositing the amount in dispute into the registry of the court or posting a bond in like amount, may reserve its right to contest the claims against it and recover that portion of any amounts not paid out of the registry of the court and obtain a discharge of any portion of the bond not required to satisfy judgments; and (5) all claims are litigated in the same court. By reserving its own right to contest claims against the deposited proceeds or bond and by filing its own affirmative claim to the proceeds, the suit remains "in the nature of interpleader" in which the surety retains an interest in the outcome rather than a straight interpleader in which the surety simply is dismissed after depositing the proceeds or its bond into court. The advantages to the surety of a suit "in the nature of interpleader," in a situation in which the surety is involved in litigation in different jurisdictions arising out of the same bond and cannot otherwise consolidate the actions, is obvious. The nationwide stay is applicable to all state and federal courts.

9. *Voluntary Stay Of Related Cases*

Where it is clear that the decision of one court in one case could resolve the same issue in related cases venued in different federal or state jurisdictions, courts in these related cases routinely will stay further proceedings in the interest of judicial economy and the fair

---

213 *See* 28 U.S.C. §§ 1335, 1397, 2361.

214 *See generally* Neel, *supra* note 210.

215 *State Farm Fire & Cas. Co. v. Tashire*, 386 U.S. 523 (1967).

administration of justice.[216] The same is true when two parties in a multi-party single case refer a portion of issues common to all parties to arbitration, since the federal arbitration act and state acts modeled on the Uniform Arbitration Act all provide for a stay in court proceedings pending arbitration.[217]

C. The Surety's Control Of Litigation By Bifurcation Of Issues

Yet another important strategic issue for early decision by the surety in complex construction cases is identification of issues for bifurcation. For instance, under Rule 42(b) of the Federal Rules of Civil Procedure, a court "in furtherance of convenience" or "to avoid prejudice" may order a separate trial of any claim or issue. Obvious issues upon which a surety should seek bifurcation in every case are the issues of bad faith and punitive damages, which can have substantial power to prejudice a jury by focusing the jury's attention on the surety's financial net worth and subjective intent rather than on basic contract and construction issues usually involving multiple parties. Such bifurcation keeps the trial of basic contract and bond issues separated from the bad faith tort and punitive damage issues. There also may be reasons not to bifurcate, such as in a "humanized" case in which the bond principal and indemnitors are sympathetic plaintiffs with a large claim against the obligee and where the obligee's counterclaim for bad faith against the surety simply appears to be heavy handed retaliation for agreeing with its bond principal on issues in dispute.

VI.
Strategic "Generalship": Marshaling The Trial To Successful Conclusion

After the strategic road map is established and litigation in multiple forums is brought under control, the final strategic issue for the surety is managing the litigation to successful conclusion. This involves continuing to develop the themes and issues clearly for trial, doing careful research on selection of jurors or arbitrators, overseeing an

216 *See* MANUAL FOR COMPLEX LITIGATION § 31.14 (3rd ed. 1995).

217 *See* Federal Arbitration Act, 9 U.S.C. § 1, *et seq.*; Unif. Arbitration Act, 7 U.L.A. § 1, *et seq.*

effective discovery process, preparing compelling exhibits, marshaling the evidence through careful selection of witnesses and demonstrative and documentary exhibits, and compelling advocacy. These are the subjects of subsequent chapters. New and valid ideas are critical even during the post-strategic road map phase of the litigation because of the need for "ever-changing commands on the battlefield."[218] Perceptions of facts may change under careful discovery or trial testimony. Legal theories may have to be revised or discarded to accommodate new realities. The fast moving flow of litigation demands a continuing flow of fresh, good ideas. Selecting the very best ideas for implementation is the highest art of litigation generalship practiced by the most knowledgeable and experienced trial lawyers.

---

218 *See Blake Const. Co. v. C.J. Coakley Co., Inc.*, 431 A.2d 569 (D.C. App. 1981).

CHAPTER 2

THE SURETY'S MANAGEMENT OF THE COMPLEX SURETY CASE

Donald G. Gavin
Vincent G. Fasano

I.
Introduction

Many of the initial management decisions which ultimately determine how a complex surety case is to be handled should be made by the surety claims department or the individual making decisions on behalf of the surety. There are a host of interrelated and subtle issues to be considered which include: determining the selection of lawyers and consultants; effectively managing the work and controlling the amounts to be spent; being cognizant of the concerns of the various other parties who have an interest in the defense including the principal, the obligee, and the reinsurers; evaluating the risks; identifying the potential for early resolution through negotiation; and utilizing mediation and other ADR techniques to encourage early settlement.

II.
Selecting Lawyers And Consultants

A. What Services Does The Surety Require

1. Does The Surety Need Its Own Independent Counsel And Consultants

The surety must first determine whether it needs to retain its own independent counsel and consultants or whether it can rely upon the counsel and consultants retained by its principal. The answer may be self-evident if the principal is no longer effectively able to retain counsel or consultants. Commonly divergent financial interest of the principal and surety suggests that there is a potential conflict of interest.[1] Even when the principal is in a position to be able to hire its own counsel and consultants, the surety may find that it requires independent counsel to monitor its own interests and to avoid what would otherwise be a conflict of interest. The surety would certainly retain its own counsel whenever there is an allegation of bad faith raised.

The difficulty for the surety is to determine when there is a conflict of interest that would necessitate retention of independent counsel by the principal and the surety, and under what circumstances the surety could recover the fees and litigation expenses incurred by its own separate counsel. To resolve the first inquiry, reference must initially be made to the bar rules in the particular jurisdiction, which may include the ABA Model Rules On Professional Conduct, the ABA Model Code of Professional Responsibility, and the state professional responsibility codes or applicable statutes pertaining to such issues.

The comment to Rule 1.7 of the ABA Model Rules provides that "[l]oyalty is an essential element in the lawyer's responsibility to a client."[2] In many circumstances it is impossible for a single lawyer or law firm to comply with the loyalty requirement if it is representing both the principal and the surety. Additionally, Rule 1.6 of the ABA Model Rules requires that: "[a] lawyer shall not reveal information relating to representation of a client . . . [other than under certain very limited circumstances called forth in the Rule,] unless the client consents after disclosure to the client."[3] This duty of confidentiality makes joint

---

1 *See* Linda A. Klein, *Ethical and Other Pitfalls of Dual Representation in Surety Bond Litigation*, Fidelity and Surety Law Committee, unpublished paper presented at Annual Mid-Winter Meeting (1991); M. Kelly Allbritton & Charles W. Langfitt, *Legal and Practical Problems With Joint Representation of the Principal and the Surety*, Fidelity and Surety Law Committee, unpublished paper presented at Annual Mid-Winter Meeting (1991); F. Malcolm Cunningham, Jr., *The Ethical Considerations of Representing a Principal and a Surety*, *in* BOND DEFAULT MANUAL 347 (Duncan L. Clore ed., 2d ed. 1995); Keith Langley, *et al.*, *Surety Obligations—Dueling Duties*, NATIONAL BOND CLAIMS ASS'N (October 3, 1996).

2 MODEL RULES OF PROFESSIONAL CONDUCT 1.7 (1994).

3 MODEL RULES OF PROFESSIONAL CONDUCT 1.6 (1994).

representation particularly difficult in many circumstances where the surety intends to seek indemnification from indemnitors who are closely related to the principal.

A surety's recovery of its fees and expenses for hiring an independent counsel depends on whether or not the principal's attorney can under the circumstances provide adequate counsel for the surety as well.[4] Today it is far more common for a surety to hire independent counsel to at least monitor its own interest, if not to undertake its own defense, given the increased recognition of the potential for a conflict of interest. If there is a reason to retain independent counsel, the question of whether or not the surety can recover these costs is usually irrelevant. This is particularly true where the surety seeks to retain the goodwill and business of a principal.

2. *What Is The Nature Of The Dispute And The Work Product Sought*

The selection of counsel and consultants is largely dependent on the nature of the dispute. Few sureties wish merely to hire a fine litigator or a good investigator. Rather, they prefer to retain individuals who have, in addition to litigation or investigatory skills, substantive expertise in the particular subject area. Litigation skills, may not be the primary concern if the attorney's role is essentially to monitor the work of the principal's counsel to preserve the surety's independent interests. Knowledge of potential surety defenses may be much more important to the surety under such circumstances. Is the case primarily a construction dispute in the context of a surety bond obligation, or does the matter raise more unusual issues involving the obligations and rights of the surety?[5] The former may require a lawyer whose practice emphasizes construction, while the latter may also require counsel to have

---

4 *Central Towers Apartments, Inc. v. Martin*, 453 S.W.2d 789, 799-800 (Tenn. Ct. App. 1969); *see also* F. Malcolm Cunningham, Jr., *supra* note 1, at 352-56.

5 *See, e.g.,* Armen Shahinian, *Strategic Use of the General Indemnity Agreement in Settling Bond Claims Over a Principal's Objections*, Fidelity and Surety Law Committee, unpublished paper presented at Annual Mid-Winter Meeting (1996).

significant surety expertise. Is the dispute really one arising out of a fidelity claim?[6] This may call for an entirely different set of skills.

What work product does the surety wish to obtain? Does it seek advice on how to proceed, how to get the matter resolved at an early stage, and what responsibility and obligations it has? The skills sought here are likely to require significant expertise and experience and the surety will benefit from innovative ideas. Does the surety wish to obtain the actual entry of a notice of appearance and the skills associated with discovery and litigation? Here the litigation skills of counsel and the cost effective staffing of the matter may be of primary importance. The surety with sizable internal expertise will likely view the outside assistance needed as requiring litigation and construction skills, as their own staff should be able to provide the necessary input on specialized surety issues. How big is the exposure, does it justify more experienced and possibly more expensive counsel and consultants? It may ultimately be wiser to retain counsel under such circumstances who the surety will be comfortable with for the long term management of this complex case. Who are the indemnitors and are they agreeable to the surety's additional expenditures as a way in which to limit their ultimate liability? Even if they are not agreeable, the surety's management usually will undertake this expenditure knowing that conflicts must be avoided and that the ultimate risk may be their's and their reinsurer's risk alone.

3. *When Are The Services Of An Expert Witness Required*

While many individuals and organizations can effectively and efficiently provide consulting services to the surety, not all may be able to effectively testify as expert witnesses. In examining the financial obligations incurred, does the consultant need to be a certified public accountant and will the consultant have more credibility if his or her employer is an accounting organization? The danger of utilizing some surety consultants as expert witnesses is that the finder of fact may not believe the consultants are sufficiently independent to be credible. In evaluating the soundness of alleged defective work, does an expert have to have an engineering degree and be a licensed professional engineer in the state in question? Even if the case law does not require this

---

6 *See* George R. Buleza, *Coordination and Working with Outside Counsel in Complex Fidelity Cases*, SURETY CLAIMS INSTITUTE (June 28, 1997).

designation, a consideration of the weight that a particular witness' testimony will have is essential in assessing the suitability of the selection. Is a time impact analysis the work product sought and does this require special expertise on the part of an expert witness whose field of expertise is in the area of critical path method (CPM) scheduling?[7] This particular subject area usually requires significant skill and experience as well as specialized education for a witness to be effective.

In hiring such consultants, can the surety rely on its own experience or should it work with or through its outside counsel? The response to that particular question depends in part on the surety's knowledge of and experience with individuals who can meet the surety's needs. It also depends upon the stage in the dispute resolution. The large surety today frequently has engineers on its own staff and will perform much of the initial investigation itself. If outside assistance is required, the scope of work will usually be developed by the surety's own personnel. The surety may utilize a consultant with whom the surety has an ongoing relationship at the inception of a potential default in order to assist in obtaining information and in investigating the facts. Additional or more specialized individuals or organizations may be retained later, and it would then be normal to coordinate selection of those individuals or organizations with outside counsel. This is particularly true where the consultant is being hired to actually fulfill the role of an expert witness, and issues of credentials, credibility, and presence on the witness stand become paramount.

4. Geographical Location

Often times the question is raised whether the attorney or consultant must be geographically located where the project or litigation is. There are large localities where a number of competent counsel and consultants are located so that hiring choices can easily be made from those at or near the project or litigation. In other geographical areas lacking such professionals, retaining someone of experience and

---

7 *See* Jon M. Wickwire, *et al.*, *Construction Scheduling: Preparation, Liability and Claims*, WILEY LAW PUBLICATIONS (1991); Jon M. Wickwire, *et al.*, *Use of Critical Path Method Techniques in Contract Claims: Issues and Developments, 1974 to 1988*, 18 PUB. CONTRACT L.J. (1989); Jon M. Wickwire & Richard F. Smith, *The Use of Critical Path Method Techniques in Contract Claims*, 7 PUB. CONTRACT L.J. (1974).

competency will more than outweigh the additional cost of travel. In particularly complex surety litigation involving significant substantive issues, retention of outside counsel and consultants from outside the jurisdiction may be necessary even though use of local counsel will be a required additional expense. Finally, with regard to specialized contracts with the federal government, the actual project location is probably less important than substantive expertise in federal procurement and that counsel is located near Washington, D.C., inasmuch as almost all board litigation and actions at the United States Court of Federal Claims will be administered out of Washington, D.C. area.[8]

B. What Methodology Is Utilized In Hiring Outside Law Firms, Consultants, And Experts And Factors To Be Evaluated

1. Methodology Of The Selection Process

Should the surety limit its hiring of law firms and consultants to entities and individuals who have worked with the surety in the past and with whose qualifications the surety is already familiar? On most occasions known counsel and consultants who are already on the surety's approved list will fulfill the surety's needs, but there are some very limited circumstances where the requirement for particular expertise may lead the surety to seek out new firms and individuals. More likely, the surety will utilize an existing approved counsel or consultant in the location where need arises even if it involves travel. Where new firms and individuals must be considered, the surety must decide whether to will seek actual proposals from a select group of pre-qualified law firms, consultants, or experts and engage in a competitive

8 *See* Donald G. Gavin, *et al., Public Works Projects, in* BOND DEFAULT MANUAL 229 (Duncan L. Clore ed., 2d ed. 1995). For additional information on selecting lawyers and consultants, see Thomas J. Acchione, *Selection and Use of the Consultants from the Surety's Perspective*, SURETY CLAIMS INSTITUTE (June 20, 1991); Thomas J. Acchione, *Selecting and Managing Outside Counsel*, Fidelity and Surety Law Committee, unpublished paper presented at Annual Mid-Winter Meeting (1992); Kenneth M. Givens, Jr., *The Surety's Use of Consultants in Connection with Contract Bond Defaults, in* BOND DEFAULT MANUAL 41 (Duncan L. Clore ed. 2d ed. 1995).

process. Once again the response will depend upon the particular surety, the nature of the case, and the number of qualified entities and individuals available. Most often this kind of competitive comparison is done on a relatively informal basis without a full statement of work or a complete written proposal.

2. *Evaluation Factors*

There are a number of particular evaluation factors beyond those already discussed that sureties will consider. The expertise of the particular individuals and team members who will be performing the work is of primary importance. An organization itself may have a wonderful reputation, but sureties are interested in which individual(s) will actually be assigned to perform the tasks involved and whether these individuals make a commitment to perform the work. While hourly rates continue to be an important limiting factor in the selection of counsel and consultants, quality and expertise are given increasingly more weight as well. This reflects what is called "best value" selection, which embodies the recognition that results are as important as the economy afforded by the hourly rate.

Costs to be incurred may ultimately be born by the principal or the indemnitors inasmuch as under a general agreement of indemnity, the principal and the indemnitors may have to reimburse such expenditures. Nevertheless, whether such costs actually can be collected under an indemnity agreement, together with the possible duty to mitigate, must be considered when committing to such expenditures. The typical surety will not only limit its approved law firms and consultants to those whose rates are within an acceptable range, but also will impose upon its outside counsel and consultants limitations on the cost of work relating to particular matters. Many sureties utilize written terms and conditions setting forth these limitations. It is increasingly common for the surety to ask counsel and consultants to consent to audits and review of their billings by organizations either independent of the surety or outside of the claims group. The purpose of such assessments are usually to provide compliance with the written billing terms and conditions. The practice is useful in getting indemnitors to recognize their obligations to pay for such costs.

The importance of personal relationships continues to remain a significant factor in the selection of lawyers and consultants. A promotional brochure or marketing call from a potential service provider does not necessarily result in the hiring of this entity over an organization whose qualifications, expertise, and record for getting results is already known to the surety. There is only so much time that a surety can devote to the selection process, and the natural tendency is to deal with those individuals and entities whom already have established relationships built on cost effective work and successful results. In an effort to improve objectivity, however, the surety industry is increasingly endeavoring to implement performance measurements and initiatives as a way of measuring past successes. Such efforts can be expected to continue.

In selecting counsel, the surety will look for individuals who are already admitted to practice in and have significant experience with the relevant specialized tribunal. Such bodies include those set up to accept claims from a governmental entity, *i.e.,* United States Court of Federal Claims, the various federal boards of contract appeals, the Ohio Court of Claims, or the Commonwealth Court of Pennsylvania. Likewise, bankruptcy is another area where specialized skills are sought by those endeavoring to hire counsel and consultants.[9] There are also executive agencies where particular expertise is required. For example, special expertise is needed in dealing with the United States Treasury Department on issues arising out of the listing or delisting on the Treasury List or agencies concerned with procurement fraud.[10] Sureties exposed to environmental liability may require specialized credentials on environmental law or engineering.[11] Finally, because sureties are not in business to promote litigation and to sponsor endless iterations of

---

9 *See* T. Scott Leo, *Bankruptcy Considerations and Bond Defaults*, *in* BOND DEFAULT MANUAL 249 (Duncan L. Clore ed., 2d ed. 1995); Bruce C. King & Maria C. Charles, *Effect of Principal's Filing for Bankruptcy on the Surety's Right of Indemnification*, LAW OF SURETYSHIP (Edward G. Gallagher ed., 1993); Jeffrey G. Gilmore, *et al., Contractor Bankruptcy—Surety's Prospective*, CONSTRUCTION BRIEFINGS 92-4 (Mar. 1992).

10 *See* Donald G. Gavin & Michael A. Gatje, *Surety Exposure to Federal Procurement Fraud Liability and Liability for Bad Faith*, Fidelity and Surety Law Committee (OCTOBER 30, 1992).

11 *See* Donald G. Gavin & Robert M. Wright, eds., *The Surety's Environmental Risk*, Fidelity and Surety Law Committee (1997); Wright & Ryan, *Hazardous Waste Liability and the Surety Revisited*, 30 TORT & INS. L.J. 739 (1995).

pleadings, discovery, and trials, those counsel and consultants who have effectively demonstrated their ability to utilize mediation and other alternative dispute resolution (ADR) processes to obtain early and successful resolution of disputes are in increasing demand.

III.
Controlling The Budget And The Work

A. Budgetary And Work Plan Requirements

The use of a budget and a work plan on a complex surety matter is becoming much more common in the surety industry. However, the advantages and limitations of the process are often not well understood. Since a complex surety matter, by definition, may be difficult to forecast, the projection of activities in a long-term budget and work plan may not be practicable. Nevertheless, a budget and work plan provides a strong incentive for rationale planning and a discipline against unbridled expenditures. In both respects there is a positive management contribution in the budgetary planning process. Likewise, it is useful in providing information to reinsurers and in giving notice of costs to indemnitors. However, the submission of a budget and work plan by outside law firms and consultants should not be viewed as a guarantee of the total cost of handling the particular matter or a complete list of activities required.

In order for the budgetary planning process to work effectively, the initial budget and work plan should encompass a reasonable period in which an accurate projection of activity can be made. It is not necessarily effective or useful to require a breakdown of each individual sub-activity inasmuch as this is neither possible nor in itself accurate. Allowance should be made for updating the budget and work plan after key events such as submission of a claim by a principal, receipt of the obligee's position in regard to the claim, completion of certain discovery milestones, completion of certain motions such as those for summary judgment, and completion of designated ADR steps. Updates at periodic time intervals should be required in order to keep the budget and work plan realistic and effective. Explanations of budgetary planning changes should be provided so that there is some accountability.[12]

---

12 *See* Donald M. Bieda, *et al., Budgets, Bad Guys, and Other Distasteful Acts*, NATIONAL BOND CLAIMS ASS'N (October 12, 1995); D.K. Wright,

B. Fixed Priced Tasks And Incentives

Fixed priced billing will become increasingly commonplace for assessments and other tasks where the scope of the work is clearly defined. The service provider may need to make clear that the level of detail will reflect the fixed priced limitation. Legal audit firms are collecting data which will be utilized to evaluate pricing for task based billing. This alternative, however, is not a panacea to the surety inasmuch as the service provider will have to include some contingency in the fixed price in order to allow for the unknown. The surety may be reluctant to pay such a sum believing that billing on an hourly basis would actually be more cost effective or that the quality of the product will suffer by virtue of the fixed amount. The reality is that certain tasks are amenable to a fixed priced form of billing, but these tasks must be judiciously chosen so as to include only work assignments where the service provider can predict or control the amount of work that must be performed and where the nature of the task is sufficiently common to allow for competition. Where the services sought are more unusual and require greater skill, expertise, and ingenuity, task based billing may not be practical except when priced at a premium.

Another alternative is for a surety to utilize reduced rates with bonuses. In a contractual context, bonuses are known as "award fees" or "success fees." Given the structure of the surety often within the larger insurance company, award or success fees may not always be feasible. Even where corporately acceptable, in order to make reduced rates and award fees workable for both parties, exceptions will have to be negotiated. The law firm may be willing to agree to a lower hourly rate in exchange for a stated percentage of a recovery where there is a reasonable estimate of this amount. In addition, the service provider might not necessarily share in all of the cost recovery where there is already an outstanding offer of a monetary payment to the surety at the time the law firm or consultant was engaged. The amount of this existing offer could be excluded from such calculation. Likewise, there would be events that would be beyond the prediction and control of the service provider that should be excluded from any reduced rate work. For example, assertion of a new counterclaim by a government obligee in the form of an allegation of fraud against the surety's principal should

---

Developing a Case Budget, Fidelity and Surety Law Committee, unpublished paper presented at Annual Mid-Winter Meeting (1992).

not affect payment to the law firm or the consultant. Few sureties utilize reduced rate billings with success fees in the defense of a claim from an obligee where there is only a down side exposure to liability, but not an estimated affirmative recovery.

IV.
Reporting To The Reinsurers

Most sureties enter into reinsurance treaties providing for some portion of their surety bond exposure to be borne by reinsurers. The reinsurance treaty itself or other supplementary understanding or correspondence usually establishes an initial reporting requirement. The surety must be sure to obtain such information from outside counsel and consultants as will be necessary to pass along to the reinsurers in order to comply with these requirements. Such information would likely include an assessment of the exposure of the surety, the costs associated with handling of the complex matter, the probability of obtaining indemnification, and some procedural status. Updating of this information may be required under the particular reinsurance treaty or under the procedures put in place. Consequently, the surety and its outside counsel and consultants will have to arrange to provide for enough periodic information to meet these requirements.

There may be other obligations that are either actually required under the reinsurance treaty or implied in law that require the surety to act in good faith in handling the particular potential bond loss to which the reinsurer may be exposed. Consequently, the surety must consider not only its obligations to its principals and the obligee, but also its obligations to the reinsurer.[13]

V.
Evaluating The Risk And Obligation To Perform An Investigation

The surety's rights and obligations to the obligee generally arise out of the bond and contract terms and conditions. An investigation is always needed to make an assessment of these rights and obligations,

---

13 *See* James A. Black, Jr., *et al., Primary/Reinsurer Relationships*, SURETY CLAIMS INSTITUTE (June 25, 1992).

but the level and intensity of the effort may vary.[14] Questions to be addressed in conducting an investigation might include: whether a termination for default has been properly taken by the obligee requiring the surety to respond under the bond; whether the surety has defenses such as the obligee's failure to mitigate damages; and whether or not the project can be completed within the penal sum of the bond, and thus is it cost effective for the surety to undertake to enter into a takeover agreement, tender another contractor, or tender the penal sum in the alternative. Additionally there may be payment bond claims where surety defenses should be identified, if possible. Finally, the surety itself has a duty of mitigation and may have obligations to the indemnitors to provide notice of claims on underlying bonds prior to settlement or compromise, and to exercise judgment in the expenditures undertaken.

Because of such obligations and the need to have affirmative information to carry out the surety's role and evaluate its defenses, sureties undertake a reasonable investigation as a matter of course. Usually the investigation is initiated by the surety's own employees, who may be engineers, and conducted with the assistance of outside counsel. Sometimes the help of consultants is obtained, many of whom have considerable expertise in the area. In the complex surety matter, both the level of sophistication and the amount of dollars involved may require significant input from outside counsel. This is particularly true in evaluating surety defenses, if the surety's own staff believes it requires assistance.[15] Consequently, the effort in the investigatory phase could be jointly undertaken with the surety's in-house claims and engineering personnel, outside counsel, and consultants. If the objective is also to protect as privileged the work product of the consultant, the consultant should be retained directly by outside counsel anticipating litigation to assist with the development of the attorney's work product.

---

14 *See* William S. Piper, *The Surety's Investigation*, *in* BOND DEFAULT MANUAL 25 (Duncan L. Clore ed., 2d ed. 1995); J. William Ernstrom, *et al., Pivotal Legal Precedent Affecting Underwriting and Claim Handling By A Surety—II Investigation of Claims*, Fidelity and Surety Law Committee, unpublished paper presented at Annual Mid-Winter Meeting (1995).

15 *See* Edward G. Gallagher, *Steps A Surety Can Take To Manage Claims Rather Than React To Them*, Fidelity and Surety Law Committee, unpublished paper presented at Annual Mid-Winter Meeting (1990).

A copy of a sample consulting agreement between outside counsel and a consultant is included in the Appendix as Exhibit 2.1.

The need for such investigatory work does not end with the initial evaluation. The responsibility and obligations continue and must be updated periodically and become part of the ongoing task of the counsel and consultants.

VI.
Analyzing The Case For Early Negotiations

A. Avoidance Of Disputes

Although the underwriting process does not often evaluate the degree of risk in a non-environmental contractual agreement, consideration might be given to devoting some effort to this analysis to avoid surety losses. As a general principal contracts should seek to avoid gross disparity in the allocation of risk and values exchanged. Public contracts are particularly vulnerable to an inequality in the bargaining process. This imbalance in bargaining power, together with terms unreasonably favorable to the stronger party, may confirm indications that the transaction involved elements of compulsion, or may show that the weaker party had no meaningful choice, no real alternative, or did not in fact consent or appear to consent to the unfair terms.[16] In a tough marketplace, some principals may enter into contractual agreements that are one-sided in the allocation of risk and that, in terms of the basic award price, do not offer sufficient contractual compensation to provide for completion of the project. Some contractors hope for a favorable buy out from the subcontractors in order to increase the amount of the contract price they may retain, while other contractors rely on the receipt of change orders from the obligee in order to increase the compensation on the project. Neither result can be relied upon. Yet in a surety market with over capacity, few underwriters will inquire too strenuously as to these factors.

The misallocation of risk creates legal uncertainties and consistently ranks as a leading cause of disputes. While contract provisions which result in oppressive risk-shifting and provide gross disparity in the values exchanged may be unenforceable in whole or in part, such

---

16 *See* RESTATEMENT 2D CONTRACTS § 208d (1979).

defenses asserted by a principal usually lead to prolonged and uncertain litigation. Consequently, some attention to the nature of the agreement in the underwriting process would go a long way in reducing the loss ratio of the surety.

An additional way of judging whether a project may avoid contractual disputes is by the obligee's willingness to use partnering and total quality management techniques. If the obligee and the principal are serious about the implementation of such concepts, there is an increased likelihood that the project will proceed to completion without prolonged disputes. The surety obviously has an interest in this process. However, most sureties do not seek to be a participant in an active way, since it will enlarge the role of the surety into areas with which their principals usually wish not to have the surety directly involved.[17]

Finally, projects that have utilized Disputes Resolution Boards and other early dispute resolution methodologies frequently resolve problems without resort to the performance bond obligations and are thus advantageous to the surety.[18]

B. When Is A Case Amendable To Early Negotiations

For the surety, complex litigation often arises where the principal, based on assertions that it has numerous defenses under the contract with the obligee, contests the obligee's decision to terminate the principal for default and demands that the surety support its position. Early negotiation of such disputes are difficult in the face of the often hotly contested and emotionally charged positions. On the other hand, where a principal lacks the financial ability to complete and does not have significant defenses, the surety can choose to either finance completion, tender a completion contractor, complete the project, or pay the obligee, and is therefore capable of bringing the matter to an early conclusion.

Even where there are disputed facts, once the investigation is completed and input from outside consultants and counsel is obtained, it may be possible for the surety to arrive at a reasoned and knowledgeable

---

17 *See* Donald G. Gavin & Leslie King O'Neal, *Partnering And the Surety*, Fidelity and Surety Law Committee, unpublished paper presented at Annual Mid-Winter Meeting (1993).

18 *See* R. M. Matyas, *et al.*, CONSTRUCTION DISPUTES REVIEW BOARD MANUAL (1996).

evaluation of the merits of the various contentions. If the dollar amounts to be compromised and resolved are within a range justifying settlement, careful consideration should be given to early negotiations.

Occasionally, there may not be disputed issues of fact, and the only questions may be issues of law. Proper evaluation of legal issues may result in a decision to resolve the matter through negotiation. Even where legal issues are not capable of being stipulated, a motion for summary judgment well prior to the trial date could assist in resolving issues sufficiently so that the remaining questions can be negotiated. The process of complex litigation is so costly that the parties should always consider the potential litigation cost via-a-vis the potential exposure to an award or judgment. The amount in issue could often be compromised for less than the cost of proceeding forward, allowing for some value for the risk of an unfavorable decision.

Finally, no matter how complex the issues may appear to be, if the parties to the dispute are willing to bring the matter to an early resolution, then negotiations should certainly be attempted.

VII.
Utilizing ADR Techniques To Encourage Early Settlement

A. Various ADR Techniques Available

Among the methodologies to be employed in resolving complex disputes, the following might be considered: negotiation; conciliation; facilitation; mediation; neutral evaluation; mini trials; arbitration; disputes review boards; court adjudication; and hybrids of the above. Although few would like to admit that they are not fully familiar with each of these alternatives, these alternatives are not necessarily the same or functionally equivalent. Although much has been written on the subject of ADR and mediation,[19] actual misunderstandings among the

---

19 *See* Thomas J. McCoy, *The Sophisticated Consumers Guide to Alternative Disputes Resolution Techniques: What You Should Expect (or Demand) From ADR Services*, SURETY CLAIMS INSTITUTE (June 25, 1992); Donald G. Gavin, *Mediation to Resolve Disputes*, WATER, ENVIRONMENT AND TECHNOLOGY (March 1993); Allan H. Goodman, *Basic Skills For The New Mediator And Basic Skills For The New Arbitrator* (Solomon Publications, 1993); John W. Cooley, *Mediation Advocacy* (NITA Practical Guide Series 1996); Robert M. Wright, *The*

users of these services still occur frequently. Most participants in an ADR process do not need someone merely to carry offers back and forth between the parties or to try to smooth the ruffled feelings of the participants. Many do not wish to have a binding mini-trial or a full blown arbitration, believing that if that alternative is necessary, they would just as well go through a court process, where rules of evidence apply and appeals are possible. A neutral nonbinding evaluation may be exactly what they wish to have, and the use of a truncated mini-trial to obtain a nonbinding neutral evaluation is an alternative that should be considered. The parties need to agree on the amount of presentation time. The third party neutral would deliver a nonbinding opinion, based on an evaluation of the positions of both sides in a joint session. After delivery of an opinion, the neutral evaluator may commence a more traditional mediation utilizing shuttle diplomacy in individual sessions with each of the parties.

It is common for the contract documents between a principal and an obligee to require binding arbitration.[20] A surety does not normally consent to participating as a party in a binding arbitration between the principal and the obligee. The surety may await the outcome of the arbitration to determine whether or not it wishes to resolve its own obligations through payment of some amount, through a negotiated settlement, or through adjudication of its own independent surety defenses in a separation proceeding. However the use of an ADR process to resolve the latter issues would in many instances effectively lead to a settlement.

B. Evaluating What Type Of Process Would Be Most Useful In Resolving The Particular Complex Litigation

Usually most parties to an ADR process wish to have the process be nonbinding. In a nonbinding procedure, the entities participating retain control and are able to engage in compromises and risk evaluation that a court may not be able to undertake as a matter of law. When a court is given an issue to decide, often one side's interpretation must be deemed to prevail, even when there is ambiguity or other lack of clarity in the division or allocation of risk. While it is said that arbitrators frequently

---

Mediation Process, Fidelity and Surety Law Committee, unpublished paper presented at Annual Mid-Winter Meeting (1997).

20 *See* American Institute of Architects Form A-201, General Conditions.

compromise issues in arbitration, their task is often as difficult as a judicial determination, and one side or the other may be exceedingly unhappy with the decision. For these reasons, many parties to disputes prefer to be able to retain control and the ability to fashion their own compromise. Nevertheless, they frequently need the assistance of neutral third parties in order to arrive at a negotiated settlement.

In choosing the kind of assistance that would fit the particular circumstances, one must remember that not all third party neutrals or mediators approach their task in the same fashion or are equally effective. What most participants hope to obtain from mediation is not necessarily provided by a conciliator or a facilitator, but usually requires an evaluator. An evaluator is an individual who will, using preexisting subject expertise, identify the issues and assess the merits of each side and predict, at least privately, what might occur should the matter ultimately be resolved through a more formalistic disputes resolution process. Evaluators come with different personalities and approaches, with some individuals obviously being more effective than others. Some parties may wish to have a low-key individual who will nonoffensively suggest in polite terms where the parties might end up if they do not resolve their disputes. There may be a need for a more forceful individual who can "arm twist" or even "bash" the parties into some resolution.

C. How Does The Outside Council Or The Surety Locate An Effective Mediator

In identifying prospective mediators, direct recommendations from individuals whose opinions you value who have previously utilized the services of effective mediators are the most reliable means of selection. There are surprisingly few individuals who excel as mediators. The mere fact that an individual is a retired jurist or a serving magistrate or judge does not necessarily mean that they will be a good mediator. Many jurists are, by temperament, unsuitable to this task. They have become so used to rendering judgments that they are not particularly skilled at manipulating and moving parties into positions that the parties do not necessarily wish to acknowledge but will ultimately have to recognize. It may be necessary to pay more for a third party neutral or mediator than one might reasonably expect, but this will be the smallest cost to be incurred in the entire dispute resolution process. The mediator's actual hours will be few and the cost is normally divided equally among the parties. Likewise, it may be necessary to travel to a locality where the mediator is available, even though it would seem sensible to have the mediator come to the parties. There are organizations such as the American Arbitration Association and JAMS/ENDISPUTE who have lengthy lists of mediators, but care should be exercised to obtain an individual with known capabilities meeting the demands of the particular dispute.

D. What Prerequisites May Have To Occur To More Effectively Utilize Mediation Or Other ADR Resolution Processes

Many complex surety matters could be effectively mediated at their inception, if the parties would be flexible. Unfortunately, each side generally believes its own rhetoric at that stage and therefore will not compromise sufficiently to allow for a settlement. However, since the alternative is exceedingly costly and time consuming, it may be worth the effort to try mediation or some other ADR alternative.

However, it is probably necessary for the parties to learn more about the facts and issues before there is a realistic possibility of success in mediation or other ADR methodologies. Thus, the parties may need to exchange documents and/or take some depositions in order to achieve some common understanding of the facts. It is not necessary to have full and unrestrictive discovery for this to occur, nor is it necessary to actually be in litigation. Agreement to engage is a document exchange and even to have some depositions can be arrived at without formal litigation. If counsel and their clients are willing to recognize that mediation or some other ADR methodology may be useful and that some discovery will help, they can structure some limited discovery that will provide the insight needed to effectuate a settlement. Counsel should first meet and identify the issues that need to be developed, and should try to agree upon a plan addressing these issues. Even where it does not appear that the participants are within range of a possible settlement, the ADR process itself can create enough momentum and commitment toward resolution to be successful. If the ADR is not fruitful, a second round of discovery can address any remaining concerns that will have to be addressed before an actual trial would take place.

E. Structuring The Actual Mediation

The first step is to obtain an agreement of the other parties that a mediation or other ADR process should be employed. Usually the parties will enter into some written agreement. The actual details of the written agreement are less important, since most mediators will have their own ideas on how the process should proceed and impose their own structure on it.

Usually the next step is to agree upon the identity of the mediator, the date when the mediation will take place, and who should be present. Someone from each participant who has actual authority and decision making power must personally participate. It is often helpful to specify someone in an organization who has authority above the participants who have been active in the dispute. Not all of the individuals who would otherwise be present as witnesses need to attend. Only such persons who are necessary to present a truncated and abbreviated presentation are really necessary. The individuals who are most emotionally involved may actually be an impediment to a settlement, and although they may have to be present, they need to be instructed on when to be silent.

Not every minor issue must be mediated. Instead, only those major issues whose resolution will effectively eliminate the need for extensive discussion of other sub-issues need to be addressed. Once the major issues are resolved, usually the parties themselves can agree on the minor ones. Frequently, the participants become amazingly flexible at that point, as the euphoria of resolution motivates them toward complete settlement.

The parties need to agree on where the mediation shall occur. It is usually best to have the proceedings at some neutral location such as a hotel or law firm office where multiple rooms can be made available and food services are obtainable. Other questions to be addressed are how long the initial presentations should be and when the actual caucuses between the mediator and the individual parties should commence. Since the mediator can usually provide some input into these issues, these latter concerns are often agreed upon after the mediator is selected. The mediator will often set a page limit on the written submissions to be provided in advance since he or she usually does not want to review extensive backup documentation and pleadings.

F. Other Reasons For Utilizing Mediation Or ADR

Even if the entire case cannot be resolved in the initial mediation or ADR proceeding, the process may still be useful in eliminating some collateral issues. Certainly mediation will narrow the focus of the remaining disputes and provide for more efficient procedures to address them. Early evaluation of a theory, claim, or defense may result in the parties themselves realizing either that the particular proposition is not supportable, or that it is far more important than the parties previously recognized. The focus that a mediation can bring can limit additional discovery and facilitate settlement either without assistance or with a mediator's aid at some future date. In cases that will go before a jury unless resolved, a neutral third party evaluator's assessment of how the case and witnesses will be perceived by a jury is often useful in deflating a party's enthusiasm for a jury trial. Once the parties become more realistic, cases tend to settle much more readily.

Finally, among the most significant reasons for utilizing mediation or other ADR procedures is to avoid the incredible expense and time associated with litigation. Recognizing that the "black or white" solution reached through litigation is too great a risk for most parties, some evaluation and appropriate allocation of risk in a mediated settlement is usually a preferable alternative.

Mediation also offers the possibility of a "creative solution." For example, the principal who would insist on a huge financial recovery might be satisfied with a negotiated settlement involving little recovery provided that the termination for default is rescinded and the contractor's reputation is thus restored. There may be further incentives that an obligee can offer, such as some future work on another project or approval of change orders on an ongoing project.

The principal itself may be capable of providing some creative assistance to a settlement. What may be really troubling the obligee is unperformed warranty work, latent defects, or other defective or nonperforming elements of the project. A principal may be able to use its relationships with subcontractors and suppliers to effectuate a solution to these problems that will go a long way in assisting in resolution of the overall dispute. The surety itself may be able to furnish a maintenance bond or other forms of performance guarantees that will alleviate the long term concerns of the obligee without the surety actually advancing any significant sums.

G. Preparing For And Participation In Mediation Or Other ADR Proceedings

Just because the mediation does not involve a full presentation of the evidence does not mean that preparation can be done on the back of an envelope the night before. Particularly in complex surety matters, the actual presentation of the case in ADR, while less costly, is more difficult in many respects than in a trial. The issues must be developed and presented in a very limited period of time. This means coming up with a cohesive and coherent presentation utilizing audio visual, graphic, and testimonial techniques in a unified fashion that will impress not only the mediator, but more particularly the individuals with authority from the other parties that the surety's position is exceedingly strong. The groundwork for this presentation must be commenced long before the written document to be provided to the mediator is actually prepared. The written submission must be sufficiently short that it will keep the mediator's attention and will be read and understood by the other parties, without detracting from the oral presentation. A short statement between five to twenty-five pages is usually the maximum that can be expected to be evaluated. Support documentation can be appended in easily accessible binders with indexes and tabs.

Finally, even though it is unlikely that the mediator will want to see most of the backup documentation, such materials must nevertheless be taken to the mediation session. Frequently the parties will end up focusing in negotiations on some particular letter or position that can be effectively addressed if the backup information is present. Thus, the documentation must be organized so that it can be accessed easily and quickly.[21] It is probably necessary to have available at the mediation session some technical personnel, including the surety's consultants and key individuals from its principal. These individuals may not necessarily provide extensive testimony, but must be accessible to answer questions and provide rebuttal that can become crucial to the resolution of the issues. In the actual mediation, the participants should be prepared to stay whatever hours are required and not be impatient with what may appear to be a lack of progress initially. It is often after

---

21 This is likewise true of transcripts from depositions; fortunately with the assistance of computer search engines, access to transcripts are now more readily available.

seemingly endless exchanges that resolution is finally achieved when the hour for the end of the mediation is finally drawing close.

Mediation can be characterized as a great deal of "hurry up and wait." But even with the relatively inefficient down time while the mediator is in caucus sessions with the other parties, the one to three days that it normally takes to resolve a complex case through mediation is only a small fraction of the time required for litigation or binding arbitration.

VIII.
Conclusion

Complex surety litigation can be one of the most trying and inefficient experiences of any surety claims executive. It can create the need for significant reserves and unpleasant reporting, and ultimately contribute to unfavorable lost ratios.

Better surety company management of complex cases can be rewarding to all the participants. It can provide for clearer understanding of each party's role, obligations, and responsibilities, with more cost-effective performance of the work through planning with accountability. Utilization of alternative dispute resolution processes that are increasingly available and recognized by sureties, principals, and obligees can lead to earlier and more efficient resolution of such disputes.

CHAPTER 3

THE ETHICAL OBLIGATIONS OF THE SURETY AND ITS COUNSEL BEFORE AND DURING LITIGATION AND THE SURETY'S BASIC DUTY OF GOOD FAITH AND FAIR DEALING[1]

Douglas M. Reimer

I. Introduction

The contract surety business has changed dramatically! The distinction between surety bonds and insurance policies has disappeared! Extra contractual liabilities are imposed upon sureties! Contract sureties are in a high risk business!

Each of these statements could be a headline on an article in a surety newsletter—and perhaps they have been. Bonding contractors has always been a risky business. Sureties have recognized the traditional financial risks and dangers in an industry where each project and each site is unique and a surety has to rely upon the contractors strength, skills and integrity to complete without a loss.

Now sureties face increased risks associated with their own performance and the manner in which they react to breaking developments in construction projects. The imposition of tort and insurance principles to the surety's performance has created a new layer of financial risk which must be recognized and appropriately managed.

---

1 The author gratefully acknowledges receipt of permission from the American Bar Association to quote extensively from an article he and Sheila M. Kalteux wrote for the ABA published in 1991.

These new risks can be managed and severe consequences can be avoided, but they must be recognized early and dealt with promptly. The key, though a gross oversimplification, is to initiate a timely and thorough investigation of the facts, research of the law, examination of potential conflicts of interest, and development of a plan of action based upon the investigation and research. Prompt implementation of the plan must be accompanied with notice to all parties in interest and the plan must remain flexible to accommodate fresh developments. If all of this sounds like good claims practice, that is exactly what it is. But how do we know that we have devised a good claims handling approach to each case? First, we must know where these problems are likely to arise.

Much has been written about this subject and there will be more to come. This chapter discusses some recent decisions, points the reader to a collection of authorities, draws some conclusions, and makes some suggestions.

The "BAD FAITH" cases can be roughly divided into those that do and those that do not allow extra contractual remedies against the surety. Not surprisingly, there are decisions in jurisdictions that allow bad faith claims where no award has been made against a surety, and others where damages have been assessed against sureties in jurisdiction where such claims have not been the subject of legislation or recognized by the courts. The facts of the cases and the reasoning of the courts are instructive in preparing a surety risk management program to avoid these problems.

II.
RISKS OUTSIDE THE BOND

A. Surety Bonds Treated As Insurance

Courts have treated surety bonds as they would insurance policies for purposes of dealing with complaints by owners, contractors, and others that sureties have handled claims poorly:

> It has frequently been held that contracts of suretyship are regarded as those of 'insurance', where a corporate surety engages in the business for a profit, and that the rights and liabilities of the parties are governed by the rules applicable to contracts of insurance. This is a rule governing the construction of such contracts, however, and is not intended to alter the normal incidents of a suretyship contract, such as

> the surety's recourse against the principal for losses paid by it. If a compensated surety's contracts were regarded as insurance for all purposes, it is apparent that the surety would not have such right of recourse, together with all the other rights and duties which devolve upon a surety as such.[2]

The reasoning quoted above, based on the supporting cases cited by Appleman in his treatise, has not been followed in every jurisdiction but it has had influence in encouraging disappointed owners and trade contracts to assert claims against surety companies for poor claims handling and other perceived flaws in the surety's performance.

Some jurisdictions have adopted legislation to regulate unfair claims settlement tactics or unfair insurance practices, and a large majority of courts have conceded that they do not create a private cause of action.[3] Other jurisdictions place these provisions in the Business Corporations Act as unfair business practices.[4]

B. Duty Of Good Faith And Fair Dealing

The tension pointed out in the Appleman treatise between the rights of the surety company described as the "normal incidents of a suretyship contract" and the right of a disappointed claimant on a surety bond to bring a bad faith claim are discussed in *Brighton School District v. Transamerica Premiere Insurance Company*.[5] In that case a bonded mechanical contractor was defaulted for delay, poor workmanship and material, and failure to correct defective performance. The contractor sued the school district and the school district sued the surety as third-party defendant for breach of the surety contract and breach of duty of good faith and fair dealing. When the surety was notified of the default it refused to make progress payments claiming the school district had not complied with the construction and surety contracts. The work was

---

2 APPLEMAN, INSURANCE LAW & PRACTICE, § 5273.

3 *See* the extensive analysis of this issue in *Farmer's Union Central Exchange Inc. v. Reliance Ins. Co.,* 675 F. Supp. 1534 (D.N.D. 1987).

4 The nature and extent of this legislation and cases decided thereunder have been exhaustively collected in CUSHMAN ET AL., HANDLING FIDELITY, SURETY AND FINANCIAL RISK CLAIMS, Ch. 12 (2d ed. 1997).

5 923 P.2d 328 (Colo. Ct. App. 1996), *aff'd* 940 P.2d 348 (Colo. 1997).

completed by a new contractor and the surety persisted in its refusal to pay.

The general assembly of the State of Colorado enacted provisions prohibiting unfair or deceptive trade practices in the insurance industry.[6] This act establishes industry standards concerning misrepresentation, unfair discrimination, unfair claim settlement practices, and other improper practices, and its provisions expressly apply to contracts of suretyship. The court found that the act provides only for state regulation of the insurance industry and does not create a private cause of action. However, the court went on to say it was not error to allow a bad faith claim to be brought against a surety based on the common law duty of good faith and fair dealing owed by an insurance company in claims adjustment. The surety argued that bonds are not insurance, but the court said that bonds are the business of insurance: "while [corporations who receive consideration for issuing bonds] may call themselves 'surety companys' their business in all essential particulars is that of insurers."[7] The court went on to say that the obligee of a surety contract is therefore in substantially the same position as the beneficiary of a first party insurance contract.[8]

The record on appeal disclosed that the surety was provided with status reports, notice of termination, a request for information on how to complete the project, and attended a meeting with the school board before refusing to pay for completion on the grounds that the owner had not complied with the construction and surety contracts. Perhaps the essence of the *Brighton* case is this:

> Finally, subjecting a surety to a claim of bad faith by a beneficiary of a surety contract does not require the surety to act in bad faith with respect to its principal. As the jury was instructed in this case, a breach of the duty of good faith and fair dealing arises *only when the surety knows its position is unreasonable or recklessly disregards the fact that its position is unreasonable.*[9]

---

6 COLO. REV. STAT. §§10-3-1101 to 10-3-1114 (1994).

7 *Brighton School Dist.*, 923 P.2d at 328 (quoting *Federal Surety Company v. White*, 295 P. 281, 290 (Colo.1930)).

8 *Id.; see also Loyal Order of Moose v. International Fidelity Ins. Co.*, 797 P.2d 622 (Alaska 1990).

9 923 P.2d at 333.

C. Duty To Investigate

The Supreme Court of Alaska allowed a bad faith claim against a surety for failure to adequately investigate an obligee's claim.[10] A contractor, who agreed to furnish a "turn key" new facility for the Moose Lodge received all but $13,000.00 of the contract price when it was substantially completed. The contractor "declined" to complete the project until he had been paid $58,000 in claimed extras. When the owner refused to pay for the extras the contractor refused to complete the project. The owner declared the contractor in default and notified the surety to complete or pay completion damages. The surety's local agent responded: "it is not anticipated that the surety will become involved . . . since Mr. Darling is certainly financially capable of completing the contract."

The owner furnished evidence of defective performance, code violations, and unpaid subcontractors to the surety. The surety received at least one subcontractor payment bond claim on the project. But the surety continued to doubt the contractor's default and urged the owner to let the contractor complete. The surety also offered to be bound by arbitration, provided no issues relating to the bond or any defenses of the surety (as distinguished from defenses of the contractor) were submitted to the arbitrator.

The surety was reassured when the trial court granted summary judgment in its favor on the owner's bad faith claim holding that there was a legitimate dispute whether the contractor was in default; and, until the dispute was resolved in arbitration, the surety was not required to complete the project or respond in damages. The court held the failure of the surety to investigate the case was not evidence of bad faith on the part of the surety.

The Supreme Court of Alaska outlined the owner's position as follows:

> Plaintiff's position is simple: sureties are insurers; insurers are subject to bad faith tort liability; therefore, sureties are subject to bad faith tort liability. The court of appeals rejected this syllogism as "too

---

10 *Loyal Order of Moose v. International Fidelity Ins. Co.*, 797 P.2d 622 (Alaska 1990).

> simplistic." Although simple, this proposition is supported by our statutes, case law and sound policy reasons.[11]

The court went on to elaborate on its opinion of the relationship between the surety and its obligee:

> In our view the relationship of a surety to its obligee—an intended creditor third party beneficiary—is more analogous to that of an insurer to its insured than to the relationship between an insurer and an incidental third party beneficiary. A surety may satisfy its duty of good faith to its obligee by acting reasonably in response to a claim by its obligee, and by acting promptly to remedy or perform the principal's duties where default is clear.[12]

Based upon this view and a showing by the owner that the surety failed to adequately investigate the owner's claims, the court reversed the summary judgment entered in favor of the surety and remanded the case for trial on the bad faith claim.

The lesson of this case is that the surety must conduct its own independent investigation of the facts and not side with the contractor unless it is clearly warranted.[13] Even then circumspection is advised.

The Supreme Court of Arizona also addressed the subject of surety bad faith in *Dodge v. Fidelity & Deposit Co. of Maryland*:[14]

> The question before this court is one of first impression in Arizona: can a surety on a contractor's performance bond be liable for the tort of bad faith? We answer the question in the affirmative.[15]

The case involved construction of a residence in Scottsdale and the owners claimed the contractor failed to complete and furnished defective workmanship and materials. The owners brought a bad faith claim against the surety for refusal to investigate and remedy their claim for the contractor's default. The trial court dismissed the bad faith claim

---

11 *Id.* at 627 (citations omitted).

12 *Id.* (citations omitted).

13 This result also was reached in *Cates Constr. Co., Inc. v. Talbot Partners,* 62 Cal. Rptr. 2d 548 (Cal. Ct. App. 1997).

14 778 P.2d 1240 (Ariz. 1989).

15 *Id.*

against the surety when the contractor paid the judgment entered confirming the arbitration award.

The Supreme Court of Arizona reversed, holding that a surety is an insurer,[16] and that a surety has a duty to act in good faith in responding to its obligees' claims that the principal defaulted. Failure to investigate by the surety is evidence of bad faith. The case was remanded for trial of the bad faith claim.

D. Vexatious Refusal To Pay Claims

What about vexatious refusal to pay a claim? Though not a construction contract bond case, the United States Court of Appeals for the Eighth Circuit decided this question under Missouri law in *State of Missouri ex rel. Penniscot County, Missouri vs. Western Surety Company.*[17] The county sued a sheriff's surety to recover sums paid to the sheriff by the United States Marshall for transporting federal prisoners who were housed in the county jail. The sheriff remitted to the county $40 per day per prisoner held in the county jail which he received from the Marshall. He retained $7 per prisoner per day for transporting them to and from the federal court. The evidence showed that his office used the $7 payments from the Marshall to pay for off duty employees and vehicles owned by the sheriff used in moving the prisoners.

The county sued the surety for vexatious and unreasonable refusal to pay the amount claimed under the bond. The court reviewed the Missouri statutes[18] and found them to be penal in nature and therefore to be narrowly construed.

> To recover penalties for vexatious refusal to pay, an insured must prove that the refusal was willful and without reasonable cause. If there is an open question of fact or law determinative of coverage, the insurer may insist upon a judicial determination of that question without incurring this statutory penalty. However, the existence of a

---

16 The court cited ARIZ. REV. STAT. § 20-257(2), which includes the definition of "surety insurance," and ARIZ. REV. STAT. § 20-106(B)(2), which includes guaranty and suretyship contracts as transaction of insurance business in the state.

17 51 F.3d 170 (8th Cir. 1995).

18 MO. REV. STAT. §§ 375.420, 375.296.

> litigable issue will not insulate the insurer from penalty if there is other evidence that the insurer's conduct was vexatious and recalcitrant.[19]

In responding to the county's demand for payment of its claim, the surety furnished the county with a copy of the sheriff's attorney's three paragraph letter disputing the county's claim. The county's more serious claim was that the surety vexatiously refused to conduct an independent investigation into the merits of the county's claim. The court responded:

> [I]t is undisputed that Sheriff Davis was primarily liable for that claim, even though the County was permitted to proceed directly against Western under the official bond. Missouri courts have expressly held that the surety for a solvent principal need not independently investigate a claim on a bond to avoid a vexatious refusal penalty.[20]

The teaching of this case is reaffirmation of the old adage: "know your principal's financial condition." The surety should always conduct an investigation to update the underwriting file and should not rely upon the principal's defense without an independent check of its validity.

E. Misleading Negotiations

How far must a surety's investigation carry? In negotiations with claimants is there a duty to follow up and to respond before the claim has been documented? *Szarkowski v. Reliance Insurance Co.*,[21] was a victory for the surety in the trial court when the subcontractor's claim against the surety was dismissed on summary judgment because the statute of limitations had run. In North Dakota surety bonds are treated as insurance policies because surety underwriting is part of the regulated business of insurance. The subcontractor claimed the surety admitted liability to pay the claim provided he could document his charges. The claimant alleged the surety failed to respond in a good faith manner to the document's he submitted and did not specify what additional material was necessary for payment to be made.

---

19 51 F.3d at 174 (citations omitted).

20 *Id.*

21 404 N.W.2d 502 (N.D. 1987).

The court found the subcontractor to be a third party beneficiary and an intended claimant under the bond having the right to sue the surety in the event of the contractor's default. As an intended claimant the subcontractor enjoyed the same relationship with the surety as a personal injury claimant enjoys with a liability insurer. Therefore the court concluded that the surety owed the subcontractor a duty of good faith and fair dealing in handling its surety claim and ruled he had a right to bring an independent tort action against the surety for a breach of that duty. The court applied the doctrine of equitable estoppel to preclude the application of the statute of limitations as a defense by the surety. The surety's actions misled the plaintiff into thinking his claim would be paid without a suit. The summary judgment was reversed.

This case points out the need to conduct negotiations with claimants in such a way that there is a full exchange of accurate information, and disclosure of the surety's defenses. Where appropriate, claimants should be informed of deadlines such as the statute of limitations.[22]

F. Bad Faith Claim Is Independent Action

In a bad faith claim case in Hawaii a claimant urged the Supreme Court to rule that a separate cause of action sounding in tort can arise against a surety for failure to properly investigate a claim. *Board of Directors v. United Pacific Insurance Co.* [23] The case involved performance and maintenance bonds for repair work done with water resistant painting material. The work was completed successfully but a year later the paint on the building exhibited a streaky, dirty appearance due to the adherence of foreign particles to its sticky surface.

The surety conducted a preliminary investigation of the claim and denied it on the grounds that the problems were not covered by the bonds. The owner filed a complaint against the surety for punitive damages, bad faith, and treble damages for unfair and deceptive trade practices. The action was stayed pending arbitration between owner and contractor.

---

22 For a more detailed discussion of the surety's obligations to payment bond claimants in another case arising out of the same default as in *Swarkowski*, see *Farmer's Union Central Exchange, Inc. v. Reliance Ins. Co.*, 675 F. Supp. 1534 (D.N.D. 1987), in which the court denied the bad faith claims of two payment bond claimants.

23 884 P.2d 1134 (Hawaii 1994).

The arbitrator found that the bonded contractors performance was of good quality, the contractor was not in breach of its contract, and the surety was not in breach of its performance bond. As to the maintenance bond, the arbitrator said the surety did not bond the architect and as the work was done according to his specifications the surety was not in breach of its maintenance bond. Subsequent to these findings, which were confirmed by the judgment of the court, the surety was awarded summary judgment as to all of the owner's claims including the bad faith claim.

The owner appealed asserting fact questions about the surety's good faith for failing to conduct a separate investigation. This caused the Supreme Court to examine the bad faith claim separately from the bond claims. After recognizing the bad faith claim as a separate cause of action, the court found that since no breach of contract on the bond occurred the bad faith claim was properly denied.

This case demonstrates the need to conduct an independent technical investigation of the quality of a principal's work. Failure to do so places the surety in the untenable position of guessing about whose expert is right. In addition the existence of the surety's expert is solid evidence the surety has honored a duty to investigate.

G. Punitive Damages, Attorneys Fees, And Interest

Not all jurisdictions recognize bad faith claims against sureties, but sureties must nevertheless be careful how they handle claims as other forms of penalty may be imposed for perceived misbehavior.

Turner Construction Co. v. First Indemnity of America Insurance Co.,[24] involved construction of a space frame atrium in downtown Philadelphia. The contractor awarded a design build contract to a French fabricator and required surety bonds to secure performance. When it became apparent that the subcontractor would be unable to complete the contractor turned to the surety to perform. The surety decided to sue the subcontractor before responding to the contractor. The contractor, faced with a completion date, arranged performance of the work by another company and paid to complete the work.

The district court judge referred the contractor's suit against the surety to a special master who was very familiar with the technology of

24 829 F. Supp. 752 (E.D. Pa. 1993).

space frames. The special master's report, among other things, found that the surety had failed to timely and responsibly investigate the claim and respond to the contractor's claims. The master also found that the contractor reasonably spent funds in excess of the subcontractor's contract balance to complete the space frame and ruled that the surety was responsible to pay these costs.

The surety asserted a number of defenses including the position that the subcontractor was in default before the bonds were executed. The evidence demonstrated that the surety never bothered to follow the progress of the subcontractor's work, was uninformed about the subcontractor's financial condition and, failed to follow its own practices and procedures for tracking its principal. The surety also claimed that the contractor waited too long to declare a default, failed to mitigate its damages, and should be estopped to assert a claim against the surety.

The special master found nothing appealing in the sureties' defenses, but did not find the surety guilty of bad faith under Pennsylvania statutes, because he found the only acts Pennsylvania courts have found to violate this statute are intentionally false, misleading, and deceptive statements made by insurance agents in order to attract and secure new business. Further, the court could not find in the master's findings evidence to support a bad faith claim under Pennsylvania statutes. The master assessed the surety attorneys' fees incurred by the contractor in completing the project, prejudgment interest, and sanctions for attempting to disqualify the contractor's attorney in the case for a purported conflict of interest.

This case demonstrates the risk of raising defenses on behalf of the surety without a full and proper investigation. It also shows that other forms of penalties may be imposed upon a surety for inappropriate conduct.

H. Is There Any Surety Bad Faith In Texas[25]

Can there be a state where a surety does not owe a duty of good faith and fair dealing to an obligee? Apparently there is: *Great American*

---

25 James A. Knox, Jr., *Is There Any Surety Bad Faith after Great American v. North Austin MUD?*

Insurance Co. vs. North Austin Municipal Utility District No. 1,[26] distinguishes surety bonds from liability insurance. The opening paragraph of the opinion of the Supreme Court's opinion reads as follows:

> The issues in this case involve the duties and liabilities of a commercial surety to its bond obligee. We hold there is no common law duty of good faith and fair dealing between the surety and the bond obligee comparable to that between a liability insurer and its insured. We further hold that article 21.21 of the Insurance Code is inapplicable to a commercial surety, and accordingly, reverse the judgment of the court of appeals in part. 850 S.W. 2d 285, 902 S.W. 2d 488. We affirm the holding of the court of appeals that the surety in this case is liable under the terms of the bond for the default of the principal.[27]

The case involved construction of a municipal waste water lift station consisting of a wet well and a dry well. The dry well, a steel tank in which pumps and electrical equipment were situated, was to be refurbished and relocated at a deeper elevation. The contractor had a performance specification to work from and a dispute arose a year after completion when the sides of the dry well collapsed inward about three inches demonstrating lack of structural strength.

The surety conducted an investigation, met with the owner, its engineers, and designers, and requested additional information from the owner. Nevertheless a jury found the surety guilty of deceptive practices under the Texas Insurance Code and judgment on the verdict was entered against the surety for the actual damages, prejudgment interest, and attorneys' fees.

The Supreme Court reduced the award of attorneys fees' to an amount which, added to the damages for the collapse of the tank, did not in total exceed the penal amount of the bond plus fees for enforcement of the bond as provided in the bond itself. But most importantly, the court held there is no duty of good faith and fair dealing imposed upon sureties in Texas. The moral of this case is: don't mess with Texas.

---

26 908 S.W.2d 415 (Tex. 1995).

27 *Id.* at 416.

I. Does The Miller Act Pre-empt Bad Faith Claims

In a Miller Act case brought in the United States District Court in Houston, *Tacon Mechanical Contractors v. Aetna Casualty & Surety Co.*,[28] the court said the dispute over a government construction project involved the dissatisfaction of a subcontractor and a subsubcontractor with the promptness of the surety's payment on the payment bond. The dissatisfaction prompted the tradesmen to assert a state law claim sounding in tort for breach of the duty of good faith and fair dealing and claims under the Texas Insurance Code for vexatious failure to pay, and tortuous interference with contract. The court reviewed the cases decided under the Miller Act and ruled that the state law claims for damages were preempted by the Miller Act.

Apparently, being uncertain of this result the court analyzed the state law claims and ruled that the duty of good faith and fair dealing arises out of the "special relationship" between the insurance company and the insured, but that Texas law does not imply this relationship into ordinary contracts. Payment bonds do not imply a special relationship at all similar to the insurance relationship and therefore there can be no recovery on a breach of good faith and fair dealing claim.

Finally, the court said attorneys' fees could only be recoverable where a party acted vexatiously or oppressively. As to that claim the judge wrote:

> While the court agrees that the certain claims should have been paid with more dispatch, both the inability of all the contractors to submit complete applications for payment as well as the animosity that developed among the lawyers mitigate Aetna's tardiness. In no event does Aetna's conduct stoop to the level required to award attorneys fees.[29]

With respect to the claim of tortuous interference with contract, the court said it fell as none of the other claims had merit.

In contrast to the decision in *Tacon Mechanical Contractors*, the decision in *United States ex rel Ehmcke Sheet Metal Works v. Wausau Insurance Companies*[30] stated that "the Ninth Circuit has ruled that the

---

28 860 F. Supp. 385 (S.D. Tex. 1994).

29 *Id.* at 388.

30 755 F. Supp. 906 (ED Cal. 1991).

Miller Act does not exclude state law claims against sureties for conduct relating to the performance of obligations arising out of Miller Act bonds." In this case, a sheet metal subcontractor on a federal government project sued the prime contractor and its surety for breach of contract and the payment bond surety for breach of the covenant of good faith and fair dealing.

The court ruled that the claim for the breach of covenant of good faith and fair dealing brought against a surety by a party who has not contracted for the surety bond is not pre-empted by the Miller Act, and that if California permitted such an action, the surety's motion to dismiss would be denied.

In reviewing California law the court found no decision in point. The court had to determine what the California Supreme Court would decide if presented with the issue. It concluded that the California courts would not permit a cause of action against a surety for breach of the covenant of good faith and fair dealing, relying upon *Moradi-Shalal v. Firemens Fund Insurance Companies*.[31] The reasons given in that decision by the California Supreme Court for not allowing such claims are: (1) that such a claim would promote multiple litigation because it encourages two lawsuits by an injured claimant; (2) granting a second cause of action against the insurer by the insured claimant may tend to encourage unwanted settlement demands and coerce inflated settlements by insurers; and (3) the court identified a serious conflict of interest created for the insurer holding that the insurer must not only protect the interests of its insured, but must also safeguard its own interests from the adverse claims of third party claimants. This conflict disrupts the settlement process and may disadvantage the insured.

The court then had to decide whether the law with respect to an insurance policy should apply to a surety. It found that the Unfair Practices Act in California itself created an administrative sanction for bad faith and appointed the insurance commissioner to enforce the law but created no private remedy. The court laid a special emphasis on the third reason in the *Moradi-Shalal* case stating: "permitting the subcontractor to sue the surety for bad faith would also create an unresolvable conflict of interest for the surety. The surety owes the

---

31 758 P.2d 58 (Cal. 1988).

party who obtained the bond a duty of good faith and fair dealing."[32] If the surety may also be liable to a third party for bad faith then it may be forced to compromise the duty to one to fulfill the duty to the other. This potential conflict could interfere with the progress of the primary law suit, create uncertainty for the surety, and disadvantage the purchaser of the surety bond. The court also found applicable the rules of construction set forth in *Bank of America v. HM Dowdy*.[33] In that case it is pointed out that a surety on an official bond undertakes no liability for anything which is not within the letter of his contract. The obligation is *strictissimi juris*; that is, he has consented to be bound only within the express terms of his contract and his liability must be found within that contract or not at all.

For all of those reasons the court found that the California Supreme Court would not permit a subcontractor on a Miller Act project to sue a surety for breach of the covenant of good faith and fair dealing.

The Supreme Court of Nevada rejected *Brighton* in *Great American Insurance Company v. General Builders, Inc.*[34] The court ruled on a claim for punitive damages brought against a surety for revoking performance and payment bonds because they were issued by an agent without authority. The contractor sued because he lost a hospital project.

The agent who issued the bonds had apparent authority to issue bonds but unknown to owner and contractor could not do so without express prior approval from the surety. The bonds were complete with corporate seal and attached powers of attorney.

The unauthorized bonds came to the attention of a senior bond claim lawyer in the surety's office and he informed the contractor and the hospital the bonds would be revoked because the agent lacked authority to issue them and not due to any fault or misconduct of the contractor.

The senior bond claims attorney apparently ignored a memorandum he received the same day from one of the lawyers under his supervision advising him that the bonds were binding on the company because the agent had apparent authority to issue them. At trial the court directed a

---

32 *Ehmcke Sheet Metal Works*, 755 F. Sup. at 906 (quoting *Pacific-Southern Mortgage Trust Co. v. Insurance Co. of N. Am.*, 212 Cal. Rptr. 754 (Cal. 1985).

33 9 Cal. Rptr. 779 (Cal. App. 1961).

34 934 P.2d 257 (Nev. 1997).

verdict for the contractor against the surety for compensatory damages and punitive damages.

The Supreme Court of Nevada affirmed the decision of the trial court that the record established that a binding surety contract had been entered into and that the surety had an obligation to honor the bonds. The surety argued that the case was one of ordinary breach of contract and that the award of punitive damages was therefore improper. The Supreme Court agreed with this argument.

> The tort action for breach of the implied covenant of good faith and fair dealing requires a special element of reliance or fiduciary duty . . . and is limited to 'rare and exceptional cases . . . .'
>
> Tort liability for breach of the good faith covenant is appropriate where 'the party in the superior or entrusted position' has engaged in 'grievous and perfidious misconduct' . . . Accordingly, we have denied tort liability in certain relationships where agreements have been heavily negotiated and the aggrieved party was a sophisticated businessman . . .
>
> [C]learly the elements of a relationship which would support tort liability are not present in this case. The facts of this case do not raise the same public policy concerns implicated where an insurance company refuses to compensate a policyholder for losses covered by the policy.[35]

The court believed the award of expected profits for mere breach of contract, rather than tort damages, was sufficient to cover the loss of the hospital's contract.

A survey of the case law of all the states would likely establish that the jurisdictions fall into one camp or the other. But for purposes of defining a surety's duties toward obligees, principals, and claimants alike the lesson is the same. Negligent and incompetent claims handling is too risky to be tolerated and probably more expensive than the cost of good investigations and adjustments by a well-trained staff.

---

35 *Id.*

III.
Ethical Obligations

> Studies of professional liability claims indicate that there is an increase in malpractice claims against lawyers in situations where a conflict of interest was undetected or ignored.[36]

In the Preamble to the Model Rules of Professional Conduct adopted by the ABA it is said that the rules "are rules of reason. [that] do not . . . exhaust the moral and ethical considerations that should inform a lawyer, for no worthwhile human activity can be completely defined by legal rules. The Rules simply provide a framework for the ethical practice of law."[37]

The ABA ethics rules are to be extensively reviewed in an ABA project designated "Ethics 2000." The contemplated completion date of the special committee's work is some time in the year 2000. The indications are that the review will address significant variations that currently exist between and among the Model Rules and the Rules of Professional Conduct that have been adopted in the states.[38]

---

36 *Legal Malpractice Claims in the 1990s*, ABA Standing Committee on Lawyers' Professional Liability (October 1997).

37 Model Rules of Professional Conduct and Code of Judicial Conduct, 1989 Edition, Center for Professional Responsibility, American Bar Association, page 7.

38 The following jurisdictions have adopted new legal ethics rules since about the time the ABA approved the Model Rules of Professional Conduct in August 1983: Alabama: Model Rules as amended, Jan. 1, 1991; Arizona: Model Rules as amended, Feb. 1, 1985; Arkansas: Model Rules as amended, Jan. 1, 1986; California: Follows neither Model Rules nor Model Code, May 27, 1989; Connecticut: Model Rules as amended, Oct. 1, 1986; Delaware: Model Rules as amended, Oct. 1, 1985; District of Columbia: Model Rules as amended, Jan. 1, 1991; Florida: Model rules as amended, Jan. 1, 1987; Idaho: Model Rules as amended, Nov. 1, 1986; Illinois: Structure of Model Rules, with substance from both Model Rules and Model Code, Aug. 1, 1990; Indiana: Model Rules as amended, Jan. 1, 1987; Kansas: Model Rules as amended, March 1, 1988; Kentucky: Model Rules as amended, Jan. 1, 1990; Louisiana: Model Rules as amended, Jan. 1, 1987; Maryland: Model Rules as amended, Jan. 1, 1987; Michigan: Model Rules as amended, Oct. 1, 1988; Minnesota: Model Rules as amended, Sept. 1, 1985; Mississippi: Model Rules as amended, July 1,

As the ABA rules grew out of older Canons of Ethics and ABA Model Standards, we may expect "Ethics 2000" to result in the reformulation of the rules in such a way that the numerous ethics opinions written by the ABA Standing Committee on Ethics and Professional Responsibility will not be mooted. Similarly, it is assumed that the larger body of ethics opinions written in the various states will remain of precedential value.

The legal profession is largely self-regulated and the emphasis upon ethical consideration is self-imposed. The requirement of ethical conduct imposed upon lawyers obviously affects the conduct of their surety client and there are many cases in which ethical questions may appear.[39]

While the respective interests of the surety, the contractor, the owner, and the guarantors may differ at the outset of a project, they are not necessarily adverse. The interests of trade contractors and vendors, while different from the core group described above, are not necessarily adverse but are at arms length from the others. But the relative

---

1987; Missouri: Model Rules as amended, Jan. 1, 1986; Montana: Model Rules as amended, July 1, 1985; Nevada: Model Rules as amended, March 28, 1986; New Hampshire: Model Rules as amended, Feb. 1, 1986; New Jersey: Model Rules as amended, Sept. 10, 1984; New Mexico: Model Rules as amended, Jan. 1, 1987.

New York: Amended Model Code incorporating substance of some Model Rules, Sept. 1, 1990; North Carolina: Takes structure and substance from both Model Rules and Model Code, Oct. 7, 1986; North Dakota: Model Rules as amended, Jan. 1, 1988; Oklahoma: Model Rules as amended, July 1, 1988; Oregon: Amended Model Code incorporating substance of some Model Rules, June 1, 1986; Pennsylvania: Model rules as amended, April 1, 1988; Rhode Island: Model Rules as amended, Nov. 15, 1988; South Carolina: Model Rules as amended, Sept. 1, 1990; South Dakota: Model Rules as amended, July 1, 1988; Texas: Model Rules as amended, Jan. 1, 1990; Utah: Model Rules as amended, Jan. 1, 1988; Virginia: Amended Model Code incorporating substance of some Model Rules, Oct. 1, 1983; Washington: Model Rules as amended, Sept. 1, 1985; West Virginia: Model Rules as amended, Jan. 1, 1989; Wisconsin: Model Rules as amended, Jan. 1, 1988; Wyoming: Model Rules as amended, Jan. 12, 1987.

39 Matthew M. Horowitz, *Ethical Considerations in Interviewing Third Parties in the Course of a Surety Investigation: Applying the Rules of Professional Conduct and the Code of Professional Responsibility to Sureties and Their Counsel*, 32 FIDELITY & SURETY NEWS, NO. 1 (1997).

relationships change over time and as these changes occur the ethical duties of the surety's lawyer may change.

The ethical considerations embodied in the ABA Rules that are commonly implicated in representing the surety are: (1) Rule 1.7: Conflict of Interest: General Rule; (2) Rule 1.8: Conflict of Interest: Prohibited Transactions; (3) Rule 1.9: Conflict of Interest: Former Client; (4) Rule 1.16: Declining or Terminating Representation; (5) Rule 3.7: Lawyer as Witness; (6) Rule 4.3: Dealing with Unrepresented Person; and (7) Rule 4.1: Truthfulness in Statements to Others.

Where the surety is represented by the contractor's lawyer for purposes of resolving unlitigated questions with subcontractors or the owner, or the surety's attorney conducts negotiations representing both contractor and surety, the parties must anticipate what might occur in the event they cannot agree with one another or the surety is called upon to perform under the bond. A conflict may arise because the surety has several rights with respect to the contractor and guarantors, including exoneration, subrogation, and reimbursement, which are adverse to the other parties.

Rule 1.7(b)(2) requires that the lawyer set forth the various differing interests that might affect his independent judgment. Rule 1.7(b) provides:

> A lawyer shall not represent a client if the representation of that client may be materially limited by the lawyer's responsibilities to another client or to a third person, or by the lawyer's own interests, unless: (1) the lawyer reasonably believes the representation will not be adversely affected; and (2) the client consents after consultation. When representation of multiple clients in a single matter is undertaken, the consultation shall include explanation of the implications of the common representation and the advantages and risks involved.

Under the old canons, a comment was made that a lawyer may not represent a conflicting interest unless he "reasonably believes his professional judgment will not be adversely affected," as follows:

> [A] lawyer may represent multiple clients if it is obvious that he can adequately represent the interest of each and if each consents to the representation after full disclosure of the possible effect of such representation on the exercise of his independent professional judgment on behalf of each.

A lawyer may not represent multiple parties to the same transaction whose interests are fundamentally antagonistic. However, a lawyer may represent multiple parties whose interests are generally aligned. The official comment to Rule 1.7 provides:

> Loyalty to a client is also impaired when a lawyer cannot consider, recommend or carry out an appropriate course of action for the client because of the lawyer's other responsibilities or interests. The conflict in effect forecloses alternatives that would otherwise be available to the client. A possible conflict does not itself preclude the representation. The critical questions are the likelihood that a conflict will eventuate and, if it does, whether it will materially interfere with the lawyer's independent professional judgment in considering alternatives or foreclose courses of action that reasonably should be pursued on behalf of the client. Consideration should be given to whether the client wishes to accommodate the other interest involved.

Relevant factors in determining whether there is potential for adverse effect include: the duration and intimacy of the lawyer's relationship with the client or clients involved; the functions being performed by the lawyer; the likelihood that actual conflict will arise; and the likely prejudice to the client from the conflict if it does arise. The question is often one of proximity and degree.

When it becomes apparent that potential conflicts will develop into actual ones (*i.e.*, that litigation will ensue), the attorney may have to withdraw from both representations in order to avoid violating Rule 1.9. Where the interests are adverse and the conflict is actual, withdrawal is mandatory, even in the nonlitigation context where there is a greater willingness to allow the attorney to engage in multiple representation.

A lawyer may be disqualified or forced to withdraw from representing a client in a litigation arising out of a prior negotiated transaction if there is a danger that the lawyer will be called as a witness, either on behalf of his client or to give testimony prejudicial to his client. If the lawyer is called to give testimony adverse to his client, then the duty of loyalty to his client is compromised. If the lawyer will be a witness on behalf of a client then the danger exists that the trier of fact will either discount the lawyer's testimony or accord it too much weight.

Where an attorney has undertaken the dual representation of contractor and surety and must terminate the dual representation because of the development of an actual conflict of interest the courts have developed a "substantial relationship test" to analyze the propriety of the continued representation of one of the clients.[40] This is the test employed in the ABA Rules.

The party seeking to disqualify counsel from representing the other party of the former dual representation on these grounds must prove that the lawyer-client relationship is substantially related to the matter at hand.

The test indulges in some presumptions, including the presumption that confidential disclosures were made to counsel in his or her dual capacity which could be used adversely when the attorney-client relationship terminates. Some courts have held that this presumption is rebuttable.[41] Other courts are uncertain about whether or not it is a rebuttable presumption,[42] and one court has held the presumption to be irrebuttable.[43]

Before litigation is commenced there are a myriad of situations in which the surety and its attorney are dealing with other parties from underwriting concerns to payment disputes. The ethics rules touch upon these situations as well. Rule 4.1 provides:

> In the course of representing a client a lawyer shall not knowingly:
>
> (a) make a false statement of material fact or law to a third person; or
>
> (b) fail to disclose a material fact to a third person when disclosure is necessary to avoid assisting a criminal or fraudulent act by a client, unless disclosure is prohibited by Rule 1.6.[44]

Hopefully, you never will need to look up and be concerned with 4.1(b). The comment on this rule in the 1989 ABA Edition says:

---

40 *Freeman v. Chicago Musical Instrument Co.*, 689 F.2d 715 (7th Cir. 1982); *Duncan v. Merrill, Lynch, Pierce, Fenner & Smith*, 646 F.2d 1020 (5th Cir. 1981).

41 *LaSalle National Bank v. County of Lake*, 703 F.2d 252 (7th Cir. 1983).

42 *Emle Industries Inc. v. Patentex, Inc.*, 478 F.2d 562 (2d Cir. 1973).

43 *Government of India v. Cook Industries, Inc.*, 422 F. Supp. 1057 (S.D.N.Y. 1976), *aff'd*, 569 F.2d 737 (2d Cir. 1978).

44 Rule 4.1.

> A lawyer is required to be truthful when dealing with others on a client's behalf, *but generally has no affirmative duty to inform an opposing party of relevant facts*. A misrepresentation can occur if the lawyer incorporates or affirms a statement of another person that the lawyer know is false. Misrepresentations can also occur by failure to act.[45]

Rule 4.3 provides:

> In dealing on behalf of a client with a person who is not represented by counsel, a lawyer shall not state or imply that the lawyer is disinterested. When the lawyer know or reasonably should know that the unrepresented person misunderstands the lawyer's role in the matter, the lawyer shall make reasonable efforts to correct the misunderstanding.

The comment on this section further provides:

> An unrepresented person, particularly one not experienced in dealing with legal matters, might assume that a lawyer is disinterested in loyalties or is a disinterested authority on the law even when the lawyer represents a client. During the course of a lawyer's representation of a client, the lawyer should not give advice to an unrepresented person other than advice to obtain counsel.

Some of the obvious problems likely to arise in the prelitigation phase of representing a surety include the situation described in the *Szarkowski* case, where the subcontractor with a payment bond claim formed the impression his claim would be paid and let the statute of limitations run. Guarantors (indemnitors) want to hear that they will not be called upon to pay losses and the owner wants to think the surety will complete the project when the contractor is in default.

When surety and contractor are co-litigants, potential conflicts implicate Rule 1.7(b). It is in the surety's interest to cooperate with the contractor and minimize the adversity between them which may increase costs and force the contractor into bankruptcy. However, the surety's ability to tender the defense of a claim to the contractor poses some potential problems. An impermissible conflict may exist when an

---

45 Rule 4.1, cmt. (emphasis added).

attorney represents parties whose interests in litigation may conflict, such as co-plaintiffs or co-defendants, where there is a substantial discrepancy in the party's testimony, an incompatibility in positions in relation to an opposing party, or substantially different possibilities of settlement of the claims or liabilities in question. However, common representation of persons having similar interests is proper if the risk of adverse effect is minimal.

In some circumstances, the attorney-client privilege may be waived.[46] In one case, a lawyer was permitted to represent both contractor and surety in an action against the owner of the project even though the surety might have future claims against the contractor for indemnity. However, the lawyer was not permitted to subsequently represent the surety in its suit against the contractor for indemnity related to the prior litigation even though both parties had consented to such representation before agreeing to the earlier joint representation.[47]

In yet another case, a lawyer was permitted to represent both a guarantor of the surety on a payment bond and the general contractor, even though the guarantor could seek indemnification from the general contractor. However, the general contractor had conceded that he was liable to the guarantor in the event of a recovery against the guarantor and had requested the representation after full disclosure of the facts and consultation with independent counsel.[48]

Dual representation may also pose problems in negotiation of settlement. Rule 1.8(g) prohibits group settlements in civil cases without disclosure to, and consent of, each client. Rule 1.8(g) provides:

> A lawyer who represents two or more clients shall not participate in making an aggregate settlement of the claims of or against the clients, unless each client consents after consultation, including disclosure of the existence and nature of all the claims or pleas involved and of the participation of each person in the settlement.

---

46 Michigan Committee on Professional and Judicial Ethics, Informal Opinion C1-555 (Sept. 22, 1980). In this opinion, attorney-client privilege applied to one who sought legal advice through his attorney from the attorney of a co-party.

47 Oregon State Bar Legal Ethics Committee Opinion 498 (Sept. 4, 1984).

48 New York County Lawyers Association Committee on Professional Ethics Opinion 523 (1970).

When the surety brings suit against the contractor to enforce its rights under the surety agreement, a lawyer cannot represent both parties because he or she would be representing directly adverse interests and would not meet the "reasonably believes" requirement under Rule 1.7(a)(1), since it would be obvious that the lawyer cannot adequately represent both interests. Rule 1.7(a) provides:

> A lawyer shall not represent a client if the representation of that client will be directly adverse to another client, unless: (1) the lawyer reasonably believes the representation will not adversely affect the relationship with the other client; and (2) each client consents after consultation.

When a disinterested lawyer would conclude that the client should not agree to the representation under the circumstances, the lawyer involved cannot properly ask for or provide representation on the basis of the client's consent. When more than one client is involved the question of conflict must be resolved as to each client. In some circumstances it may be impossible to make the disclosure necessary to obtain consent. For example, if one client refuses to consent to the disclosure necessary to permit the other client to make an informed decision, the lawyer cannot properly ask the latter to consent.

If the attorney had previously represented both surety and contractor, he or she would be prohibited from representing either party unless there is full disclosure and waiver under Rule 1.9. Rule 1.9(a) provides:

> A lawyer who has formerly represented a client in a matter shall not thereafter represent another person in the same or a substantially related matter in which that person's interests are materially adverse to the interests of the former client unless the former client consents after consultation."

Rule 1.9(c)(1)(2) further provides:

> A lawyer who has formerly represented a client in a matter or whose present or former firm has formerly represented a client in a matter shall not thereafter: (1) use information relating to the representation to the disadvantage of the former client except as Rule 1.6 or Rule 3.3 would permit or require with respect to a client, or when the information has become generally known; or (2) reveal information

> relating to the representation except as Rule 1.6 or Rule 3.3 would permit or require with respect to a client.

The comment to Rule 1.9 provides that the lawyer's involvement is a question of degree:

> The underlying question is whether the lawyer was so involved in the matter that the subsequent representation can be justly regarded as a changing of sides in the matter in question.

Many cases hold that where a lawyer is prohibited from representing multiple parties, the taint of representing conflicting interests may not be eliminated by dropping one client in favor of another. Consent and waiver will not avoid the prohibition against suing a former client according to these cases.[49] Other courts will allow counsel to continue to represent one of the parties, but will treat it as a simultaneous representation of conflicting interests which requires the representation to pass a more rigorous test than the substantial relationship test.[50]

Interviewing unrepresented contractors, indemnitors, or potential claimants is proper as long as the persons know that the statements are being taken by the attorney in his role as counsel for the surety. However, it is improper for the attorney to advise the unrepresented person as to the course of conduct the attorney thinks the claimant should pursue.[51]

A lawyer's presentation of settlement documents to unrepresented potential claimants for execution, in most instances, will not constitute giving advice (*i.e.* the advice that signing the documents is appropriate) as long as the lawyer makes clear that he or she is representing only the interests of the client and does not mislead the unrepresented person as to the law.[52] The lawyer should explain the contents of the documents

---

49 *See e.g., Picker International Inc. v. Varian Assocs., Inc.*, 670 F. Supp. 1363, 1365-66 (N.D. Ohio 1987) (citing cases).

50 *United Sewerage Agency v. Jelco, Inc.*, 646 F.2d 1339 (9th Cir. 1981).

51 ABA Committee on Professional Ethics Informal Opinion 1034 (1968) (letter sent by plaintiff's attorney to prospective defendants outlining claim and containing advice on the course of conduct the attorney thought the prospective defendant should pursue was improper).

52 ABA Committee on Professional Ethics and Grievances, Formal Opinion 102 (1933) (an attorney representing an employer may draft settlement papers for workmen's compensation settlements with unrepresented

in order to avoid a charge of overreaching and should allow the claimants to reach an independent decision on whether or not to sign them. The same is true of a lawyer's presentation of the indemnity agreement to an unrepresented indemnitor for execution.

In one case,[53] the court held that obtaining written authority from an unrepresented party to examine otherwise confidential papers was improper when accompanied with the implied advice that signing the authorizations would be in the party's best interest. The court found that the lawyer had acted inappropriately in obtaining the authorizations, despite the fact that he had explained precisely what the documents contained, because he had implied that the party's best interest would be served by signing the authorizations. The court apparently viewed the following statement by the lawyer as objectionable advice to an unrepresented party: "I think this would be the best way for you to help the company."

In another case,[54] the court held that the acts of drafting documents and presenting them for execution, without more, do not constitute "advice" and are proper as long as the attorney does not engage in misrepresentation or overreaching. The court found that the mere possibility that the unrepresented party had trusted the attorney and relied on his expertise in permitting him to draft the documents in a real estate transaction did not establish a relationship of confidence sufficient to characterize the attorney as a fiduciary.

As the surety client may risk a bad faith claim or other penalty by raising frivolous claims or defenses, so the lawyer risks violation of Rule

---

employees as long as the attorney does not advise or mislead the employees as to the law); ABA Committee On Ethics and Professional Responsibility, Informal Opinion 1269 (1973) (attorney in a domestic relations case may submit for signature to an unrepresented party a waiver of the issuance and service of summons and the entry of an appearance as long as the documents are not coupled with the giving of any advice); ABA Committee on Ethics and Professional Responsibility, Formal Opinion 84-350 (1984) (withdrew two ABA informal opinions that had prohibited lawyers from presenting documents to opposing unrepresented parties for their signatures).

53 *Haines,* 531 F.2d at 675-76.

54 *Dolan v. Hickey*, 431 N.E.2d 229, 231 (Mass. 1982); *see also Gold v. Vasileff*, 513 N.E.2d 446 (Ill. 1987) (buyers of grocery business cannot bring an action against seller's attorney based on breach of fiduciary duty or confidential relationship).

3.1 by agreeing to do so on behalf of the surety. Rule 3.1 Meritorious Claims and Contentions provides:

> A lawyer shall not bring or defend a proceeding, or assert or controvert an issue therein, unless there is a basis for doing so that is not frivolous, which includes a good faith argument for an extension, modification or reversal of existing law . . .

The comment on this rule further provides:

> The advocate has a duty to use legal procedure for the fullest benefit of the client's cause, but also a duty not to abuse legal procedure. The law, both procedural and substantive, establishes the limits within which an advocate may proceed. However, the law is not always clear and never is static. Accordingly, in determining the proper scope of advocacy, account must be taken of the law's ambiguities and potential for change.
>
> The filing of an action or defense or similar action taken for a client is not frivolous merely because the facts have not first been fully substantiated or because the lawyer expects to develop vital evidence only be discovery. Such action is not frivolous even though the lawyer believes that the client's position ultimately will not prevail. The action is frivolous, however, if the client desires to have the action taken primarily for the purpose of harassing or maliciously injuring a person, or, if the lawyer is unable either to make a good faith argument on the merits of the action taken or to support the action taken by a good faith argument for an extension, modification or reversal of existing law.

In *Turner Construction Co. v. First Indemnity of America Insurance Co.,* the court entered sanctions against the surety and its attorney for filing motions to disqualify plaintiff's counsel and for leave to file an amended answer. The court spent three pages of its opinion discussing these motions and reluctantly concluded that sanctions were warranted as the motion to disqualify counsel had no basis in fact. Moreover, the surety violated Rule 11 by submitting false information to support the motion. The court also found the amended pleading to have been filed for purpose of delay. The court acknowledged that the surety's lawyer had accepted on faith what his client gave him in connection with these motions but said:

> Similarly, the test as to whether a pleading has been used for "an improper purpose, such as to cause harassment, undue delay, or needless increase in litigation expense" is also objective. This means that a "pleader may not escape liability because he did not intend to bring about additional delay or expense." . . . In other words, 'There is no room for a pure heart, empty head defense under Rule 11.

IV.
Conclusion

Heads up surety underwriting and claims management and good lawyering are the answer to these problems. Knowing about the risks and taking prompt and appropriate action is the key. Simple but most effective.

CHAPTER 4

DECIDING TO LITIGATE: THE SURETY'S RECOURSE AGAINST THE PRINCIPAL AND INDEMNITORS

Jeffrey S. Osgood
Paul Goldean

I. Introduction

A. Overview

Surety bonds are devices which encourage timely completion of projects and the full payment for labor and material associated with those projects. Issuance of a surety bond creates a tripartite relationship between the "principal," "obligee," and "surety."[1] These bonds are meant to function as credit accommodations. As such, the surety, upon completion of a full underwriting analysis of the principal, anticipates no loss. However, given the relationship of the parties, whenever trouble arises, the surety faces potential threats from not only the obligee to complete obligations under the bond, but also from subcontractors, suppliers, and, quite often, the defaulting or recalcitrant principal himself.

What is the surety to do? This paper suggests that the surety, in deciding to litigate, plan to do so from the very beginning of the relationship, not necessarily as the first choice, but at the proper time. Often, a surety will not consider its rights until faced with "weathering

---

1 *See* RESTATEMENT (THIRD) OF SURETYSHIP & GUARANTY § 1 (1996) [hereinafter RESTATEMENT OF SURETYSHIP].

the default storm."[2] But by then the surety may have diminished or potentially lost valuable opportunities to force an advantage in its favor. Because there is no crystal ball which will tell the surety how parties will act when faced with financial adversity, the surety should understand its legal rights and obligations while conducting the initial investigation of the principal. More importantly, in the context of disputes between surety and principal, it is essential the surety understand how and when to use its common law and contractual rights to avoid losses, settle its claims, and settle the claims of a recalcitrant principal.

B. Surety's Basic Rights

In deciding to litigate, the surety must have a thorough understanding of its rights and the means for enforcing those rights. This requires review of both the facts and the law of the case. More exactly, the facts of each situation should be examined to determine the actions the surety may take before litigation and those causes of action which it can assert to its advantage. As a general proposition, surety liability is measured by that of the principal.[3] Thus, where the principal is not liable, neither is the surety.[4] Of course, however, it is not that simple. Before the surety determines the principal is not liable, it is likely both parties will have developed differing views of their rights and obligations.

---

2 *See* David C. Dreifuss, *Bond, Contractual and Statutory Provisions and General Agreement of Indemnity, in* BOND DEFAULT MANUAL 1, 1 (Duncan L. Clore ed., 2d ed. 1995).

3 *Morse/Diesel, Inc. v. Trinity Indus., Inc.*, 67 F.3d 435 (2d Cir. 1995); *United Aluma Glass v. Bratton Corp.*, 8 F.3d 756 (11th Cir. 1993); *President & Dirs. of Georgetown College v. Madden*, 505 F. Supp. 557 (D. Md. 1980), *aff'd in part*, 660 F.2d 91 (4th Cir. 1981); *American Sur. Co. of N.Y. v. Axtell Co.*, 36 S.W.2d 715 (Tex. Comm'n App. 1931).

4 RESTATEMENT OF SURETYSHIP, *supra* note 1, §§ 39, 48; *see also Tri-State Ins. Co. v. United States*, 340 F.2d 542, 545 (8th Cir. 1965); *National Union Fire Ins. Co. of Pittsburgh, Pa. v. Alexander*, 728 F. Supp. 192, 198 (S.D.N.Y. 1989); *Tudor Dev. Group, Inc. v. United States Fidelity & Guar. Co.*, 692 F. Supp. 461, 463 (M.D. Pa. 1988) *Cole v. Wadsworth*, 376 S.W.2d 13 (Tex. Civ. App. 1964).

Through this conflict, sureties enjoy two sets of substantive rights: (1) those created by contract; and (2) those created by common law and equity.[5] The primary common law/equity based rights that a surety may enforce against a principal are reimbursement, subrogation, exoneration, and contribution. These rights may be created or expanded by agreements between the surety on one hand, and the principal and indemnitor(s) on the other.

1. Non-Contractual Rights

Non-contractual, common law rights allow the surety to protect itself either by offensive or defensive use of those rights. Guidance for offensive and defensive use of a surety's rights is found in the *Restatement (Third) of Suretyship & Guaranty* (*Restatement of Suretyship*).[6] The non-contractual, common law rights listed below are

---

5 Section 17 of the RESTATEMENT OF SURETYSHIP provides:

(1) The rights of the secondary obligor against the principal obligor are (i) those existing as a result of any contract between them, and (ii) those that arise out of suretyship status (§§ 18, 21-31).

(2) The duties of the secondary obligor to the obligee, and of the obligee to the secondary obligor, are those existing pursuant to the contract creating the secondary obligation, subject to the secondary obligor's defenses arising out of suretyship status (§§ 19, 32-49).

RESTATEMENT OF SURETYSHIP, *supra* note 1, § 17.

6 Section 18(2) of the RESTATEMENT OF SURETYSHIP lists the surety's rights against the principal:

Recourse against the principal obligor to cause the principal obligor to perform the underlying obligation or bear the cost of performance may be effectuated by:

(a) enforcement of the principal obligor's duty of performance (§ 21); or

(b) enforcement of the principal obligor's duty to reimburse the secondary obligor (§§ 22-24); or

(c) enforcement of the secondary obligor's right of restitution (§ 26); or

(d) subrogation of the secondary obligor to the rights of the obligee (§ 27-31).

best understood in their temporal relations. First, prior to the time for performance of the underlying obligation, the surety may enforce its *quia timet* right requiring the principal to provide adequate assurance of performance. Second, at the time performance of the underlying obligation is due, the surety may enforce its exoneration right and require the principal to perform. Finally, after performance of the underlying obligation is due and the surety has performed, the surety will be able to seek reimbursement from the principal and/or contribution from cosureties. Thus, there is a certain time and place when each of these rights become ripe.

a. Right Of Exoneration

The equitable right of exoneration allows a surety to proceed by action in the courts to compel the principal to satisfy an obligation. Thus, a surety may require the principal to perform the underlying obligation which was originally bonded.[7] The surety may, before the underlying obligation comes due, compel performance of the principal through the equitable action of *quia timet*.[8] However, it is only after the obligation matures that the surety can compel performance through an action for exoneration. Although the surety will have a right to reimbursement or restitution in the event it satisfies the underlying obligations, exercise of those rights presupposes some expense and inconvenience to the surety which it would ordinarily want to avoid. A surety's right to exoneration attempts to avoid such expense.

b. Right Of Subrogation

A surety is equitably subrogated to the obligee's rights against the principal where the surety satisfies the obligee's claims.[9] Thus, equitable subrogation is akin to the surety purchasing the obligee's claim

---

7 *See* RESTATEMENT OF SURETYSHIP, *supra* note 1, §§ 18(2), 21; *Filner v. Shapiro*, 633 F.2d 139 (2d Cir. 1980).

8 For a more detailed discussion of exoneration and *quia timet,* see *infra* Section VII.

9 *Pearlman v. Reliance Ins. Co.*, 371 U.S. 132, 141-42 (1962); *Transamerica Ins. Co. v. United States*, 989 F.2d 1188 (Fed. Cir. 1993); *Federal Ins. Co. v. Community State Bank*, 905 F.2d 112, 114 (5th Cir. 1990).

against the principal.[10] Once the surety satisfies the obligation, it is then subrogated to the obligee's rights as of the time of issuance of the bond.[11] As a result, the surety may be able to increase its priority position in relation to other creditors of the principal or indemnitor(s).[12] Although the scope of the subrogation right is broad, it is most commonly asserted by the surety against the principal.[13] It is important to note that the surety, by obtaining subrogation, takes no greater rights

---

10 *See* RESTATEMENT OF SURETYSHIP, *supra* note 1, § 27.

11 *See Balboa Ins. Co. v. United States*, 775 F.2d 1158 (Fed. Cir. 1985).

12 *See, e.g., Argonaut Ins. Co. v. United States*, 434 F.2d 1362, 1370 (Ct. Cl. 1970); *Kansas City, Mo. v. Tri-City Constr. Co.*, 666 F. Supp. 170 (W.D. Mo. 1987).

13 Section 28 of the RESTATEMENT OF SURETYSHIP discusses the rights obtained through subrogation and provides:

> (1) To the extent that the secondary obligor is subrogated to the rights of the obligee, the secondary obligor may enforce, for its benefit, the rights of the obligee as though the underlying obligation had not been satisfied:
> (a) against the principal obligor pursuant to the underlying obligation;
> (b) against any other secondary obligor for the same underlying obligation, unless the other secondary obligor is a subsurety for the subrogated secondary obligor;
> (c) against any interest in property securing either the obligation of the principal obligor or that of any other secondary obligor against whom the rights of the obligee may be enforced; and
> (d) against any other persons whose conduct has made them liable to the obligee with respect to the default on the underlying obligation.
>
> (2) Recovery under this section is limited as follows:
> (a) the total recovery of the secondary obligor pursuant to subsection (1) may not exceed the secondary obligor's cost of performance of the secondary obligation;
> (b) the enforcement of the obligee's rights against a cosurety, and against any interest in property securing performance of the obligation of a cosurety, is limited to the amount that will satisfy the cosurety's duty of contribution (§ 55).

RESTATEMENT OF SURETYSHIP, *supra* note 1, § 28.

than the obligee had against the principal.[14] And, if the surety also pursues its reimbursement right, the surety is subject only to those defenses of the principal of which the obligee had notice.

c. Right Of Contribution

Where a surety is compelled to pay more than its proportionate share of a debt, that surety will have a right of contribution against any and all cosureties.[15] But the question arises whether a surety can assert contribution rights against cosureties before paying more than its share of the debt. As a general rule, a surety, before paying more than its fair share, may proceed in equity against its cosureties.[16] Thus, a surety can sue for contribution in an action of implied assumpsit or where the principals' debt is due, and a surety may pay the debt and demand contribution without deprivation of its contribution rights against cosureties.[17] The *Restatement of Suretyship* also provides that in cases of cosuretyship each cosurety is liable to the other.[18]

---

14 *See, e.g.,* CAL. CIV. CODE § 2848 (West 1994) (upon satisfying obligations of principal, surety may enforce every remedy that creditor has against principal); LA. CIV. CODE Ann. art. 3048 (West 1994) (surety who pays principal obligation is subrogated to the rights of the creditor).

15 *See In re Wetzler*, 192 B.R. 109 (Bankr. D. Md.), *aff'd sub nom., Wetzler v. Cantor*, 202 B.R. 573 (D. Md. 1996); *Alberding v. Brunzell*, 601 F.2d 474 (D. Nev. 1979); *Roberson v. Tonn*, 13 S.W. 385 (Tex. 1890).

16 *See* Roland F. Chase, Annotation, *Right of Guarantor or Surety, in Order to Avoid Paying Amount in Excess of His Proportionate Share, to Compel Coguarantors or Cosureties to Pay Their Share to Creditor*, 38 A.L.R. 3d 680 (1972).

17 *See Atalla v. Abdul-Baki*, 976 F.2d 189 (4th Cir. 1992); *Alberding*, 601 F.2d at 474; *In re Wetzler*, 192 B.R. at 109; *e.g., Hollis v. Winfree*, 216 S.W.2d 625 (Tex. Civ. App. 1948).

18 The RESTATEMENT OF SURETYSHIP provides:

(1) As between cosureties for the same underlying obligation, each cosurety is a principal obligor to the extent of its contributive share (§ 57), and a secondary obligor as to the remainder of its duty pursuant to its secondary obligation.

(2) To the extent that, as between themselves, one cosurety is a secondary obligor and the other is a principal obligor, the former has rights of contribution against the latter. The rights of

d. Right Of Reimbursement

When a surety satisfies the underlying obligation in whole or in part, it has a right of reimbursement from the principal.[19] Where the surety undertakes an obligation at the principal's request, the principal has impliedly authorized the surety to pay the debt if the principal fails to do so. This also implies a promise by the principal to reimburse the surety. In practice, reimbursement is akin to an implied indemnification.[20] Thus, the principal, once it has notice, has a duty to reimburse the surety.[21] However, a surety should not be concerned with the notice requirement because the principal's signature on the surety bond gives the principal actual notice of the obligation.[22]

The surety is entitled to reimbursement of the reasonable costs of performance, costs incurred in investigation, and expenses for asserting defenses.[23] However, cost associated with bringing an action to collect reimbursement does not constitute a "reasonable cost" and is not recoverable.[24] Also, there are several situations in which a right to reimbursement will not arise.[25] Although the duty of reimbursement is an important right within the suretyship relationship, it is a reactive duty because the surety must first suffer a loss before it can exercise its right of reimbursement. For this reason a surety will be interested in any rights it can acquire or use without first suffering a loss. These issues will be reviewed further in the next section.

> contribution are the same as the rights of a secondary obligor against a principal obligor as set forth in §§ 21-31.

RESTATEMENT OF SURETYSHIP, *supra* note 1, § 55.

19 *See id.* §§ 21, 22, 23.

20 *See id.* § 22 cmt. a ("Just as the principal obligor impliedly agrees that it will perform the underlying obligation so that the secondary obligor will not have to perform, the principal obligor also agrees that it will reimburse the secondary obligor to the extent the secondary obligor does perform, thereby fulfilling all or part of the underlying obligation."); *see also In re J.T. Moran Fin. Corp.*, 124 B.R. 926 (Bankr. S.D.N.Y. 1991).

21 *See* RESTATEMENT OF SURETYSHIP, *supra* note 1, § 18.

22 *See* James A. Black, *Symposium: The Restatement of Suretyship: Miscellaneous Surety Bonds and the Restatement*, 34 WM. & MARY L. REV. 1195, 1204 (1993).

23 *See, e.g.*, RESTATEMENT OF SURETYSHIP, *supra* note 1, § 19.

24 *See Burr v. Lichtenheim*, 460 A.2d 1290 (Conn. 1983).

25 *See, e.g.*, RESTATEMENT OF SURETYSHIP, *supra* note 1, § 20.

2. *Contractual Rights*

Besides non-contractual rights, sureties may also provide for heightened protection through contractual agreement. This practice requires the principal and/or other interested parties to agree to indemnify the surety in the event the surety is called upon to answer for the principal's default. Unlike the surety's noncontractual rights, which may only be enforced against the principal, the surety's rights under an indemnity agreement may be enforced against any signatory to the agreement. This remains the fundamental advantage of an indemnity agreement. Additionally, the surety may require an agreement to post collateral. In any event, whether collateral or additional indemnity is required, these agreements generally give the surety the right to settle claims and recover losses, costs, and expenses from the principal and the indemnitors.

II.
Statutory And Common Law Indemnity

Indemnity is a well-settled common law right compelling the principal to reimburse the surety to the extent of its reasonable outlay.[26] Thus, once the underlying obligation is paid the surety obtains a right to recover from the principal.[27] The basis for this right relates to the relationship of the parties under the contract. The party who receives and retains the benefit of the contract is the principal.[28] The party who secures benefit for another (*i.e.*, the obligee) is the surety. Because of the benefit received, the principal impliedly agrees to "indemnify" the surety for losses arising from this relationship so that when the surety

---

26 *See, e.g., Lake Charles Elec. Co. v. Globe Indem. Co.* 128 So. 2d 780 (La. Ct. App. 1961); *Lori-Kay Golf, Inc. v. Lassner*, 460 N.E.2d 1097 (N.Y. 1984); *Central Sur. & Ins. Corp. v. Martin*, 224 S.W.2d 773 (Tex. Civ. App. 1949).

27 The principal-surety relationship is analogous to the master-servant or principal-agent relationship. In each case once the underlying liability is paid the principal, master, or surety may recover the amounts paid from its agent, servant, or principal.

28 *See* AM. JUR. 2D *Suretyship* § 3 (1974).

pays the debt its indemnity is immediate.[29] Thus, suretyship is simply a relationship between two parties who may become obligated to the same creditor. This right is generally expanded by contract under the General Indemnity Agreement (GIA) to include a right against not only the principal but also any indemnitors. Indemnity agreements usually provide the surety with indemnification for all losses, costs, and expenses, including actual attorneys' fees incurred as a result of issuing the bond.[30]

Absent a written indemnity agreement, however, the surety's common law indemnity right exists only when there is actual liability, so that a surety who settles a claim must prove actual liability before the principal is required to indemnify. And, where a written indemnity agreement exists between the parties, the surety need only show potential liability to invoke its rights.[31] Since the surety does not have to suffer a loss before its rights attach, written indemnity agreements are commonplace in the industry. Still, a surety's common law rights remain important, because a general indemnity agreement may not have been executed by an indemnitor or it may suffer from a procedural defect. In such a case the surety is wholly dependent on its common law rights. Additionally, understanding the surety's common law rights is important because the language of a standard indemnity agreement is usually a written expression of common law.

Several states provide a statutory right of indemnification for sureties.[32] These statutes are nothing more than a codification of the

---

29 *Western Coach Corp. v. Roscoe*, 650 P.2d 449 (Ariz. 1982); *Frank Lerner & Assocs., Inc. v. Vassy*, 599 N.E.2d 734 (Ohio Ct. App. 1991); *Barger v. Gething*, 52 N.E.2d 94 (Ohio Ct. App. 1943); *Wisconsin Inv. Bd. v. Hurst*, 410 N.W.2d 560 (S.D. 1987); *Fulton v. South Oak Cliff State Bank*, 439 S.W.2d 730 (Tex. Civ. App. 1969); *First Interstate Bank of Wash., N.A. v. Nelco Enters., Inc.*, 822 P.2d 1260 (Wash. Ct. App. 1992).

30 For a broader discussion on recovery of attorneys' fees, see *infra* Section III.C.

31 *American Motorists Ins. Co. v. United Furnace Co., Inc.*, 876 F.2d 293 (2d Cir. 1989); *United States ex rel. Trustees of Elec. Workers Local Pension Fund v. D Bar D Enters., Inc.*, 772 F. Supp. 1167 (D. Nev 1991); *Safeco Ins. Co. of Am. v. Gaubert*, 829 S.W.2d 274 (Tex. App. 1992).

32 *See, e.g.*, Arkansas: ARK. CODE ANN. §§ 303, 304, 305 (Michie 1987); California: CAL. CIV. CODE §§ 2847-2848 (West 1991); Georgia: GA. CODE ANN. §§ 10-7-40,-41, (1989); Indiana: IND. CODE § 34-1-55-5

common law. As a general matter, the indemnity statutes are not as helpful as the terms of the written indemnity agreement. The important thing to remember about these statutory enactments is that they do not allow the terms of the written indemnity agreement to contravene the statute.

III.
General Agreement Of Indemnity

The surety issues a bond to function as a credit accommodation. But before issuing the bond, the surety conducts an investigation into the principal's capacity to perform and its financial ability. Satisfied of the principal's ability, the surety will not anticipate a loss because of its common law rights of indemnity, subrogation, and exoneration. But a surety intending to continue as a profit making entity will also require the principal and the persons who control it to enter into a written indemnity agreement. In this way the surety expressly buttresses its contractual position and strengthens its common law rights.

It is accepted practice that sureties require the principal and its owners or shareholders to sign a GIA before issuance of a bond. The specific terms of the GIA will vary from surety to surety, but its basic purpose of facilitating settlements and obviating unnecessary costly litigation remains the same.[33] This is accomplished by recognizing that the common objectives of such agreements are to: (1) provide the surety with a contractual right of recovery against the principal and other named indemnitors; (2) facilitate the handling of bond claims; (3) require the deposit of collateral; (4) ease burdens of proof in actions to recover losses and expenses; (5) provide a security interest in the principal's equipment, machinery, and receivables; and (6) establish the surety's right to pay or otherwise settle its bond obligations.

(West 1982); Louisiana: LA. CIV. CODE ANN. 3052 (West 1994); Montana: MONT. CODE ANN. §§ 28-11-413 to 28-11-420 (West 1996); North Dakota: N.D. CENT. CODE §§ 22-03-10,-11 (1991); Ohio: OHIO REV. CODE ANN. §§ 1341.18-1341.21 (Anderson 1997); Oklahoma: OKLA. STAT. ANN. tit. 15, §§ 380, 381, 382 (West 1993); South Dakota: S.D. CODIFIED LAWS §§ 56-2-13,-14 (Michie 1988).

33 *See Transamerica Inc. Co. v. Bloomfield*, 401 F.2d 357, 362 (6th Cir. 1968).

This section of the article explores the rights and remedies of the surety under provisions which should be and which are customarily included in the GIA. These include the: (1) indemnity clause; (2) right to settle provision; (3) prima facie/conclusive evidence clause; (4) collateral security requirement; and (5) assignment clause. Because the GIA enables the surety to avoid unnecessary and costly litigation while protecting its rights of indemnity against the principal and named indemnitors, the GIA itself spawns litigation by principals and indemnitors seeking to avoid paying the surety. As a result, the surety must have a keen appreciation and full working knowledge of the ins and outs of the above mentioned clauses.

A. The Indemnity Clause

As discussed above, the equitable right of indemnification arises when the surety discharges an obligation for which the principal is liable. However, an express indemnification contract governs the rights of the parties by its own terms.[34] Thus, the parties can extend the

---

34 An example of a common indemnity provision is as follows:

> The Contractor and Indemnitors shall exonerate, indemnify, and keep indemnified the Surety from and against any and all liability for losses and/or expenses of whatsoever kind or nature (including, but not limited to, interest, court costs, and counsel fees) and from and against any and all such losses and/or expenses which the surety may sustain and incur: (1) by reason of having executed or procured the execution of the bonds, (2) by Indemnitors to perform or comply with the covenants and conditions of this agreement, or (3) in enforcing any of the covenants and conditions of this agreement. Payment by reason of the aforesaid causes shall be made to the Surety by the Contractor and Indemnitors as soon as liability exists or is asserted against the Surety, whether or not the Surety shall have made any payment therefor. Such payment shall be equal to the amount of the reserve set by the Surety. In the event of any payment by the Surety the Contractor and Indemnitors further agree that in any accounting between the Surety and the Contractor, or between the Surety and the Indemnitors, or either or both of them, the Surety shall be entitled to charge for any and all disbursements made by it in good faith in and about the matters herein contemplated by this Agreement under the belief that it is or was liable for the sums and amounts so disbursed, or that it was necessary so expedient to make such disbursements, whether or not

principal's and indemnitor's obligations beyond those at common law. As a general rule, the indemnity clause expressly requires indemnification for all losses sustained in good faith by the surety.[35] Where the surety is liable for immediate payment of the principal's debt, it may pay the debt and resort at once to any funds which the surety holds to satisfy its indemnity right. However, the surety will usually have no remedy at law or equity until a loss actually occurs.[36]

The indemnity clause usually extends indemnification coverage to any named indemnitors who must indemnify the surety for all losses regardless of the surety's actual liability under the bond. The validity of the indemnity clause is ordinarily unassailable absent fraud or bad faith on the part of the surety.[37] Even where the surety settles claims while the principal appeals an earlier ruling, the indemnity agreement remains valid absent bad faith.[38] Thus, indemnitors can only defeat a surety's right to recover by demonstrating that the surety failed to act in good faith in discharging its obligations.[39] Because of the limited defense of fraud or bad faith, most courts will consider principal and indemnitor defense claims on summary judgment.[40]

---

such liability, necessity, or expediency existed; and that the vouchers or other evidence of any such payments made by the Surety shall be prima facie evidence of the fact and amount of the liability to the Surety.

35 *See Commercial Ins. Co. of Newark, N.J. v. Pacific-Peru Constr. Corp.*, 558 F.2d 948 (9th Cir. 1977). The court recognized that the indemnity agreement did not limit the principal's liability for indemnification. The broad terms of the indemnity clause would provide indemnity for any loss by the surety by reason of having executed the bond. *Id.* at 953.

36 74 AM. JUR. 2D *Suretyship* § 178 (1974).

37 *See, e.g., Fidelity & Deposit Co. of Md. v. Bristol Steel & Iron Works, Inc.*, 722 F.2d 1160, 1163 (4th Cir. 1983); *Commercial Ins. Co. of Newark, N.J. v. Pacific-Peru Constr. Co.*, 558 F.2d 948, 953 (9th Cir. 1977); *Bloomfield*, 401 F.2d at 362-363; *Engbrock v. Federal Ins. Co.*, 370 F.2d 784, 786 (5th Cir. 1967); *Insurance Co. of N. Am. v. Bath*, 726 F. Supp. 1247, 1252 (D. Wyo. 1989).

38 *See Transamerica Ins. Co. v. Avenell*, 66 F.3d 715 (5th Cir. 1995).

39 *Engbrock,* 370 F.2d at 784 (holding that neither a lack of diligence or negligence is the equivalent of bad faith.).

40 *See, e.g., Continental Cas. Co. v. American Sec. Corp.* 443 F.2d 649 (D.C. Cir. 1970); *Continental Cas. Co. v. Guterman*, 708 F. Supp. 953 (N.D. Ill. 1989); *Hartford Accident & Indem. Co. v. Millis Roofing & Sheet Metal, Inc.*, 418 N.E.2d 645 (Mass. App. Ct. 1981).

1. Conditions Precedent To Right Of Pay

Before moving into a full blown analysis of bad faith and fraud it is important to note that most of the surety's rights under the indemnity clause are conditioned upon the occurrence of particular events. These events are usually stated in each clause. For example, conditions in the collateral security clause are usually different from those in the assignment clause. Thus, under most of the provisions in indemnity agreements, the surety may invoke its rights without the principal being adjudicated in default or the surety's payment to bond claimants.[41] To be able to establish that the conditions precedent to contractual rights have occurred, good practice requires the surety to document relevant events and decisions, and compare those to the express and implied conditions within each clause of the contract. Failure to do so allows the principal and indemnitors a better opportunity to plead and successfully argue defenses of fraud and bad faith.

2. No Bad Faith[42]

Unlike liability insurance, in suretyship the obligation to perform and pay is squarely upon the principal. This obligation does not change simply because a payment is secondarily required by the surety. Rather, in consideration for acting as a guarantor, a surety has the right of indemnification from the principal. This right is a critical factor in a surety underwriter's decision to issue a bond and in setting a fair premium. Without this fundamental right the character of risk would be altered and surety bonds would either not be written or premiums would be prohibitive. Yet the right to indemnification under the GIA is conditional upon the surety acting in good faith.

To understand the potential liability of sureties for bad faith claims, it is necessary to distinguish and review the general principles involved. The basis for bad faith is the general principle that every contract imposes on each party a duty of good faith and fair dealing in its

---

41 *See, e.g., American Motorists Ins. Co. v. United Furnace Co.,* 876 F.2d 293, 299-300 (2d Cir. 1989); *General Ins. Co. of Am. v. Howard Hampton, Inc.*, 8 Cal. Rptr. 353 (Cal. Dist. Ct. App. 1960).

42 For an extensive review of bad faith, see Gary Rouse, *Extra-Contractual Damages Considerations*, *in* BOND DEFAULT MANUAL 305 (Duncan L. Clore, ed., 2d ed. 1995).

imposes on each party a duty of good faith and fair dealing in its performance and enforcement.[43] The duty of good faith is one arising by law and emphasizes faithfulness to the agreed common purpose and consistency with the justified expectations of the other party so that when a party's conduct violates community standards of decency, fairness, or reasonableness it will be characterized as bad faith.[44]

Good faith arises in both the contractual and common law contexts. Thus, the nature of the surety relationship creates a divided duty of good faith and fair dealing. Under the contractual context, where a surety fails to act in good faith it loses its indemnification rights and is subject to liability for the stated value of the contract. Among the factors courts will consider in determining whether the surety has acted in good faith under the bond are: (1) whether the obligation of the surety is provided by the terms of the bond; (2) whether the principal demanded that the surety deny the claim; (3) whether there was cooperation or lack thereof by the principal in dealing with the obligee; (4) whether a judgment or other administrative proceeding has determined the issue of default; (5) whether there was undue or unusual pressures brought to bear on the surety by the obligee; (6) whether the surety incurred expenses or bad faith penalties in resisting the obligee's claim when liability was clear; (7) whether the surety failed to protect or preserve the indemnitor's collateral in the hands of the surety; (8) whether a suit against the surety was allowed to go into default; and (9) the thoroughness of the investigation performed by the surety.[45]

However, the bad faith cause of action, allowing recovery in tort for breach of what would normally be considered a contractual duty, recognizes that the contract measure of damages is inadequate to compensate a party who has suffered damages not in contemplation of the contract. To overcome the general rule that the surety is liable only for contract damages, courts imposing a common law duty of good faith and fair dealing refer to the surety relationship as one based on principles of insurance.[46] Thus, the normal tripartite relationship analysis is not considered or is not given predominance.[47]

---

43 RESTATEMENT (SECOND) OF CONTRACTS § 205 (1981).

44 *Id.* § 231.

45 *See* Bernard L. Balkin & Keith Witten, *Current Developments in Bad Faith Litigation Involving the Performance and Payment Bond Surety*, 28 TORT & INS. L.J. 611, 623 n.57 (1993).

46 *See, e.g., General Ins. Co. of Am. v. Mammoth Vista Owners Ass'n, Inc.*,

Yet courts have often recognized that the surety is placed in a "good faith dilemma."[48] The surety has a divided duty of good faith and fair dealing toward both the obligee and principal. Under the liability imposed by contract, the surety faces tough decisions concerning application of its rights under the GIA. Even more disturbing, though, is the potential liability for the tort duty of good faith and fair dealing. In deciding to enforce its rights in jurisdictions allowing the tort of bad faith, the surety must be prepared to face this increased potential for litigation.

How can the surety protect itself from this increased risk? Before the surety is entitled to its rights under common law or the GIA, it must make an effort to act in good faith with respect to the obligee's demands and those of the principal and indemnitors. As discussed above, this is the good faith dilemma. Thus, the surety wants to act, and appears to act, in good faith to protect its rights. To the extent practicable, this requires leaving a significant paper trail of investigation and communication with both principal and obligee. Inherent in this mix is the retention of knowledgeable claim representatives and consultants to conduct the investigation.[49] Not doing so leaves opportunities for principals, indemnitors, and obligees to assert claims of bad faith.

For example, in *Rush-Presbyterian-St. Luke's Medical Center v. Safeco Insurance Co. of America*,[50] the surety entered into a settlement agreement with the obligee over the objections of the principal. The principal and indemnitors claimed that the surety could not recover under the GIA because it failed to properly investigate. The court expressly rejected the surety's argument that a bad faith claim must be

---

220 Cal. Rptr. 291 (Cal. Ct. App. 1985).

47 To separate surety from fidelity and insurance it is critical to recognize that suretyship creates a tripartite relationship between and among the party secured (obligee), the principal (obligor), and the party secondarily liable (surety). For a judicial recognition of this critical distinction, see *Meyer v. Building & Realty Serv. Co.*, 196 N.E. 250 (Ind. 1935).

48 Webster, *Performance Bond Surety Dilemma When Both Bond Principal and Obligee Claim the Other is in Default: A Guide Through No Man's Land*, 12 FORUM 238 (1976).

49 For a complete discussion on the proper investigation techniques for a surety, see William S. Piper, *The Surety's Investigation*, *in* BOND DEFAULT MANUAL 25 (Duncan L. Clore, ed., 2d ed. 1995).

50 712 F. Supp. 1344 (N.D. Ill. 1989).

based in fraud, dishonest, or spiteful purpose, conscious wrongdoing, or some other self-interested motive. Instead, the court stated that "negligence and bad faith are synonymous" in the surety context.[51] Thus, a surety could be liable for negligently entering into a settlement agreement with an obligee where it fails to give equal considerations to all parties. *Rush-Presbyterian* is instructive, even in jurisdictions not applying to negligence standard, because it reinforces the idea that good investigative habits and techniques can make all the difference in a bad faith scenario.

B. Right-To-Settle Provisions

One of the most frustrating problems for a surety is the "good faith" dilemma posed when attempting to settle a bond claim and the principal unreasonably refuses to compromise or settle claims it has against the obligee. The competing claims of both parties, obligee and principal, means the surety can achieve settlement only by simultaneous resolution of the principal's affirmative claims. If the principal refuses to compromise or release its claims, the surety is at a decisional crossroads: going left, the surety agrees with the principal and bears the risk of not settling; going right, the surety seeks to override the principal's decision and attempts to settle all the claims. The decision to go left or right is often confused because any dispute between the parties usually arises long after a bond claim is made. Compound this with the complexity of construction cases, the general lack of sufficient information in the initial stages of dispute, the fact that discovery is not yet available, and the potential liability for "bad faith," and it becomes easy to see why a surety would rather stand in the middle of the road waiting for oncoming traffic than appear to choose a side.[52]

More importantly, in most cases the surety and principal will have undertaken a joint defense against the obligee's claims. The surety undoubtedly will develop a close, cooperative working relationship with the principal. As a result, the surety will likely make decisions by

---

51 *Id.* at 1346. The court also went on to cite *Cernocky v. Indemnity Ins. Co. of N. Am.*, 216 N.E.2d 198 (Ill. App. Ct. 1966), for the proposition that "the words 'good faith' and 'bad faith' are not particular words of art . . . . They mean either being faithful or unfaithful to the duty or obligation that is owed." *Rush-Presbyterian*, 712 F. Supp. at 1346.

52 Consider the *Rush-Presbyterian* discussion in Section III.A.2, *supra*.

agreement with the principal and will be reluctant to take any action which conflicts with that agreement. Thus, a surety often feels powerless to settle a case over the principal's objections even where the principal is unable to secure the surety against loss. To avoid this dilemma, a surety must utilize proactive clauses incorporated within the GIA.

1. The Basic Format

One of the most important provisions in helping the surety decide which road to take is the right-to-settle provision. Depending on the particular GIA used, the right-to-settle provision may be in the indemnity clause or set forth separately. This clause allows the surety to: (1) discharge the principal's obligations before suit;[53] (2) effectively and efficiently resolve claims; (3) seek immediate reimbursement from the principal and indemnitor; and (4) avoid unnecessary litigation. Thus, the right-to-settle provision gives the surety broad discretion in determining claims to compromise and/or settle under the terms of the bond. However, this provision alone may not allow the surety to settle the principal's affirmative claims against the obligee.[54]

Compromising or settling an insolvent or recalcitrant principal's affirmative claims against an obligee requires the GIA to contain complimentary right-to-settle, assignment, and attorney-in-fact clauses.[55] Combined, these three clauses will allow the surety maximum protection when it attempts to settle the claims of the obligee and override the principal's refusal to settle or compromise. Typically, the right-to-settle provision provides:

> The surety shall have the right to adjust, settle, or compromise any claim, demand, suit, or judgment upon the bonds, unless the Contractor

---

53 *Transamerica Ins. Co. v. Bloomfield*, 401 F.2d 357, 362 (6th Cir. 1968); *Commercial Union Ins. Co. v. Melikyan*, 430 So. 2d 1217 (La. Ct. App. 1983); *Hess v. American States Ins. Co.*, 589 S.W.2d 548 (Tex. Civ. App. 1979).

54 *See Hutton Constr. Co., Inc. v. County of Rockland*, 52 F.3d 1191, 1192-93 (2d Cir. 1995).

55 Thomas J. Demski & Lester Chanin, *The Surety's Rights of Assignment and Power of Attorney Under General Indem. Agreements*, 31 TORT & INS. L.J. 17, 26 (1995).

> and the Indemnitors shall request the surety to litigate such claim or demand, or to defend such suit, or to appeal from such judgment, and shall deposit with the surety, at the time of such request, cash or collateral satisfactory to the surety in kind or amount, to be used in paying any judgment or judgments rendered or that may be rendered, with interest, costs, expenses, and attorneys' fees, including those of the surety.

The pertinent part of the assignment clause provides:

> [the principal] will assign to the surety, as collateral, to secure the obligations in any and all paragraphs of the agreement and any other indebtedness and liabilities of [the principal] to the surety, whether heretofore or heretoafter incurred, the assignment to become effective only in the event of (1) any abandonment, forfeit or breach of any contracts referred to in the bonds or (2) of any breach of the provisions of any of the paragraphs of this agreement (a) all the rights of [the principal] in, and growing in any matter out of, all contracts referred to in the bonds . . . (e) any and all percentages retained and any sums that may be due or hereafter become due on account of any and all contracts referred to in the bonds . . . .[56]

This type of assignment clause, coupled with the surety's common law subrogation rights, allows the surety to dispose of claims in a manner that makes economic sense to the surety.[57] The assignment becomes effective on the date of execution of either the GIA or the bond, but since execution of the GIA creates a security interest, the assignment rights do not transfer until default. Once default occurs, though, the assignment clause is automatically triggered.

The final piece to the puzzle is found in the attorney-in-fact clause, which provides:

> the Contractor and Indemnitor hereby irrevocably nominate, constitute, appoint and designate the surety as their attorney-in-fact with the right,

---

56 *Hutton Constr. Co., Inc. v. County of Rockland*, Nos. 93 Civ. 2465, 87 Civ. 4027, 1993 U.S. Dist. LEXIS 18046 (S.D.N.Y. Dec. 22, 1993).

57 *See* Stephen J. Trecker et al., *The Agreement of Indemnity—The Surety's Handling of Contract and Bond Problems: Administration and Resolution of Performance Bond and Payment Claims*, *in* THE AGREEMENT OF INDEMNITY: PRACTICAL APPLICATIONS BY THE SURETY (1990).

but not the obligation, to exercise all of the rights of the Contractor and Indemnitors assigned to the surety in this Agreement.[58]

The attorney-in-fact clause is irrevocable and gives the surety broad power to exercise all of the rights assigned. This clause is an important complement to the surety's right-to-settle powers under the GIA because it enables the surety to (1) take title and possession of assigned property, and (2) enter into settlement agreements with the obligee and release any of the principal's affirmative claims. As with the other clauses previously mentioned, a default under the GIA automatically transfers the principal's rights to the surety.

2. Surety's Good Faith In Settling Claims

Although the surety has broad discretion to determine which claims should be compromised and settled, its decision will be binding on the principal and indemnitors only as long as the surety acts in good faith.[59] The determination of good faith raises the same issues under a right-to-settle provision as under an indemnity clause. The primary focus is on the surety's handling of claims.[60]

---

58 *Id.* at 5-6.

59 *Massachusetts Bonding & Ins. Co. v. Gautieri*, 30 A.2d 848, 850 (R.I. 1943); *Hess v. American States Ins. Co.*, 589 S.W.2d 548, 551 (Tex. Civ. App. 1979); *Central Sur. & Ins. Corp. v. Martin*, 224 S.W.2d 773, 776 (Tex. Civ. App. 1949). Further, as stated in the discussion of indemnification clauses, in order to prove lack of good faith, the principal or indemnitor must prove that the surety acted with fraud, ill will, or some other improper motive. *Safeco Ins. Co. of Am. v. Criterion Inv. Corp.*, 732 F. Supp. 834, 841 (E.D. Tenn. 1989); *City of Portland v. George D. Ward & Assocs., Inc.*, 750 P.2d 171, 175 (Or. Ct. App. 1988); *cf. Rush-Presbyterian-St. Luke's Med. Ctr. v. Safeco Ins. Co. of Am.*, 712 F. Supp. 1344 (N.D. Ill. 1989) (principal need only show negligence in establishing bad faith).

60 Among the factors considered by courts in evaluating the surety's good faith in settling a claim are: (1) whether the surety exceeded the face amount of the bond, *Commercial Union Ins. Co. v. Melikyan*, 430 So. 2d 1217, 1222 (La. Ct. App. 1983); (2) whether the principal has demanded that the surety deny the claim and/or provided documentation supporting denial of the claim, *M-Pax, Inc. v. Dependable Ins. Co.*, Inc., 335 S.E.2d 591, 593 (Ga. Ct. App. 1985); *Glens Falls Indem. Co. of Glens Falls v.*

Courts, when trying to determine if the surety has settled the claims in good faith, focus on whether the surety reasonably believed that the claim should be paid. Thus, where a surety reasonably considers the claims and possible defenses to those claims, it will have acted in good faith.[61] Also, where the surety's attorney continually informs the principal of the steps taken to discharge a claim, receives no objections or cooperation in response, and hires an expert when it becomes apparent that substantial overruns will result, the surety will be acting in good faith.

3. *Theory In Practice*

a. *Hutton Construction*: The Blueprint

Given the tripartite relationship of suretyship, situations arise where a surety will seek settlement with an obligee which encompasses not only the claims under the bond but also the affirmative claims of the principal against the obligee. Such was the case in *Hutton Construction Co., Inc. v. County of Rockland*,[62] where the United States Court of Appeals for the Second Circuit enforced a settlement between a surety and the obligee to the detriment of an insolvent principal. *Hutton Construction* is instructive for sureties who wish to take the road to the right and proactively control an insolvent or recalcitrant principal.

In *Hutton Construction*, Hutton, the principal, entered into a contract with the county of Rockland, the obligee, to install 40,000 feet of sewer lines. While performing the work, Hutton encountered subsurface rock and water conditions which were different from those originally disclosed in the contract. Hutton demanded a series of price increases

---

Carobine, 36 N.Y.S.2d 253 (N.Y. City Ct. 1942); (3) whether the obligee has placed undue or unusual pressure on the surety by the obligee, *Fidelity & Deposit Co. of Md. v. Bristol Steel & Iron Works, Inc.*, 722 F.2d 1160, 1164 (4th Cir. 1983); and (4) the thoroughness of the investigation performed by the surety, *Maryland Cas. Co. v. R. & L. Constr. Co.*, 368 S.W.2d 134 (Tex. Civ. App. 1963); *see also Rush-Presbyterian*, 712 F. Supp. at 1344 (Illinois requires only a showing of negligence in actions alleging breach of the surety's duty of good faith).

61 *United States ex rel. Trustees of Elec. Workers Local Pension Fund v. D Bar D Enters., Inc.* 772 F. Supp. 1167 (D. Nev. 1991).

62 52 F.3d 1191 (2d Cir. 1995).

based on these unforeseen conditions. The county, through its engineers, rejected Hutton's price increase requests. According to the county, Hutton then failed to complete portions in a timely fashion. In response, Hutton alleged a failure to negotiate and investigate in good faith by the county. Soon afterward the county terminated the contract and sought performance by the sureties under the performance bond.[63]

The cosureties investigated the county's claims, declined to complete or pay any difference in completion costs under the contract, and exercised their right to litigate the issue of Hutton's breach of contract. Hutton's subcontractors and suppliers then sent claims to the cosureties under the payment bond. Hutton advised the cosureties of its inability to pay and indemnify them for the amounts owed to the suppliers and subcontractors, and sought advancement of attorneys' fees for the suit with the county. The cosureties advanced Hutton the proceeds.[64]

Hutton then filed a $4,800,000 suit against the County for wrongful termination of the construction contract and for additional costs incurred because of the unanticipated subsurface conditions. The County, in response, filed a $4,100,000 counterclaim for costs to complete and remedy defective work against Hutton and its cosureties. The County also sued the engineering firm for indemnification and contribution. During the pretrial phase the cosureties expended more than $1,500,000 for Hutton's legal fees and like payments under the bond.[65] As the costs

---

63 At the time Hutton entered into the contract it purchased performance and payment bonds which provided for a total penal sum of $4,226,347.00 binding three separate sureties as follows: (1) Union Indemnity Insurance Company of New York in the amount of $2,500,000; (2) Indemnity Insurance Company of North America in the amount of $1,126,347; and (3) International Fidelity Insurance Company in the amount of $600,000. Prior to the County's demand Union Indemnity had been placed in liquidation leaving the remaining two as cosureties. *See Brief for Counterclaim-Defendants-Appellees Indemnity Insurance Company of North America and International Fidelity Insurance Company* at 2, *Hutton Constr. Co. v. County of Rockland*, 52 F.3d 1191 (2d Cir. 1995) [hereinafter *Brief of Counterclaim-Defendants-Appellees*] (on file with author).

64 Hutton and its principal owner entered into a standard General Indemnity Agreement pursuant to which they agreed to indemnify the cosureties from and against any and all losses which might be incurred.

65 *See Brief of Counterclaim-Defendants-Appellees*, *supra* note 63, at 5.

mounted, the cosureties demanded payment from Hutton and its indemnitors. Hutton failed to make payment and instead instructed the cosureties to consent to a settlement proposal whereby Hutton would receive a recovery of $1,700,000 and the cosureties would recover a mere fraction of their losses.[66]

After five years of numerous court appearances, depositions, and expert reports, the parties recognized that the actual trial was upon them. They engaged in extensive settlement negotiations, but Hutton continued to make unreasonable demands.[67] With the court's involvement, the County's engineers made an offer which provided a fund to be shared by the cosureties and the County.[68] Even though this proposal provided the cosureties and Hutton a full release from the $4,100,000 counterclaim, Hutton refused to agree to settle and instead renewed its same unreasonable demands. The cosureties advised Hutton and its indemnitor that their refusal to accommodate a settlement and continued refusal to reimburse and secure the cosureties would leave them with no other recourse than to exercise their powers under the GIA. Hutton did not comply and the cosureties settled their bond claims under the right-to-settle provision and Hutton's affirmative claims based on the cosureties assignee and attorney-in-fact powers under the GIA.

Later, the cosureties moved to enforce the settlement over Hutton's objections and to dismiss all of Hutton's affirmative claims. The cosureties argued that Hutton had assigned all of its rights to its claims to the cosureties by virtue of the assignment and attorney-in-fact clauses. Hutton's only response was that the right-to-settle clause only allowed

---

66 *Id.* at 6. It is also interesting to note that Hutton actually increased its settlement demand to $1,750,000 with no reimbursement of the cosureties and complete forgiveness of its $1,500,000 debt to the cosureties. From a surety's point of view, this should define the term recalcitrant principal.

67 On the eve of trial, Hutton proposed that the cosureties contribute $1,000,000 to a total settlement fund of $2,900,000 from which $1,900,000 would be paid to Hutton and the remainder to the County. Hutton also insisted that the cosureties waive their indemnification rights against Hutton and its indemnitors. *Id.* at 7.

68 The engineers agreed to pay a total of $1,500,000. The County to receive $825,000 in full satisfaction of its claims against Hutton, the cosureties, and the engineers. The cosureties, as Hutton's assignees, would receive $675,000 in full satisfaction of Hutton's claims against the County and the engineers. *Id.* at 8.

the cosureties "to adjust, settle or compromise any claim . . . upon the Bonds" and thus did not give them the right to settle Hutton's affirmative claims growing out of the construction contract.[69] The court agreed this clause alone did not give the cosureties that power, but concluded that all three clauses combined gave the cosureties the authority to settle all claims on behalf of Hutton, including not only claims against the cosureties "upon the bonds," but also Hutton's affirmative claims growing out of its construction contract.[70]

Interestingly, the court determined that all the relevant GIA provisions were properly bargained for and enforceable. It also found that Hutton's failure to comply with the cosureties demands for collateral to cover expenses and a potential adverse judgment constituted a breach of the indemnity agreement.[71] Furthermore, it was this breach which triggered the cosureties' powers under the assignment and attorney-in-fact clauses. As assignees of all of Hutton's rights under the bond and construction contract, the cosureties properly settled the lawsuit.[72]

For a surety seeking to control an unreasonable, recalcitrant, and/or insolvent principal, *Hutton Construction* is a powerful blueprint for a proactive approach to settlement. Where, as in *Hutton Construction*, a principal attempts to hold its surety hostage by maintaining a stance on unreasonable affirmative claims, the surety with a properly drafted GIA will find its powers to settle greatly enhanced. However, the surety must always keep in mind its duty to settle claims in good faith.

b. *Matthews Constructors*: Applying The Blueprint

Matthews Constructors v. Carothers Construction, Inc.[73] adds one layer of complexity to the *Hutton Construction* fact pattern. In

---

69 *Hutton*, 52 F.3d at 1192-93.

70 *Id.* at 1192.

71 *Id.*

72 Hutton, not satisfied with this result, is seeking alternate means to recover. More specifically, Hutton, owing over $432,000, filed suit against one of its law firms for malpractice. The moral of the story may be that where a surety cannot trust its principal, then the attorneys for the principal should be equally weary.

73 *Matthews Constructors* is an unreported case originally filed in the United States District Court for the Western District of Texas, San

Matthews Constructors, Carothers Construction, Inc. (CCI) was awarded a general contract from the United States Army Corps of Engineers (Corps) to construct a tank repair and maintenance facility at the United States Military installation in Fort Hood, Texas. CCI, as part of the contract, received surety bonding from St. Paul Fire and Marine Insurance Company (St. Paul). At this point there was a complete suretyship relationship between CCI the principal, the Corps the obligor, and St. Paul the surety. CCI, in turn, entered into two subcontracts with Matthews Constructors, Inc. (MCI) for mechanical contracting work. CCI required MCI to obtain bonding. MCI contracted with Fidelity and Deposit Company of Maryland (F&D) as surety. At this point, another level of suretyship was established with CCI as obligee, MCI as principal and F&D as the surety.

Almost immediately the project developed problems. First, CCI was challenged for failure to produce an acceptable project schedule.[74] As a result, before MCI began any work under the subcontracts, the entire project was arguably seventy-three days behind even the revised schedule. Second, the Corps issued an evaluation of CCI, indicating that CCI failed to properly staff and manage the site, thus creating a log jam of problems for all subcontractors. As a result, CCI entered into an accelerated completion date agreement to avoid penalties under the general contract. Consequently, the entire project required acceleration by CCI and all of its subcontractors.

Soon thereafter, CCI sent a letter to F&D claiming that MCI was in default due to its failure to (1) properly manage the job; (2) make satisfactory progress; and (3) discharge its responsibilities under the subcontract. MCI initiated a claim against CCI for wrongful termination, and CCI responded with a counterclaim and third-party complaint against F&D alleging (1) breach of contract; (2) bad faith and violation of the Texas Insurance Code; and (3) Texas Deceptive Trade Practices Act claims.

Unlike the tripartite relationship in *Hutton Construction*, *Matthews Constructors'* double tripartite relationship added different pressures to

---

Antonio Division, where settlement was reached using the formula set forth in *Hutton Construction*. *See Matthews Constructors, Inc. v. Carothers Constr., Inc.*, Case No. SA94CA0649 (on file with the author).

74 The Corps rejected the critical path method schedule adopted by CCI. The Corps cited logic problems and sequencing errors that it claimed made the model inaccurate.

the mix. As the Corps pressed CCI and its surety, St. Paul, on the general contract, CCI sought to roll the proverbial ball down hill onto MCI and its surety, F&D. Thus CCI, as both principal and obligee, occupied a position to relieve pressure from the general contract by applying pressure to the subcontract through that status as obligee. Settlement negotiations ensued, but MCI rejected all proposals. F&D, utilizing the same provisions outlined in *Hutton Construction*, settled all the claims, to include MCI's affirmative claims. Also like *Hutton Construction*, the district court approved the settlement and granted F&D's and CCI's joint motion to dismiss.[75]

c. Overview

Hutton Construction and *Matthews Constructors* suggest that a surety has several viable options which require certain steps when attempting to unilaterally settle the affirmative claims of a recalcitrant principal. These steps include investigation, demand for exoneration, demand for collateral, financing of the principal, and retention of separate counsel.[76] However, the most important step comes from implementing a standardized GIA which encompasses a right-to-settle provision, an assignment of rights clause, and an attorney-in-fact clause.

C. Prima Facie/Conclusive Evidence Clauses

To ease the surety's burden of proving the propriety of its payments and exact amount of the indemnitors' liability, many GIAs include a prima facie/conclusive evidence clause which states:

> In the event of any payment by the Surety the Contractor and Indemnitors further agree that in any accounting between the Surety and the Contractor, or between the Surety and Indemnitors, or either or both of them, the Surety shall be entitled to charge for any and all disbursements made by it in good faith in and about the matters herein contemplated by this Agreement under the belief that it is or was liable for the sums and amounts so disbursed, or that it was necessary or

---

75 *Order, Matthew Constructors, Inc. v. Carothers Constr., Inc.*, Case No. SA94CA0649 (on file with the author).

76 For a broad discussion of the proper steps, see Demski & Chanin, *supra* note 55, at 23-26.

> expedient to make such disbursements, whether or not such liability, necessity or expediency existed; *and that the vouchers or other evidence of any such payments made by the Surety shall be prima facie evidence of the fact and amount of liability to the Surety.*

This clause operates in conjunction with the right-to-settle, attorney-in-fact, and assignment clauses. Courts have upheld these clauses to establish the propriety of the surety's payment and the amount of the indemnitors' liability to the surety.[77] Thus, where a surety seeks a motion for assessment of damages and provides affidavits and supporting documentation of payment, courts will consider this prima facie proof of the indemnitors' liability and will often grant summary judgment to the surety.[78]

Since courts recognize the enforceability of prima facie evidence clauses, a surety resolving claims under its right-to-settle clause can and should use vouchers and affidavits to obtain summary judgment for indemnification of losses and expenses incurred in discharging bond obligations. In instances where a court does not grant summary judgment, the prima facie evidence clause shifts the burden of proof to the indemnitors to prove that the surety did not act in good faith while settling claims and incurring expenses. Since sureties make these payments only when they believe such payments are necessary and

---

77 *See, e.g., Continental Cas. Co. v. American Sec. Corp.*, 443 F.2d 649 (D.C. Cir. 1970), *Transamerica Ins. Co. v. Bloomfield,* 401 F.2d 357, 362 (6th Cir. 1968); *Engbrock v. Federal Ins. Co.*, 370 F.2d 784 (5th Cir. 1967); *United States Fidelity & Guar. Co. v. Napier Elec. & Constr. Co., Inc.*, 571 S.W.2d 644, 646 (Ky. Ct. App. 1978); *Hartford Accident & Indem. Co. v. Millis Roofing & Sheet Metal, Inc.*, 418 N.E.2d 645, 647 (Mass. App. Ct. 1981).

78 *See Curtis T. Bedwell & Sons, Inc. v. International Fidelity Ins. Co.*, 1989 WL 55388 (E.D. Pa. May 23, 1989); *Buckeye Union Ins. Co. v. Boggs*, 109 F.R.D. 420, 423-424 (S.D.W. Va. 1986) (surety's affidavit as to expenses incurred was sufficient evidence on which to grant summary judgment); *Continental Cas. Co. v. American Sec. Corp.*, 443 F.2d 649 (D.C. Cir. 1970) (where the surety presents prima facie evidence of payment by way of affidavit and supporting documentation the case is ideally suited for summary judgment); *Ford v. Aetna Ins. Co.*, 394 S.W.2d 693, 700 (Tex. Civ. App. 1965) (affidavit from surety's attorney was prima facie evidence of the payments made by the surety).

expedient, it is highly unlikely that an indemnitor will be able to meet this burden in the normal case.[79]

Fallon Electric Co., Inc. v. Cincinnati Insurance Co.[80] is a recent case demonstrating use of this clause. In *Fallon*, an unpaid subcontractor sued the principal and surety for nonpayment. Before conclusion of trial, the surety and principal settled the subcontractor's claims. Immediately thereafter, the surety sought indemnification for $87,752 in attorneys' fees.[81] The district court awarded the surety only $53,429 and placed the burden of justifying the fees on the surety. The court disagreed and held that the prima facie evidence clause shifts to the principal/indemnitor the burden of proving that the fees were not recoverable.[82] The court did not remand, but instead entered judgment in favor of the surety for the full $87,752. In so deciding, the court made it clear that where a GIA between a surety and principal specifies that (1) a voucher is prima facie evidence of payment; and (2) the surety is entitled to reimbursement of expenses it reasonably believed were owed, then the burden is on the principal/indemnitor to prove that indemnification is not warranted under the contract.[83] This requires a showing that the surety either (1) did not believe its payment was reasonable; or (2) its payment was unreasonable. In such a circumstance, the principal/indemnitor would establish fraud or bad faith.[84]

It is important to note, however, that many courts draw a clear distinction between prima facie evidence clauses and conclusive evidence clauses.[85] Generally, courts are less likely to enforce indemnity agreements which state that evidence of payment will be

---

79 But just because the indemnitor cannot meet its burden does not mean it will automatically pay. It is likely that the indemnitors will always assert a lack of good faith. Also, it is likely that no indemnitor will ever agree that the amount they owe results from a normal case scenario.

80 121 F.3d 125 (3rd Cir. 1997).

81 *Id.* at 127.

82 *Id.* at 129.

83 *Id.* at 130.

84 *But see Rush-Presbyterian-St. Luke's Med. Ctr. v. Safeco Ins. Co. of Am.*, 712 F. Supp. 1344 (N.D. Ill. 1989).

85 Conclusive evidence clauses provide that: "vouchers and other evidence of any such payments made by the surety shall be *conclusive evidence of the fact and amount of liability of the Surety*."

conclusive rather than prima facie evidence of the principal and indemnitor's liability to the surety. Courts refusing to enforce these clauses do so out of a belief that such clauses deprive indemnitors of the opportunity to rebut the surety's evidence.[86] Where a conclusive evidence clause precludes an indemnitor from contesting any of the surety's evidence, courts will view the clause as against public policy.[87]

However, several courts do recognize the validity of conclusive evidence clauses and enforce them to their fullest extent with the surety's good faith as the only restriction on the validity of the clause.[88] Where the principal and indemnitors sign a GIA which includes a conclusive evidence provision, those parties are bound by its express terms and are liable to the surety if the surety settles the claims in good faith.[89]

Because courts are split on the issue of whether conclusive evidence clauses are valid, a surety must be aware of the jurisdiction controlling the GIA's enforceability. The safest bet is to use the prima facie evidence clause since it is enforced in every jurisdiction. Procedurally, the conclusive evidence clause operates very much like the prima facie evidence clause and both are limited by good faith. Be aware, however that the conclusive evidence clause is more of a gamble, and a surety wishing to use it should provide a clear choice of law provision within the GIA to insure that the law of a favorable jurisdiction applies. Of course, if the conclusive evidence clause is found to be against public policy and there is no prima facie evidence clause, the surety needlessly increases its burden to prove the validity of its expenditures under the bond.

---

86 *Fidelity & Deposit Co. of Md. v. Davis*, 284 P. 430 (Kan. 1930); *Fidelity & Cas. Co. of N.Y. v. Eickhoff*, 65 N.W. 351 (Minn. 1895); *see also International Fidelity & Ins. Co. v. Jones*, 682 A.2d 263 (N.J. Super. Ct. App. Div. 1996).

87 *See, e.g., Sork v. United Benefit Fire Ins. Co.*, 161 So. 2d 54 (Fla. Dist. Ct. App. 1964).

88 *See e.g. Engbrock v. Federal Ins. Co.*, 370 F.2d 784 (5th Cir. 1967); *Home Indem. Co. v. Wachter*, 496 N.Y.S.2d 252 (N.Y. App. Div. 1985).

89 *See, e.g., Home Indem. Co.*, 496 N.Y.S.2d at 252.

D. The Collateral Security Clause

1. Collateral Posted

In addition to the clauses already mentioned, most GIAs contain what is known as a collateral security clause. The purpose of this clause is to force a protesting principal and/or indemnitor to provide the surety with a reserve of funds when faced with anticipated bond liabilities. Under this provision, once a surety receives a demand on the bond, the principal and/or indemnitors must deposit sufficient funds with the surety to secure the claim. Then, if necessary, the surety will pay the claim from this fund. If no payment is necessary the fund will be returned.[90] Under the typical collateral security clause, all the surety must do to trigger an obligation by the principal and indemnitors to deposit funds is establish a liability reserve and make demand for the payment.[91] A typical collateral security provision may provide as follows:

> If for any reason the surety shall be required to or shall deem it necessary to set up a reserve in any amount to cover any (a) judgment, actual or contingent, with interest and costs, in any action instituted against the principal and/or the surety of (b) unadjusted claims or (c) losses, costs, attorneys' fees, and disbursements and/or expenses in connection with said bond or (d) default(s) of the principal or (e) abandonment of the contract or (f) liens filed or (g) dispute with the owner or obligee or (h) for any reason whatever regardless of any proceedings contemplated or taken by the principal or the pendency of any appeal, the undersigned shall immediately upon demand, deposit with the surety funds in the amount of such reserve and any increase thereof, to be held by the surety as collateral with the right to use such fund or any part thereof, at any time, in payment or compromise of any judgment, claim, liability, loss, damages, attorneys' fees, and disbursements and/or other expenses.

It is interesting to note that the surety does not need to be found liable for any debt prior to making demand for collateral. The surety must only be faced with possible liability or deem it necessary that

---

90 *See Safeco Ins. Co. of Am. v. Schwab*, 739 F.2d 431, 433 (9th Cir. 1984).

91 *See United Bonding Ins. Co. v. Stein*, 273 F. Supp. 929 (E.D. Pa. 1967).

collateral be posted.[92] Thus, the net effect of the collateral security clause is that the surety, after payment of the collateral demand by the indemnitors, is secured against any loss it may subsequently sustain if a claim on the bond matures.[93]

This contractual right to collateral security allows the surety to proceed against the signatories to the GIA. It also gives the surety a contractual right to choose the triggering event which compels the deposit of funds by the principal and/or indemnitors.[94] However, not all collateral security provisions use the creation of a reserve as a triggering event. This is because indemnity agreements, in general, are not

---

92 The court in *Standard Sur. & Cas. Co. of N.Y. v. Caravel Indus. Corp.*, 15 A.2d 258 (N.J. 1940), recognized this, explaining that:

> [Indemnitors] did not agree to post collateral only if a valid claim were made, but if any claim were made and [the surety] found it necessary in consequence thereof to set up a reserve . . . .What the parties had in mind when the [indemnitors] agreed to deposit collateral was that at some distant day the liability of the [surety] might be established by judgment and the principal debtor, as well as the [indemnitors], be at the time insolvent. To guard against that contingency, the [indemnitors] agreed to make the deposit as soon as a serious question of liability should arise. Such situation has now arisen.

Id. at 260.

93 *See National Sur. Corp. v. Titan Constr. Co.*, 26 N.Y.S.2d 227 (N.Y. Sup. Ct.), *aff'd*, 24 N.Y. S.2d 141 (N.Y. App. Div.1940).

94 For example, a typical provision in a General Indemnity Agreement provides:

> If the Surety shall set up a reserve to cover any liability, claim asserted, suit or judgment under any such bond, the Indemnitors will, immediately upon demand and whether or not the Surety shall have made any payment therefor, deposit with the Surety a sum of money equal to such reserve and any increase thereof as collateral security on such bond, and such sum and other money or property which shall have been or shall hereafter be pledged as collateral security on any such bond shall be available, at the discretion of the Surety, as collateral security on all bonds coming within the scope of this instrument or for any other indebtedness of the Indemnitors to the Surety; and any such collateral security shall be held subject to the terms of the Surety's regular form of receipt of collateral, which is by reference made a part thereof.

uniform and standardized in practice.[95] Though most GIAs contain a promise to deposit collateral upon establishment of a reserve, others require collateral security upon any demand made against the surety.[96] Furthermore, not every collateral security clause conditions the right to require deposit of collateral upon the creation of a reserve.

Before moving further into the discussion, it is important to distinguish the right to indemnify from the right to be "collateralized" against a possible future loss.[97] Making the distinction that the surety has bargained for the right to be collateralized in addition to its ultimate indemnity rights helps the surety establish that it is entitled to specific performance of its indemnity agreement.

Short of filing suit for damages or specific performance, a demand for collateral is a drastic measure for an aggressive surety responding to either a principal's or an indemnitor's imminent default. When a surety makes a demand, it sends a message that it is serious. The demand is a warning and reminds both principal and indemnitors that they will be responsible for the cost of resolving any obligee claims. In a tactical sense, demand for collateral should be combined with an invitation to enter settlement negotiations. As in cases like *Hutton Construction* and *Matthews Constructors*, where a principal acts unreasonably, demand for collateral establishes the surety's position. Couple the demand with the right-to-settle, assignment, and attorney-in-fact clauses, and the principals and indemnitors should interpret an offer of settlement by the surety as an indication of reasonableness, rather than a sign of weakness. If that fails, the surety can still resort to its remaining options, seek enforcement of the collateral security clause through a claim for specific performance, and simultaneously utilize its powers under the other GIA provisions to seek an ultimate settlement with the obligee.

---

95 *See generally* Meeker, *Surety's Right to Specific Performance of Indemnity Agreements*, CONSTR. LAW., Spring 1982, at 1.

96 *See Safeco Ins. Co. of Am. v. Schwab*, 739 F.2d 431, 431-32 (9th Cir. 1984).

97 In *Northwestern Nat'l Ins. Co. of Milwaukee, Wis. v. Alberts,* the court, in discussing the issue of why a remedy at law was inadequate to protect a surety, noted the difference between indemnification and collateralization when it stated "a vindication of [this] right—to be collateralized while the litigation is pending, [is] not the ultimate settling of accounts." 741 F. Supp. 424, 431 (S.D.N.Y. 1990), *vacated in part*, 937 F.2d 77 (2d Cir. 1991).

2. *Actions Available To A Surety To Enforce Its Collateralization Rights*

a. Action For Specific Performance

As recently as 1982, it was unclear whether a surety could seek specific performance of a collateral security clause under the GIA.[98] In 1982, Charles Meeker pushed sureties down the road to specific performance of collateral security claims when he published an article recognizing that sureties probably did not utilize their equitable action enough to enforce their rights or mitigate their damages.[99] Since then many courts have agreed with Mr. Meeker, and today "it is well settled that a surety may sue for specific performance for deposit of security to cover the surety's reserve if such a provision is in the indemnity agreement."[100] Thus, sureties are now ordinarily entitled to specific performance of collateral security clauses.

As discussed above, a surety can enforce the collateral security clause by instituting an action for specific performance. Courts will generally grant the remedy if the surety can prove that: (1) it has no adequate remedy at law; (2) the indemnity agreement is just and reasonable and supported by adequate consideration; (3) the terms of the indemnity agreement are sufficiently definite to allow enforcement by the court; and (4) the performance sought is substantially identical to that promised in the indemnity agreement.[101] From the surety's standpoint, the most important element to prove is lack of an adequate

---

98 Before 1982, only two cases expressly granted such relief. *See Milwaukee Constr. Co. v. Glens Falls Ins. Co.*, 367 F.2d 964 (9th Cir. 1964); *United Bonding Ins. Co. v. Stein*, 273 F. Supp. 929 (E.D. Pa. 1967).

99 *See* Meeker, *supra* note 95. Meeker's conclusion was that the use by sureties of an equitable action of specific performance of an indemnity agreement had not been considered as it should have been. If the surety expects to mitigate its losses, it should and must expedite fulfillment of the promises made by indemnitors. Delay in seeking demand and enforcement is detrimental to the sureties. *See* Meeker, *supra* note 95, at 22.

100 *Employers Ins. of Wausau v. Bond*, U.S. Dist. Lexis 951 (D. Md. 1991).

101 *See* Meeker, *supra* note 95.

remedy at law.[102] Courts usually consider this the deciding factor in whether they will specifically enforce the collateral security provision.

However, where the amount of the bond claim is not definite, courts recognize that the surety has no legal remedy available because the surety cannot yet institute an action for indemnification.[103] Also, where an indemnitor fails to post collateral promptly after a demand, courts will deem this sufficient evidence of no adequate remedy.[104] But a surety's right to sue for specific performance on the collateral security clause is extinguished once the surety's debt becomes certain.[105] Once the debt is certain, the surety has an adequate legal remedy to secure an award of indemnification under other provisions of the GIA. It is also important to consider, in an action to compel posting of collateral, that a

---

102 *See BIB Constr. Co. v. Fireman's Ins. Co. of Newark, N.J.*, 625 N.Y.S.2d 550 (N.Y. App. Div. 1995) (surety may demand security collateral from its principal even though the amount of damages are in doubt. The surety is then entitled to specific performance of its GIA because future damages are not ascertainable, therefore, its legal remedies are inadequate).

103 *Milwaukee Constr. Co.*, 367 F.2d at 964; *Safeco Ins. Co. of Am. v. Criterion Inv. Corp.*, 732 F. Supp. 834, 843 (E.D. Tenn. 1989) (court granted specific performance of the collateral security provision upon the surety's showing that it had obtained a judgment against the principal after the principal had defaulted on its bond); *United Bonding Ins. Co.*, 273 F. Supp. at 929.

104 The court in *National Sur. Corp. v. Titan Constr. Co.* recognized this fact and stated:

> The very purpose of Clause 2 (the collateral security clause) of the agreement is to entitle plaintiff (surety), the moment of danger of possible future liability appears, to security against any loss plaintiff may subsequently sustain if such liability matures, and to save plaintiff from being in a position of a general creditor of defendants, on a party with their other creditors. The inadequacy of the legal remedy is obvious. Plaintiff, therefore, is entitled to have Clause 2 of the agreement specifically enforced.

26 N.Y.S.2d 227, 231 (N.Y. Sup. Ct.), *aff'd*, 24 N.Y.S.2d 141 (N.Y. App. Div. 1940).

105 *See American Motorist Ins. Co. v. United Furnace Co.,* 876 F.2d 293, 300 (2d Cir. 1989); *Commercial Ins. Co. of Newark, N.J. v. Pacific-Peru Constr. Corp.*, 558 F.2d 948, 955 (9th Cir. 1977).

surety may meet difficulties if the collateral security clause contains ambiguous language. Also, the surety faces certain pitfalls if it seeks relief beyond specific performance to post collateral.[106]

The prime advantage of seeking specific performance is that it allows the surety to realize the benefit of the GIA while the principal is still solvent. But even if the principal is insolvent, the surety may enforce the provision against the assets of indemnitors and, by utilizing the power of specific performance, the surety can often prevent recalcitrant indemnitors from hiding assets which should be used to offset bond losses.

b. Injunctive Relief

Where an indemnitor or principal is attempting to dissipate assets to avoid collateralizing the surety, an action for specific performance should be accompanied by a temporary restraining order or an injunction. The standard for obtaining injunctive relief on the collateral security clause is essentially the same as for specific performance. However, the procedure and elements are somewhat different and include: (1) demonstration of irreparable injury and no adequate remedy at law; (2) lack of harm to the principal and indemnitors; (3) probability of the surety's success on the merits; and (4) that the public interest favors issuance of injunctive relief under the circumstances.[107]

Courts recognize the inherent overlap between specific performance and an injunction ordering the deposit of collateral. This is because the surety's rights may extinguish if not asserted prior to final payment. Thus, a surety has a window in which to enforce its rights and unless some type of injunctive relief is available, the surety will not be adequately secured and will forever and irreparably lose its rights. This loss of rights is irreparable harm. As a result, courts recognize that the irreparable injury requirement for injunction overlaps with the lack of adequate remedy at law necessary to establish equitable rights.[108] Also, with respect to the probability of the surety's success on the merits, the

---

106 *See Safeco Ins. Co. of Am. v. Schwab*, 739 F.2d 431, 433 (9th Cir. 1984).

107 *See* Jay M. Mann, *Exoneration and Quia timet Remedies of the Surety, in The Law of Suretyship* (1993); *see also Blackwelder Furniture Co. of Statesville v. Seilig Mfg. Co.*, 550 F.2d 189 (4th Cir. 1977).

108 *See Northwestern Nat'l Ins. Co. of Milwaukee, Wis. v. Alberts*, 741 F. Supp. 424 (S.D.N.Y. 1990), *vacated in part*, 937 F.2d 77 (2d Cir. 1991).

issue is whether the surety can show that it has a likelihood of success in its right to compel a deposit of collateral. Therefore, if the claim is valid and the surety is acting in good faith, the likelihood of success is established.

c. Alternative Use Of The Collateral Security Clause

The GIA should also contain a collateral security clause allowing the surety to demand the posting of collateral by the principal and/or indemnitors when either of them insist that the surety refrain from paying the obligee's claim. The surety would enforce this provision by requiring the indemnitors to post enough collateral to cover (1) the amount of the obligee's claims; (2) the amount of estimated accrued interest; (3) estimated expenses; and (4) estimated attorneys' fees. By enforcing this clause, principals and indemnitors are less likely to raise spurious defenses and claims because they are directly on the hook for the amount of the claim and the surety's costs of defense. As a result, where a surety anticipates an attack by the principal or indemnitors based on "good faith," the surety can use this provision to counter their efforts.

d. Recovering Attorneys' Fees As A Result Of The Principal Or Indemnitor's Failure To Collateralize

In cases where the surety must force, by an action for specific performance or injunction, the principal and indemnitors to post collateral, the surety must provide for a means to recover attorneys' fees. This can be accomplished by the attorneys' fees clause found in most GIAs which provides:

> The undersigned shall indemnify and keep indemnified the Company against any and all liability for losses and expenses of whatsoever kind or nature, including fees and disbursements of counsel, and against any and all said losses and expenses, which the Company may sustain or endure (i) by reason of having executed or procured execution of the bond or bonds hereinabove applied for; (ii) by reason of the failure of the undersigned to perform or comply with the covenants and conditions of this Agreement; or (iii) enforcing any of the covenants and conditions of this Agreement.

> bond or bonds hereinabove applied for; (ii) by reason of the failure of the undersigned to perform or comply with the covenants and conditions of this Agreement; or (iii) enforcing any of the covenants and conditions of this Agreement.

This clause allows the surety to recover attorneys' fees and other reasonable costs which are usually not recoverable under the common law right of subrogation.[109] Common sense dictates that a surety should retain separate counsel even if it feels confident that the principal will cooperate under the bond and in the course of litigation. This is because unanticipated rifts usually occur as a claim develops or a lawsuit progresses and in such cases, inherently, a conflict of interest will arise. Also, where a principal has affirmative claims against the obligee, the surety's use of a separate counsel gives it the excess powers detailed in *Hutton Construction*. Finally, the attorneys' fees clause will potentially offset any expenses for the retaining of separate counsel.

E. Assignment Clause

Complimenting the other provisions in the GIA, the assignment clause operates to assign to the surety various rights. These generally include rights to (1) the principal's equipment; (2) the principal's accounts receivable; and (3) the principal's rights to pursue claims under the bond.[110] The assignment clause allows the surety to recover any and all funds owed to the principal under bonded or unbonded contracts and is valid even if the GIA has not yet been filed under the Uniform Commercial Code (UCC).[111]

---

109 A surety's claim for reimbursement of attorneys' fees under the GIA should make sure to include an unequivocal written demand for collateral on the principal and indemnitor. An expressed written demand, clearly marked, is essential. If there is a failure to collateralize, courts will consider this as evidence of the surety's reasonable fear that the indemnitor or principal is unable to mount an effective defense. *See, e.g., Perkins v. Thompson*, 551 So. 2d 204 (Miss. 1989); *United Riggers & Erectors, Inc. v. Marathon Steel Co.*, 725 F.2d 87 (10th Cir. 1984).

110 *See* discussion of *Hutton Construction* and *Matthews Constructors, supra* Section III.B.3.

111 The assignment clause generally provides:

> The Contractor, the Indemnitors hereby consenting, will assign, transfer,

As detailed by its express language, the purpose of the assignment clause is to: (1) enable the surety to complete performance and payment of the obligations of the principal covered under the bond; and (2) provide the surety with collateral security. For all general purposes, the assignment becomes effective on the date of the bond or GIA execution, but there is no transfer of assignment rights to the surety until a default.

---

of the Bond covering such contract, but only in the event of (1) any abandonment, forfeiture, or breach of any contracts referred to in the Bond or of any breach of any said Bonds; or (2) of any breach of the provisions of any of the paragraphs of this Agreement; or (3) of a default in discharging such other indebtedness or liabilities when due; or (4) of any assignment by the Contractor for the benefit of creditors, or of the appointment, or of any application for the appointment of a receiver or trustee for the Contractor whether insolvent or not; or (5) of any proceeding which deprives the Contractor for the use of any of the machinery, equipment, plant, tools, or material referred to in section (b) of this paragraph; or (6) of the Contractor's dying, absconding, disappearing, incompetency, being convicted of a felony, or imprisoned if the Contractor be an individual: (a) All the rights of the Contractor in and growing in any manner out of all contracts referred to in the Bonds, or in, or growing in any manner out of the Bonds; (b) All the rights, title, and interest of the Contractor in and to all machinery, equipment, plant, tools, and materials which are now or may hereafter be, about or upon the site or sites of any and all of the contractual work referred to in the Bonds or elsewhere, including materials purchased for or chargeable to any and all contracts referred to in the Bonds, materials which may be in process of construction, in storage elsewhere, or in transportation to any and all of said sites; (c) All the rights, title, and interest of the Contractor in and to all subcontracts let or to be let in connection with any and all contracts referred to in the Bonds, and in and to all surety bonds supporting such subcontracts; (d) All actions, causes of actions, claims, and demands whatsoever which the Contractor may have or acquire against any subcontractor, laborer, or materialman, or any person furnishing or agreeing to furnish or supply labor, material, supplies, machinery, tools, or other equipment in connection with or on account of any and all contracts referred to in the Bonds; and against any surety or sureties of any subcontractor, laborer, or materialman; (e) Any and all percentages retained and any and all sums that may be due or hereafter become due on account of any and all contracts referred to in the Bonds and all other contracts whether bonded or not in which the Contractor has an interest.

This does not mean that the principal must be declared in default.[112] Once an event of default occurs on any obligation under the GIA, the assignment clause is automatically triggered. This automatic vesting of rights to the surety obviates the need for immediate court action to enforce assignment rights. Thus, once the principal is in default, the surety's assignment right extends even to contract funds for projects not in default. Usually, a surety does not exercise this right until the principal and indemnitors fail to post collateral.

Although the surety can usually rely upon its subrogation rights to obtain contract balances, this right is somewhat limited in that a surety retains priority in contract funds only upon actual payment of a claim and only for the amount of that claim. In short, the surety has no subrogation right power to offset losses on one project against contract proceeds due from a different owner or another project. But, since the surety's assignment rights arise from default or the mere receipt of a claim, the surety may obtain superior rights to the contract funds and, more importantly, allow the surety to offset losses on different projects.

It is extremely important that a surety recognize the timing difference between its assignment and subrogation rights. Subrogation rights relate back to the date of bond issuance. Assignment rights relate to the event of default. Also, subrogation rights which vest on payment of a claim are usually stronger than assignment rights. But the key feature of assignment rights is their ability to protect the surety before it has paid on any losses. The clause gives the surety a right to immediate possession of all contract funds due the principal and gives the surety all rights, title, and interest to job-related supplies, tools, equipment, and materials in order to complete work on the project with minimal excess expense. Thus, the surety has a right to possess not only supplies and materials on-site or in storage, but also those items in transit, on order, or even in the process of being produced. If the surety decides not to use the assigned material it may also have the right to sell the assigned property. Because the execution of an assignment clause creates a security interest, a surety should consider filing a UCC financing statement to buttress its priority rights over any materials or equipment.

To aid in completion of the defaulted project, the assignment clause also authorizes the surety to execute any necessary instruments in the principal's name to secure assigned property. The clause assigns to the

---

112 *Gray v. Travelers Indem. Co.*, 280 F.2d 549, 552 (9th Cir. 1960).

surety the principal's interest in all project subcontracts and gives the surety the power to use some or all of the principal's subcontractors to complete the originally bonded contract. This is a major advantage of utilizing the assignment clause because where a principal refuses or can no longer perform, the surety can minimize expenses by seizing control of the overall scene and directing resources to the area of its exposure.

Also, any and all claims or causes of action which the principal has against subcontractors are assigned to the surety. This assignment right also allows the surety of a general contract to bring any cause of action the principal may have against a surety of a subcontractor, laborer or material supplier. Thus, in a case like *Matthews Constructors* where the subcontract surety recognizes the potential of the general contract surety's inheriting its principal's claims, the assignment right by the general contract surety is an extremely powerful negotiating tool. On the other hand, a general contract surety may want to avoid utilizing its rights under this clause for very practical reasons. Primarily, a surety's acceptance of assignment rights may expand its liability to subcontractors beyond that originally contemplated under the bond. For example, in a case like *Matthews Constructors*,[113] if St. Paul had availed itself to CCI's rights against MCI, there was the potential that MCI could sue St. Paul for consequential damages resulting from CCI's breach of the contract. If this were the case, St. Paul could have become liable for consequential damages which are generally not covered under a payment bond because availment under the GIA may require the surety to succeed to the burdens as well as benefits of the contract. This fact itself likely explains why a surety in the position of St. Paul would prefer to operate from behind its principal. As a result, a surety must be careful in invoking its assignment rights. The situation must be clearly advantageous with little or no risk of assumed burden.

113 *See* discussion *supra* Section III.B.3.b.

IV.
Indemnity Agreement As Security Agreement

A. Basic Security Rights Under The GIA

Most GIAs constitute not only a security agreement, but also a financing statement under the UCC.[114] The surety has the option to file the GIA as a standard UCC-1 form or rely exclusively on the express terms of the GIA and the surety's equitable subrogation rights. Before deciding whether or not to file, a surety needs to recognize that its attorney-in-fact clause gives it the power to execute a UCC-1 form over the objections of an uncooperative and unwilling principal. However, before any filing occurs, the surety must also realize the distinction between its subrogation rights and those rights achieved by filing under the UCC. For example, equitable subrogation protects the surety's claims to contract balances, while a perfected security interest may be necessary to allow the surety access to the principal's materials and equipment. Given this basic difference, the question of whether the surety should file the GIA depends largely on the facts relevant to the case.

The key distinction to note in whether or not to file is that the UCC protects only "contract rights" and does not have any effect on rights which arise by operation of law. A surety's equitable subrogation rights arise by operation of law. As a result, the surety is not required to file a financing statement under the UCC to preserve its subrogation right because of the well settled rule that a surety who makes payments or undertakes to complete the principal's obligations is entitled to be reimbursed.[115] Therefore, a surety relying on its subrogation rights is

---

114 Usually, the GIA will contain a provision which states:

> This agreement shall constitute a security agreement to the Surety and also a financing statement, both in accordance with the provisions of the Uniform Commercial Code of every jurisdiction wherein such code is in effect and may be so used by the Surety without in any way abrogating, restricting or limiting the rights of the surety under this agreement or under law, or in equity.

115 *Pearlman v. Reliance Ins.* Co., 371 U.S. 132, 141 (1962); *National Shawmut Bank of Boston v. New Amsterdam Cas. Co.*, 411 F.2d 843 (1st Cir. 1969); *McAtee v. United States Fidelity & Guar. Co.*, 401 F. Supp.

entitled to stand in (1) the shoes of the bond obligee; (2) the shoes of the contractor; and (3) the shoes of the subcontractor or material supplier.[116] However, subrogation is limited to the recovery of contract balances to reimburse costs of completing performance. In situations where a surety does not face a loss, its subrogation rights do not vest and it must rely exclusively on its GIA to protect its security interest. If the GIA is not filed, then the surety has no rights to the security against superior claims. Though subrogation remains a viable theory of recovery as to contract proceeds, it is not foolproof when a surety is faced with a defaulting contractor situation. The UCC may supplement and help in solving these problems, but the issue remains whether the surety should rely solely on equitable subrogation or instead pursue available UCC remedies.

B. Filing The GIA As A Security Interest

Under the UCC, a security interest is an "interest in personal property or fixtures which secures payment or performance of an obligation."[117] The interest "attaches" when (1) either collateral is in possession of the secured party or the debtor has signed an agreement containing an adequate description of the collateral; (2) value has been given; and (3) the debtor has rights to the collateral.[118] Once the interest attaches, it is enforceable against the debtor, but it is not superior to the claims of third parties unless it is "perfected." Perfection means that the secured party has taken all the necessary steps to enforce its security interest against subsequent lienholders or purchasers for value.[119] As a result, perfection is extremely important when multiple creditors claim rights in the same property because it allows the holder, upon repossession and sale of collateral, to satisfy its own secured debts in full after paying expenses.

---

11 (N.D. Fla. 1975); *Mid-Continent Cas. Co. v. First Nat'l Bank & Trust Co. of Chickasha*, 531 P.2d 1370 (Okla. 1975).

116 *Transamerica*, 540 So. 2d at 116.

117 *See* U.C.C. §§ 1-201(37).

118 *See* U.C.C. § 9-203.

119 *See* U.C.C. §§ 9-203, 9-301, 9-303, 9-312.

1. Obtaining Priority Over Previously Perfected Claims: The Purchase Money Security Interest

Given the basic filing rule, a surety may be able to use the UCC to achieve the same result it would achieve under its subrogation rights. If the surety perfects before any other security interest, the surety will have priority as to all the principal's assets described within the GIA.[120] Even in situations where a prior perfected security interest exists on all of the principal's assets, the surety may still obtain priority on the contract proceeds if it can obtain a purchase money security interest (PMSI).[121] Classification as a PMSI allows the surety to take priority over prior perfected security interests if the surety perfects the PMSI "at the time the debtor receives possession of the collateral or within ten days thereafter."[122] In this limited sphere of the PMSI, a surety may be able defeat prior perfected security interests because it furnishes a bond on a construction project and incurs an obligation which gives value to the principal/contractor (the debtor) to obtain a contract right to payment (collateral) that is necessary in order to obtain financing for the project. Because the PMSI is attached, the surety's perfection of the GIA before or within ten days of the execution of the construction contract may give the surety priority.

However, a surety's claim to other collateral, such as materials and equipment, will remain junior to prior perfected security interests. This is because the surety's bond is an obligation incurred to enable the principal to obtain the construction contract and will not be given for value to purchase materials or equipment. Thus, a PMSI cannot help the surety gain priority position as to the principal's existing assets. Still, a

---

120 *See* U.C.C. § 9-312. The description of collateral provided for in the GIA does not use the UCC description system. Because of this difference, a junior creditor with nothing to lose will likely challenge the surety's priority over the collateral even if the surety filed in an appropriate and timely manner.

121 The UCC provides that a security interest is a purchase money security interest to the extent that it is: (1) taken or retained by the seller of the collateral to secure all or part of its price; or (2) taken by a person who by making advances or incurring an obligation gives value to enable the debtor to acquire rights in or the use of collateral if such value is in fact so used. U.C.C. § 9-107.

122 U.C.C. § 9-312(4).

PMSI is a valid mechanism for obtaining first right to contract balances without reliance on the surety's equitable subrogation rights.

2. *Advantages Of Filing Under The UCC*

Aside from its PMSI potential, a surety may find filing to be advantageous. First, the surety will receive enforceable rights to materials, equipment, and other personal property owned by the principal and described in the GIA. Therefore, a surety can assert claims against owners and subcontractors to collect accounts receivables and amounts due on other bonded or unbonded contracts in an effort to reduce or eliminate losses.[123] Second, a proper filing discourages disputes with other creditors for contract balances and assets. Third, courts addressing the issue have concluded that a filing under the UCC does not waive or relinquish the surety's equitable subrogation rights.[124] But prudence is warranted under this line of reasoning because these holdings are not consistent with the general principle that equity is not available where a party has adequate remedies at law.

3. *Disadvantages Of Filing Under The UCC*

Despite its advantages, filing retains numerous disadvantages to the surety. First, if the surety files subordinate to a prior perfected interest, most of its reasons for filing disappear. Of course, there is the narrow PMSI situation, but for the most part the surety becomes a junior lienholder. Second, filing requires added time, effort, and expense. Determining where to file adds to the expense because the UCC is not uniform in every state. Thus, in a multi-state transaction, determining where to file is an extremely complex and costly process. Third, a surety's filing may have an effect on other sources of financing for the principal. Under pre-existing loans, the surety may create an instant default where a bank expressly provides. Also, since banks generally require a first priority position on collateral before a loan is made, a

---

123 Also, under the bankruptcy context, filing gives the surety specific recourse to specific property. The surety avoids the unenviable situation of being an unsecured creditor.

124 *See, e.g., American Oil Co. v. L.A. Davidson, Inc.*, 290 N.W.2d 144 (Mich. Ct. App. 1980); *Canter v. Schlager*, 267 N.E.2d 492 (Mass. 1971).

surety's filing will likely inhibit needed financing. In situations where filing will result in a principal's loss of needed financing, the surety must be aware of the potential for a bad faith claim.

4. *When Should A Surety File*

As demonstrated above, there are differences between equitable subrogation rights and liens imposed under the UCC. Given both the advantages and disadvantages, a surety should not always file the GIA as a financing statement. But when is filing appropriate? As a general rule, the surety should file the GIA to establish priority when the surety seeks to recover contract proceeds on bonded or unbonded projects where it has suffered no loss. Also, the surety must file when it is seeking to obtain the principal's equipment. Equitable subrogation gives the surety no power over previously perfected security interests in these situations.

Where the surety seeks contract funds and tangible property on a bonded contract, the surety will usually rely exclusively on its subrogation rights, believing that there is no need to file its GIA. However, by filing, the surety augments its recovery prospects for contract funds on unbonded jobs or jobs in which there may be no losses. Also, since the surety's subrogation rights arise only upon payment of claims, filing the GIA during the investigation of unpaid claims may offer substantial benefits. Since the risk of waiver to subrogation rights is minimal,[125] many sureties perfect GIAs as financing statements when it is advantageous to do so.

V.
Principal's Failure To Perform: Default

A. Performance By Surety Of Bonded Obligations

In situations where the principal defaults on an underlying contract, the surety is obligated to the obligee to complete the contract. If the surety performs the underlying contract pursuant to the performance bond, the principal and indemnitors remain ultimately liable for any

---

125 *See Old Kent Bank-Southeast v. City of Detroit*, 444 N.W.2d 162 (Mich. Ct. App. 1989).

loss.[126] Also, the surety that completes the work under the contract is entitled to the contract balance in the hands of the obligee. The surety's right to these proceeds is free from any obligee setoff for taxes or other obligations owed by the principal to the obligee.[127] The rationale is that upon principal's default the obligee retains the right to use unpaid contract funds to complete the work, thus the surety who completes the work is subrogated to the obligee's right.[128]

Performance bonds are issued with "penal sums" which limit the monetary amount of the surety's liability under the bond. In most cases, the surety will not be responsible for completion costs which exceed the bond's penal sum.[129]

1. What Type Of Bond

In some circumstances, the surety's obligations to complete the project are dictated by the type of bond at issue. The modern trend in the industry is to rely on either general performance bonds, completion

---

126 Put simply, this is because of the surety's contractual, statutory, and/or equitable rights of indemnification and reimbursement. *See* RESTATEMENT OF SURETYSHIP, *supra* note 1, §§ 22-23; *Fidelity & Deposit Co. of Md. v. Bristol Steel & Iron Works, Inc.*, 722 F.2d 1160, 1163-65 (4th Cir. 1983); *Hays Livestock Comm'n Co., Inc. v. Maly Livestock Comm'n Co., Inc.*, 498 F.2d 925, 932 (10th Cir. 1974); *In re J.T. Moran Fin. Corp.*, 124 B.R. 926, 930 (Bankr. S.D.N.Y. 1991).

127 *Prairie State Nat'l Bank v. United States*, 164 U.S. 227 (1896); *Aetna Cas. & Sur. Co. v. United States*, 435 F.2d 1082 (5th Cir. 1970); *Security Ins. Co. of Hartford v. United States*, 428 F.2d 838 (Ct. Cl. 1970); *National Shawmut Bank of Boston v. New Amsterdam Cas. Co.*, 411 F.2d 843 (1st Cir. 1969).

128 The court in *Trinity Universal Ins. Co. v. United States* explained this relationship of rights as follows: "The surety who undertakes to complete the project is entitled to the funds in the hands of the government [obligee] not as a creditor and subject to setoff, but as a subrogee having the same rights to the funds as the government." 382 F.2d 317, 320 (5th Cir. 1967).

129 *See Simmons, Inc. v. Pinkerton's, Inc.*, 762 F.2d 591, 608 (7th Cir. 1985); *Coyne-Delany Co., Inc. v. Capital Dev. Bd.*, 717 F.2d 385, 393 (7th Cir. 1983); *cf. In re Technology for Energy, Corp.*, 123 B.R. 979, 982-83 (Bankr. E.D. Tenn. 1991) (surety that takes over performance of contract has a duty to complete the contract without regard to penal sum of bond).

bonds, or indemnity bonds. Under the terms of a general performance bond, the surety does not usually have an absolute duty to complete the project, but will likely be bound to pay completion costs up to the penal limit of the bond. Thus, the generic performance bond gives the surety an option to either sit back and face paying out the full penal sum or completing the project itself to mitigate costs. Completion bonds are different in that upon default by the principal, the surety has only one option: to take over and complete the contract work. On the other hand, indemnity bonds only require the surety to indemnify the obligee for any losses suffered by the obligee as a result of the principal's default.

Before the surety can determine the extent of its liability resulting from the principal's default, it must construe the terms and conditions of the underlying contract with those of the bond. This is because the obligee's right of recovery against the surety typically extends no further than the obligee's right of recovery against the principal.[130] Therefore, as a general rule the surety's liability in the event of default is defined by the underlying construction contract.

2. *Surety's Options Upon Principal's Default*

Prior to defining its duties under the bond, a surety should be mindful that it assumes only the duties of the principal to the obligee, and that separate duties run directly to the surety from the obligee. As a result, there may be a number of ways a surety can limit or even alleviate its obligations under the bond.[131] Also, before a surety

---

130 *See National Union Fire Ins. Co. of Pittsburgh, Pa. v. D&L Constr. Co.*, 353 F.2d 169, 175 (8th Cir. 1965); *Trustees of Bricklayers & Allied Craftsmen Local No. 3 Health & Welfare Trust v. Reynolds Elec. & Eng'g Co.*, 747 F. Supp. 606, 614 (D. Nev. 1990); *H.H. Robertson Co. v. Lumbermen's Mut. Cas. Co.*, 94 F.R.D. 578, 583 (W.D. Pa.), *aff'd*, 696 F.2d 982 (3rd Cir. 1982).

131 For example a surety may put forth defenses such as impossibility of performance or defective plans/specifications provided by the principal. *See, e.g., United States v. Spearin*, 248 U.S. 132 (1918); RESTATEMENT (SECOND) OF CONTRACTS § 261 (1981). The obligee's duties owed directly to the surety which may alleviate liability include (1) material alteration of the contract without the surety's knowledge; (2) substantial extensions of time by the obligee to the principal; (3) overpayment or failure to mitigate by the obligee; or (4) improper or inadequate notice of default. *United States. v. Freel*, 186 U.S. 309 (1902); *Texaco Refining &*

determines to perform under the bond it should confirm that the principal is in default and is unable or unwilling to cure the default. It is a sure bet that if a surety takes over a project when the principal is not in default, the principal will assert that the surety is not able to recover its losses or expense because it acted as a volunteer.[132] Thus a surety would be wise to determine whether the obligee has complied with the contract, or in some situations, the surety should obtain a voluntary letter of default from the principal. Once the surety has defined its duties and confirmed default it has a variety of options available to it in fulfilling its bond obligations. Most often, a surety will have five basic options to choose from, including: (1) buying back the bond; (2) financing the principal; (3) tendering a new contractor; (4) taking over and completing the project; or (5) doing nothing and letting the obligee complete the project. These options all assume that the obligee timely terminated the principal for good cause.

a. Buying Back The Bond

If default occurs early in the construction contract and completion will exceed the penal sum of the bond, the surety should consider tendering payment of the penal sum in exchange for a full and final release from the obligee. This is a viable option because most obligees will have interest in the infusion of cash and the surety will have a quick resolution to a potentially costly and drawn out affair. The one drawback or risk is that principals and/or indemnitors may argue that tender of the full penal sum was premature and in error, thus eliminating the surety's right of indemnification.

Mktg., Inc. v. Aetna Cas. & Sur. Co., 895 F.2d 637 (9th Cir. 1990); *United States ex rel. Army Athletic Ass'n v. Reliance Ins. Co.*, 799 F.2d 1382 (9th Cir. 1986); *Argonaut Ins. Co. v. Town of Cloverdale*, 699 F.2d 417 (7th Cir. 1983); *Keene Corp. v. International Fidelity Ins. Co.*, 736 F.2d 388 (7th Cir. 1984); *Bank of Nova Scotia v. St. Croix Drive-In Theatre, Inc.*, 728 F.2d 177 (3rd Cir. 1984).

132 *Terra Resources, Inc. v. Lake Charles Dredging & Towing, Inc.*, 695 F.2d 828 (5th Cir. 1983); *Commercial Ins. Co. of Newark, N.J. v. Pacific-Peru Constr. Corp.*, 558 F.2d 948 (9th Cir. 1977).

b. Financing The Principal

Where the principal was unable to perform solely because of a lack of financial resources, a surety may finance the principal to complete the project by paying the principal's bills as they become due. This option is most viable where a shutdown of the project will result in substantial liquidated damages or where the project is substantially complete. However, a principal's default is not usually just a cash flow problem but is more often based on poor business judgment and management. Therefore, a surety looking to finance a principal must conduct a thorough investigation because as a financier, the surety is a lender no longer protected by the penal sum ceiling under the bond.

c. Tendering A New Contractor

Tendering a new contractor is a viable option where the underlying project is not yet near completion. If the contract is near completion, the costs and time delay associated with obtaining new bids will outweigh risks associated with other surety options. However, where it is plausible to tender a new contractor, the surety will be able to fix its loss. This is because the completion contractor, once accepted, will directly enter into a separate contract with its obligee. The new contractor will obtain a new performance bond and the original, tendering surety will pay the obligee the difference in price between the original contract and the completion contract, limited only by the penal sum. For this payment the surety is fully released of liability under the bond.

d. Completion By The Surety

Completion by the surety is probably most appropriate where the project is near completion. In this situation the surety will enter into a takeover agreement with the obligee to complete the project in accord with the original contract. In return the surety will seek the obligee's agreement to pay all remaining contract balances, including items like retainage, to the surety. Once this agreement is reached, the surety will either negotiate a completion contract with a new contractor or utilize its own consultants to complete the project.

Most often the surety will obtain a new contractor to complete the project because it obtains greater control over the work, can monitor costs and expenses more effectively, and can add another level of suretyship by requiring the new contractor to obtain a performance bond for the completion contract. However, absent specific contractual language to the contrary, a surety that decides to take over and complete the project assumes full responsibility for completion of the project. This potentially extends liability beyond the penal sum of the bond. As a result, a surety should only choose to complete the project when it is certain that the cost is less than other options, and the risks of completion are manageable.

e. Do Nothing And Let Obligee Complete

A surety may elect to let the obligee complete or contract to complete the original obligations of the underlying contract. Doing nothing allows the surety to subsequently pay the difference between the cost of completion and the original contract balance. Also, since a surety is not a contractor, doing nothing allows the surety to avoid risks associated with other options. However, the surety will have no control of costs incurred in completion and by doing nothing the surety tempts the obligee into suing for breach of the bond and bad faith. As a result, this option is rarely utilized.

B. Recovery Of Funds Expended

Depending on the mix of contractual or non-contractual rights, the surety will generally have a right to recover funds expended to cure the principal's default from either the principal or, in most cases, the indemnitor. Simply stated, the ability to recover expended funds arises largely because of the surety's right of subrogation. Is it therefore firmly established that "a surety who pays the debt of another is entitled to all the rights of the person he paid to enforce his right *to be reimbursed*."[133] Thus, after the surety is called upon to perform its obligations, subrogation operates as a vehicle for reimbursement to the surety to reduce or eliminate loss. For the most part, a surety in a construction context will be targeting not only the principal and indemnitors, but more importantly the unpaid contract funds and retainage held by the obligee. However, the surety will likely not be the

---

133 *Pearlman v. Reliance Ins. Co.*, 371 U.S. 132, 137 (1962).

only party seeking these funds.[134] To recover from the principal and/or indemnitor for sums expended to cure default, the surety need only show certain essential elements including: (1) the existence of an obligation of the principal to the obligee; (2) failure of the principal to perform that obligation; (3) rights in the obligee arising from this principal's failure to perform; and (4) the performance by the surety, pursuant to the suretyship, of the obligation that the principal failed to perform. Thus, a surety establishing these elements can recover its expended funds from the principal and/or indemnitor. The only restriction on recovery is that the payments were reasonable and made in good faith.

Also, as stated previously, a surety performing fully upon its principal's default has a right of contribution from cosureties. However, in cases where the principal and/or indemnitor are insolvent, a surety may attempt to recover expended funds from the obligee. This situation arises where the obligee releases the funds to the principal prior to the time a release was mandated by the construction contract. Thus, where an obligee issues unauthorized payments to a principal and that principal later defaults, the surety may want to take the position that the unauthorized payments result in a total or partial discharge of the surety's obligation. The surety will achieve discharge to the extent of the prejudice shown to result from the unauthorized payments.[135]

C. Recovery Of Funds On Bonded And Closely Related Unbonded Projects

In certain situations, the surety may be able to recover assets and contract funds on both bonded and unbonded contracts of its principal. This is accomplished by extending equitable subrogation rights and showing that the bonded contracts are interdependent with unbonded contracts so that failure to honor unbonded contracts will result in a total inability to honor the bonded ones. The blueprint for this maneuver is

---

134 Competing claimants for these funds include the obligee, the principal, third party beneficiary claimants under the payment bond, principal's lender, bankruptcy trustees, taxing authorities, and other general creditors of the principal.

135 *See Transamerica Ins. Co. v. City of Kennewick*, 785 F.2d 660 (9th Cir. 1986).

found in the unpublished opinion of *Bank One v. Highlands Insurance Co.*[136]

In *Bank One*, the principal (JHG) was a construction firm specializing in curtainwall, glass jobs, and roof decking. Bank One's (bank) predecessor made a number of loans to JHG, taking perfected security interests in JHG's accounts, contract rights, and inventory. The surety (Highlands) issued payment and/or performance bonds on a number of JHG projects. In early 1988, JHG suffered financial problems and sought additional loans from the bank. However, the infusion of capital was not sufficient to keep JHG in operation. JHG informed Highlands of its default on fourteen bonded contracts and assigned all its rights under those contracts to Highlands.

During the three months following the default, Highlands advanced $8,846,812 for claims and costs to complete the bonded and unbonded contracts of JHG. These funds were expended while Highlands conducted a full assessment and evaluation of its options. During its investigation, Highlands decided not to finance JHG, having no confidence JHG could complete the bonded contracts. Highlands also determined that the interdependence of the bonded and unbonded JHG contracts with respect to materials, manpower, supplies, equipment, and office and warehouse space made it highly likely that suppliers and subcontractors common to several contracts would refuse to perform the bonded contracts if Highlands did not help with the unbonded contracts as well. As a result, Highlands and JHG entered into a number of agreements with the end result being that Highlands was assigned all rights to the bonded and unbonded contracts, previously mentioned, as well as JHG's rights to certain new unbonded contracts. Thereafter, Highlands contracted with two new contractors to complete the bonded, unbonded, and new unbonded contracts.[137] In 1989, one year later, the bank foreclosed on JHG's assets and made demand on Highlands for an accounting and payment of contract proceeds. When Highlands refused to pay, the bank filed suit to recover proceeds from the bonded and unbonded contracts.

---

136 No. 04-95-00201-CV, 1996 WL 656697 (Tex. App. Nov. 13, 1996).

137 Highlands hired one contractor to complete work on the bonded and new unbonded contracts. Highlands entered into an agreement with a second contractor to take an assignment from JHG on the unbonded contracts and agreed to bear up to $250,000 of losses on those contracts.

The bank alleged that Highlands' acts as to the unbonded contracts were those of a volunteer and that the bank's perfected creditor status allowed it to recover on the bonded and unbonded contracts. The court disagreed. First, the court recognized that Highlands' right to subrogation was based on its payment of sums which should have been paid by JHG. Thus, on the bonded contracts, Highlands clearly had an obligation, and payments under those bonds could not have been voluntary. Second, there was ample evidence to prove that any refusal by Highlands to honor unbonded contracts would result in the common suppliers' and subcontractors' refusal to honor the bonded contracts. Thus, the court concluded from the evidence that Highlands' payment on the unbonded contracts was done in order to protect the bonded contracts. The court further concluded that Highlands was not a volunteer, because to fulfill its obligations under the bonded contracts, it had to fulfill JHG's obligations under the unbonded contracts. Third, even if Highlands had mistakenly believed that its bonded contracts would be adversely affected by a failure to complete the unbonded contracts, that mistake would still not have rendered it a volunteer.

There are several key points to glean from this case: (1) Highlands acted in good faith in its payment of funds immediately after default and its completion of all bonded and unbonded contracts; (2) Highlands conducted a full assessment and evaluation of its rights and obligations after default; (3) Highlands retained competent replacement contractors who completed all the projects; (4) the interdependence between the bonded and unbonded contracts allowed Highlands to extend its equitable subrogation rights beyond just the bonded contracts; and (5) Highlands presented extensive evidence of this interdependence. As a result, in situations where a surety is faced with a principal's default on several bonded and unbonded contracts, it should consider following the blueprint set forth in *Bank One Taxas, N.A. v. Highlands Insurance*. Doing so may allow the surety to defeat a creditor's superior perfected claim and allow recovery on both bonded and unbonded contracts.

VI.
Principal's Bankruptcy

Bankruptcy entanglements are a way of life for a surety. Entanglements generally occur when the principal defaults and files a bankruptcy petition. The surety is immediately obligated to compensate unpaid subcontractors and suppliers and faces the likely possibility that multiple claimants will battle for the remaining contract proceeds. As discussed above, the doctrine of equitable subrogation entitles the surety to priority over other secured creditors in the principal's contract retainage funds. Thus, bankruptcy has little affect on this area. But the principal's bankruptcy can affect the surety's interest in progress payments and indemnification rights to collateral in which the surety has not made a UCC filing. Furthermore, the principal's bankruptcy subjects the surety to various UCC provisions, Bankruptcy Code provisions regarding claims, preferences, and priorities, and state law governing priority of creditors. Given the extent of the entanglement, it is not enough that the surety simply understand its bankruptcy rights. Rather, the surety must use those rights in an aggressive manner to avoid excessive costs with the potential of minimum returns, and it is essential that the surety utilize its available rights during the window in which they are most useful. In the bankruptcy arena, windows of opportunity are divided in two distinct areas: (1) the rights available before the principal files bankruptcy; and (2) the rights available afterward.

A. Surety's Actions Before Bankruptcy

1. Investigation

Entering into a suretyship relationship without a full investigation of the principal, obligee, and underlying contract is analogous to Russian Roulette. A surety in that situation may be able to avoid trouble for awhile, but chances are that somewhere down the line the gun will go off. Putting it quite simply, investigation is a necessity because a financially stable principal is probably less likely to file bankruptcy than a principal currently in, or with a history of, financial trouble. But even financially sound principals can face bankruptcy when the contracts they enter into are unworkable and/or the obligee does not facilitate the relationship. Thus, in order to prepare itself for all available options, the surety must conduct a thorough investigation of the principal, obligee, and underlying contract.

2. *Filing The GIA To Perfect A Security Interest*

As discussed above, the surety's equitable subrogation rights arise independently of the UCC. However, these rights are not foolproof in protecting the surety from loss. When the surety holds a GIA with an assignment of the principal's equipment and assets, it must perfect its rights by filing the GIA as a security interest in order to obtain a position ahead of unsecured creditors. Moreover, the surety's rights to machinery, tools, and materials are not enforceable in bankruptcy absent perfection. Even where perfection is achieved, the order of filing, which establishes priority of rights, will play a part in determining which party recovers collateral. Also, if the surety's filing is within ninety days of the principal's bankruptcy filing, any transfers of property or payments to a creditor will be considered a "preference" and may be avoided by the debtor or the bankruptcy trustee.[138]

The end result is that filing the GIA as a security interest is recommended but in most cases is of little effect. In the best case scenario, a surety who files first, outside of ninety days from the principal's bankruptcy filing, and obtains first priority will find filing the GIA a valuable tool. Any other situation results in some level of lost rights or increased battling over collateral.

3. *Utilizing Financing Agreements*

Where the surety has decided to help finance the principal through a difficult time, it must utilize finance agreement provisions which can protect the surety from a preference attack. As before, all transfers to creditors within ninety days of the debtor's bankruptcy may be avoided by the Bankruptcy Code.[139] Obviously, the financing agreement provisions cannot be made one hundred percent preference-proof. However, the Bankruptcy Code contains numerous exceptions to the preference rule.[140] For example, one exception requires the surety, as creditor, to give "new value" contemporaneously with the debtor's transfer.[141] A surety, in defense to a preference action, could argue that

---

138 *See* 11 U.S.C. § 547.

139 11 U.S.C. § 547(b).

140 *See* 11 U.S.C. § 547(c).

141 *See* 11 U.S.C. § 547(c)(1); *see also* 11 U.S.C. § 547(a)(2) (providing "'New Value' means money or money's worth in goods, services, or new

its financing and subsequent transfer of assets was a contemporaneous exchange for new value.[142]

Even if it is concluded the transfer was not contemporaneous, a surety could assert the subsequent advance exception.[143] Under this exception, the surety's commitment to finance and cash advances are considered new value.[144] Important under both of these exceptions is the rule that a surety is not required to finance its principal to complete the bonded project.[145] This rule will defend an argument by the bankrupt principal that the financing was not new value because the surety had a pre-existing duty to pay the expenses.

4. Encouraging An Involuntary Petition

A financially troubled principal will first try to take care of its own by transferring assets without consideration, or by repaying loans to insiders or favored creditors. By involuntarily commencing a bankruptcy petition against the principal, a surety is attempting to prevent or "recapture" fraudulent conveyances.[146] The one big limiting factor in using this procedure is the risk of a bad faith claim.

B. Actions After Bankruptcy

Immediately upon the filing of a petition for bankruptcy, any and all efforts to recover from the principal are automatically stayed.[147] However, the automatic stay does not effect the surety's obligation

---

credit, or release by a transferee of property previously transferred to such transferee in a transaction that is neither void nor voidable by the debtor or the trustee under any applicable law, including proceeds of such property, but does not include an obligation substituted for an existing obligation.").

142 *See In re Spada*, 91 B.R. 668 (Bankr. M.D. Pa. 1988), *aff'd*, 115 B.R. 796 (M.D. Pa. 1989), *aff'd in part, rev'd in part*, 903 F.2d 971 (3rd Cir. 1990).

143 *See* 11 U.S.C. § 547(c)(4).

144 *See In re Amarex, Inc.*, 88 B.R. 362 (W.D. Okla. 1988).

145 *See Fischer Constr. Co. v. Fireman's Fund Ins. Co.*, 420 F.2d 271 (10th Cir. 1969).

146 For the rules on filing an involuntary petition, see 11 U.S.C. § 303.

147 *See* 11 U.S.C. § 362(a).

under a bond. As a result, the surety may have an obligation to continue performance or payment while the principal seeks shelter in bankruptcy.

1. Seeking Information

Assuming the principal gave no warning of its imminent bankruptcy, it is likely that information concerning the bonded contract's status, unpaid suppliers and subcontractors, and job progress will be slow coming from the principal.[148] Without this information, maneuvering through the bankruptcy maze will be difficult. To alleviate the information blackout, the surety should utilize Rule 2004 of the Federal Rules of Bankruptcy Procedure.[149] The rule allows the surety to compel

---

148 After filing bankruptcy the principal will have a number of different issues to attend. Where the principal views the surety as an adversary in the upcoming proceeding, it is likely that information will flow at a trickle.

149 Rule 2004 of the Federal Rules of Bankruptcy Procedure provides:

> (a) Examination on Motion. On Motion of any party in interest, the court may order the examination of any entity.
>
> (b) Scope of Examination. The examination of an entity under this Rule . . . may relate to the acts, conduct, or property or to the liabilities and financial condition of the debtor, or to any manner which may effect the administration of the debtor's estate, or to the debtor's right to a discharge. In . . . a reorganization case under Chapter 11 of the Code, . . . the examination may also relate to the operation of any business and the desirability of its continuance, the source of any money or property acquired or to be acquired by the debtor for purposes of consummating a plan and the consideration given or offered therefore, and any other matter relevant to the case or to the formulation of a plan.
>
> (c) Compelling Attendance and production of Documentary Evidence. The attendance of an entity for examination and the production of documentary evidence may be compelled in the manner provided in Rule 90167 for the attendance of witnesses at a hearing or trial.
>
> (d) Time and Place of Examination of Debtor. The court may for cause shown and on terms as it may impose order the debtor to be examined under this Rule at any time or place it designates, whether within or without the district wherein the case is pending.

FED. R. BANKR. P. 2004.

the principal to appear and to produce and explain its documentation. The scope of the rule is extremely broad and allows the surety the opportunity to examine the principal with respect to all aspects of bonded job performance, subrogation issues, and any other related matters.[150] In situations where information is essential in determining the proper action to take, Rule 2004 is an essential tool for the surety.

2. *Obtain Bonded Contract Funds*

As discussed above, a performing surety's right to contract proceeds is firmly rooted in its equitable subrogation rights. For the most part, a surety relying entirely on its subrogation rights will have a superior right to contract retainages in bankruptcy situations.[151] However, unpaid progress payments related to ongoing projects under a Chapter 11 case may provide difficulty.[152]

3. *Protect Contract Funds*

Once the principal files bankruptcy, the bonded contract proceeds become "cash collateral."[153] To protect its interest in the cash collateral

---

150 Courts have stated the scope of the Rule 2004 examination to be "so all encompassing as semantically to include and encourage harassment on every subject." *In re Georgetown of Kettering*, 17 B.R. 73, 75 (Bankr. S.D. Ohio 1981).

151 *See In re Alliance Properties, Inc.*, 104 B.R. 306 (Bankr. S.D. Cal. 1989); *In re Pacific Marine Dredging & Constr.*, 79 B.R. 924 (Bankr. D. Or. 1987); *In re Massart Co.*, 105 B.R. 610 (Bankr. W.D. Wash. 1989).

152 *See In re Glover Constr. Co., Inc.*, 30 B.R. 873 (Bankr. W.D. Ky. 1983); *In re Ram Constr. Co., Inc.*, 32 B.R. 758 (Bankr. W. D. PA 1983).

153 The surety's interest in the bonded contract funds is deemed the property of the principal's bankrupt estate and is termed "cash collateral." A comprehensive definition is found in 11 U.S.C. § 363(a) which provides:

> '[C]ash collateral' means cash, negotiable instruments, documents of title, securities, deposit accounts, or other cash equivalents whenever acquired in which the estate and an entity other than the estate have an interest and includes the proceeds, products, offspring, rents, or profits of property subject to a security interest as provided in section 552(b) of this title, whether existing before or after commencement of a case under this title.

and limit its liability under the bond, a surety should consider filing motions: (1) for sequestration of the cash collateral; (2) for adequate protection, if the court permits the principal to use the cash collateral; (3) to lift the automatic stay; and/or (4) to compel its principal to assume or reject executory contracts.[154]

a. Sequestration Or Adequate Protection

A trustee or debtor-in-possession must segregate and account for cash collateral in its possession, custody, or control[155] and cannot use it unless (1) the surety consents; or (2) the court authorizes its use.[156] To avoid any lapse or statement of mistake, the surety should file a brief with the court to notify the principal it does not consent to use of the cash collateral. The surety should also file a motion to prohibit or condition or, alternatively, to adequately protect the surety if the court allows the principal to use the cash collateral.[157] If the court prohibits the use of funds, the surety is protected. However, if the court does not prohibit use of the funds, the surety can obtain adequate protection.[158] Thus, the surety can require the principal, as debtor-in-possession completing the bonded contract to: (1) limit its disbursements to necessary bonded contract expenses; (2) permit surety control of expenditure payments to ensure payment on bonded contracts; (3) provide an accounting for each project; (4) limit salary payments to the principal's stockholders, officers, or their relatives; and (5) deny payment for unnecessary equipment or expenses.[159]

---

154 Schexnayder & Berens, *The Surety on Attack in The Principal's Bankruptcy: Proven Strategies and Brave New Tactics*, Torts and Insurance Practice Section Annual Meeting (ABA 1992).

155 *See* 11 U.S.C. § 363(c)(4).

156 *See* 11 U.S.C. § 363(c)(2).

157 *See* 11 U.S.C. § 363(e).

158 11 U.S.C. § 361 defines adequate protection.

159 *See In re Ram Constr.*, Inc., 32 B.R. 758, 759 (Bankr. W.D. Pa. 1983). The court in *In re Glover Constr. Co, Inc.* required progress payments to be used first to pay all bona fide claims against the bonded projects that the surety may become liable, and any surpluses could be used for the debtor's job-related operating and overhead expenses "most stringently viewed." 30 B.R. 873, 882 (Bankr. W.D. Ky. 1983). *But see In re Universal Bldrs., Inc.*, 53 B.R. 183 (Bankr. M.D. Tenn. 1985) (court denied surety's action for adequate protection as concerns unpaid

b Relief From Automatic Stay

Where the surety wants to maintain control over the bonded contract funds or where it decides to complete the project and wants to terminate the debtor-in-possession from controlling the bonded projects, it should file a motion for relief from the automatic stay.[160] To obtain control of the contract funds the surety must prove the funds are not "property of the estate," and to terminate the debtor-in-possession the surety must show that the construction contract is not assumable as an executory contract because the debtor-in-possession (principal) cannot cure the existing default or provide adequate assurance of its future performance.[161]

---

progress payments finding that the surety had not perfected its interest in the funds.).

160 11 U.S.C. § 362(d) allows the surety to file this motion and provides:

On request of a party in interest and after notice and a hearing, the court shall grant relief from the stay provided under subsection (a) of this section, such as by terminating, annulling, modifying, or conditioning such stay-

(1) for cause, including the lack of adequate protection of an interest in property of such party in interest; or

(2) with respect to a stay of an act against property under subsection (a) of this section, if-

(A) the debtor does not have an equity in such property; and

(B) such property is not necessary to an effective reorganization.

161 *See* 11 U.S.C. § 362(d)(1); *see also* 11 U.S.C. § 365(b)(1).

VII.
Exoneration And *Quia Timet*

A. Distinctions

1. *Quia Timet*[162]

Quia timet is an extraordinary right that can be used to "protect a party against an anticipated future injury when it cannot be avoided by an action at law."[163] It allows the surety to compel the principal to post collateral for an anticipated liability. For all practical purposes, *quia timet* is the common law version of the GIA's collateral security provision.[164] As a result, where contractual rights to collateral fail or are non-existent, *quia timet* rights may be utilized to protect the surety.[165] Given the relationship of *quia timet* rights with those found in a collateral security provision, a wise surety would better off trying to persuade the court that its contractual rights are consistent with principles of equity.

---

162 *Quia timet* is a Latin term meaning "because he fears." BLACK'S LAW DICTIONARY 1247 (6th ed. 1990).

163 *Northwestern Nat'l Ins. Co. of Milwaukee, Wis. v. Alberts*, 741 F. Supp. 424, 429 (S.D.N.Y. 1990).

164 For a full discussion, see *supra* Section III.D.

165 In *Milwaukee Constr. Co. v. Glen Falls Ins. Co.*, 367 F.2d 964 (9th Cir. 1966), the court was asked to specifically enforce a collateral security provisions of a GIA. However, the court did recognize *quia timet* as a prominent reason for granting relief when it stated:

> Not only does a bill *quia timet* be to compel the principal to pay the debt after it has become due, but its use has also been extended to compel the principal to furnish the surety to indemnity against possible loss where the surety has reasonable grounds for anticipating that his rights are being jeopardized and that he will incur a liability by threatened conduct of the principal.

Id. at 966; *see also Alberts*, 741 F. Supp at 430, a surety, "even absent a contractual provision, is entitled to invoke the equitable and common law rights of exoneration and *quia timet*.").

2. *Exoneration*

Exoneration gives the surety the right to "compel its principal to pay for debt in which the surety's liability already matured."[166] To invoke this right the surety must show its liability is "absolute and must in the end rest with the principal."[167] Herein lies the distinguishing feature between the claim of exoneration and a bill of *quia timet*. Exoneration requires the exact liability be established where as *quia timet* allows relief when the surety can show "reasonable grounds" that it will suffer some future injury due to the principal's conduct which cannot currently be determined.

3. *Distinctions Illustrated*

The key distinction is illustrated in the case of *Northwestern v. Alberts*, where the court granted relief under both doctrines.[168] The surety issued a bond guaranteeing payment by certain limited partners. These partners defaulted and the surety made payments on their behalf. The surety then sued to recover exoneration relief for the payments it had already made and *quia timet* for collateralization for future payments which were not yet due.

Note that with either exoneration or *quia timet*, the principal's ultimate liability is not certain. However, if the principal is ultimately liable and the surety must perform the principal's obligation, that loss should be the principal's loss. Exoneration is thus the surety's right to compel the principal to indemnify it for amounts actually paid. Since damages resulting from a principal's default are not absolute, known, or determinable until the surety performs, *quia timet* exists to compel payment to the surety so it can prepare for the potential loss.

B. *Quia Timet* Used To Protect Contract Funds

Upon the principal's default, the surety is entitled to (1) have contract proceeds applied to contract obligations; or (2) be reimbursed from contract proceeds to the extent of payment made to discharge bond

---

166 *Alberts*, 741 F. Supp. at 429.

167 *Id.* at 430.

168 *Id.*

obligations. The surety's right to have contract funds applied to contract obligations is meaningless unless the surety has some recourse to prevent dissipation or fraudulent conveyance of the funds. Because at this early stage the surety's obligations are not clearly defined and absolute, the surety technically has no rights to contract proceeds. However, if the surety has a reasonable basis to fear dissipation, it may use its *quia timet* power to prevent the diversion of contract proceeds.[169]

However, *quia timet* is not an unlimited right in this context. Some courts will not allow *quia timet* to be asserted where the principal is not yet in default of the bonded contract.[170] Accordingly, a surety should not seek to enforce *quia timet* rights unless it can show probable bond liability resulting from the principal's default.

C. Collateral Requirements From Indemnitors

A standard GIA will often provide that indemnitors must post collateral to secure anticipated bond liabilities. This contractual right of indemnitor collateralization is independent of, and in addition to, the surety's common law rights. Thus, *quia timet* may be exercised under the appropriate circumstances to require the indemnitors to collateralize the surety for anticipated losses.[171]

Generally, the surety will not have to show a fraudulent disposition of property or any other special reason for fearing a loss before it can

---

169 *See Morley Constr. Co. v. Maryland Cas. Co.*, 90 F.2d 976 (8th Cir. 1937) (contractor's insolvency provided the surety with a reasonable apprehension of diversion); *National Sur. Corp. v. Barth*, 89 A.2d 104 (N.J. Super. Ct. Ch. Div. 1952), *aff'd*, 95 A.2d 145 (N.J. 1953) (right of surety to have contact funds paid to bond claimants was superior to principal's rights to the funds, and the United States government's rights for back taxes); *Western Cas. & Sur. Co. v. Biggs*, 217 F.2d 163 (7th Cir. 1954) (court seized contract funds due to principal's failure to pay subcontractor on bonded project); *Lambert v. Maryland Cas. Co.*, 418 So. 2d 553 (La. 1982) (surety of an insolvent principal is entitled to take action to freeze contract funds for subsequent payment of bond claims.

170 *See Fireman's Fund Ins. Co. v. S.E.K. Constr. Co.*, 436 F.2d 1345 (10th Cir. 1971).

171 *See Doster v. Continental Cas. Co.*, 105 So. 2d 83 (Ala. 1958). (although the surety has a signed, written GIA with a collateral provision, court's issuance of injunctive relief was based entirely on common law principals); *see also Ellis v. Phillips*, 110 N.W.2d 772 (Mich. 1961).

invoke its *quia timet* rights. However, there are some essential elements to establish in requiring the principal or indemnitors to transfer assets as collateral. These include: (1) alleging the relationship between the surety and principal/indemnitors; (2) showing that the principal, in substance, declared itself in default; (3) showing that the obligee, subcontractor, or material providers have made demand on the surety; (4) showing that the principal retains control and/or possession of funds related to the bonded project and refuses to make payments on the project; and (5) the surety assures the court it will not avoid its obligations.[172] Aside from these elements, the surety must also show that its fear of bond liability is reasonably based on objective facts.[173] This requires more than a showing that the principal is insolvent and bond claims have been made.

D. Procedural Considerations

In seeking either to prevent dissipation of contract funds or collateralization from indemnitors, the surety must take prompt action. This requires injunctive relief at the commencement and for the duration of litigation to prevent the transfer of assets. The standards for injunctive relief in *quia timet* and exoneration situations are the same as in any other case.[174]

The proper order for securing funds and collateral is to first seek a temporary restraining order (TRO) to enjoin indemnitor transfers. A TRO may actually be issued if it clearly appears from specific facts that immediate and irreparable injury will result,[175] and will only stay in effect for ten days unless the parties agree to extend or the surety makes a good cause showing for an additional ten days. If the TRO is granted, the assets will be preserved pending a preliminary injunctive hearing. Once the preliminary injunction is issued an order for posting of collateral will arise to secure anticipated bond liability. Obtaining this

---

172 *Doster*, 105 So. 2d at 87.

173 *See, e.g., Transamerica Premium Ins. Co. v. Calvary Constr., Inc.*, 552 So. 2d 225 (Fla. Dist. Ct. App. 1989) ("without proof that a surety realistically faces loss under the performance bond and is in jeopardy, the trial court correctly determined that additional *quia timet* relief was not appropriate").

174 For a full discussion see *supra* Section III.D.2.b.

175 *See* FED. R. CIV. P. 65(b).

type protection by utilizing *quia timet* and exoneration rights at the earliest possible stage will often be the difference between bond loss and full indemnification for the surety.

VIII.
Conclusion

A surety's rights against principals and indemnitors vest at different times. These windows of opportunity require the surety to understand the specific facts surrounding each of its bonds. Thus, the surety should closely investigate and obtain a thorough understanding of its principal, indemnitors, and the obligee, and of the underlying contract. Failure to do so retards the surety's rights from the outset. However, through up-front planning and investigation, the surety is allowed to consider its full array of rights in any decision to litigate. Timely, informed, and decisive action with respect to the principal and indemnitor(s) can minimize, or even eliminate, losses in connection with default.

CHAPTER 5

DECIDING TO LITIGATE: THE SURETY'S RECOURSE AGAINST THIRD PARTIES

James F. Crowder, Jr.
Kenneth C. Borden
Guy W. Harrison
Russell A. Yagel
Esperanza Diaz Briscoe

I. Introduction

The fact patterns suggesting third party liability generally emerge in the form of anomalies in the picture most sureties develop regarding their principal's construction competence and financial strength. If something which contradicts the known profile of the principal crops up on a job, the likelihood is great that it is due to misinformation of a technical nature inherent in the plans and specifications or financial difficulty resulting from third party misinformation or actions.

These anomalies, like the tip of an iceberg, may surface on a project at anytime. Early detection depends upon the vigilance of the principal. The sooner problems are recognized by both principal and surety, the better the planning may be on how to deal with them. As is sometimes the case, a principal may not want to acknowledge the problem for fear of upsetting the owner and the surety by eroding the confidence of both in its technical or financial ability. It may not be until the problem exceeds the principal's technical and/or financial ability that it comes to the attention of the surety. At that point, the surety must make a decision to become involved in less than a "default situation" to get the job done and to later seek recovery from third parties. As these are generally suits to recover against design professionals, accountants, subcontractors, material suppliers, or lenders whose conduct has contributed to the problems of the principal on the job for which the surety is called upon to respond, the surety's actions from its first involvement should be directed toward documenting its claim which may not find its way to court or mediation for months or years. The traditional claims of a surety against third parties are based on subrogation to the rights of the obligee or tort theories arising from the breach of a duty or standard of care owed by a third party to the surety. A "new" weapon in the arsenal of the surety and its counsel is the *Restatement (Third) of Suretyship & Guaranty (Restatement of Suretyship).*[1] The concepts of the *Restatement of Suretyship* are not new, but their assembly into an updated set of rules designed to deal comprehensively with the rights and duties of those with "suretyship status" is new and powerful as a source of the action to take and the authority to sustain that action in subsequent litigation.

II.
Subcontractors And Material Suppliers

In considering actions against third parties, sureties should look to recover from subcontractors and material suppliers in situations where: (1) the surety completes the contract of its defaulting principal and the default was caused in part by the improper performance of a subcontractor; (2) the surety rescinds the subcontract as a result of the subcontractor's breach, thereby causing the surety to incur additional completion costs in hiring another completing subcontractor; (3) the surety satisfies a payment bond claim as a result of the subcontractor's misappropriation of contract funds and failure to pay its suppliers; or (4) the obligee looks to the surety under its performance bond for replacement or repair of defects in work performed by the subcontractor. Although these are the prevalent circumstances suggesting subcontractor or material supplier claims by sureties, any fact pattern which suggests that a subcontractor or material supplier is the party that should have performed

---

1 RESTATEMENT (THIRD) OF SURETYSHIP AND GUARANTY (1995) [hereinafter RESTATEMENT OF SURETYSHIP].

or should bear the cost of the performance is a proper circumstances for a claim by a performing surety.

A. Assignment

The facts calling for a surety's performance will dictate whether the surety brings its claim pursuant to its subrogation rights or the assignment rights given by the principal pursuant to a contractual agreement of indemnity.

Assignment is distinctly different from subrogation. The existence of subrogation, although growing out of a contractual setting and frequently articulated by a contract, does not depend on a grant in a contract, but is created by law to avoid injustice.[2] An assignment, by contrast, arises strictly by contract, and generally confers legal rights or a security interest to the assignee in exchange for valuable consideration.

Typically, a surety's General Indemnity Agreement transfers or assigns the principal's rights in a contract upon the occurrence of an event, such as a payment bond claim or a declaration of default by the obligee. Once assignment occurs, the surety is empowered with the same rights conferred by the assignor and is free to act as if the surety stands in the shoes of the assignor.

Whether a surety pursues subrogation or assignment may determine the applicable statute of limitations. Whether a surety pursues subrogation or assignment will determine the surety's priority in recovered assets where a priority contest with other claimants is likely. In situations where a surety may have a choice between subrogation and assignment, the statue of limitations and priority are therefore considerations of primary importance in planning the surety's recovery strategy.

B. Subrogation Under The *Restatement of Suretyship*

The right of subrogation in the suretyship context is set forth in Section 27 of the *Restatement of Suretyship*.[3] Although subrogation may

---

2 *Pearlman v. Reliance Ins., Co.*, 371 U.S. 132, 136 n.12 (1962).

3 Section 27 provides:

(1) Upon total satisfaction of the underlying obligation, the secondary obligor is subrogated to all rights of the obligee with respect to the

be confirmed or limited by contract, it is a creature of equity, enforced solely for the purpose of accomplishing the ends of substantial justice.[4] In simple terms, subrogation is the substitution of one party (the surety) in place of another (the creditor or obligee) with respect to the second party's lawful claim or right.[5] Typically, this substitution occurs when the surety pays or performs on behalf of the principal, thereby extinguishing the principal's obligations to creditors or the obligee.[6] Upon complete performance by the surety, the surety stands in the shoes of the obligee to enforce any rights the obligee may possess to reimbursement for loss incurred.

The essential elements of subrogation are: (1) the existence of an obligation from the principal to the obligee; (2) the failure of the principal to perform the obligation; (3) the obligee's rights arising from the principal's failure to perform; and (4) the performance by the surety of the obligation the principal failed to perform.[7] The absence of one or more of these elements may extinguish a surety's subrogation rights. For example, a surety acting outside the scope of its bond obligations in a pre-default financing situation is merely a volunteer who has no right of subrogation.

A common circumstance where a surety would pursue recovery from a subcontractor or supplier is when a surety is called upon to complete the general contractor/principal's construction contract in accordance with the terms of a performance bond based upon the defective performance of subcontractors. This situation is covered by Section 59 of the *Restatement of Suretyship*.[8]

---

underlying obligation to the extent that performance of the secondary obligation contributed to the satisfaction.

(2) For purposes of subsection (1), an underlying obligation that has been totally satisfied except to the extent of discharge of the secondary obligor from secondary obligation pursuant to §§ 39-46 is treated as totally satisfied.

RESTATEMENT OF SURETYSHIP, *supra* note 1, § 27

4 *Pearlman,* 371 U.S. at 136 n.12.

5 *Ammerman v. Miller*, 488 F.2d 1285, 1298-99 (D.C. Cir. 1973).

6 *Pearlman,* 371 U.S. at 138; *Trinity Universal Ins. Co. v. United States*, 382 F.2d 317, 320 (5th Cir. 1967).

7 *District of Columbia v. Aetna Ins. Co.*, 462 A.2d 428, 430 (D.C. 1983).

8 Section 59 provides:

If the relationship between two secondary obligors is that of subsuretyship,

The distinction between "Subsuretyship" and "Cosuretyship" is addressed by Section 53 of the *Restatement of Suretyship.*[9] Because the rights to which a surety is subrogated are the rights of the obligee for whose benefit the surety has performed, the *Restatement of Suretyship* recognizes a second layer of obligation by which the subcontractor and or its surety become the principal surety and the general contractor and its surety become the subsurety. The effect of this rule is that the subcontractor, and its surety, if there is one, becomes the principal and the general contractor and its surety become their obligee. Thus, once the general contractor's surety completes performance of its general contractor/principal'sobligations, it is subrogated to its defaulting general contractor's rights to pursue any subcontractor responsible for contributing to the default.[10]

C. Contingent Third Party Subrogation

Some states, including New York, recognize the viability of contingent third party claims based on subrogation.[11] In *Menorah Nursing Home, Inc. v. Zukov*,[12] for example, the court explained how a contractor's completing surety had standing to sue third party subcontractors for losses sustained under the surety's performance bond to the extent the subcontractors contributed to the principal contractor's default. The surety in *Menorah Nursing Home* sued multiple subcontractors to indemnify the surety for losses incurred as a result of the subcontractor's failure to properly perform the subcontracts. The subcontractors defended the third

---

as between them the principal surety occupies the position of a principal obligor and the subsurety occupies the position of a secondary obligor, and the rights and duties between them are those set forth in §§ 21-31.

RESTATEMENT OF SURETYSHIP, *supra* note 1, § 59.

9 RESTATEMENT OF SURETYSHIP, *supra* note 1, § 53.

10 *See, e.g., St. Paul Ins. Cos. v. Fireman's Fund Am. Ins. Cos.*, 245 N.W.2d 209, 213-14 (Minn. 1976).

11 *See e.g., American Home Assurance Co. v. Flushing Sav. Bank*, 416 N.Y.S.2d 591, 594 (N.Y. App. Div. 1979) (Murphy, P.J., concurring), *aff'd*, 420 N.E.2d 92 (N.Y. 1981); *Consolidated Edison Co. of N.Y., Inc. v. Royal Indem. Co.*, 340 N.Y.S.2d 991, 993 (N.Y. App. Div. 1973); *see also Town of Wappinger v. Republic Ins. Co.*, 452 N.Y.S.2d 674 (N.Y. App. Div. 1982).

12 548 N.Y.S.2d 702 (N.Y. App. Div. 1989).

party action by asserting that the surety was subrogated only to the rights of the obligee, not to the rights of its principal, and that the surety thus lacked standing to sue because it had not paid any money on the bond prior to instituting the action. The court, contradicting the general rule adopted by Section 59 of the *Restatement of Suretyship*, held that the surety's failure to fully complete its principal's obligations as required under the general rules of subrogation was not a bar to the surety's action against the subcontractors.[13]

D. Defenses To Subrogation

A surety's subrogation rights are subject to any defenses a subcontractor may have against the defaulting principal. For example, in *National Surety Corp. v. Allen-Codell Co.*,[14] a completing surety brought an action against a subcontractor for failing to properly perform the terms of the subcontract. The court determined that because the principal failed to honor invoices from the subcontractor, failed to adequately prepare the site for the subcontractor's work, and removed employees from the site, the principal had abandoned its rights under the subcontract which was binding on the surety.[15] Weighing the equitable positions of the surety and the subcontractor under the theory of subrogation, the court held that the surety was prevented from recovering due to the actions of its principal and the surety's substantial delay in securing services of a new completing contractor after receiving notice of termination and default from the obligee.[16]

The court in *Sentry Insurance v. Lardner Elevator Co.*[17] reached a different result. In *Sentry Insurance*, a subcontractor continued to perform after the bonded principal had defaulted. The completing surety, dissatisfied with the subcontractor's progress, rescinded the subcontract and hired a substitute contractor to complete the work. The surety then brought suit against the original subcontractor for the cost of completion. The court allowed the surety to recover, rejecting the subcontractor's argument that the surety breached the subcontract by failing to pay certain

---

13 *Id.* at 710-11.
14 70 F. Supp. 189 (E.D. Ky. 1947).
15 *Id.* at 192.
16 *Id.* at 192-93.
17 395 N.W.2d 31 (Mich. Ct. App. 1986).

invoices.[18] The court further held that because the subcontract provided for attorneys' fees in the event of a party's default, the surety was entitled to recover the entire amount of its fees, even though only a portion of the subcontractor's delay resulted in penalties against the surety.[19]

A common situation sureties face is a material supplier's payment bond claim where the subcontractor received payment for work on the project, but applied those monies to debts previously accrued. In *Story v. American States Insurance Co.*,[20] the subcontractor admitted that it withheld payment to material suppliers because the money it received was paid to previous creditors. Upon obtaining this information, the surety denied liability under its principal's payment bond and was promptly sued by the material supplier. The surety brought a third party action against the subcontractor for which the court entered judgment in favor of the material supplier on its claim and in favor of the surety on the third party action. The Arkansas Supreme Court affirmed the lower court's award, but declined the award of attorneys' fees to the surety, stating that the surety could have avoided litigation by confessing judgment as to the material supplier and bringing a separate action against the subcontractor.[21]

E. As Contractual Indemnitors

An interesting question arises whether a contractor's effort to insure its protection against any loss occurring as a result of the subcontractor's non-performance of the subcontract by including an indemnity provision in the subcontract is necessary in light of the remedy recognized by Section 59 of the *Restatement of Suretyship*. Given the right circumstances, a subcontractor can be held liable for breach of an indemnity provision contained within the subcontract even where the

---

18 *Id.* at 34-35.

19 *Id.* at 35-36.

20 704 S.W.2d 162 (Ark. 1986).

21 *Id.* at 164. Comment g of Section 28 of the RESTATEMENT OF SURETYSHIP permits recovery not only of the amount paid to the obligee, but also includes expenses in performance of the secondary obligation as part of the surety's cost of performance. Thus, the RESTATEMENT provides not only the rule of recovery, it also provides the rule of damages in subrogation cases. Accordingly, the result in *Story* may have been different under the RESTATEMENT rule.

subcontractor is not in breach of material terms of the subcontract. In *Continental Heller Corp. v. Amtech Mechanical Services, Inc.*,[22] for example, the court concluded that the subcontractor's refusal to accept the tender of defense by the contractor was a breach of the indemnity provisions of the subcontract, even without a showing that the subcontractor was at fault in performing its work.[23]

F. Contractual Warranties

Contained within the obligations of the subcontract may be terms of warranty by the subcontractor for work performed or material supplied in furtherance of the subcontract. For example, in *Rhone Poulenc Rorer Pharmaceuticals, Inc. v. Newman Glass Works*,[24] the United States Court of Appeals for the Third Circuit held a subcontractor liable for damages in failing to honor the express warranties provided for in the subcontract, even though the subcontractor performed in accordance with the specifications of the subcontract.[25] The owner in *Rhone*, through its architect agent, specified the exact manufacturer and type of window the subcontractor was to install in performing its contract, which the contractor supplied accordingly. The subcontract further provided that the subcontractor would repair or replace any defective work or materials.[26]

During the construction phase of the project, the owner detected imperfections in the windows caused by the manufacturer's misapplication of a protective coating which covered the glass. The subcontractor refused to replace the windows, citing the specifications in the subcontract and the subcontractor's strict adherence thereto. The court recognized that the subcontractor "had virtually no discretion in carrying out its contractual obligations in light of the exacting specifications in the subcontracts."[27] The court further recognized that the subcontractor "entered into subcontracts that require it to remove and to replace any defective materials at its own cost and expense."[28] The court accordingly held in favor of the owner.

---

22 61 Cal. Rptr. 2d 668 (Cal. Ct. App. 1997).
23 *Id.* at 671.
24 112 F.3d 695 (3d Cir. 1997).
25 *Id.* at 697.
26 *Id.* at 696-97.
27 *Id.* at 698.
28 *Id.*

G. Implied Warranties Under The Uniform Commercial Code

Sureties may look to implied warranties as provided for under the Uniform Commercial Code (UCC) in seeking recovery from the material suppliers that contribute to the default of its principal. Even in the absence of express warranties, courts have held that the UCC provides that a supplier's product carries an implied warranty of fitness for the particular use for which it was intended.[29]

III.
Design Professionals, Accountants, And The Economic Loss Rule

A. Claims Against Design Professionals

1. Introduction

Many "anomalies" on construction projects for which the surety may become responsible arise from errors by design professionals, *i.e.*, architects and engineers. Design professionals are involved in every phase of a project, including initial cost and quantities estimates; investigating, testing, and determining site conditions; preparation of plans and specifications; evaluation of bids; supervision and inspection of work in progress; certification of progress payments; and final acceptance and release of retainage. Errors in any of these activities can have grave consequences for the contractor and its surety. How can the surety, who is not in privity of contract with the design professionals, obtain a remedy against them?

A growing body of case law permits sureties to directly sue architects and engineers who improperly perform their contractual duties. These cases recognize tort actions for negligence in favor of the surety, but look to the contract documents to determine the scope of liability. The courts generally hold that the contract documents define the activities of the design professional, and tort law imposes upon the professional the duty of reasonable care and competence in performing those activities. This duty

---

29 *See, e.g., Certain-Teed Prods. Corp. v. Goslee Roofing & Sheet Metal, Inc.*, 339 A.2d 302, 308-309 (Md. 1975) (allowing roofing subcontractor to sue material supplier based on UCC implied warranty of fitness for a particular purpose).

extends to anyone who may be foreseeably harmed by incompetent performance, and courts have little difficulty in recognizing that a surety is such a person. Tort principles, however, will not impose upon the professionals the duty to perform any activities not specified by their contract. Thus, the contract documents remain the starting point for evaluating any negligence suit against design professionals.

Additionally, several of the cases allowing direct suit against design professionals discuss the interlocking nature of construction contracts, especially the plans and specifications, the prime contract, and the surety's bond. Several courts have noted that these documents often refer to or incorporate one another, and that the relationship between the design professionals, owner, contractor, and surety can be "tantamount" to privity.[30] This notion of a "privity equivalent" is often not developed, because the courts, in recognizing a tort duty, ultimately do not require privity at all. However, this concept may prove useful to sureties trying to overcome the obstacles of the economic loss rule and the privity doctrine.

The relatively few cases involving sureties are part of a much larger body of law dealing with the existence and nature of claims against design professionals by participants in a construction project who are not in contractual privity. Attempting to predict which jurisdictions may allow such an action by a surety partakes of crystal-gazing, as this area is one of the most unsettled, conflicted, and rapidly evolving areas of construction law. Most jurisdictions allow such claims, but on different theories, and, occasionally, to some construction participants but not others. Some jurisdictions still deny claims based on the privity doctrine, and a growing number have adopted the economic loss rule in the construction context.

The economic loss rule permits torts for personal injuries or property damage apart from the property which is the subject of the contract. The economic loss doctrine does not allow recovery for "disappointed economic expectations" of third parties related to the contract. This doctrine also would effectively bar a direct surety suit against a design professional whose negligence causes or increases a surety's loss on a project.

Because of the breadth of these areas, far beyond suretyship, only a brief survey, limited to construction cases, is possible here. Even with this restriction, no survey of this area can be definitive for several reasons.

---

30 *Calandro Dev., Inc. v. R.M. Butler Contractors, Inc.*, 249 So. 2d 254, 265 (La. Ct. App. 1971).

First, the economic loss rule often arises in products liability or other non-construction contexts. The rule may therefore be established in a jurisdiction, but may not have been applied into the construction context. Second, many states have internal conflicts regarding the scope of the economic loss rule, carving out exceptions, which, again, may or may not have appeared in the construction context. Finally, this area is in constant flux. States continually adopt, refine, or reject the economic loss rule. Thus, for every jurisdiction, detailed, state-specific research is necessary in order to determine exactly what causes of action may be available to the surety.

2. Direct Suit By Sureties Against Design Professionals

Several courts have allowed sureties to sue design professionals directly. The seminal case establishing the surety's right to bring a direct action against a design professional is *State ex rel. National Surety Corp. v. Malvaney*.[31] *Malvaney* offers two conceptual bases for liability by sureties against design professionals. The first conceptual basis may be called "quasi privity." This contractual theory of recovery is based on: (1) the interlocking nature of most construction documents; and (2) the identity of obligation between contractor and surety.[32] Because the bond usually incorporates the prime contract by reference, the design professionals' obligations as defined in the prime contract (and any other documents incorporated into the prime contract by reference) then become obligations to the surety, as if the surety did have privity with the professionals. This analysis could apply equally to bonded subcontracts or material supply contracts. Also, the scope of the duties is defined by the terms of the incorporated documents, not by the formal parties to those documents. In *Malvaney*, for example, nothing in the prime contract incorporated the owner/architect contract by reference. The prime contract, however, explicitly defined the architect's duties to the owner with regard to certifying payment. Incorporation of the prime contract into the bond extended these obligations to the surety as well.[33]

The second *Malvaney* basis for liability by sureties against design professionals is pure tort negligence. Yet, even here, the contracts play an

---

31 72 So. 2d 424 (Miss 1954).

32 *Id.* at 431.

33 *Id.*

important role. The terms of the contract determine the precise "acts" the design professionals will perform. Tort concepts, however, translate these acts into a "duty" to use due care, and then extend that duty to any person for whom it is "apparent" that negligent performance will cause harm. Defining "apparent" simply calls for applying the tort concept of foreseeability. Thus, while this tort analysis relies in part on the contracts, it extends liability beyond the formal parties to any third persons who may foreseeably be harmed by improper performance. And so the *Malvaney* court held that the prime contract created duties which extended to the non-privy surety.[34]

Several cases illustrate the *Malvaney* principles. The court in *Peerless Insurance Co. v. Cerny & Associates, Inc.*[35] held that privity of contract was not a prerequisite to recovery by the surety against a design professional. According to the court, the architect's duty to exercise due care arose at common law; the contracts simply defined the specific scope of that duty and the types of persons who could foreseeably be harmed by its breach.[36] The *Peerless* court's discussion had a dimension not present in *Malvaney*, however. The *Peerless* court noted that the surety "would be entitled to be subrogated to the rights of the Owner and against defendant for loss sustained or met with by reason of defendant's negligence."[37] Thus, the *Peerless* court put the subrogated surety in the shoes of the owner who was in privity. Because Minnesota law ultimately did not require privity to permit recovery, the court did not discuss this aspect any further. However, the court's statement could be very useful in jurisdictions still following the privity rule or who have adopted the economic loss rule.

In *Seaboard Surety Co. v. Walker*,[38] the surety brought an indemnity action against the architect for negligence in failing to discover and report defects and negligent certification for payment. The court held that the surety could bring the indemnity action, noting that a supervising architect owes a common law duty to any persons who might foreseeably be harmed by the architect's negligent performance.[39]

---

34 *Id.*
35 199 F. Supp. 951 (D. Minn. 1961).
36 *Id.* at 955.
37 *Id.* at 954.
38 260 Cal. Rptr. 924 (Cal. Dist. Ct. App. 1963).
39 *Id.* at 927.

In *Westerhold v. Carroll,*[40] the court ruled that privity of contract was not required for the surety to bring suit against the architect. The court reasoned that contractual provisions calling for architect supervision and certification reduced the hazards to the owner and to the surety, stating that such provisions in a contract are "as much for the protection of the surety as of the owner."[41] The interlocking construction contracts, by defining the economic expectations to be met and identifying the interested parties, eliminated any concern about abolishing privity of unlimited liability to an unlimited class. Instead, once the duties and class of interested persons were defined, traditional negligence standards of reasonable care governed liability.[42]

Aetna Insurance Co. v. Hellmuth, Obata & Kassabaum, Inc.,[43] took *Westerhold* one step further. *Westerhold* and preceding cases had dealt only with the architect's negligent certification for release of funds. *Aetna* extended the architect's potential liability to any acts of negligence in supervising the construction process as well. *Aetna* permitted the surety to advance claims for damages due to delays and improper construction practices which resulted in corrective work caused by negligent inspection and supervision.[44]

In *Calandro Development, Inc. v. R.M. Butler Contractors, Inc.*,[45] the surety sued the engineering firm for negligence in inspecting the work, authorizing payments for defective work, failing to suggest that the owner withhold sufficient funds to correct defective work, and failing to halt construction pending correction of defects. The *Calandro Development* court adopted the reasoning of *Malvaney, Westerhold, Peerless,* and *Aetna,* and concluded that the surety had a claim notwithstanding lack of direct privity. The court ruled that, because of the nature of the construction documents, "the relationship between the surety and engineer is tantamount to one of privity" and that "the degree of care owed by an engineer to a surety is the same as that owed to the owner."[46]

---

40 419 S.W.2d 73 (Mo. 1967).

41 *Id.* at 78-79 (internal quotation omitted).

42 *Id.* at 80-81.

43 392 F.2d 472 (8th Cir. 1968).

44 *Id.* at 475.

45 249 So. 2d 254 (La. Ct. App. 1971).

46 *Id.* at 265.

City of Houma v. Municipal Industrial Pipe Service, Inc.[47] involved a surety on a sewer project. The contractor failed to comply with specifications and the surety was held liable for completion costs of over $400,000.00. The surety sought indemnity from the engineer for negligently inspecting the work and certifying compliance with the specifications. The court upheld the tort judgment for indemnity. The court, looking to *Calandro,* held that the surety was one who would foreseeably rely on the engineer properly performing its duties.[48]

Sureties have not been uniformly successful in asserting claims in their own right against design professionals, however. Cases rejecting such claims do so not on theoretical grounds, but simply rely on exculpatory language in the owner/design professional contract.

American Casualty Co. v. Town of Shattuck[49] and *American Casualty Co. v. Board of Education of Independent School District No. 2*[50] were companion cases involving basically identical facts. In each case, the surety's claim against the architect failed because of the lack of any contractual obligation on the part of the architect to the surety. Because of this, the court distinguished *Malvaney* and *Peerless*, and found that tort concepts of duty would not add to the scope of the architect's contractual duties.[51]

---

47 884 F.2d 886 (5th Cir. 1989).

48 *Id.* at 889-90 n.3.

49 228 F. Supp. 834 (W. D. Okla. 1964).

50 228 F. Supp. 843 (W. D. Okla. 1964).

51 *Shattuck*, 228 F. Supp at 841; *Board of Education*, 228 F. Supp. at 851. The court also concluded, after a detailed analysis of Oklahoma's statutory payment bond scheme, that the surety's interest in the design professional's certifications did not coincide with the owner's interest. Because a public body cannot be sued, and its property cannot be liened by unpaid subcontractors or material suppliers, the public body's only interest in the architect's certification is whether the physical work is complete. The courts viewed payment of project bills as the exclusive concern of the payment bond surety. The payment bond surety cannot "piggyback" its interest in payment onto the owner's interest in completion. As a result, the design professional's duties to the owner did not extend to the payment bond surety. *See Shattuck*, 228 F. Supp. at 838; *Board of Education*, 228 F. Supp. at 850. This distinction has not been followed in other cases, and seems contrary to general principles of equitable subrogation, which puts the surety in the shoes of the owner upon either performance of project work or payment of project debts.

In *J & J Electric, Inc. v. Gilbert H. Moen Co.,*[52] an electrical subcontractor's surety sued the project architects for negligence in failing to discover defects in the subcontractor's work. The court did recognize a body of state law upholding an architect's liability to third parties who may foreseeably be harmed by negligent performance. The court even assumed, without deciding, that sureties might be included in this class of third persons.[53] However, the court looked to language in the owner/architect contract stating that the architect had no duty to make "exhaustive or continuous" inspections and had no liability for the contractor's "failure to carry out the construction work in accordance with the Contract Documents."[54] The court viewed this language as clear and unambiguous, absolving the architects of any liability for the contractor's or subcontractors' failure to properly perform the work.[55]

The unpublished opinion in *Great American Insurance Co. v. Wade-Trim/Associates, Inc.*[56] expands on this analysis. In *Wade-Trim,* the surety brought a third party complaint against the engineer for negligence. In its analysis, the United States Court of Appeals for the Sixth Circuit recognized a Michigan state court case allowing a contractor to sue an engineer for negligence, even in the absence of privity.[57] Nonetheless, the Sixth Circuit Court flatly refused to extend this cause of action to the contractor's surety. With no explanation at all, the court stated that the relationship between a surety and an engineer is "different" from that between a contractor and an engineer.[58] The Sixth Circuit rejected the decision in *Calandro* finding that the interlocking construction contracts create duties that flow to contractors and sureties. Instead, the court focused on multiple provisions in the contract sharply limiting the responsibilities of the engineer.

The cases allowing a surety's direct action against design professionals apply three basic rationales. First, these cases recognize a common law duty of competence and reasonable care on the part of design professionals toward persons who may foreseeably be harmed by their

---

52 516 P.2d 217 (Wash. Ct. App. 1973).

53 *Id.* at 227.

54 *Id.*

55 *Id.*

56 No. 90-71311, 1992 WL 1679 (6th Cir. Jan. 7, 1992).

57 *Id.* at *2 (citing *Bacco Constr. Co. v. American Colloid Co.*, 384 N.W.2d 427 (Mich. Ct. App. 1986)).

58 *Id.*

negligence. Because this duty exists at law, it is separate and independent from the design professional's contract. The contract, however, does bear on liability, because, absent contract, design professionals have no obligation to act at all. The contract defines exactly how the design professionals will act, and the common law will not require any additional activities. The duty only attaches to those specific actions which the design professionals perform. The duty, however, is imposed by law and is non-delegable.

Next, the courts, examining the surety's role in the construction project, recognize that the surety's economic interests are most closely allied, and in fact are identical with, the owner's interests, *i.e.,* proper completion of the work for the contract price. The harm a surety suffers from the principal's non-performance is the same as that of an owner who, absent a surety, would bear the same expense. This is a crucial point, for it allies the surety with the owner's expectations and not the principal's. The principal has its own independent contractual obligation to comply with project specifications and is compensated accordingly. Thus, the principal may have limited rights against third parties for damages caused by its own breach. However, the economic damages a principal suffers from non-performance (lost profit, delay expense, wasted labor and materials, etc.) are not the same as the losses the surety suffers (completion expenses). Thus, the surety is not considered to be in the traditional role of a third party, but instead is considered to be in the role of the owner. This analysis may give the surety an even stronger position against design professionals than contractors have.

This leads to the third rationale for allowing a direct action by a surety against a design professional: the notion that the surety's relationship to the design professional is "tantamount" to privity.[59] The close interrelationship of construction documents, where each contract often refers to or incorporates other contracts, gives every participant actual knowledge that the other participants rely on that participant's proper performance. Thus, courts find that it is "apparent" that a breach by design professionals can harm the surety, especially since the surety's economic interests are so closely tied to those of the owner in direct privity with the design professionals. In *Malvaney*, the court recognized that these contractual interrelationships by themselves created duties

---

59 *Calandro Dev., Inc. v. R.M. Butler Contractors, Inc.*, 249 So. 2d 254, 265 (La. Ct. App. 1971).

between the participants in the construction process, even absent direct privity. This duty to those whose reliance, by contract, is "apparent" appears to be stronger than the traditional tort duty based on "foreseeability" of harm.

Seen in light of these general principles, it can be suggested that the courts in *J & J Electric* and *Wade-Trim* confused the nature of the design professional's duty. The contract merely describes the activities the design professional is to perform, it does not set the duty. The common law of tort requires the duty of competence and due care when performing the contractual activities, and this duty is non-delegable. Thus, contractual language limiting the scope of the architect's activities, and making other persons also responsible for compliance with project specifications, do not negate the legal duty of competence and due care when the design professional performs its requisite activities under the contract. For example, the contract may only call for periodic, "non-exhaustive" inspections. The design professional, however, must still perform these inspections competently, and must discover what a reasonable professional would discover. This important distinction seems to have escaped the *J & J Electric* and *Wade-Trim* courts.

Additionally, these cases overlooked other state law decisions which should have supported the surety's claim. In both Washington and Michigan, state courts had allowed third party contractors, to bring direct suits against architects for negligence. However, the *J & J Electric* and *Wade-Trim* courts felt that allowing a cause of action to the surety would be an unwarranted "expansion" of liability. The courts reasoned that because the contractor had the exclusive responsibility to comply with project specifications, the contractor, and by extension, the surety, could not reasonably rely on the performance of the design professionals. This reasoning is flawed in several respects.

First, state law in Washington and Michigan had given a contractor a direct action against a design professional notwithstanding the fact of the contractor's own independent obligation to comply with project specifications. If the contractor, therefore, can reasonably rely on the design professional's duty of competent performance, then so can the surety. The surety, after all, is ultimately responsible for the same contractual obligations as the principal.

The more critical flaw in these decisions, however, is the courts' failure to recognize that the surety's economic and reliance interests in the design professionals' proper performance is the same as the owner's, not the contractor's. Because of this, allowing a cause of action to a contractor is actually more of an expansion of liability than allowing a cause of action to the surety. The courts' failure to recognize this distinction puts them at odds with the general weight of authority allowing a cause of action to the surety.

Finally, the *J & J Electric* and *Wade-Trim* courts did not consider that the interlocking nature of construction documents gives the surety, like the contractor, a basis in contract for relying on the reasonable and competent performance of the design professionals. In sum, the decisions denying the surety's claim for relief rely on reasoning at odds with the general weight of authority, and may be subject to challenge even in the jurisdictions in which they were rendered.

3. Jurisdictions That Permit Third Parties Not In Privity To Sue Design Professionals

Courts from California, Louisiana, Minnesota, Mississippi, and Missouri have recognized sureties' direct remedies against design professionals. Courts in Oklahoma, Michigan, North Carolina, and Washington have denied such claims, but on fact-specific or legally questionable grounds. While the body of law involving sureties is becoming well-developed, sureties in most jurisdictions are still without direct guidance.

For this reason, cases permitting or refusing suits by other participants in the construction process who lack privity with the design professionals are extremely useful as predictors for whether sureties may do the same. This is one of the most contentious areas of construction law today. States have many varied views on whether and under what conditions they will allow such suits. The subject has generated lengthy litigation and numerous articles. A detailed analysis, which could fill a book in itself, is therefore beyond the scope of this chapter. Instead, this section will present a brief description of the various legal theories employed by the courts and a survey of cases in various jurisdictions addressing the issue. Since this area is continually changing, state-specific research is essential before a surety pursues a claim against a design professional.

The majority of jurisdictions addressing the issue have upheld suits against design professionals even in the absence of privity. Most jurisdictions simply recognize common law duties of due care, reasonableness, and competence on the part of design professionals, and allow suit by anyone foreseeably injured by negligent performance.[60]

Other jurisdictions take a more limited view, as expressed in the Section 552 of the *Restatement (Second) of Torts* regarding negligent misrepresentation.[61] These cases focus primarily on limiting the scope of potential claimants to those persons who are members of a class whom the design professionals know will rely on their product. This class would, seemingly without doubt, include the owner and the prime contractor, and, by extension, the surety as well. Indeed, *Malvaney* and its progeny provide the ideal conceptual basis for this extension. The class also could,

---

60 Alabama: *Berkel & Co. Contractors, Inc. v. Providence Hosp.*, 454 So. 2d 496 (Ala. 1984) (subcontractor allowed to sue architect); Alaska: *City of Dillingham v. CH2M Hill N.W., Inc.*, 873 P.2d 1271, 1277 n.12 (Alaska 1994) (indicating that third parties may have direct action against design professionals); Arkansas: *Carroll-Boone Water Dist. v. M. & P. Equip. Co.*, 661 S.W.2d 345 (Ark. 1983) (contractor allowed to sue engineer); Massachusetts: *Parent v. Stone & Webster Eng'g Corp.*, 556 N.E.2d 1009,1012 (Mass. 1990) (court indicating in dicta that "a claim in tort may arise from a contractual relationship" and may be available to persons not in privity); New Jersey: *Conforti & Eisele, Inc. v. John C. Morris Assocs.*, 418 A.2d 1290 (N.J. Super. Ct. App. Div. 1980), *aff'd,* 489 A.2d 1233 (N.J. 1985) (allowed claim by contractor against architect); North Carolina: *Browning v. Maurice B. Levien & Co.*, 262 S.E.2d 355, 358 (N.C. Ct. App. 1980) (owner allowed to sue inspecting architect hired by lender; allows suit by "anyone who can be reasonably foreseen as relying" on the design professional's information); *Davidson & Jones, Inc. v. New Hanover County*, 255 S.E.2d 580 (N.C. Ct. App. 1979) (allowing subcontractors to sue architect); Oklahoma: *Keel v. Titan Constr. Corp.*, 639 P.2d 1228 (Okla. 1981) (owner allowed to sue architect hired by contractor in absence of privity); South Carolina: *Griffin Plumbing & Heating Co. v. Jordan, Jones & Goulding, Inc.*, 463 S.E.2d 85, 88-89 (S.C. 1995) (contractor allowed to sue engineer based on the special relationship between contractor and design professional); South Dakota: *Mid-Western Elec., Inc. v. DeWild Grant Reckert & Assocs. Co.*, 500 N.W.2d 250 (S.D. 1993) (subcontractor allowed to sue engineer); Texas: *Associated Architects & Eng'rs, Inc. v. Lubbock Glass & Mirror Co.*, 422 S.W.2d 942 (Tex. Civ. App. 1967) (subcontractor allowed to sue architects); Wisconsin: *A.E. Inv. Corp. v. Link Builders, Inc.*, 214 N.W.2d 764 (Wis. 1974) (allowing suit against engineer for failing to properly determine subsoil conditions).

61 Section 552 is discussed in greater detail in Section III.B.2, *infra*, regarding accountants' liability.

with little conceptual effort, be extended to subcontractors and material suppliers. The cases construing Section 552 of the *Restatement (Second) of Torts* in the construction context have extended its reach this far.[62]

Finally, a few jurisdictions allow third party claims against design professionals under even more restrictive conditions than those of Section 552. One jurisdiction has, for policy reasons, allowed third party claims only where the negligence of the design professionals presents a clear and substantial risk of serious personal injury.[63] Another jurisdiction has allowed claims only when the professionals have "actual knowledge" that a *specific* person will be relying on their work.[64] New York courts have used a position between "actual knowledge" and Section 552 of the *Restatement*, allowing third parties to sue design professionals when their reliance is foreseeable and their relationship is the functional equivalent of privity.[65]

---

62 Colorado: *Wolther v. Schaarschmidt*, 738 P.2d 25 (Colo. Ct. App. 1986) (purchaser allowed to sue construction lender's inspecting architect); Delaware: *Guardian Constr. Co. v. Tetra Tech Richardson, Inc.* 583 A.2d 1378 (Del. Super. Ct. 1990) (contractor allowed to sue design engineer); Georgia: *Robert & Co. Assocs. v. Rhodes-Haverty Partnership*, 300 S.E.2d 503 (Ga. 1983) (purchaser allowed to sue inspecting engineer under Section 552; however, court limited claim to the misrepresentation, supply, or omission of information, not supervisory activities); Montana: *Jim's Excavating Serv., Inc. v. HKM Assocs.*, 878 P.2d 248 (Mont. 1994) (contractor allowed to sue engineer); Tennessee: *John Martin Co. Inc. v. Morse/Diesel, Inc.*, 819 S.W.2d 428 (Tenn. 1991) (subcontractor allowed to sue construction manager).

63 Maryland: *Council of Co-Owners, Atlantis Condominium, Inc. v. Whiting-Turner Contracting Co.*, 517 A.2d 336, 345 & n.5 (Md. 1986) (allowing claims by purchasers not in privity for expense of repairs only where defects pose "clear danger of death or personal injury;" other defects "will not suffice" to support claim).

64 Indiana: *Essex v. Ryan*, 446 N.E.2d 368, 372 (Ind. Ct. App. 1983) (third party claims against surveyors only when surveyor has "actual knowledge" that the third party will rely).

65 New York: *Ossining Union Free Sch. Dist. v. Anderson LaRocca Anderson*, 539 N.E.2d 91 (N.Y. 1989) (allowed suit by owner against consulting engineers hired by architect, because the relationship was the "functional equivalent of contractual privity"); *Reliance Ins. Co. v. Morris Assocs., Inc.*, 607 N.Y.S.2d 106, 107 (N.Y. App. Div. 1994) (surety subrogated to claims of contractor acceded to functional equivalent of privity with design professionals).

In sum, in addition to the six jurisdictions which allow surety suits outright, an additional twenty-one jurisdictions allow suits against design professionals under conditions which may extend to sureties under the appropriate factual circumstances.

4. Jurisdictions That Do Not Permit Third Parties Not In Privity To Sue Design Professionals

While two states still recognize a privity requirement for suit,[66] the most widespread grounds for denying recovery is the economic loss rule. The economic loss rule limits the extent of tort recovery when the losses are purely economic. Courts describe such losses as disappointed economic expectations arising from contracts in which the person has an interest, but lacks privity.[67] Such losses explicitly include "damages for inadequate value, costs of repair and replacement of the defective product, or consequent loss of profit."[68] Plainly, these are the quintessential losses a surety suffers in the construction setting.

The conceptual underpinning of the economic loss rule is to preserve the distinction between contract and tort remedies. Tort remedies traditionally protect a person's interest in freedom from physical harm or the destruction of his or her property. Contract remedies traditionally protect a person's economic expectations created through dealings with other persons. Because persons, in creating contracts, can bargain for allocation of risk, allowing tort remedies to intrude could undermine the contracts themselves.[69] Although, again, a detailed explication of the

---

66 Nebraska: *John Day Co. v. Alvine & Assocs., Inc.*, 510 N.W.2d 462, 466 (Neb. Ct. App. 1993); Pennsylvania: *Linde Enters., Inc. v. Hazelton City Auth.*, 602 A.2d 897, 901 (Pa. Super. Ct. 1992).

67 *See, e.g., Casa Clara Condominium Ass'n, Inc. v. Charley Toppino & Sons, Inc.*, 620 So. 2d 1244, 1246 (Fla. 1993).

68 *Id.* (internal quotation omitted).

69 Two recent articles, both of which provided valuable source information for this chapter, provide an excellent discussion of the theoretical underpinnings of the economic loss rule and its ramifications in the construction setting. *See* Martha Crandal Coleman, *Design Professionals' Liability for Negligent Design and Project Management*, unpublished paper presented at the Annual Midwinter Meeting of the Fidelity & Surety Law Committee (1997); J. William Ernstrom, *Sureties and the Economic Loss Rule, Can We Salvage Something from Nothing?*, National Bond Claim Association Seminar (Sept. 1991). These articles in turn cite many of the

economic loss doctrine could fill volumes, the doctrine has had a distinct impact in the construction setting.

The best example is Florida. The Florida Supreme Court, in *A.R. Moyer, Inc.. v. Graham,*[70] produced one of the seminal decisions establishing the right of a contractor to sue an architect for negligent supervision of a project. *Graham* based the architect's liability on his close relationship with the contractor and on the foreseeability that his negligent actions might cause harm to the contractor.[71] However, Florida subsequently adopted the economic loss rule,[72] and the rule found its way into the construction setting in *Casa Clara Condominium Ass'n v. Charley Toppino & Sons, Inc.*[73] In *Casa Clara,* the court refused to permit home purchasers to sue a supplier of defective concrete due to their lack of privity with the supplier. The court did not overrule *Moyer*, but stated that the *Moyer* decision was limited strictly to its facts.[74] Subsequent to *Casa Clara,* Florida courts have, without discussion, refused to allow a contractor to sue a design professional,[75] and have stated that, unless all facts, including the exact terms of the contracts, are identical to those in *Moyer*, no action will lie.[76] Thus, the rise of the economic loss rule in Florida has seriously undermined, if not eliminated outright, one of the key authorities establishing the right of a third party to sue a design professional.[77]

---

numerous and extensive articles concerning the economic loss rule, both generally and in the construction context.

70 285 So. 2d 397 (Fla. 1973).

71 *Id.* at 402.

72 *AFM Corp. v. Southern Bell Tel. & Tel. Co.*, 515 So. 2d 180 (Fla. 1987).

73 620 So. 2d 1244 (Fla. 1993).

74 *Id.* at 1248 n.9.

75 *D.I.C. Commercial Constr. Corp. v. Broward County ex rel. Bd. of County Comm'rs*, 668 So. 2d 697 (Fla. Dist. Ct. App. 1996).

76 *Spancrete, Inc. v. Ronald E. Frazier Assocs., P.A.,* 630 So. 2d 1197, 1198 (Fla. Dist. Ct. App. 1994) (refusing subcontractor's suit and requiring total identity of facts with *Moyer*).

77 Florida is not the only state to exhibit confusion between holdings allowing third party suits and the adoption of the economic loss rule. In Wisconsin, established law allows third parties to sue design professionals directly in tort for negligent acts. However, Wisconsin adopted the economic loss rule in *Sunnyslope Grading, Inc. v. Miller, Bradford & Risberg, Inc.*, 437 N.W.2d 213, 217-18 (Wis. 1989), a products liability case. Federal courts, however, have predicted that the rule will apply in all contractual contexts,

Relatively few jurisdictions actually apply the economic loss rule in the construction context.[78] Nevertheless, the economic loss rule may appear in state law in non-construction contexts before it appears in the construction arena.[79] Although a full survey of every state which has adopted the doctrine, in any context, is beyond the scope of this chapter, sureties should engage in thorough state-specific research, and be wary of their chances of recovery in any jurisdiction following the rule.

---

and have prohibited tort claims against engineers. *See Wausau Paper Mills Co. v. Chas. T. Main Co.*, 789 F. Supp. 968 (W.D. Wis. 1992). Despite this prediction, Wisconsin appellate courts have squarely held that "a negligent defendant is liable for economic loss." *Anderson v. Regents of Univ. of Cal.*, 554 N.W.2d 509, 517 (Wis. Ct. App. 1996). In Missouri, long recognizing a surety's right to sue design professionals, the court of appeals, using the economic loss rule, denied a construction manager's suit against an architect for defective specifications. *Fleischer v. Hellmuth, Obata & Kassabaum, Inc.*, 870 S.W.2d 832, 834 (Mo. Ct. App. 1993). The court did expressly reaffirm the surety's rights against design professionals, based on the surety's identity of interest with the owner. This may provide a conceptual basis for sureties in other economic loss jurisdictions to argue for an exception.

78 Illinois: *2314 Lincoln Park West Condominium Ass'n v. Mann, Gin, Ebel & Frazier, Ltd*, 555 N.E.2d 346 (Ill. 1990) (subsequent purchaser could not sue architect); Ohio: *Floor Craft Floor Covering, Inc. v. Parma Community General Hosp.*, 560 N.E.2d 206 (Ohio 1990) (subcontractor not permitted to sue architect); Pennsylvania: *Palco Linings, Inc. v. Pavex, Inc.*, 755 F. Supp. 1269 (M.D. Pa. 1990) (under Pennsylvania law subcontractor could not sue project engineer); Utah: *Maack v. Resource Design & Constr., Inc.*, 875 P.2d 570 (Utah Ct. App. 1994) (subsequent purchaser could not sue builder); Virginia: *Sensenbrenner v. Rust, Orling & Neale Architects, Inc.*, 374 S.E.2d 55 (Va. 1988) (must be privity of contract to sue for economic losses); West Virginia: *National Steel Erection, Inc. v. J.A. Jones Constr. Co.*, 899 F. Supp. 268 (N.D. W.Va. 1995) (predicting that West Virginia state courts will adopt economic loss rule to bar third parties from suing for economic losses); Washington: *Berschauer/Phillips Constr. Co. v. Seattle Sch. Dist. No. 1*, 881 P.2d 986 (Wash. 1994) (contractor could not sue architect directly for economic losses).

79 *See e.g., AFM Corp. v. Southern Bell*, 515 So. 2d 180 (Fla. 1987) (involving a claim for damages due to an error in a telephone book advertisement).

5. *Strategies For Avoiding The Economic Loss Rule And The Privity Doctrine's Barriers To Recovery*

Even in jurisdictions following the economic loss rule or the privity doctrine, however, participants in a construction project, including sureties, have available certain strategies for pursuing claims against third parties. The first strategy is to obtain an assignment of claim from the person who is in privity with the design professional. In the surety's case (absent subrogation), this would mean an assignment from the owner. In *Berschauer/Phillips Construction Co. v. Seattle School District No. 1*,[80] the court, at the same time it barred an action in tort under the economic loss rule, held that the contractor could obtain an assignment of the owner's cause of action, and so pursue the architect directly. The court rejected a contention that an "anti-assignment" clause in the contract prohibited such a strategy, noting that there is a distinction between assigning the right to perform the contract with assigning a cause of action for non-performance.[81]

Besides assignment, the third party can allege a direct cause of action for intentional torts. In *BIB Construction Company, Inc. v. City of Poughkeepsie*,[82] the contractor alleged that the architect tortiously interfered with its contract by falsely blaming the contractor for deficiencies in the architect's own designs, in order to divert attention from the architect's fault.[83] The court upheld the claim. In reality, this situation is rare. However, since many projects which go bad do so in part because of conflicts between the contractor and the supervising architects and engineers, it is an option that the surety should keep in mind.

6. *Actions Against Design Professionals Based On Subrogation*

In addition to direct claims in their own right, sureties who fully perform a project and pay all claims have equitable subrogation rights. Comment F of Section 28 of the *Restatement of Suretyship* deals with rights against other persons. The comment sets forth the rule that if a person other than the principal obligor is liable to the obligee for default on the underlying obligation, the surety may enforce the subrogation rights

---

80 881 P.2d 986 (Wash. 1994).

81 *Id.* at 995.

82 612 N.Y.S.2d 283 (N.Y. App. Div. 1994).

83 *Id.* at 285.

of the obligee against such other person performs the underlying obligation.[84] These rights place sureties in the same position as an owner or a prime contractor on a project, making the contractual and tort remedies of the owner or contractor available to the surety. Privity no longer becomes a defense, nor does the economic loss rule, for the surety obtains "privity" through subrogation. Generally, the subrogated surety obtains the owner's traditional tort action for negligence.[85] However, in certain cases, the undertakings of the design professionals may give rise to warranty claims.[86]

In *Unity Telephone Co. v. Design Service Co.*,[87] the surety paid the owner of a telephone building project all costs and expenses which the owner incurred in completing the building after the contractor defaulted. The surety then sued the supervising engineer for negligent supervision. The court held that through the doctrine of equitable subrogation, the surety became subrogated to the owner's contractual rights against third persons.[88] The court in *Peerless Insurance Co. v. Cerny & Associates, Inc.*[89] recognized the same principle.

Not all cases, however, have considered the surety as subrogated to the rights of the owner. In *Reliance Insurance Co. v. Morris Associates, Inc.*,[90] the court considered a performance bond surety's suit as subrogee of the contractor. The court concluded that the relationship between the contractor and architect was the functional equivalent of privity, and allowed the surety's suit to proceed.[91] While this was a good result for the

---

84 Subrogation is discussed in greater detail in section II.B, *supra*, and section III.B.4, *infra*. Comment F of Section 28 of the *Restatement of Suretyship* cites to extensive authorities generally describing a surety's subrogation rights.

85 *City of Mounds View v. Walijarvi*, 263 N.W.2d 420 (Minn. 1978) (due to necessity to exercise judgment, architect's duties limited to negligence standard, not warranty).

86 *See C.L. Maddox, Inc. v. Benham Group, Inc.*, 88 F.3d 592, 600 (8th Cir. 1996) (applying Missouri law, court held that, where design professional held itself out as holding special expertise in technical work, and where it knew bidder relied exclusively on professional's calculations in preparing bid, assurances of design professional held to create implied warranty).

87 201 A.2d 177 (Me. 1964).

88 *Id.*

89 199 F. Supp. 951 (D. Minn. 1961).

90 607 N.Y.S.2d 106 (N.Y. App. Div. 1994).

91 *Id.* at 107.

surety, the court omitted from its analysis a discussion of how a subrogated surety stands in the shoes of an owner.

Additionally, in jurisdictions that follow the economic loss rule, even subrogation may not provide a surety relief, unless it places the surety in the shoes of someone in privity with the design professional. In *Fireman's Fund Insurance Co. v. SEC Donohue, Inc.*,[92] for example, a subcontractor's surety who, as subrogee, sued the project architect for damages due to defective specifications was denied relief based on the economic loss rule. The subcontractor did not have privity or the equivalent thereof sufficient to support the surety's claim.

Once the surety has become subrogated, it has the full panoply of actions against the design professionals as does the owner. These claims are limited only by the terms of the contracts, and may be of almost infinite variety. The "typical" claims include negligent preparation of designs and specifications, negligent supervision and inspection of work, and negligent certification of payment. Additional types of claims courts have recognized in favor of owners include: negligent estimation of construction costs;[93] negligent observation and acceptance of the work;[94] negligent failure to respond to contractor inquiries;[95] negligent rejection of proposed substituted materials;[96] inadequate review of shop drawings;[97] and delay in shop drawing review. Again, a complete list of possibilities is virtually impossible, as claims are limited only by the language of the contract itself.

7. *Contractual Limitations On Tort And Subrogation Claims*

A major obstacle to both tort claims and subrogation claims against design professionals comes from the design professionals' attempts in recent years to completely disclaim any responsibility in their contracts.

---

92 666 N.E.2d 881 (Ill. App. Ct. 1996), *aff'd as modified*, 679 N.E.2d 1197 (Ill. 1997).

93 *Brock Bridge Ltd. Partnership, Inc. v. Development Facilitators, Inc.*, 689 A.2d 622 (Md. 1997).

94 *Shephard v. City of Palatka*, 414 So. 2d 1077 (Fla. Dist. Ct. App. 1981).

95 *Miller v. City of Broken Arrow*, 660 F.2d 450 (10th Cir. 1981).

96 *Waldor Pump & Equip. Co. v. Orr-Schelen-Mayeron & Assocs., Inc.*, 386 N.W.2d 375 (Minn. Ct. App. 1986).

97 *Duncan v. Missouri Bd. for Architects, Prof. Eng'rs & Land Surveyors*, 744 S.W.2d 524 (Mo. Ct. App. 1988).

Beginning in 1976, the standard American Institute of Architects form construction contracts began to contain language purporting to sharply curtail the responsibility of design professionals.[98] The consequences of this limiting language, however, differ depending on whether the surety's claim is brought directly in tort or is based upon subrogation.

In tort, the limiting language should be of little or no effect. As discussed previously, tort duties arise from common law and public policy and extend to all those foreseeable persons who may be injured by negligent conduct. In a subrogation situation, however, the contractual limitations become much more of a problem to the surety. Because the surety now is treated the same as the owner who bargained for the contract, the surety is bound by any lawful disclaimers or limitations of liability. For example, in *Marbro v. Borough of Tinton Falls*,[99] the city's consulting engineer had a limitation of liability clause restricting its total liability to the total amount of fees it charged for services. The engineer's negligence caused the city to have to resurface an entire park. Nevertheless, the court upheld the clause and limited the engineer's liability to the amount of fees charged.[100]

Similarly, in *Wausau Paper Mills Co. v. Chas. T. Main, Inc.*,[101] the owner sued the engineering firm who provided services for modernization and expansion of a paper mill. The contract had a provision under which the engineer disclaimed all liability for delay, lost profits, lost use of equipment, increased expense of operation, or cost of purchased or replacement equipment. The engineer raised the contractual disclaimer as a defense. The court noted that, if the defense was valid, it would leave the owner without a remedy for its breach of contract claims. Notwithstanding this harsh result, the court found that the contract terms were not unconscionable, and entered summary judgment in favor of the engineer.[102]

---

98 *See Great Am. Ins. Co. v. Wade-Trim/Assocs. Inc.*, No. 90-71311, 1992 WL 1679 (6th Cir. Jan. 7, 1992).

99 688 A.2d 159 (N.J. App. 1996).

100 *Id.* at 164; *see also Peter Kiewit Sons' Co. v. Iowa Southern Utilities Co.*, 355 F. Supp. 376 (S.D. Iowa 1973) (upholding a "no damages for delay" clause even though it had the effect of barring, in its entirety, the contractor's claims for extensive delay damages).

101 789 F. Supp. 968 (W.D. Wis. 1992).

102 *Id.* at 975.

Not all disclaimers are completely effective, however. In *Watson, Watson, Rutland/Architects, Inc. v. Montgomery County Board of Education,*[103] the court held that the standard exculpatory language as cited in *Wade-Trim* did not operate as a total disclaimer of all liability. If during inspections the architect observed a known defect, he had a duty to notify the owner: "an architect cannot close his eyes on the construction site and refuse to engage in any inspection whatsoever and then disclaim liability for construction defects that even the most perfunctory monitoring would have prevented."[104]

In *U.R.S. Co. v. Gulfport-Biloxi Regional Airport Authority,*[105] the owner sued the architect for damages resulting from roof leaks. During construction, a roofing expert had warned the inspecting architect about problems with improper roof construction. Despite the warnings, the architect certified the roof as properly completed. The court quoted extensive disclaimers and limitations in the form contract. Nevertheless, the court held that an architect "is required to exercise ordinary professional skills and diligence and such duty is non-delegable."[106] Since the architect had knowledge of the defective construction, and did nothing to stop it, the court found a breach of the contract.[107]

In *City of Houma v. Municipal & Industrial Pipe Service, Inc.,*[108] the city alleged that an engineering firm did not properly inspect and monitor the project. As a result, the contractor committed extensive fraud. The engineering firm's contract had provisions requiring regular inspections but stated that the engineer would only "endeavor" to protect against defects and deficiencies.[109] The evidence indicated that the engineers who performed the inspections were "untrained, unskilled" personnel who were unable to perform any reasonable inspection method to detect performance deficiencies. Based on this evidence, the city prevailed on its claim.

Thus, while contractual disclaimers may not relieve design professionals of all responsibility on the project, such disclaimers can drastically reduce the scope of a subrogated surety's remedies. This is particularly true where disclaimers are accompanied by express limitations

---

103 559 So. 2d 168 (Ala. 1990).
104 *Id.* at 174.
105 544 So. 2d 824 (Miss. 1989).
106 *Id.* at 827.
107 *Id.*
108 884 F.2d 886 (5th Cir. 1989).
109 *Id.* at 889 n.2.

on damages. Subrogated sureties, therefore, have no greater guarantee of success than those sureties who must cast their fortunes on the shifting sea of tort law.

B. Accountants

1. Introduction

As part of establishing and maintaining a bonding line of credit for a contractor, a surety will usually request the principal to provide it with a current financial statement, audited by a certified public accountant (CPA). The CPA's participation in the preparation of these materials gives the surety comfort and the statement credibility. All too often, however, after paying a claim, the surety is left with little or no recovery potential from its principal which supplied it with a glorious financial statement bearing the signature of a CPA.

This odd turn of events almost always calls into question the work of the CPA. If the CPA was negligent, does the surety have any recourse as a third party, as the assignee, or as the subrogee of the principal?

Courts have been grappling with the issue of accountants' liability to third parties for their negligence since the seminal case on the issue, *Ultramares Corp. v. Touche,*[110] was decided in 1931. In *Ultramares*, Justice Cardozo opined that an accountant's liability should be limited to those with whom the accountant was in privity.[111] In the post-*Ultramares* era, many jurisdictions continue to adhere to the requirements of *Ultramares* and its progeny.[112] Under this rule of law, a surety is

---

110 174 N.E. 441 (N.Y. 1931).

111 *Id.* at 448.

112 *See, e.g., Colonial Bank of Ala. v. Ridley & Schweigert*, 551 So. 2d 390 (Ala. 1989); *Selden v. Burnett*, 754 P.2d 256 (Alaska 1988); *Robertson v. White*, 633 F. Supp. 954 (W.D. Ark. 1986) (applying Arkansas law); *MacNerland v. Barnes*, 199 S.E.2d 564 (Ga. Ct. App. 1973); *Idaho Bank & Trust Co. v. First Bancorp of Idaho*, 772 P.2d 720 (Idaho 1989); *Toro Co. v. Krouse, Kern & Co.*, 827 F.2d 155 (7th Cir. 1987) (applying Indiana law); *Thayer v. Hicks*, 793 P.2d 784 (Mont. 1990); *Citizens Nat'l Bank of Wisner v. Kennedy & Coe*, 441 N.W.2d 180 (Neb. 1989); *Pennine Resources, Inc. v. Dorwart Andrew & Co.*, 639 F. Supp. 1071 (E.D. Pa. 1986) (applying Pennsylvania law); *Hartford Accident & Indem. Co. v. Parente, Randolph, Orlando, Carey & Assocs.*, 642 F. Supp. 38 (M.D. Pa.

precluded from recovery against the principal's accountant on a mere negligence theory.[113] However, at least two other distinct lines of authority expanding the liability of CPAs have developed.

2. *Restatement (Second) Of Torts*

In recent years a growing number of states and federal courts applying state law have expanded the accountant's liability to third parties by adopting Section 552 of *Restatement (Second) of Torts*[114] which provides:

> (1) One who, in the course of his business, profession or employment, or in any other transaction in which he has a pecuniary interest, supplies false information for the guidance of others in their business transactions, is subject to liability for pecuniary loss caused to them by their justifiable reliance upon the information, if he fails to exercise reasonable care or competence in obtaining or communicating the information.
>
> (2) Except as stated in Subsection (3), the liability stated in Subsection (1) is limited to loss suffered

---

1985) (applying Pennsylvania law).

113 *Dworman v. Lee*, 441 N.Y.S.2d 90 (N.Y. App. Div. 1981), *aff'd*, 438 N.E.2d 103 (N.Y. 1982) (holding that a surety could not recover under the *Ultramares* rule).

114 *See, e.g.*, *Boykin v. Arthur Andersen & Co.*, 639 So. 2d 504 (Ala. 1994); *First Fla. Bank, N.A. v. Max Mitchell & Co.*, 558 So. 2d 9 (Fla. 1990); *Badische Corp. v. Caylor*, 356 S.E.2d 198 (Ga. 1987); *Merit Ins. Co. v. Colao*, 603 F.2d 654 (7th Cir. 1979) (applying Illinois law); *Seedkem, Inc. v. Safranek*, 466 F. Supp. 340 (D. Neb. 1979) (applying Indiana and Nebraska law); *Pahre v. Auditor of State*, 422 N.W.2d 178 (Iowa 1988); *Ryan v. Kanne*, 170 N.W.2d 395 (Iowa 1969); *First Nat'l Bank of Commerce v. Monco Agency, Inc.*, 911 F.2d 1053 (5th Cir. 1990)(applying Louisiana law); *Bonhiver v. Graff*, 248 N.W.2d 291 (Minn. 1976); *Aluma Kraft Mfg. Co. v. Elmer Fox & Co.*, 493 S.W.2d 378 (Mo. Ct. App. 1973); *Spherex, Inc. v. Alexander Grant & Co.*, 451 A.2d 1308 (N.H. 1982); *First Interstate Bank of Gallup v. Foutz*, 764 P.2d 1307 (N.M. 1988); *Rusch Factors, Inc. v. Levin*, 284 F. Supp. 85 (D.R.I. 1968) (applying Rhode Island law); *ML-Lee Acquisition Fund, L.P. v. Deloitte & Touche*, 489 S.E.2d 470 (S.C. 1997); *Bethlehem Steel Corp. v. Ernst & Whinney*, 822 S.W.2d 592 (Tenn. 1991); *Federal Land Bank Ass'n v. Sloane*, 825 S.W.2d 439 (Tex. 1991); *First Nat'l. Bank of Bluefield v. Crawford*, 386 S.E.2d 310 (W.Va. 1989).

(a) by the person or one of a limited group of persons for whose benefit and guidance he intends to supply the information or knows that the recipient intends to supply it; and

(b) through reliance upon it in a transaction that he intends the information to influence or knows that the recipient so intends or in a substantially similar transaction.

(3) The liability of one who is under a public duty to give the information extends to loss suffered by any of the class of persons for whose benefit the duty is created, in any of the transactions in which it is intended to protect them.[115]

The *Restatement* approach seems to be a compromise between the broad protection accountants enjoy from third party liability afforded by *Ultramares* and the almost unrestricted liability to which accountants would be exposed under a traditional tort standard. The *Restatement* appears to include an actual knowledge element, thus preserving some of the *Ultramares* doctrine, while expanding liability to any member of a class of persons.

A recent Florida decision involving a surety is an excellent example of the application of the Section 552 to a common factual scenario. In *Amwest Surety Insurance Co. v. Ernst & Young*,[116] a surety brought an action seeking damages for losses it suffered as a result of issuing payment and performance bonds to its principal. The surety claimed that the principal's auditor had failed to disclose certain liabilities of the principal to the Internal Revenue Service.

Prior to issuing the financial statements, the principal had sent its accountants a letter stating that the financial statements would be used for bonding requirements, banking requirements, suppliers' requirements, and pre-qualification requirements.[117] In addition, the client advised the accountant via correspondence that the "bonding underwriters are requiring a current audited statement so that a line of bonding can be issued with reasonable rates and adequate limits."[118]

The accountants moved for summary judgment, arguing that under the *Restatement* position, the surety had failed to establish that it was one of a limited group of persons for whose benefit the accountant supplied the

115 RESTATEMENT (SECOND) OF TORTS § 552 (1976).

116 677 So. 2d 409 (Fla. Dist. Ct. App. 1996).

117 *Id.* at 410.

118 *Id.* (emphasis omitted).

information or knew that its client intended to supply the information.[119] The trial court granted the accountants' motion and entered judgment in favor of the accountants. The surety appealed.

On appeal, the accountants claimed that the client/principal identified such a multitude of groups that to hold the accountants liable would be tantamount to making them liable to a limitless group of foreseeable users, which is contrary to the provisions of Section 552 of the *Restatement*.[120] The appellate court disagreed and held that liability was not determined by analyzing the size of the class of users; rather, it was dependent only upon the accountants having notice of the intended use of the work.[121]

Amwest reaffirms an obvious conclusion one formulates when analyzing the Section 552 of the *Restatement (Second) of Torts*: liability is not dependent upon the party's status (*i.e.*, surety, shareholder, privity, or third party); instead, it is dependent upon the accountant having notice of the intended use of the work. It is also important to note that the accountant's liability in *Amwest* was due to actual notice of the use of the audited financial statements. The comments and illustrations to the *Restatement* reveal that constructive notice is insufficient to attach liability.[122]

---

119 *Id.*

120 *Id.* at 411.

121 *Id.* at 412.

122 Illustration 10 to Section 552 of the RESTATEMENT (SECOND) OF TORTS provides:

> A, an independent public accountant, is retained by B company to conduct an annual audit of the customary scope for the corporation and to furnish his opinion on the corporation's financial statements. A is not informed of any intended use of the financial statements; but A knows that the financial statements, accompanied by an auditor's opinion, are customarily used in a wide variety of financial transactions by the corporation and that they may be relied upon by lenders, investors, shareholders, creditors, purchasers and the like, in numerous possible kinds of transactions. In fact B Company uses the financial statements and accompanying auditor's opinion to obtain a loan from X Bank. Because of A's negligence, he issues an unqualifiedly favorable opinion upon a balance sheet that materially misstates the financial position of B Company, and through reliance upon it X Bank suffers pecuniary loss. A is not liable to X Bank.

RESTATEMENT (SECOND) OF TORTS § 552, ill. 10 (1976).

3. *Standard Negligence*

Another line of authority has rejected both *Ultramares* and the *Restatement* position by adopting traditional tort concepts and extending liability to any person the accountant reasonably could have foreseen would rely upon his work.[123] Under this approach, an insurer that issues a fidelity bond to a bank has been held to fall within the purview of foreseeable users of an audit.[124]

Many accounting firms, particularly the larger, sophisticated international firms, claim in their brochures to have particular expertise in auditing contracting firms. As "experts" they are well aware that contractors are required to obtain payment and performance bonds, and that their financial statements are used to establish and maintain the contractors' bonding line of credit. In such circumstances, the surety should clearly constitute a foreseeable user of the CPA's work under the *Restatement* view.

4. *Claims Against The Accountant Based Upon The Surety's Rights As Assignee And Subrogee*

Routinely the performing surety will obtain an assignment of any claims the insured has against third parties. While many jurisdictions prohibit the assignment of a personal chose in action for legal malpractice[125] based on the uniquely close and personal relationship between the attorney-client, the relationship of the accountant to his client appears to lack the same close and personal nature.

In *United States v. Arthur Young & Co.*,[126] the United States Supreme Court refused to recognize a CPA work-product privilege. The Court distinguished the attorney-client relationship from that of the accountant, finding that unlike the attorney that serves as the advisor and advocate of

---

123 *Touche Ross & Co. v. Commercial Union Ins. Co.*, 514 So. 2d 315 (Miss. 1987); *H. Rosenblum, Inc. v. Alder*, 461 A.2d 138 (N.J. 1983) (superseded by statute); *Citizens State Bank v. Timm, Schmidt, & Co.*, 335 N.W.2d 361 (Wis. 1983).

124 *Touche Ross & Co.*, 514 So. 2d at 323.

125 *See* Francis M. Dougherty, Annotation, *Assignability of Claim for Legal Malpractice*, 40 A.L.R. 4TH 684 (1986).

126 465 U.S. 805 (1984).

its client, the CPA has a public responsibility because it assumes the role of a disinterested party certifying the financial condition of an entity.[127]

Addressing the assignment issue directly, the court in *First Community Bank & Trust v. Kelly, Hardesty, Smith & Co.*[128] the court held that an accounting malpractice case was assignable to bank directors that relied upon the CPA's financial statements, and the existence of a statutorily created accountant-client privilege did not create the close, personal relationship between the accountant and the client that is the foothold of the attorney-client relationship. The court found that the attorney-client privilege is derived not only from the duty of confidentially owed to the client, but also from the attorneys' duty to be a zealous advocate.[129] The accountant-client privilege, on the other hand, was premised only upon the duty of confidentiality and its only value was to the client.

In the few reported cases involving actions against accountants as subrogees of the injured party, the courts have made no mention of the policy issues that preclude the assignment of legal malpractice cases.[130] Accordingly, a subrogation claim bolstered by Section 28 of the *Restatement of Suretyship*[131] should be the least problematic cause of action the surety may present.[132]

---

127 *Id.* at 817-18.

128 663 N.E.2d 218 (Ind. Ct. App. 1996).

129 *Id.* at 221-22.

130 *See Dantzler Lumber & Export Co. v. Columbia Cas. Co.*, 156 So. 116 (Fla. 1934); *Western Sur. Co. v. Loy*, 594 P.2d 257 (Kan. Ct. App. 1979); *Federal Ins. Co. v. Arthur Anderson & Co.*, 552 N.E.2d 870 (N.Y. 1990).

131 Section 28 provides:

> (1) To the extent that the secondary obligor is subrogated to the rights of the obligee, the secondary obligor may enforce, for its benefit, the rights of the obligee as though the underlying obligation had not been satisfied:
>
> ***
>
> (d) against any other persons whose conduct has made them liable to the obligee with respect to the default on the underlying obligation.

RESTATEMENT OF SURETYSHIP, *supra* note 1, § 28(1).

132 Support for the position that when a surety stands in the shoes of the wronged party that privity or lack thereof is not an issue can be found in

A defense that has been raised to a subrogation action against accountants is that such claims are barred by the superior equities doctrine. However, in *Federal Insurance v. Arthur Anderson & Co.,*[133] the court rejected this argument, concluding that the superior equities doctrine deals with the equitable allocation of a loss as between two innocent parties, and where there has been an assertion of negligence against the CPAs, the doctrine is inapplicable.[134]

IV. Lender Liability

In considering potential third parties the surety may pursue, there are some persuasive theories under which the surety may assert a claim against the lender. As the overriding concern for the surety is to recover the amount expended, it will generally look to the "deepest pocket" for recovery. In most instances, that will be the lender. The typical scenario may involve the financial failure of the owner which causes the lender to cut off construction funds or the lender to assume control over the construction funds. Additionally, the lender may have somehow misrepresented the financial condition of the owner to the detriment of the surety. In certain cases the claim may also be valid against a governmental lender. Each instance assumes that the lender has somehow acted (or failed to act) to the detriment of the surety, or that the lender is otherwise obligated to the surety under contractual or equitable principles.

A. Privity Not An Issue

As the theories of recovery by third parties to the lending agreement are based on equitable principles and implied contract duties, they are not barred by traditional contractual privity concepts. It is rather an assessment of fairness that supports the surety's position. The engine which drives the application of these principles of fairness is that the lender exercises control over the project, control in drafting the loan and construction documents, and control over advancing funds, both during the course of construction and during completion if the owner/borrower

---

Western Sur. Co., 594 P.2d at 257.

133 552 N.E.2d 870 (N.Y. 1990).

134 *Id.* at 874.

defaults on the construction loan agreement. This places the lender in a superior bargaining position which also carries with it some risks.

B. Obligatory vs. Optional Advances

Periodic payments made under a loan agreement by the lender are either obligatory or optional advances, depending on the terms of the agreement.[135] The distinction brings with it a balance of risks and benefits to the lender. For instance, the lender may face liability where it structures the loan agreement by designating the advances made by the lender as "obligatory." The obligatory payment is conditioned upon the completion of predetermined stages of construction,[136] contrasted with the optional payment which is conditioned upon the lender's subjective satisfaction as to the progress of the construction or depends upon other conditions separate from the project.[137] The obligatory structure protects the lender inasmuch as it preserves the viability of the project, but removes from the lender the discretion it would otherwise enjoy. A surety will have a stronger position against a lender who has made obligatory advances because the courts seem to find that this demonstrates the intention that the loan funds benefit the job by giving less control to the lender. The theories espoused throughout this section will presuppose a factual scenario in which the lender is constrained by obligatory advances and that it has either failed or refused to make those advances.

C. Theories Of Lender Liability To The Surety

The surety's claim against the lender generally arises when the surety has had to perform the principal's contract with the owner. The situation can develop as a domino effect: the lender's refusal to advance funds under the loan agreement with the owner causes the owner's inability to pay the principal which, in turn, causes the principal's inability to perform and requires the surety to step in for the principal. The following discussion addresses some of the theories under which the surety can seek redress and pursue the lender for recovery.

---

135 *See* Craig D. Tindall, *The Obligatory Advance Rule in the Construction Lending Context*, 12 CONSTR. LAW. (Jan. 1992).

136 Leidner, *An Overview Of Mechanics Lien & Future Advance Mortgages*, 3 PRAC. REAL EST. LAW. 39, 42-44 (May 1987).

137 *Id.* at 42-43.

1. Third Party Beneficiary Theory

The intent of both the lender and owner that third parties benefit from some aspect of the loan transaction is the basis for third party beneficiary approach to lender liability. The Section 302 of the *Restatement (Second) of Contracts* confers third party beneficiary status where it is appropriate and either (1) the performance of the promise will satisfy an obligation of the promisee to the third party; or (2) the circumstances indicate that the promisee intended to give the third party the benefit of the promise.

a. The Construction Loan Agreement

While there is certainly case law to suggest that third party beneficiary status will not be liberally applied,[138] several cases have interpreted the loan agreements to confer third party beneficiary status to those who were not parties to the agreement.

The courts essentially look for an indication in the loan agreement that lender and the owner intended the funds to be used to pay the would-be beneficiary. That intention was found, and third party beneficiary rights were created in the contractor, in *Ralph C. Sutro Co. v. Paramount Plastering, Inc.*[139] and *Whiting-Mead Co. v. West Coast Bond & Mortgage Co.*,[140] where the construction loan agreements provided that labor and materials had to be fully paid by owner as a condition precedent to the

---

138 *See* Curtis R. Reitz, *Construction Lender's Disability to Contractors, Subcontractors, and Materialmen*, 130 U. PA. L. REV. 416, 423-428 (1981); *see also Burns v. Washington Savings*, 171 So. 2d 322 (Miss. 1965) (where contracts did not expressly include building contractor, contractor was not a third party beneficiary and could not sue lender for failure to carry out agreement with owner); *Volpe Constr. Co., Inc. v. First Nat'l Bank of Boston*, 567 N.E.2d 1244 (Mass. App. Ct. 1991) (where a loan agreement provides that nothing contained in the agreement will confer any right upon or benefit in any other party other than the borrower and the lender, the courts should deny the third party any direct remedy on the contract); *compare Twin City Constr. Co. of Fargo, N.D. v. ITT Indus. Credit Co.*, 358 N.W.2d 716 (Minn. Ct. App. 1984) (contractor entitled to recover against lender as third party beneficiary regardless of disclaimer in loan agreement).

139 31 Cal. Rptr. 174 (Cal. Dist. Ct. App. 1963).

140 152 P.2d 629 (Cal. 1944).

lender releasing final payment. Also, a provision in the construction loan agreement requiring construction loan funds to be used for only one project was used in *In re Gebco Investment Corp.*[141] to show third party beneficiary rights in the contractor. In *Kreimer v. Second Federal Savings & Loan Association of Pittsburgh,*[142] a clause requiring that complete releases of liens be provided by the contractor before performing work on the project was held to create third party beneficiary status.[143]

b. Direct Statements

A lender's statement directly to a contractor or subcontractor has served to create third party beneficiary rights in that entity. In *Jim Walter Corp. v. Laperouse,*[144] a lender sent a commitment letter directly to the contractor to guarantee the credit of the owner and confirm the lender's commitment to finance all of the contractor's construction. That action on the part of the lender was held to create third party beneficiary rights in the contractor later when the owner could not make full payment.[145]

2. Misrepresentation As To Financial Condition

Another theory of recovery for the surety lies in fraud or misrepresentation. This form of "direct statement" case arises where the lender misrepresents facts related to project funding or to the principal's financial status. If a surety suffers loss in reliance on the lender's inaccurate or misleading statements and can show that it would not have suffered a loss if the true facts had been known (not just that it would not have issued a bond), it can recover against the lender on theories of fraud or misrepresentation.

---

141 641 F.2d 143 (3d Cir. 1981).

142 176 A.2d 132 (Pa. Super. Ct. 1961)..

143 *Id.*

144 196 So. 2d 539 (La. Ct. App. 1967).

145 *Id.*; *see also Rudick Co., Inc. v. First Nat'l Bank of Lafayette*, 643 So. 2d 846 (La. Ct. App. 1994) (liability attached to bank by virtue of commitment letter issued to contractor); *A.F. Blair Co., Inc. v. Haydel*, 504 So. 2d 1044 (La. Ct. App 1987) (construction contractor was third party beneficiary of contract between bank and debtor for construction loan, finding that contractor had alleged facts sufficient to establish claim in detrimental reliance).

In *Associated Indemnity Corp. v. Del Guzzo*,[146] the court held the lender liable for intentionally misleading parties to the project, including the surety. The court recognized that parties must exercise good faith in their dealings and that any concealment of facts which induces a party to enter a contract which it would not otherwise enter avoids the contract. In *Associated Indemnity*, the court held the bank liable on the ground that the bank, through misleading statements concerning the contractor's financial standing, induced the surety to execute a performance bond. Similarly, in *Marine Midland Bank v. Smith*,[147] the court held that the lender must not be inaccurate or incomplete in any factual statement upon which the surety may rely to its detriment in issuing a bond.[148]

Not all jurisdictions place a duty to provide accurate financial data on those in such business, however. This issue was addressed by the Supreme Court of Louisiana in *Greene v. Gulf Coast Bank*.[149] There the court reversed the appellate court's holding that the bank had committed fraud when it failed to disclose to the guarantor the poor credit history of the borrower. The court noted that the guarantor was not an unsophisticated or inexperienced investor who had to rely on the bank to advise him of the risks of his investment. Finding that the guarantor was in a position to know as much or more than the bank about the financial condition of the company of which he was a major stockholder and officer, the court held that the bank did not owe a duty to disclose under those circumstances.

In *Popp v. Dyslin*,[150] the court held that a bank does not have a legal duty to a third party creditor for negligently investigating the financial

---

146 81 P.2d 516 (Wash. 1938).

147 482 F. Supp. 1279 (S.D.N.Y. 1979).

148 *Id.* at 1286-87; *see also Cooper v. Wesco Builders, Inc.*, 253 P.2d 226 (Idaho 1953) (false representations by officers of bank would constitute fraud and deceit, and bank would be liable for the tort of its managing officers if made of behalf of bank); *Loyola Fed. Sav. & Loan Ass'n v. Galanes*, 365 A.2d 580 (Md. 1976); *First-Citizens Bank & Trust Co. v. Akelaitis*, 214 S.E.2d 281 (N.C. Ct. App. 1975) (where there is a duty to speak, fraud can be practiced by silence as well as positive misrepresentation; guarantor sufficiently stated affirmative defense on theory that bank had breached duty to disclose to defendant such facts as were material to risk assumed by defendant).

149 593 So. 2d 630 (La. 1992).

150 500 N.E.2d 1039 (Ill. App. Ct. 1986).

qualifications of a borrower and found that a bank's only obligation flows between the bank and its borrower.[151] While the text of the holding seems to contradict the broader argument in favor of lender liability toward sureties, a persuasive argument can be made that *Popp* is distinguishable. The court in *Popp* expressed concern that the imposition of liability upon the bank in favor of any creditor that receives information regarding the credit-worthiness of its borrower would "create a situation of overwhelming potential liability."[152] This concern is wholly inapplicable to the case where the bank specifically provides information to the surety (or contractor) with respect to a specifically contemplated project to induce the surety to execute a bond. The argument rests upon the idea that the bank owes a duty not to all creditors at large, but instead to the individual creditors that the lender knows will rely on its representations.

3. *Negligent Disbursement*

In jurisdictions that do not restrict recovery to those in privity, courts have imposed a duty on lenders to properly disburse funds.[153] This has most often been the case where the lender, as opposed to a third party agent such as a title company or escrow agent, is making the disbursements after inspections. In the leading case of *Commercial Standard Insurance Co. v. Bank of America*,[154] a lender's liability to properly disburse funds was identified as arising in tort. The court established a test for imposing independent tort liability upon a third party in favor of a surety. It noted that even if a bank did not intend to affect the contractor's surety when it undertook to make the disbursements, the lender was aware of the surety's bonds and it was foreseeable that negligent disbursements could cause a loss to the surety. Thus, the surety was held to have an independent tort action against the bank.[155]

---

151 *Id.* at 1043.

152 *Id.*

153 *See, e.g., Cook v. Citizens Sav. & Loan Ass'n*, 346 So. 2d 370 (Miss. 1977) (lender paying out construction funds must discharge its duty using reasonable diligence to see that the funds are actually used in payment of materials or other costs of construction).

154 129 Cal. Rptr. 91 (Cal. Ct. App. 1976).

155 *Id.*; *see also Community Nat'l Bank v. Fidelity & Deposit Co. of Md.*, 563 F.2d 1319 (9th Cir. 1977) (following *Commercial Standard* and holding that the policy of preventing future harm is more clearly served when the

4. Rights Against Lender Obtained Through Subrogation Under Sections 27 And 28 Of The Restatement Of Suretyship

Once a surety steps in to satisfy the total obligation to third parties under a payment bond, Section 27 of the *Restatement of Suretyship* provides that the surety is subrogated to all rights of the obligee, to the extent the payment satisfies the obligation.[156] As the obligee in the cases discussed here is usually the owner, Sections 27 and 28 of the *Restatement of Suretyship* give the surety, as subrogee, a claim against "any other persons whose conduct has made them liable to the obligee with respect to the default on the underlying obligation."[157] Thus, if the surety on a payment bond must respond for the contractor and pay subcontractors or material suppliers because the lender has wrongfully failed or refused to disburse construction loan funds to the owner, these sections give the surety a claim against the lender. The rule in the *Restatement of Suretyship* is supported by general case law of equitable subrogation and does not depend upon contractual privity.[158]

---

third party conduct is intentional); *Gartner v. Atlantic Nat'l Bank of Jacksonville*, 350 So. 2d 495 (Fla. Dist. Ct. App. 1977) (issues of fact existed on the question of the bank's alleged negligence in the management of the construction loan fund and whether such conduct materially affected the guarantor's liability under a guaranty agreement).

156 RESTATEMENT OF SURETYSHIP, *supra* note 1, § 27.

157 RESTATEMENT OF SURETYSHIP, *supra* note 1, § 28(1)(d).

158 *Pearlman v. Reliance Ins. Co*, 371 U.S. 132, 136 n.12 (1962) (the right of subrogation is not founded on contract, but is a creature of equity, enforceable solely for the purpose of accomplishing the ends of substantial justice); *Balboa Ins. Co. v. Bank of Boston Conn.*, 702 F. Supp. 34 (D. Conn. 1988) (the doctrine of equitable subrogation is fully applicable where a party seeks reimbursement for a debt paid under the compulsion of a payment bond).

CHAPTER 6

DECIDING TO LITIGATE: THE SURETY'S RIGHTS AGAINST PROPERTY AND LIABILITY INSURERS OF THE OBLIGEE, PRINCIPAL, AND SUBCONTRACTORS

Patrick J. O'Connor, Jr.

I. Introduction

Sureties and principals have an interest in insurance coverage for a number of reasons. If a loss or default for which the obligee has called upon the surety to pay or perform is covered also by insurance, then the surety may look to its principal's rights under the policy for reimbursement. Similarly, if a principal has a right to a defense for an owner's claim, then the surety benefits from such defense coverage. Sureties are well advised to analyze all performance bond claims for purposes of determining whether a liability insurer has either a defense or indemnity obligation with respect to the claim. Oftentimes the surety can convince the liability carrier to participate in defense costs, and still permit the surety to exercise a significant amount of control over the litigation.

Insurance coverage, generally in the form of property coverage, coupled with standard contractual language involving waivers of subrogation rights up to the extent of insurance coverage, can provide the surety with a valuable defense from owner claims that are covered by such insurance.[1]

Insurance coverage, then, can provide the surety not only with an additional pool of assets from which to lessen any loss exposure but also provide a defense to certain claims asserted against its bond.

II.
Commercial General Liability Coverage

A. Introduction

Nearly every party involved in a construction project will carry a CGL policy. This is the most common liability policy issued today. The policy protects insureds against loss arising out of "bodily injury" or "property damage" caused by an "occurrence." Most of the issues that arise with respect to the scope of this coverage in a construction setting involve policies issued to contractors and subcontractors. A number of issues of interest to sureties and principals recur with some frequency. They include:

1. To what extent does a contractor have coverage for losses due to defective work—either performed by itself or by one of its subcontractors?

2. Does a contractor have coverage for losses incurred as a result of agreeing to indemnify another (*e.g.*, an owner or design professional) from loss?

---

1 This may be the case even though the owner either failed to obtain the proper coverage or elected not to secure the benefits of existing coverage. *See, e.g., Richmond v. Grabowski*, 781 P.2d 192 (Colo. Ct. App. 1989) (holding that the party who agrees to procure the insurance and fails to do so assumes the position of the insurer and, thus, the risk of loss.); *Christiansen v. Holiday Rent-A-Car*, 845 P.2d 1316, 1321-22 (Utah Ct. App. 1992); *Sutton v. First Nat'l Bank of Crossville*, 620 S.W.2d 526 (Tenn. Ct. App. 1981) (upholding discharge of debt due to bank's failure to procure credit insurance); 44 C.J.S. *Insurance* § 225 (1955) ("The measure of damages for breach of the contract to procure insurance is the amount which might have been recovered under such insurance.").

3. What is the extent of coverage afforded to owners and design professionals when listed as additional insureds under a contractor's CGL policy?

4. How is coverage effected when loss results from a latent defect that deteriorates over time?

5. How does the common industry practice of waiving subrogation rights in construction contracts affect an owner's claims against a surety and its principal?

An examination of these issues touches upon many of the operative portions of the CGL policy relevant to construction practice.

B. Coverage For Losses Arising Out Of Defective Work

1. The "Occurrence" Requirement

The first issue that one often confronts when analyzing insurance coverage for defective work is the CGL policy's requirement that losses arise as a result of an "occurrence." The 1986 CGL policy form defines a "occurrence" as: "[A]n accident, including continuous or repeated exposure to substantially the same general harmful conditions."[2] Is faulty work an accident?[3] An accident has been defined as "an

---

2 *See, e.g.*, Susan J. Miller & Philip Lefebvre, I *Miller's Standard Insurance Policies Annotated*, at 419 (4th ed. 1995).

3 While policy definitions such as "occurrence" and "property damage" tend to work to limit coverage, their application to a particular factual situation is often treated different than whether a particular exclusion applies to bar coverage. "The burden is on the insurer to establish the applicability of an exclusion." *O'Shaughnessy v. Smuckler Corp.*, 543 N.W.2d 99, 102 (Minn. Ct. App. 1996); *see also Johnson v. Insurance Co. of N. Am.*, 350 S.E.2d 616, 619 (Va. 1986) ("where an insured has shown that his loss occurred while an insurance policy was enforced, but the insurer relies on exclusionary language in the policy as a defense, the burden is upon the insurer to prove that the exclusion applies to the facts of the case."); *White v. State Farm Mut. Auto Ins. Co.*, 157 S.E.2d 925 (Va. 1967) (exclusions are to be strictly interpreted against the insurer); *see, e.g.*, *Hennings v. State Farm Fire & Cas. Co.*, 438 N.W.2d 680, 683

unforeseen occurrence, usually of an untoward or disastrous character or an undersigned, sudden or unexpected event of an inflictive or unfortunate character."[4] This analysis, often engaged in by the courts, quickly leads to unproductive results given its circularity. On the one hand, if an "occurrence" is given an expansive meaning, then little coverage is afforded. If any act of human activity containing a volitional element is considered to be something other than "accidental," then few situations resulting in loss would trigger coverage. Take for example the following hypothetical: a construction worker places a piece of plywood over an opening on a construction site, which happens to be too thin to support the weight of a worker who falls and injures himself. If one focuses upon the *act* rather then on the *consequences* of the act, then no coverage results. Yet this common situation routinely results in coverage.

a. Intention

A finding of no coverage simply because "poor workmanship" is not an accident is an insufficient analysis of the issue.[5] In reality, the "occurrence" element should not pose a significant impediment to coverage if all other policy requirements are met. The reason for this is two-fold. First, in order for the policy to provide any meaningful coverage, the focus of the inquiry regarding what should be deemed "accidental" must be upon the injury resulting from the act rather then

---

(Minn. Ct. App. 1989); *Reinsurance Ass'n of Minn. v. Patch*, 383 N.W.2d 708, 711 (Minn. Ct. App. 1986). The contra-insurer rule does not apply with such strict application in interpreting the assuring agreement. *See, e.g.*, *Stanford Ranch, Inc. v. Maryland Cas. Co.*, 89 F.3d 618, 627 (9th Cir. 1996).

4 *Indiana Ins. Co. v. Hydra Corp.*, 615 N.E.2d 70, 73 (Ill. App. 1993); *see also Aetna Cas. & Sur. Co. v. Freyer*, 411 N.E.2d 1157 (Ill. App. 1980).

5 *See, e.g.*, *Oak Crest Constr. Co. v. Austin Mut. Ins. Co.*, 905 P.2d 848 (Or. Ct. App. 1995); *United States Fidelity & Guar. Corp. v. Advance Roofing & Supply Co.*, 788 P.2d 1227 (Ariz. Ct. App. 1989) (no coverage existed for contractor that installed negligent roofs as poor workmanship did not amount to an occurrence and to hold otherwise would be tantamount to converting the CGL policy into a performance bond.).

the act itself.[6] Second, the question of whether the resulting loss was unexpected or unforeseeable is generally viewed subjectively, *i.e.,* from the insured's standpoint. The "general trend appears to require an inquiry into the actor's subjective intent to cause injury."[7] Even where the act seems foolhardy, it is not unusual for the courts to require an inquiry into the actor's subjective intent.[8]

Some courts have concluded that the term "occurrence" is ambiguous and, accordingly, do not find the term a limitation on coverage. For example, an Illinois Court of Appeals concluded that the use of the word "occurrence" in insurance policies broadens coverage and eliminates the need to find the exact cause of damages as long as they are neither intended nor expected by the insured.[9] Along similar lines, a Louisiana court in *Western World v. Paradise Pools & Spas, Inc.*[10] concluded that defective construction resulting in cracks in a

---

6 *See, e.g.*, Voorhees v. *Preferred Mut. Ins. Co.*, 607 A.2d 1255, 1263 (N.J. 1992); *CTC Dev. Corp., Inc. v. State Farm Fire & Cas. Co.*, 1977 WL 525237 (Fla. App. 1 Dist., Aug. 26, 1977); *Economy Lumber Co. of Oakland, Inc. v. Insurance Co. of N. Am.*, 204 Cal. Rptr. 135 (1984) (damage caused by the application of defective siding was an occurrence, however, if more siding applied after knowing of the damage it caused then damage is not unforeseeable and no "occurrence.").

7 *Voorhees v. Preferred Mut. Ins. Co.*, 607 A.2d 1255, 1264 (N.J. 1992).

8 *See, e.g.*, *Frankenmuth Mut. Ins. Co. v. Masters*, 570 N.W.2d 134 (Mich. Ct. App. 1997) (as a matter of law, one could not conclude no "occurrence" existed where an insured intentionally started fire, but claimed that it only intended to cause damage to its own structure and not to adjoining buildings. The standpoint by which one determines whether the resulting damage was unforeseen or unexpected is a subjective one, analyzing what the insured anticipated.) *Voorhees v. Preferred Mut. Ins. Co.*, 607 A.2d 1255, 1264 (N.J. 1992) ("even when the actions in question seem foolhardy and reckless, the courts have mandated an inquiry into the actor's subjective intent to cause injury); *but see, R.N. Thompson & Assocs. v. Monroe Guar. Ins. Co.*, 686 N.E.2d 160, 165 (Ind. Ct. App. 1997) (rotting roofs "natural and ordinary" consequence of the defective work and thus "occurrence element not met.).

9 *Bituminous Cas. Corp. v. Gust K. Newberg Constr. Co.*, 578 N.E.2d 1003, 1009 (Ill. Ct. App. 1991).

10 633 So. 2d 790 (La. Ct. App. 1994).

swimming pool constituted an "occurrence" because the definition of the term was ambiguous.[11]

A question sometimes arises as to whether coverage may exist where the insured claims that it did not intend or foresee the resulting damage but that there is little dispute that the conduct was reckless. The Eighth Circuit Court of Appeals recently discussed this issue in *Koch Engineering Co., Inc. v. Gibraltar Casualty Co. Inc.*,[12] where an insured was hired to design and supply a chemical company with certain equipment which proved defective resulting in a $7,000,000 judgment against the insured based upon a breach of warranty claim.[13] The problem with the equipment was that it permitted a rust-like substance known as mill scale to develop. The first issue examined by the court was whether the plugging of the distribution system with this mill scale constituted an occurrence. In the underlying action the district court found that the insured had been "reckless with respect to its design and supervision of the filtration system."[14] As a result, the trial court concluded that the plugging was not unexpected. The Eighth Circuit, while not taking issue with the court's finding of reckless behavior, determined that the ramifications of this finding did not, however, preclude the conclusion that an occurrence existed for purposes of coverage. In applying Missouri law, the court stated "regardless of the reckless character of behavior, the insurer must show that Koch [the insured] intended the results of its actions."[15] In this case, the project was highly technical and complex, and there was simply no evidence in the record that the insured actually intended that the equipment would be plugged with debris. Therefore, the court's finding of recklessness did not alone support the inference of intent.[16]

---

11 *Id.* at 794.

12 78 F.3d 1291 (8th Cir. 1996).

13 *Id.* at 1293.

14 *Id.* at 1294.

15 *Id.* at 1294.

16 *See also American Family Mut. Ins. Co. v. Pacchetti*, 808 S.W.2d 369, 371 (Mo. 1991) ("although recklessness is sometimes a legal equivalent of intention, . . . [i]t remains for the insurer to show that this particular insured expected or intended the result which occurred."). Obviously, in the context of construction activities, few acts are so egregious so as to render irrelevant motive or intent. *See, e.g., Shell Oil Co. v. Winterthur*

b. Continuous Or Repeated Exposures

A number of cases have focused upon that part of the definition of "occurrence" that pertains to the "continuous or repeated exposure to conditions which result in property damage." In discussing this aspect of the definition, the Supreme Court of Illinois in *United States Fidelity & Guarantee Co. v. Wilkin Insulation Co.*,[17] found a duty to defend based upon claims that the insured's incorporation of asbestos-containing materials into buildings met the requirement of an "occurrence:"

> The insurance policy's definition of "accident" [referring to the standard policy's definition of "occurrence" as an "accident"], however, includes the "continuous or repeated exposure to conditions." In the instant action, the underlying complaints all allege that the asbestos-containing products continuously release toxic asbestos fibers into the air. This continuous release of asbestos fibers constitutes a harmful condition. By virtue of the "continuous or repeated exposure to the condition" of asbestos fiber release, the buildings and their contents become contaminated. It is this continuous exposure of the buildings and their contents to release asbestos fibers which constitutes the "accident," as defined by the insurance policies.[18]

Swiss Ins. Co., 15 Cal. Rptr. 2d 815 (Cal. Ct. App. 1993) (in the context of an environmental claim a willful act under Section 533 [California statute that provides insurers are not liable for losses caused by the willful acts of their insureds] must mean an act deliberately done for the express purpose of causing damage or intentionally performed with knowledge that damage is highly probably or substantially certain to result.).

17 578 N.E.2d 926 (Ill. 1991).

18 *Id.* at 932. The court goes on to conclude that the accident of the release of asbestos fibers into the building results in "property damage" thereby satisfying the requirement that an occurrence result in property damage. The insurer's arguments that there was no duty to defend because the insured "knew or should have known" of the propensity of the product to release toxic asbestos fibers was rejected on the grounds that the pleadings made no such allegation that the insured expected or intended to contaminate the buildings and the contents therein. *Id.* The court went on to discuss a variety of exclusions including the pollution exclusion, sistership exclusion, and the various business risk exclusions and found that they did not exculpate the insurer.

c. The Liability Policy As Performance Bond

A number of courts refuse to find poor workmanship to be "accidental" within the meaning of an insurance policy because to do so would be tantamount to turning the policy into a bond.[19] Light is seldom shed on a subject by referring to another even more misunderstood subject. This has traditionally been the problem with attempting to determine the scope of coverage under a CGL policy by referring to another instrument, such as a performance bond. Many courts mistakenly believe that if a contractor had desired coverage for its poor workmanship then the proper instrument is a performance bond. This reasoning, however, arises from a fundamental misunderstanding as to the nature of suretyship. Performance bond sureties do not provide contractors with insurance. If a surety performs under a performance bond it is entitled to reimbursement from the contractor. This is not insurance. In other words, contractors do not secure performance bonds in order to protect themselves, but rather owners require contractors to secure performance bonds to protect against the credit risk that the contractor will not be able to perform its contractual obligations.

Once the nature of performance bonds and general liability policies is understood it is possible to see that a certain class of losses may trigger contractual obligations of both the surety and the insurer. In fact, it is not unheard of that a performance bond surety will commence suit seeking to recover under its principal's (contractor's) general liability policy.[20] The interplay between a performance bond and CGL policy

---

19 *See, e.g.*, *Reliance Ins. Co. v. Mogavero*, 640 F. Supp. 84, 85 (D. Md. 1986) ("however, this case turns upon one overriding principle: that Reliance issued a general liability policy, not a performance bond, . . ."); *Solcar Equip. Leasing Corp. v. Pennsylvania Mfrs. Ass'n Ins. Co.*, 606 A.2d 522 (Pa. Super. Ct. 1992) ("the insurance contract at issue is not a performance bond or any type of construction malpractice insurance."); *Bor-Son Bldg. Corp. v. Employers Commercial Union Ins. Co. of Am.*, 323 N.W.2d 58 (Minn. 1982); *Knutson Constr. Co. v. St. Paul Fire & Marine Ins. Co.*, 396 N.W.2d 229 (Minn. 1986).

20 *See, e.g.*, *Western World Ins. Co. v. Travelers Indem. Co.*, 358 So. 2d 602 (Fla. Dist. Ct. App. 1978). In this case a surety under performance bond brought suit against the principal's general liability carrier to recovery monies the surety paid to cover the principal's liability. The general liability carrier argued that the surety had no right to pursue a claim

was intelligently discussed in the New York decision of *Continental Insurance Co. v. Colangione.*[21] In this case, Continental issued both a performance bond and a general liability policy to a roofing subcontractor (Skyway) in connection with a construction project for Syracuse University. Sometime after the project was complete, the roof began to leak and the University sued its architect and general contractor. The general contractor, in turn, sued Skyway. The claims against Skyway included breach of contract, breach of warranty, and negligence.

The general contractor also sued Continental, as surety for Skyway. The surety then brought fourth-party actions against its indemnitors. Three years after the suit initiated by the University was commenced, the surety posted a reserve on its performance bond and instituted an action against the indemnitors requesting the posting of collateral. In this separate action, the indemnitors counterclaimed for declaratory relief, seeking judgment that Continental was obligated under its general liability policy to defend an insure Skyway against liability in the suit instituted by the University. The surety prevailed in its action, requiring the indemnitors to post collateral. A few days after prevailing on its posting of collateral claim, the surety settled the performance bond suit by paying $100,000 "on the surety bond in settlement of the third-party actions, and stipulated to a discontinuance of its fourth-party action against the Skyway companies."[22]

Upon settling the action on its performance bond, Continental then moved for summary judgment, seeking dismissal of all claims against its liability policy. Continental's arguments and the court's response

---

against the principal's general liability insurer for a loss that the surety had paid under the bond. The Florida court disagreed noting that suretyship is not insurance and that if a surety pays a loss, it is entitled to be indemnified by its principal. Accordingly, if the principal has claims that some or all of a judgment or settlement should be paid by a general liability carrier rather then by a principal, the principal's surety is entitled to pursue such a claim against the general liability carrier. *Id.* at 603-04; *see also American Ins. v. Ohio Bureau of Workers' Comp.*, 577 N.E.2d 756 (Ohio Ct. App. 1991) (Court noted that general liability carrier should not be absolved of its insurance defense and indemnity obligations simply because a surety steps in and pays the principal's loss).

21 463 N.Y.S.2d 619 (N.Y. App. Div. 1983).

22 *Id.* at 917, 921.

provide guidance as to the interplay between the performance bond and liability policy:

> The main thrust of Continental's appeal, however, is that defendants are conclusively precluded from asserting that claim [that the insurer was obligated to defend and indemnify against the defective work claims], either by reason of the terms of the final settlement . . . in which Continental paid in $100,000 in satisfaction of its and Skyway's liability on the performance bond and Skyway was given a general release, or because Continental was awarded judgment against the Colangiones as [indemnitors] . . . Continental reasons that since neither Skyway nor the Colangiones were held liable in or required to contribute to the settlement of the third-party action, neither of them was damaged by any alleged failure on the part of Continental to defend and insure them against liability in that action. This argument, while superficially appealing, does not withstand critical analysis in which Continental's role as surety must be separated from its role as a possible insurer of Skyway . . .. [A]s between Continental as surety and Skyway as primary obligor on the performance bond, Skyway was primarily liable. On the other hand, if there was coverage under the liability policy for the claims against Skyway . . . Skyway would have claimed over to compel Continental, as insurer, to defend and insure against its primary liability. It was only because Continental's payment in settlement . . . was made in its role as surety on the performance bond, rather then as insurer under the policy, that the Colangiones became liable under the indemnification agreement. Thus, if there was coverage and the coverage extended to the Colangiones, they did indeed suffer damages by reason of Continental's wrongful refusal to defend and insure Skyway . . . [23]

---

23 *Id.* at 918-19. What makes the *Colangione* decision as relevant as it is on the point is the fact that the same company issued both the performance bond and liability policies. Other courts have, however, understood the differences between liability policies and performance bonds. The Florida Court of Appeals decision in *Western World Ins. Co., Inc. v. Travelers Indem. Co.*, 358 So. 2d 602, 604 (Fla. Dist. Ct. App. 1978), reveals a keen appreciation of the differences between performance bonds and insurance policies:

> Clark and Moore [the insureds], by paying a premium which resulted in the purchase of a liability insurance policy from Western World, in effect made themselves financially responsible to meet their obligations under

The argument that there exists no coverage under a general liability policy because to find coverage would, in effect, turn the policy into a performance bond, is premised upon a misunderstanding regarding what the two instruments are designed to accomplish.

2. *The Requirement Of "Property Damage"*

"Property damage" means:

a. Physical injury to tangible property, including all resulting loss of use of that property. All such loss of use shall be deemed to occur at the time of the physical injury that caused it; or

b. Loss of use of tangible property that is not physically injured. all such loss of use shall be deemed to occur at the time of the "occurrence" that caused it.[24]

a. Diminution In Value Damages

This definition has, for the most part, been successful in slowing the spread of decisions permitting recovery of diminution in value losses.[25]

---

the terms of the bond. The maximum amount of coverage afforded to Clark and Moore through their liability insurance policy was far in excess of the total judgments recovered by Thomas. Since it is clear under the law of general suretyship that the surety, Travelers, may be indemnified from its principal, Clark and Moore, for the judgment paid to Thomas, Travelers has the right to be subrogated to any rights which Clark and Moore have against their insurance carrier.

24 Susan J. Miller & Philip Lefebvre, *Miller's Standard Insurance Policies Annotated*, Vol. I at 420 (4th ed. 1995).

25 *See, e.g.*, *Wyoming Sawmills v. Transportation Ins. Co.*, 578 P.2d 1253 (Or. 1977); *New Hampshire Ins. Co. v. Vieira*, 930 F.2d 696 (9th Cir. 1991); *Federated Mut. Ins. Co. v. Concrete Units, Inc.*, 363 N.W.2d 751 (Minn. 1985); *R.N. Thompson & Assoc., Inc. v. Monroe Guar. Ins. Co.*, 686 N.E.2d 160 (Ind. Ct. App. 1997) (contractor denied coverage for claims made by homeowners association alleging damages to deterioration of plywood roughing material as there was no physical

A few courts, however, continue to find diminution in value loss to be covered. For example, in *W.E. O'Neil Construction Co. v. National Union & Fire Ins. Co.,*[26] a general contractor's insurer refused to pay its insured's share of a $1.8 million dollar settlement payment to an owner for cracks in the floors and walls of a parking garage. The insurer contended that the owner's claims of the property's diminution in value did not amount to "property damage" under the CGL policy. Because "property damage" was defined as "physical injury to tangible to property," the insurer argued that the owner's claims of economic loss were not covered. The court rejected this contention holding that property damage had occurred. It held there were a number of types of economic loss including diminution in value and loss of use, which might be considered "property damage," as opposed to lost profits, which did not constitute "property damage."[27]

---

injury to any property apart from the deterioration of the roof decking itself and therefore a claim limited to remedying faulty workmanship of materials does not involve "physical injury to tangible property" or "property damage."); *Mapes Indus., Inc. v. United States Fidelity & Guar. Co.*, 560 N.W.2d 814 (Neb. 1997) (CGL carrier had no obligation to defend or indemnify a manufacturer of insulated panels used in building against claims that the panels deteriorated and needed replacement.); *Hawaiian Ins. & Guar. Co., Ltd. v. Blair, Ltd.*, 726 P.2d 1310, 1915 (Haw. Ct. App. 1986) (we hold that as applied to the facts of this case, "diminution in value" does not constitute "loss of use" within the meaning of the policy provision."); *Kartridg Pak Co. v. Travelers Indem. Co.*, 425 N.W.2d 687, 689-90 (Iowa Ct. App. 1988) (". . . requirement of "physical injury or destruction of tangible property" is clear and unambiguous . . . We conclude in tangible damages, such as diminution of value, do not constitute physical injury to or destruction of tangible property.").

26 721 F. Supp. 984 (N.D. Ill. 1989).

27 *See also Marathon Plastics, Inc. v. International Ins. Co.*, 514 N.E.2d 479 (Ill. App. Ct. 1987) (defective polyvinyl chloride pipe incorporated into water system diminished its value and was "property damage."); *Missouri Terrazzo Co. v. Iowa Nat'l Mut. Ins. Co.*, 740 F.2d 647 (8th Cir. 1984) (diminution in value of building resulting from destruction of supermarket floor covered as such damages were merely a means of measuring the damage sustained as a result of property damage.).

b. Economic Loss

Lost profits or losses in the nature of increased construction costs are difficult to recover under a CGL policy. The definition's use of the term "physical injury" eliminates many economic loss claims. In *Aetna Casualty & Surety Co. v. McIbs, Inc.,*[28] a supplier delivered cement block manufacturing equipment to a cement block manufacturer. The manufacturer used the supplier's equipment to produce certain blocks sold to a construction company for use on a specific project. The blocks, however, were either too long or too wide and as a result the construction company incurred increased labor costs to cut the blocks and re-plaster interior walls at the job site. After the block manufacturer reimbursed the contractor for some of its increased costs, it sued the supplier of the equipment, who sought coverage under its CGL policy. The insurer denied coverage arguing that there was no evidence of any physical injury to or destruction of any property caused by the blocks. Indeed, the blocks themselves were not even damaged, they were simply the wrong size. The court agreed with the carrier's position.[29]

---

28 684 F. Supp. 246 (D. Nev. 1988), *aff'd*, 878 F.2d 385 (9th Cir. 1989).

29 *Batson-Cook Co. v. Aetna Ins. Co.*, 409 S.E.2d 41 (Ga. Ct. App. 1991) (contractor's claim that insured construction manager's poor performance cost additional expenses in the construction of the work was excluded from coverage as it was not property damage); *Tobi Eng'g, Inc. v. Nationwide Mut. Ins. Co.*, 574 N.E.2d 160 (Ill. App. Ct. 1991) (delayed damages incurred by contractor because of defective materials supplied by insured were not property damage losses); *Selective Ins. Co. of S.E. v. J.B. Mouton & Sons, Inc.*, 954 F.2d 1075 (5th Cir. 1992) (loss of anticipated revenues from a failed partnership due to insured's actions do not state a claim for property damage); *Lazzara Oil Co. v. Columbia Cas. Co.*, 683 F. Supp. 777 (M.D. Fla. 1988), *aff'd*, 868 F.2d 1274 (11th Cir. 1989) (economic damages such as loss profits and loss of good will do not qualify as "property damage"); *Hommel v. George*, 802 P.2d 1156 (Colo. Ct. App. 1990) (economic loss is sustained by investors in condominium development caused when contractor failed to complete project did not result from loss of use of tangible property and were not covered "proper damage"); *but, see, American Home Assurance Co. v. Libbey-Owens-Ford Co.*, 786 F.2d 22 (1st Cir. 1986). This decision involving disputes arising out of construction problems of the John Hancock building in Boston is an expansive reading of coverage under

c. Incorporation Of Defective Products

Apart from the issue of whether loss measured as diminution in value is property damage, is the question of coverage where the insured's work or product is incorporated into a larger product or structure. Because most subcontractors' and suppliers' work is incorporated into a building, these incorporation cases are particularly significant. While not a pure incorporation case, one of the more celebrated decisions on the question of coverage for defective work is *Maryland Casualty Co. v. Reeder*,[30] where condominium owners and their association brought suit as a result of cracks in the interior and exterior walls and foundations of their structure caused by soil subsidence and a defective roofing system. The court found that there existed "property damage under the general contractor's CGL policy because the work of various subcontractors caused collateral or derivative damage to the structure. For example, there was damage to a

---

the standard CGL policy. The following is a discussion of coverage afforded the glass curtain wall manufacturer due to massive failures with that system:

> [W]e conclude that the court erred in its initial interpretation of the American Home policy. Although the plain language of the policy does require that some physical injury to tangible property must occur, other then for loss-of-use claims, the policy does not preclude the possibility that physical injury to the insured's own property can constitute property damage. Under this reading of the policy, consequential damages suffered as a result of physical injury to any product, including that of the insured, would be covered. Such a reasoning is a reasonable construction of the policy and, under general principals of insurance law, a reasonable construction that affords coverage for the insured should be adopted. Indeed, any alternative reading resulting in non-coverage for the insured would actually require us to add language to the policy against the insured, standing on its head the general insurance principal that policies are to be strongly construed against the insurer . . . .

Id. at 25-26.

30 270 Cal. Rptr. 719 (Cal. Ct. App. 1990).

four story garage resulting from improper placement of steel mesh in concrete levels by one of the subcontractors."[31]

In another case, Judge Posner, writing for the Seventh Circuit, construed a CGL policy containing the definition of "property damage" as "physical injury to tangible property" in the context of damages resulting from allegedly defective plumbing systems which were incorporated into homes and apartments. In *Elger Mfg., Inc. v. Liberty Mutual Insurance Co.,*[32] where coverage was afforded to the manufacturer even for systems where no leaking had yet to occur, Judge Posner noted:

> The central issue in the case—when if ever the incorporation of one product into another can be said to cause physical injury pivots on a conflict between the connotations of the term "physical injury" and the objective of insurance. [After an extensive discussion of the everyday meaning of "physical injury" in contrast with the drafter's intent of the new policy forms, the court continues.] We now can see more clearly that two senses of "physical injury" are competing for our support. One, which the insurers want us to adopt, is an injury that causes a harmful physical alteration in the thing injured. The other, which is what the draftsman of the comprehensive liability insurance policy apparently intended and what rational parties to such a policy would intend in order to make the policy's coverage real and not illusory, is a loss that results from physical contact, physical linkage, as when a potentially-dangerous product is incorporated into another and, because it is incorporated and not merely contained (as a piece of furniture is contained in a house but can be removed without damage to the house), must be removed, at some cost, in order to prevent the danger from materializing . . . [T]he drafting history of the property-damage clause, and the probable understanding of the parties to liability insurance contract, persuade us that the incorporation of a defective product into another product inflicts physical injury in the relevant sense on the latter at the moment of incorporation here, the moment when the defective quest [plumbing] systems were installed in homes.[33]

---

31 *Id.* at 723.

32 972 F.2d 805 (7th Cir. 1992).

33 *Id.* at 808, 810, 814. The Elger Manufacturing decision is an aggressive expression of coverage as it is reminiscent of the line of authority finding diminution in value damages covered under policy language. It is, however, not alone. For example, the Washington Supreme Court in

Many courts have similarly adopted the position that the presence of a defective product within a larger whole, without some attendant physical injury to the larger structure, does not constitute "property damage."[34] A

Yakima Cement Prods. Co. v. Great Am. Ins. Co., 608 P.2d 254 (Wash. 1980), commented the following with regard to coverage for a manufacturer which delivered defective concrete panels to a project that had to be ripped out, repaired and reinstalled:

> We . . . decline to limit the words "accident" and "occurrence" to include only damage arising from the "active malfunctioning" of the insured's product such as the breaking of a defective panel with resulting injury to a person or damage to property by its fall. An accident has been found where defective doors were installed in buildings; where the planting of the wrong type of seed resulted in a reduction in the following season's yield; and where the insured failed to properly complete an automatic sprinkler system which, if it had been operational, would have prevented a subsequent fire in a building. In each case a finding that an accident had occurred did not turn on the "active malfunctioning" of the insured's product. Rather, it turns on the failure of the product to perform the function for which it was sold or manufactured. The term "accident" must be defined in the context of the product being manufactured by the insured and the kind of losses sustained in the even of a defect in manufacturer . . . [T]he hazards to be insured against are the mishaps or unintended consequence that can result from the use of the product. Clearly the mismanufacture of the panels and their subsequent incorporation into that operation's building constituted the use of a product which failed to perform the function for which it was intended.

Id. at 257-58 (quoting *Ramco, Inc. v. Pacific Ins. Co.*, 439 P.2d 1002, 1005 (Or. 1968)). *But see Hamilton Die Cast, Inc. v. United States Fidelity & Guar. Co.*, 508 F.2d 417, 419-20 (7th Cir. 1975).

34 *See, e.g.*, *Diamond State Ins. Co. v. Chester-Jensen Co.*, 611 N.E.2d 1083 (Ill. App. Ct. 1993) (defective equipment incorporated into HVAC system did not constitute "property damage"); *Wyoming Sawmills, Inc. v. Transportation Ins. Co.*, 578 P.2d 1253 (Or. 1978) (en banc) (defective lumber studs incorporated into building did not constitute "property damage" to the building); *Aetna Life & Cas. v. Patrick Indus.*, 645 N.E.2d 656 (Ind. Ct. App. 1995) (defective particle-board incorporated into furniture products resulting in $200,000 of repairs was a claim for

few courts have focused upon the amount of destruction that will occur to the larger whole in order to repair the defective component. Some courts have concluded that if repairs necessitate significant destruction then property damage does occur for coverage purposes.[35]

d. Loss Of Use Injury

Perhaps the greatest source of confusing over the meaning of "property damage" arises in connection with the "loss of use" component of this term. As a general rule, the courts have been fairly expansive in applying the definition to factual settings in which "loss of use" claims are present. Because the policy expressly provides coverage for loss of use of tangible property which has not been physically injured, there is some question as to whether "diminution in value" loss is covered where the structure's value is diminished in part because of an impairment to its use.

The Maryland Court of Appeals decision in *Woodfin Equities Corp. v. Hartford Mutual Insurance Company*,[36] is illustrative of some of the issues that "loss of use" claims raise under a CGL policy. Woodfin, an

---

diminution in value and not physical injury to property with the meaning of the policy).

35 *See, e.g.*, *Aetna Cas. & Sur. Co. v. M & S Indus., Inc.*, 827 P.2d 321 (Wash. Ct. App. 1992); *Globe Indem. Co. v. People*, 118 Cal. Rptr. 75 (Cal. Ct. App. 1974); *Intel Corp. v. Hartford Accident & Indem.*, 692 F. Supp. 1171 (N.D. Cal. 1988), *aff'd in part, rev'd in part*, 952 F.2d 1551 (9th Cir. 1991). *But, see, New Hampshire Ins. Co. v. Vieira*, 930 F.2d 696, 701-702 (9th Cir. 1991), where the court rejected a claim for coverage arising out of the defective installation of dry wall in three housing projects:

> We hold that the nature of the repairs cannot create coverage where none exists. diminution in value and cost of repair are not two separate harms—they are two different ways of measuring the same harm. If the harm—the Vierira's defective work—is not covered as measured by diminished value, it is not covered as measured by cost of repair. The owner's decision to make repairs which necessitated damaging the roofs years after the settlement does not affect New Hampshire's liability.

36 678 A.2d 116 (Md. Ct. App. 1996), *aff'd in part, rev'd in part*, 687 A.2d (Md. 1997).

owner of a hotel in Rockville, Maryland, brought an action against the CGL insurer of a subcontractor that installed a defective HVAC system in the hotel. The HVAC units failed repeatedly and ultimately had to be replaced. The replacement caused damage to the walls and carpeting of the hotel. Prior to their replacement, the malfunctioning of the units caused the hotel to have to close off rooms serviced by malfunctioning units. The court determined that there was no coverage for the costs associated with tearing out walls, molding and carpeting in order to repair and remove the HVAC units. Nor was there any coverage for the replacement of the units. The reasoning for this was explained as "courts uniformly hold that when property damage arising out of the insured's defective workmanship is confined to the insured's own work product, the damage is not caused by an "occurrence" within the meaning of the CGL policy."[37] Nevertheless, the court did find that there was "loss of use" coverage:

> [W]e affirm the circuit court to the extent that it determined that the appellants were not entitled to damages resulting from the property damage to the HVAC systems. The property damage to the HVAC systems, however, is not the only property damage involved in this case. Appellants put forth undisputed evidence that they lost the use of guest suites as a result of the breakdown to the HVAC system. At this stage of the trial, appellee had not challenged appellant's position or denied that the hotel lost the loss of use of suites because of the HVAC failures. Therefore, under the plain reading of the CGL policy, appellants have established the "loss of use of tangible property." As noted above, in order for the loss of use of the guest suites to be compensable under the general coverage provision of the CGL policy, such loss of use must have been caused by an "occurrence."

---

37 *Id.* at 131. As the discussions on "occurrence" and the "business risks exclusions" contained in this chapter point out, the court's reasoning is subject to question. In fact, the Maryland Supreme Court as well as the United States Supreme Court have concluded that Maryland law adopts the majority position that whether a loss arises from an accident is to be determined from the standpoint of the insured's subjective intent. *See, e.g., Sheets v. Brethren Mut. Ins. Co.*, 679 A.2d 540 (Md. 1996); *Lords Landing Village Condominium Council of Unit Owners v. Continental Ins. Co.*, 117 S. Ct. 1731 (1997).

> Unlike the HVAC systems, the guest rooms are not the work product of the insured, but are the property of appellants. In other words, at least with respect to the loss of use of guest suites, the damage is to property other than the product or completed work of the insured. To the extent that the insured's defective workmanship causes damage to such other property, courts uniformly hold that such damage is caused by an "occurrence" and is, therefore, compensable under the CGL policy. . . . The loss of use of the guest suites, therefore, is "property damage" caused by an "occurrence" under the CGL policy. As a result, in light of the evidence thus far introduced, under the general coverage provision, appellee is obligated to cover the "damages" associated with the loss of use of the guest suites.[38]

3. *Exclusions Affecting Coverage For Defective Work*

The standard CGL policy contains a number of exclusions sometimes collectively referred to as the "business risk" or "work product" exclusions. The wording and numbering of these exclusions has changed over the years. The particular exclusions falling within this rubric in the 1986 policy form are exclusions (j), (k), (l), (m), and (n).[39]

---

38 *Woodfin Equities Corp. v. Hartford Mut. Ins. Co.*, 678 A.2d 116, 133 (Md. Ct. Spec. App. 1996), *aff'd in part, rev'd in part*, 687 A.2d 652 (Md. 1997); *but see, Bituminous Cas. Corp. v. Gust K. Newberg Constr. Co.*, 578 N.E.2d 1003 (Ill. App. Ct. 1991), *appeal denied*, 587 N.E.2d 1011 (1992) (faulty HVAC system resulting in loss of use of building not covered because the complaints of too hot and too cold temperatures were nothing more than the natural and ordinary consequences of installing an inadequate HVAC system and, therefore, did not result from an "occurrence").

39 *See* Susan J. Miller & Philip Lefebvre, *Miller's Standard Insurance Policies Annotated*, Vol. I at 411-12 (4th ed. 1995). Exclusion (k) under the 1986 policy form, by its terms, does not apply to real estate. It, therefore, is not particularly relevant to construction except where the work can in some way be considered a "product" unrelated to real estate. *See, e.g., McKellar Dev. of Nev. v. Northern Ins. Co. of N.Y.*, 837 P.2d 858 (Nev. 1992) (construction services are not a "product").

a. Exclusion (j) – Damage To Property

Exclusion (j) actually is six separate exclusions. Exclusion (j) provides that insurance does not apply to:

> "property damage" to:
>
> (1) Property you own, rent, or occupy;
> (2) Premises you sell, give away, or abandon, if the "property damage" arises out of any part of those premises;
> (3) Property loaned to you;
> (4) Personal property in the care, custody or control of the insured;
> (5) That particular part of real property on which you or any contractors or subcontractors working directly or indirectly on your behalf are performing operations, if the "property damage" arises out of those operations; or
> (6) That particular part of any property that must be restored, repaired or replaced because "your work" was incorrectly performed on it.
>
> Paragraph (2) of this exclusion does not apply if the premises are "your work" and were never occupied, rented or held for rental by you.
> Paragraphs (3), (4), (5), and (6) of this exclusion do not apply to liability assumed under a side-tracked agreement.
> Paragraph (6) of this exclusion does not apply to "property damage" included in the "products-completed operations hazard."[40]

Exclusions (j)(5) and (6) apply most directly to construction activities.

Exclusion (j)(5) is one of the primary "business risk" exclusions. This exclusion has been around in the standard forms for many years. Prior to 1986 the exclusion applied to all property, both personal and real. Exclusion (j)(5), however, applies only to real property.[41] While the exclusion has been around in one form or another for many years, it is still the subject of a fair amount of interpretive discussion. As a general rule, this exclusion has been applied to precluded coverage for

---

40 Susan J. Miller & Philip Lefebvre, *Miller's Standard Insurance Policies Annotated*, Vol. I at 411 (4th ed. 1995).

41 Exclusion (j)(4) is intended to exclude this risk with respect to personal property.

damages to real property resulting from or arising out of the insured's operations.[42] Still there has been a fair amount of controversy as to over what "that particular part" means and whether the exclusion applies to completed operations.

Some of the issues created by the language "that particular part" are concisely stated in the California decision of *Economy Lumber Co. of Oakland v. Insurance Company of North America.*[43] In this case a supplier of defective siding sought coverage against claims that its product damaged several residences. In discussing the application of exclusion (j)(5) the court opined:

> If we interpret "that particular part of any property" under [the exclusion] to refer to the siding, coverage for the sided is excluded by subparagraph (ii), "out of which any property arises," and subparagraph (iii), because its "restoration, repair and replacement" had been made necessary by faulty milling on behalf of [the insured]. However, the property damage to the houses itself remains covered.
>
> If we were to interpret "that particular part of any property" to refer to the houses themselves, coverage for property damage to them would still result. As the supplier, [the insured] did not perform any faulty workmanship on the houses. The only reasonable interpretation of the exclusion would be to hold it applicable if [the insured] had itself performed work on the houses, or if such work was done on its behalf. Such is not the cases.
>
> We conclude, therefore, that while there is no coverage for the defective siding as such, the exclusion does not apply to the loss of value to the 8 houses for which there is policy coverage.[44]

It is obviously to the insured's advantage to define "that particular part" to apply to as small a part of the project as possible. For the most part, insureds have had a difficult time narrowing the scope of this exclusion. For example, in *Vandivort Construction Corp. v. Seattle*

---

42 *See, e.g.*, *George A. Fuller Co. v. United States Fidelity & Guar. Co.*, 613 N.Y.S.2d 152 (N.Y. App. Div. 1994); *Brawdy v. National Grange Mut. Ins. Co.*, 616 N.Y.S.2d 846 (N.Y. App. Div. 1994).

43 204 Cal. Rptr. 135 (Cal. Ct. App. 1984).

44 *Id.* at 140-41.

Tennis Club,[45] the insured contractor's work resulted in an earth slide causing damage to the site. The contractor claimed that the "particular part" of the property it was working on at the time of the accident was a small portion of the site and, therefore, the only damage subject to the exclusion was that discrete part of the project on which its operations were being performed. The court rejected the contractor's contention noting:

> Vandivort argues that because the slide occurred at Seattle Tennis Club's north property line and damage is claimed beyond that point, the exclusion which it argues applies only to the particular part of any property upon which work is being performed is not applicable. We reject the argument. The plain meaning of the language covers the situation here. Vandivort was performing operations on the property and the injury here for which damages are claimed arose out of those operations.[46]

Another issue that arises with regard to the application of exclusion (j)(5) to construction activities involves a question of timing. The language of the exclusion which reads "are performing operations" has been interpreted to mean that the exclusion applies to damages involving

---

45 522 P.2d. 198 (Wash. App. Ct. 1974).

46 *Id.* at 201; *see also C.D. Walters Constr. Co. v. Fireman's Ins. Co. of Newark, N.J.*, 316 S.E.2d 709 (S.C. Ct. App. 1984); *Jet Line Serv. v. American Employers Ins. Co.*, 537 N.E.2d 107 (Mass. 1989) (insured's operations being performed on entire tank not just the bottom of the tank); *Continental Graphic Serv. v. Continental Cas. Co.*, 681 F.2d 743 (11th Cir. 1982) (insured's operations on entire printing press and not just one gear); *Goldsberry Operating Co. v. Cassity, Inc.*, 367 So. 2d 133 (La. Ct. App. 1979) (insured's operations on entire well); *but see, Continental Ins. Co. v. Asarco, Inc.,* 738 P.2d 368 (Ariz. Ct. App. 1987) (bents and trusses supplied by owner for conveyor system which collapsed during insured's construction activities were not "that particular part" of the property being worked on by the insured.); *Abramson v. Florida Gas Transmission Co.*, 908 F. Supp. 1389 (E.D. La. 1995) (policy language that excluded property damage to property being installed, erected or worked upon by insured did not apply to real property damaged in pipeline project as insured's work was limited to the pipeline.).

"works in progress." In other words, the exclusion, by its terms, does not apply to completed operations.[47]

Exclusion (j)(6) is commonly known as the "faulty workmanship" exclusion. This exclusion has no analogue in the 1966 or 1973 versions. The exclusion's inception occurred with the creation of the broad form property damage endorsement which narrowed the standard care, custody, or control and work performed exclusions contained in the 1973 standard form. The exclusion has been carried forward into the 1986 policy form.[48] The most significant difference between the faulty workmanship exclusion as contained in the BFPD endorsement and that set forth in (j)(6) of the 1986 policy is that the (j)(6) exclusion does not apply to property damage included in the products-completed hazard.[49]

---

47 *See Spears v. Smith*, No. 15864, 1996 WL 933931 (Ohio Ct. App. 1996). In this case, the insured contractor was retained to build a residence in Centerville, Ohio in September of 1990. The contractor constructed the home's I-beam floor system itself and utilized subcontractors to complete much of the remaining work. After the home's completion, the homeowners noted "sponginess" and depressions and slanting which effected portions of the floor. As a result of the floor sinking, the home's wall cracked in several places. The contractor had finished its work before these defects were noticed by the homeowners. The court concluded that the exclusion did not apply because it is written in the "present tense" and because the damages were discovered after the contractor's work was complete, coverage would not be excluded under this provision. *See also Robert J. Franco, Insurance Coverage for Faulty Workmanship Claims Under Commercial General Liability Policies*, 30 TORT & INS. L.J. 785, 796 ("the intent of exclusion (j)(5) is to bar coverage for the work being done by a contractor when claims arise at the time the work is being performed".)

48 The exclusion was known as (2)(d)(iii) of the broad form property damage endorsement (including completed operations). *See, e.g.*, Susan J. Miller & Philip Lefebvre, *Miller's Standard Insurance Policies Annotated*, Vol. I at 452.6 (4th ed. 1995).

49 This can be seen from the last paragraph of exclusion (j) which expressly states this proposition. The policy defines "products-completed operations hazard" as including:

All "bodily injury" and "property damage" occurring away from premises you own or rent and arising out of "your product" or "your work" except:

Thus, if the claim arises from defective work that is discovered after the contractor has completed its work, exclusion (j)(6) does not apply. The industry, instead, relies upon exclusion l or the "work performed" exclusion to address situations involving latent defective work.[50]

The primary purpose of exclusion (j)(6) is to preclude coverage for the costs to repair or replace an insured's faulty work.[51] Because the exclusion contains the language "that particular part," much of the same analysis as was discussed with respect to exclusion (j)(5) applies with equal force to (j)(6). For example, in *Frankel v. J. Watson Co., Inc.*,[52] the insured was retained to move a farmhouse and to construct a new concrete foundation for it. The insured, however, negligently constructed the foundation and as a result the farmhouse sustained damage. The faulty workmanship exclusion applied only to preclude coverage for the foundation itself and not for damages to the rest of the property.[53]

---

(1) Products that are still in your physical possession; or
(2) Work that has not yet been completed or abandoned.

See, e.g., Susan J. Miller & Philip Lefebvre, *Miller's Standard Insurance Policies Annotated*, Vol. I at 419 (4th ed. 1995).

50 *See, e.g.*, Patrick J. Wielinski & Jack P. Gibson, *Broad Form Property Damage Coverage*, 3d Ed. International Risk Management Institute, Ch. 8 (1992).

51 *See, e.g.*, *LISN, Inc. v. Commercial Union Ins. Co.*, 615 N.E.2d 650 (Ohio Ct. App. 1992) (exclusion precluded coverage for costs to repair and replace live telephone cable erroneously cut by one of insured's employees.); *Brawdy v. National Grange Mut. Ins. Co.*, 616 N.Y.S.2d 846 (N.Y. App. Div. 1994) (exclusion precluded costs for repairing partially completed wall that collapsed due to insured's negligence); *Transcontinental Ins. Co. v. Ice Sys. of Am., Inc.*, 847 F. Supp. 947 (M.D. Fla. 1994) (exclusion precluded costs associated with repairing ice rink so as to make it suitable for hockey).

52 484 N.E.2d 104 (Mass. Ct. App. 1995).

53 *See also Blackfield v. Underwriters at Lloyd's of London*, 245 Cal. App. 2d 271, 53 Cal. Rptr. 838 (Cal. Ct. App. 1966); *Baugh Constr. Co. v. Mishen Ins. Co.*, 836 F.2d 1164 (9th Cir. 1988); *Dorchester Dev. Corp. v. Safeco Ins. Co.*, 737 S.W.2d 380 (Tex. App. 1987). Not all courts have, however, found the language "that particular part" to be particularly restrictive. For example, in *Bond Bros., Inc. v. Robinson*, 471 N.E.2d

b. Exclusion l – Work Performed Exclusion

Exclusion l of the 1986 policy form, otherwise known as the "work performed" exclusion performs the lion's share of the labor in reducing coverage for losses arising from a contractor's poor workmanship in a completed operations context. Under the 1986 version of this exclusion, insurance does not apply to:

> l. "Property Damage" to "your work" arising out of it or any part of it and included in the "products-completed operations hazard."
>
> This exclusion does not apply if the damaged work or the work out of which the damage arises was performed on your behalf by a subcontractor.[54]

1332 (Mass. 1984) a subcontractor negligently installed reinforcing steal and wire mesh over which a general contractor poured concrete to complete the wall. The general contractor sought costs of redoing the wall from the subcontractor who, in turn, sought coverage under its CGL policy. Coverage was denied under the faulty workmanship exclusion contained in the BFPD endorsement. This was the case even though the subcontractor's work did not include the concrete portion of the wall. For a contrary view on similar facts *see W.E. O'Neil Constr. Co. v. National Union Fire Ins. Co. of Pittsburgh, Pa.*, 721 F. Supp. 984 (N.D. Ill. 1989).

54 *See, e.g.*, Susan J. Miller & Philip Lefebvre, *Miller's Standard Insurance Policies Annotated*, Vol. I at 411 (4th ed. 1995). Earlier versions of this exclusion included exclusion (o) of both the 1966 and 1973 CGL forms, which stated that insurance did not apply:

> (o) to property damage to work performed by or on behalf of the named insured arising out of the work or any portion thereof, or out of materials, parts or equipment furnished in connection therewith;

See, e.g., Susan J. Miller & Philip Lefebvre, *Miller's Standard Insurance Policies Annotated*, Vol. I at 451.5 (4th ed. 1995). The broad form property damage endorsement modified exclusion (o) under these previous forms by deleting the language "or on behalf of." *See, e.g.*, Susan J. Miller & Philip Lefebvre, *Miller's Standard Insurance Policies*

This exclusion is often raised by insurers in connection with claims against contractors for latent defective work. It is a common refrain within the insurance industry as well as by the courts that insurance is not intended to pay the costs of repairing a contractor's poor workmanship. The insurance industry does not intend to cover contractors for the costs of making good their own bad work as this is a risk of doing business rather than an insurance risk. This is, in essence, the heart of the "business risk" doctrine.[55]

Under the 1986 policy revisions, any confusion over the scope of the "work performed" exclusion arising out of subcontractor operations has been put to rest. Even the Minnesota courts now acknowledge that under the policy language of the 1986 and subsequent forms, contractors are afforded coverage for property damage which either arises out of or is to the work of a subcontractor.

In *O'Shaughnessy v. Smuckler Corporation,*[56] a general contractor was engaged to construct a luxury home on one of the lakes outside of the Twin Cities. The general contractor, in turn, retained the services of numerous subcontractors to perform most of the actual construction work. After the house was completed and the homeowners took

---

Annotated, Vol. I at 452.6 (4th ed. 1995). There were two lines of authority over whether the language changed to this exclusion contained in the BFPD endorsement was significant. One line of authority represented by two cases out of the Supreme Court of Minnesota held that the language change was insignificant and did not alter the scope of the exclusion. *See*, *Bor-Son Bldg. Corp. v. Employers Commercial Union Ins. Co. of Am.*, 323 N.W.2d 58 (Minn. 1982); *Knutson Constr. Co. v. St. Paul Fire & Marine Ins. Co.*, 396 N.W.2d 229 (Minn. 1986). The other line of authority held that the change was significant and the purchase of the BFPD endorsement did narrow the scope of the work performed exclusion. *See*, *Maryland Cas. Co. v. Reeder*, 270 Cal. Rptr. 719 (Cal. Ct. App. 1990); *Fire Guard Sprinkler Sys., Inc. v. Scottsdale Ins. Co.*, 864 F.2d 648 (9th Cir. 1988). This issue has largely gone away under the 1986 policy form as the policy expressly states that the scope of the exclusion does not apply to subcontractor work. *See, e.g.*, *O'Shaughnessy v. Smuckler Corp.*, 543 N.W.2d 99 (Minn. Ct. App. 1996).

55 *See, e.g.*, Tinker, *Comprehensive General Liability Insurance–Perspective and Overview*, 25 F.I.C.C. 217, 224 (1975)

56 543 N.W.2d 99 (Minn. Ct. App. 1996).

possession a number of problems surfaced with the construction. Some of the more serious problems included widespread cracking of the homes gypsum walls and granite marble flooring due to an improperly constructed floor support system. The exterior brick masonry was also improperly constructed thus allowing water to leak into the wall system at various locations.[57] The carrier argued that to permit coverage for these defects "would be to transform the policy into a performance bond."[58] The insurer argued that the claims for the cost of repairing the "defects involve [d] a business risk assumed by Smuckler Corporation for which there is no coverage under Minnesota's Business Risk Doctrine."[59]

The court rejected the insurance company's position, as the language of the 1986-form policy afforded coverage for the claimed damages:

> We cannot conclude that the exception to exclusion (l) has no meaning or effect. The CGL policy already covers damage to the property of others. The exception to the exclusion, which addresses "property damage" to "your work" must therefore apply to damages to the insured's own work that arise out of the work of a subcontractor. Thus, we conclude that the exception at issue was intended to narrow the business risk doctrine . . . . Under our holding in the current language of the policy, the business risk doctrine would still apply to work performed by the general contractor and to other deficiencies in a subcontractor's work that do not constitute "property damage." Absent policy language to the contrary, the business risk doctrine would also preclude a subcontractor from making a claim against its own insurer to recover the cost of the repair or replacement of its own defective work. The insurance industry is, of course, free to alter this result in future cases by seeking to change the language of its policies.[60]

Nevertheless, the "work performed" exclusion will apply to preclude coverage for the costs associated with repairing the insured's own defective work. Thus, a developer was denied coverage when all it

---

57 *Id.* at 100-101.

58 *Id.* at 105.

59 *Id.* at 101.

60 *Id.* at 105.

proved at trial were the costs to repair its own defective road work.[61] One insurer sought to avoid the subcontractor exception to the "work performed" exclusion by claiming that the defective work in question was performed not by a subcontractor but by a "materialman." In *National Union Fire Insurance Co. of Pittsburgh. v. Structural Sys. Technology, Inc.*,[62] a steel fabricator performed the allegedly defective work. After examining the nature of the work in question, the court upheld coverage. Given that exclusions are generally strictly construed against the insurer, it would appear that this argument would find little support in the courts.[63] Indeed, suppliers have been deemed to be "subcontractors" within the meaning of insurance policies.[64]

c. Exclusion (m) – The "Impaired Property Exclusion

The concept of "impaired property" first surfaced in 1986. Before that time, however, the standard CGL policy did contain an exclusion intended to address claims for losses were incurred as a result of the insured's failure to perform its obligations. For example, exclusion (m) contained in the 1973 CGL form and known as the "failure to perform" exclusion, precluded coverage for loss of use damages where no physical injury was present resulting from (1) a delay in or lack of performance by the insured, or (2) the failure of the insured's products

---

61 *See, J.Z.G. Resources, Inc. v. Shelby Ins. Co.*, 84 F.3d 211 (6th Cir. 1996).

62 756 F. Supp. 1232 (E.D. Mo. 1991), *aff'd*, 964 F.2d 759 (8th Cir. 1992).

63 *See, e.g., Reinsurance Ass'n of Minnesota v. Patch*, 383 N.W.2d 708 (Minn. Ct. App. 1986) (policy exclusions are to be strictly interpreted against the insurer); *National Union & Fire Ins. Co. v. Evenson*, 439 N.W.2d 394 (Minn. App. 1989) (insurance policies to be strictly construed against the insurance company as the party drafting the language.); *Caledonia Community Hosp. v. St. Paul Fire Marine Ins. Co.*, 239 N.W.2d 768 (Minn. 1976) (insurers have the burden of proving that policy exclusions apply).

64 *See, e.g., State ex rel. Regents of N.M. State Univ. v. Siplast, Inc.*, 877 P.2d 38 (N.M. 1994) (supplier of roofing materials was a "subcontractor" within the meaning of a builder's risk policy.)

or work to meet a level of performance, quality, fitness or durability warranted or represented by the insured.[65]

Exclusion (m) of the 1973 form, however, did not apply in *Aetna Casualty & Surety Co. v. M & S Industries, Inc.*,[66] where an insured supplier of plastic-coated plywood panels was sued by one of its customers, a manufacturer of concrete forms systems. The manufacturer alleged that a defect in the plastic-coated plywood panels caused it to incur expense to replace defective forms and resulted in lost profits because it had to divert production to make replacement forms. The "failure to perform" exclusion did not apply because the concrete forms, as property of the manufacturer, were physically injured by the defects in the plastic-coated plywood panels.[67]

Given the interpretation challenges posed by the 1973 form exclusion, many hoped that the revisions to the exclusion would provide a clearer picture. Unfortunately, the exclusion has not become simpler. The introduction of the concept of "impaired property" has created a new set of challenges. A number of commentators have opined that the new exclusion is "too complex to receive a uniform interpretation.[68] Others have found the new exclusion to be "difficult"[69] or "tricky."[70] One commentator has gone so far as to conclude that the exclusion is

---

65 Susan J. Miller & Philip Lefebvre, *Miller's Standard Insurance Policies Annotated*, Vol. I at 452.2 (4th ed. 1995). By its terms, this exclusion did not apply to loss of other tangible property resulting from the sudden and accidental physical injury to or destruction of the named insured's products or work performed by or on behalf of the named insured after such products or work have been put to use by any person or organization other than an insured.

66 827 P.2d 321 (Wash. Ct. App. 1992).

67 *See also Marathon Plastics, Inc. v. International Ins. Co.*, 514 N.E.2d 479 (Ill. Ct. App. 1987) (exclusion did not apply to water system containing insured's defective piping because the system had sustained physical injury).

68 *See* Pete Ligeros & Donald S. Malacki, *Impaired Property Exclusion: Using Discretion to Make it Work*, Claims 58, 60 (Nov. 1994).

69 *See* P. Magarick, 2 *Casualty Insurance Claims* 26-15 (3d Ed. 1995).

70 *See* Eugene F. Wolters, *Impaired Property CGL Definition Can Prove to be Tricky*, National Underwriter (Property & Casualty/Risk & Benefits Management Edition) 25 (1992).

subject to attack as "unintelligible or at least ineffective to overcome the insured's reasonable expectations of coverage."[71]

The 1986 version of exclusion (m) states that insurance does not apply to:

> (m) "Property damage" to "impaired property" or property that has not been physically injured arising out of:
>
> (1) a defect, deficiency, inadequacy, or dangerous condition in "your product" or "your work;" or
>
> (2) a delay or failure by you or anyone acting on your behalf to perform a contract or agreement in accordance with its terms.
>
> This exclusion does not apply to loss of use or other property arising out of sudden and accidental injury to "your product" or "your work" after it has been put to its intended use.[72]

In order to begin to understand the scope of this exclusion one has to understand the meaning of "impaired property." Unfortunately, the policy definition of "impaired property" is just as lengthy as the exclusion itself:

> "impaired property" means tangible property, other than "your product" or "your work" that cannot be used or is less useful because:
>
> a. it incorporates "your product" or "your work;" that is known or thought to be defective, deficient, inadequate or dangerous; or
>
> b. you have failed to fulfill the terms of a contract or agreement;
>
> If such property can be restored to use by:
>
> a. the repair, replacement, adjustment or removal of "your product" or "your work;" or

---

71 *See* Scott C. Turner, *Insurance Coverage of Construction Disputes*, 307 (1992).

72 *See* Susan J. Miller & Philip Lefebvre, *Miller's Standard Insurance Policies Annotated*, Vol. I at 411 (4th ed. 1995).

b. your fulfilling the terms of the contract or agreement.[73]

Because the definition of "impaired property incorporates several other defined terms including, "property damage," "your product" and "your work," the full scope of the exclusion cannot be fully understood until all the definitional terms are taken into account.[74]

At least one court has found the impaired property exclusion to be ambiguous. In *Serigne v. Wildey*,[75] the exclusion was found to be "ambiguous and/or in conflict with other segments of the policy defining coverage." Because the new exclusion carries forward the references to "contract or agreement" from the 1973 exclusion, one can anticipate a continuation of those cases finding the exclusion not applicable to claims sounding and tort.[76] This choice of language is unfortunate as it

---

73 *See* Susan J. Miller & Philip Lefebvre, *Miller's Standard Insurance Policies Annotated*, Vol. I at 418 (4th ed. 1995).

74 For purposes of the exclusions application to the construction industry, the term "your work" has significance. It is defined to mean:

a. Work or operations performed by you or on your behalf; and
b. Materials, parts or equipment furnished in connection with such work or operations.

"Your work" includes warranties or representations made at any time with respect to the fitness, quality, durability or performance of any of the items included in a. or b. above. *See* Susan J. Miller & Philip Lefebvre, *Miller's Standard Insurance Policies Annotated*, Vol. I at 419 (4th ed. 1995). It is also important to note that the standard policy introductory paragraphs define the terms "you" and "your" to refer to the named insured shown in the declarations, and any other person or organization qualifying as a named insured under the policy. *See, e.g.*, Susan J. Miller & Philip Lefebvre, *Miller's Standard Insurance Policies Annotated*, Vol. I at 409 (4th ed. 1995).

75 612 So. 2d 155 (La. Ct. App. 1992).

76 *See, e.g.*, *Geurin Contractors, Inc. v. Bituminous Cas. Corp.*, 636 S.W.2d 638, 642 (Ark. Ct. App. 1992) (store owner's claims against contractor for delays in finishing work such that road access restricted was brought on negligence theory and not breach of contract and, therefore, exclusion for "delay in or lack of performance . . . of any contract or agreement" did not preclude coverage); *Tobi Eng'g, Inc. v. Nationwide Mut. Ins. Co.*,

tends to reinforce the already rampant impression in the courts that insurance coverage is afforded for tort claims and not contract claims. In reality, coverage is afforded based upon the terms of the policy and not upon how a particular claimant styles its specific claims.

If the loss of use of the impaired property cannot be corrected by the "repair, replacement, adjustment or removal of "your product" or "your work" or by "[y]our fulfilling the terms of the contract or agreement," then the property does not qualify for the exclusion.[77] Because the 1986 version of the exclusion also contains the "sudden and accident" language, one can anticipate more of the same interpretation issues arising under the new exclusion. In *Riley Stoker Corp. v. Fidelity & Guar. Ins. Underwriters, Inc.,* [78] the insured built two coal-fired steam generators for a power plant. The generators incorporated two "ball tube mills" which were defective and shut down the power plant. The court, however, did not apply the 1973 version of exclusion (m) as it concluded that the damage arose after the insured's product had been "put to use" and thus fell within the exclusions exception.[79]

d. Exclusion (n) – The Sistership Exclusion

This exclusion applies primarily to product recalls. It states that insurance does not apply to:

> Damages claimed for any loss, cost or expense incurred by you or others for the loss of use, withdrawal, recall, inspection, repair, replacement, adjustment, removal or disposal of:

---

574 N.E.2d 160 (Ill. Ct. App. 1991); *Hartford Ins. Co. v. City of Sanibel,* 500 So. 2d 581 (Fla. Dist. Ct. App. 1986); *Alert Centre, Inc. v. Alarm Protection Serv., Inc.*, 967 F.2d 161 (5th Cir. 1992) (applying Louisiana law); *Glens Falls Ins. Co. v. Donmac Golf Shaping Co.,* 417 S.E.2d 197 (Ga. Ct. App. 1992).

77 *See, Action Auto Stores, Inc. v. United Capitol Ins. Co.*, 845 F. Supp. 417, 426 (W.D. Mich. 1993) (exclusion did not apply to insured who was sued for negligent instillation of gasoline containment system which polluted surrounding property as there was no evidence that the injury done to surrounding property could be remedied by the repair or the replacement of the insured's work.).

78 26 F.3d 581 (5th Cir. 1994).

79 *Id.* at 588 and 589.

(1) "your product;"
(2) "your work;" or
(3) "impaired property;"

if such product, work, or property is withdrawn or recalled from the market or from use by any person or organization because of a known or suspected defect, deficiency, inadequacy, or dangerous condition in it.[80]

The purpose behind the sistership exclusion is to protect insurers against liability for cost of recalls. These costs can be quite significant as was demonstrated by the cost incurred by Johnson & Johnson Companies in connection with a recall of its product known as Tylenol.[81] This exclusion, as one might expect, applies with more force to manufacturers than to contractors. In the construction setting, this exclusion can present coverage issues to material suppliers.

The case of *Superior Steel, Inc. v. Bituminous Casualty Corp.*,[82] illustrates an attempted application of the exclusion to a construction situation. In this case, the insured was a manufacturer of steel bolts. The bolts failed to meet project specifications and, as a result, the manufacturer agreed to pay for the repairs and sought recover at the value of the defective bolts and the amount it paid in damages as a result of the out of tolerance bolts. The insurers sought to avoid coverage under several exclusionary clauses, including the sistership exclusion. The court held that the sistership exclusion did not apply because the manufacturer did not seek a withdrawal from the market of all similar products because of a defect.[83]

---

80 *See, e.g.*, Susan J. Miller & Philip Lefebvre, *Miller's Standard Insurance Policies Annotated*, Vol. I at 412 (4th ed. 1995).

81 For cases discussing the purpose and policy behind the sistership exclusion *see Four City Dillon, Inc. v. Aetna Cas. & Sur. Co.,* 852 F.2d 168 (6th Cir. 1988); *Honeycomb Sys., Inc. v. Admiral Ins. Co.*, 567 F. Supp. 1400 (D. Me. 1983); *Todd Shipyards Corp. v. Turbine Service, Inc.*, 674 F.2d 401 (5th Cir. 1982), *aff'd in part, rev'd in part,* 763 F.2d 745 (5th Cir. 1985).

82 415 So. 2d 354 (La. Ct. App. 1982).

83 *Id.* at 358. *See also United States Fidelity & Guar. Co. v. Nevada Cement Co.*, 561 P.2d 1335 (Nev. 1977) (exclusion did not apply to

C. Contractual Liability Coverage

There is no insuring language or endorsement that expressly provides "contractual liability" coverage. In fact, the case law is replete with examples of holdings finding no coverage based upon the observation that a CGL policy does not afford coverage to an insured for damages incurred for breaches of contract. Sometimes the court's coverage analysis proceeds no further. For example in *Silk v. Flat Top Construction, Inc.*,[84] a construction manager was denied coverage for claims that it failed to properly supervise and inspect work, failed to insure that construction of a house conformed with the contract documents and failed to avoid excessive cost overruns. The court reasoned that because the damages sought were related to cost overruns and delays in moving and, thus, were breach of contract damages rather than property damages, no coverage existed under the policy.[85]

There is a policy exclusion which directly addresses the coverage afforded for certain types of contract liability. Exclusion b of the 1986

---

defective concrete as the claim for damages did not involve costs associated with the withdrawal, inspection, repair and replacement of the insured's product.); *Forest City Dillon, Inc. v. Aetna Cas. & Sur. Co.*, 852 F.2d 168 (6th Cir. 1988) (applying Ohio law) (exclusion did not apply to claim of defective aluminum siding where manufacturer did not withdrawal product from market); *Imperial Cas. & Indem. Co. v. High Concrete Structures, Inc.*, 858 F.2d 128 (3d Cir. 1988) (applying Pennsylvania law) (exclusion did not apply to damage to washers stamped out of steel supplied by insured as damage claimed by hot rolled carbon steel buyer was unrelated to withdrawal from the market of insured's supplier's steel.).

84 453 S.E.2d 356 (W. Va. 1994).

85 *See also Stanford Ranch, Inc. v. Maryland Cas. Co.*, 883 F. Supp. 493 (E.D. Cal. 1995), *aff'd* 89 F.3d 618 (9th Cir. 1996) (land developer brought suit against its CGL carrier, claiming entitlement to coverage for suits brought by subdevelopers that alleged that the insured misrepresented and failed to disclose the amount and significance of wetlands. Court held no coverage existed because the insured's obligation to provide accurate information arose only because it had entered into a contract with the subdevelopers and thus, the claims were dependent upon contract and not tort.).

CGL policy form addresses coverage for contractual liability. Under the exclusion, coverage is not afforded for:

> "Bodily injury" or "property damage" for which the insured is obligated to pay damages by reason of the assumption of liability in a contract or agreement. This exclusion does not apply to liability for damages:
>
> (1) assumed in a contract or agreement that is an "insured contract," provided the "bodily injury" or "property damage" occurs subsequent to the execution of the contract or agreement; or
>
> (2) That the insured would have in the absence of the contract or agreement.[86]

As was seen with the "impaired property" exclusion, the defined terms contained in the exclusion can sometimes be more lengthy than the exclusion itself. This is the case with the definition of "insured contract." The term has a six-part meaning, only the last of which is of general interest to participants in the construction industry. Part (f) of the definition of an "insured contract" means:

> That part of any other contract or agreement pertaining to your business (including an indemnification of a municipality in connection with work performed for a municipality) under which you assume that tort liability of another party to pay for "bodily injury" or "property damage" to a third person or organization. Tort liability means a liability that would be imposed by law in the absence of any contract or agreement.[87]

---

86 Susan J. Miller & Philip Lefebvre, *Miller's Standard Insurance Policies Annotated*, Vol. I at 409 (4th ed. 1995).

87 Susan J. Miller & Philip Lefebvre, *Miller's Standard Insurance Policies Annotated*, Vol. I at 418 (4th ed. 1995). (f), however, does not include that part of any contract or agreement:

> (1) that indemnifies a railroad for "bodily injury" or "property damage" arising out construction or demolition operations, within 50 feet of any railroad property and effecting any railroad bridge or trestle, tracks, road-beds, tunnel, underpass or crossing;

Before 1986, the contractual liability exclusion spoke in terms of "incidental contracts" instead of "insured contracts." The basic definition of an "incidental contract" under the 1973 policy form did not include the indemnity agreements one often finds in construction contracts. The broad form comprehensive general liability endorsement first developed in 1978 and amended in 1981, however, did broaden the definition of incidental contract to include such indemnity agreements.[88] One of the problems with the 1973 contractual liability exclusion was the exception for warranty of fitness or quality claims. A number of courts found this exception to the contractual liability exclusion to create an ambiguity with regard to the scope of coverage for warranty claims.[89]

The courts that did not find an ambiguity between the "work performed" and "contractual liability" exclusion focused upon the purpose of exclusionary language. These courts did not find credible the

---

(2) that indemnifies an architect, engineer, or surveyor for injury or damages arising out of:

(a) preparing, proving or failing to prepare or approve maps, drawings, opinions, reports, surveys, change orders, designs or specifications; or

(b) giving directions or instructions or failing to give them, if that is the primary cause of the injury or damage; or

(3) under which the insured, if an architect, engineer or surveyor, assumes liability for an injury or damage arising out of the insured's rendering or failure to render professional services, including those listed in (2) above and supervisory, inspection or engineering services.

Id.

88 *See* Susan J. Miller & Philip Lefebvre, *Miller's Standard Insurance Policies Annotated*, Vol. I at 452.4 (4th ed. 1995).

89 *See, e.g.*, *Simon v. Shelter Gen. Ins. Co.*, 842 P.2d 236 (Colo. 1992); *Bond Bros. Inc. v. Robinson*, 471 N.E.2d 1332 (Mass. 1984) (case contains a list of jurisdictions finding work product exclusion and contractual liability exclusion, when read in tandem, ambiguous and therefore affording coverage to the insured.).

argument that an ambiguity created between two exclusions could somehow create a grant of coverage.[90]

The deletion of the warranty language from the 1986 contractual liability exclusion should put an end to such policy challenges. The new language also contains an express reference that the type of liability being assumed by the insured is the "tort liability" of another. This change in language may prove significant to a particular subset of indemnity claims which often exist when surety bonds are issued in connection with construction projects. It is standard practice for sureties to obtain indemnity agreements from their principal and the individual owners of the principal protecting it against any losses incurred as a result of issuing performance or payments on behalf of the principal/contractor. When a surety experiences a loss and makes a claim under these indemnity agreements, an issue arises as to whether such claims fall within the "incidental contract" exception under the broad form endorsement.

One court answered in the affirmative. In *Natchitoches Parish School Board v. Shaw,*[91] a surety brought suit against its contractor/principal seeking indemnity for losses incurred as a result of issuing bonds to the contractor for a roofing project. The owner sued the surety for performance, and the surety tendered the defense to the contractor who, in turn, tendered the surety's request to its liability carrier. The Louisiana court concluded that the liability carrier was obligated to defend the owner's action against the surety on the ground that the contractor's indemnification obligation was an incidental contract under the terms of the general liability policy.[92]

It is thought by some that the express reference to tort liability will restrict coverage for such indemnity obligations as discussed in *Natchitoches*.[93]

Yet, the 1986 version contains language not found in the 1973 form which may bring such indemnity agreements back into coverage.

---

90 *LaMarche v. Shelby Mut. Ins. Co.*, 390 So. 2d 325 (Fla. 1980).

91 620 So. 2d 412 (La. Ct. App. 1993).

92 *Id.* at 413. *But see Gerrity Co. v. CIGNA Property & Cas. Ins. Co.*, 860 P.2d 606 (Colo. Ct. App. 1993) (breach of guaranty agreement not covered under incidental contract exception.).

93 *See, e.g.*, Gregory G. Scholtz, *Commercial General Liability Coverage of Faulty Construction Claims*, 33 TORT & INS. L.J. 257, 268 (1997).

Exception (2) of the contractual liability exclusion covers liability for damages "that the insured would have in the absence of the contract or agreement." If the insured can argue that its liability to the surety would exist even in the absence of a written indemnification agreement, then arguably coverage might lie under the policy.

A case from the Florida appellate courts several years before the 1986 policy language was adopted is of interest on this point. In *Western World Insurance Company, Inc. v. Travelers Indemnity Company,*[94] a performance bond surety and its principal brought suit against the principal's CGL insurer. The insured/principal was a security service that was sued by a person after an assault and battery was committed against them. The surety provided a bond pursuant to a Florida law that requires such instruments in order for security services to obtain the necessary licensing.[95] The security service and its surety sued the CGL insurer for its failure to defend and indemnify them against the third-party's claims. The trial court found that had the security service paid the judgment directly to the claimant, it would have been entitled to recover from its insurer. The fact that the surety made payment, allowed the surety to be subrogated to the security service's rights under its policy. In rejecting the contractual liability exclusion defense offered by the insurer, the court held:

> Western World additionally argues that since the policy expressly excludes liability assumed by Clark & Moore [the security service] under any contract or agreement, it is thereby insulated from liability to the amount of $5,000, which Clark & Moore were obligated to their surety in the event of payment. We agree with the lower court's determination that the assumption of liability by a principal to its surety must be pursuant to an express contract or agreement. While the bond furnished by travelers is a contract or agreement, it did not contain any express assumption of liability by the insureds. Clark & Moore's liability arose as a matter of law from the theory of indemnification. A general rule is that an exclusion, such as the one here, refers only to contractually-created liability, and has no affect upon the general liability of the insured arising by operation of law.[96]

---

94 358 So. 2d 602 (Fla. App. Dist. Ct. 1978).

95 *Id.* at 603.

96 *Id.* at 604; *see also Ohio Cas. Ins. Co. v. Flanagan*, 210 A.2d 221, 231 (N.J. 1965); *Larson Constr. Co. v. Oregon Auto. Ins. Co.*, 450 F.2d 1193

Because the duties principals owe their sureties (performance, reimbursement and restitution) arise out of operation of law rather than contract, the issue with respect to insurance coverage in these situations is not necessarily settled by the 1986 policy language.[97]

Contractual liability coverage under the 1986 policy form was examined in some detail in the decision of *Gibson & Associates, Inc. v. Home Insurance Company*,[98] wherein a contractor brought suit against its CGL carrier for breach of the duty to defend and indemnify it against claims brought by a municipal owner for breach of contract and indemnity. Gibson contracted with the city of Dallas to perform certain street, sidewalk and public underground utility upgrades along Main Street in downtown Dallas.[99] During the course of the work, several store owners and tenants along Main Street brought suit against the city claiming poor or inadequate planning had caused them a loss of the use of their properties. The city responded to these law suits by filing a third-party complaint against Gibson seeking indemnity and damages for breach of contract. The contractor forwarded these actions to its liability carrier who denied coverage. Gibson then brought suit against the carrier, who intern, moved for partial summary judgment on the issue of whether it was under a duty to defend the contractor against the City's claims.

In ruling upon the insured's motion, the court addressed many of the typical coverage issues that arise under CGL policies including (1) the scope of the insured's duty to defend, (2) the existence of an "occurrence", (3) the existence of "property damage" and such policy exclusions as the "impaired property" exclusion.[100] The court, in analyzing these issues, concluded that under Texas law the policy would not provide coverage for the city's breach of contract action against the

---

(9th Cir. 1971); *Home Ins. Co. v. Southport Terminals, Inc.*, 240 So. 2d 525 (Fla. App. Dist. Ct. 1970); *see also* 12 COUCH, INSURANCE § 44.446 et seq. (2d ed. 1964).

97 *See, e.g.*, RESTATEMENT (THIRD) SURETYSHIP AND GUARANTY §§ 21 (Duty of Performance), 22-25 (Duty of Reimbursement), and 26 (Restitution).

98 966 F. Supp. 468 (N.D. Tex. 1997).

99 *Id.* at 470.

100 *Id.* at 472-475.

contractor. The city's breach of contract claim involved only economic loss and was therefore not property damage as defined by the policy.[101] Nor did the court find that the breach of contract claim met the policy's "occurrence" requirement.[102] Finally, the court concluded that the impaired property exclusion applied to the breach of contract claim.[103]

In light of the court's findings relative to the city's breach of contract claim, its decision on the city's indemnity claim is somewhat surprising. As one might expect, the insurer raised the contractual liability exclusion as a defense to the indemnity claim. The contract between the city of Dallas and Gibson, however, contained an indemnification provision which the court construed fell within the "insured contract" exception to the exclusion.[104] While the court did not

---

101 *Id.* at 474.

102 *Id.*

103 *Id.* at 474-475.

104 *Id.* at 476. On this score the court noted:

> By creating an exception [the insured contract exception] from exclusion (b) for property damage for which the insured becomes obligated to pay due to an indemnification agreement assumed in an "insured contract," Home thus has expressly agreed to provide a limited range of contractual liability coverage.

> Home also argued that because the indemnity agreement between Gibson and the City of Dallas was based in contract and CGL policies provide coverage for only tort-based liabilities, it had no duty under the policy to defend Gibson on the indemnity claim. With respect to this argument, the court found:

> . . . that this reasoning misconstrues the nature of indemnity obligations: "the assumption by contract of the liability of another is distinct conceptually from the breach of one's contract with another. Liability on the part of the insured for the former is triggered by contractual performance; for the latter liability is triggered by contractual breach." Contractual liability coverage provisions are specifically designed to provide coverage for indemnification actions, not withstanding the fact that the policy may offer no coverage for breach of a contractual duty.

Id. at 476; *see also Musgrove v. Southland Corp.*, 898 F.2d 1041, 1044 (5th Cir. 1990) (applying Louisiana law.); *White Mountain Cable Constr.*

find "property damage" with respect to the city's breach of contract claim, the situation was different with respect to the indemnity claim. Citing cases from Arkansas and Wisconsin, the court agreed with the insured that the store owner's claims for financial losses created by construction-imposed restrictions on access to their property and by its consequent loss of use constituted "loss of use of tangible property that is not physically injured" within the meaning of the policy's definition of "property damage."[105] The court also found the "occurrence" requirement met with respect to the indemnity claim. Under Texas law, injuries are "accidental" and within coverage of an insurance policy if from the viewpoint of the insured, the injuries are not the natural and probably consequences of the action or occurrence which produced the injury.[106] The focus is not on whether the insured's conduct or actions were intentional, but on whether the insured intended the damages or injuries which are the subject of the underlying claims.[107] The store owner's loss of their leases, occasioned by the reduction in the flow of customers, was regarded by the court as consequential damage that was neither intended nor certain to occur. As such, the extent of damages exceeded the harm normally foreseeable thus rendering the defective performance and "accident" within the meaning of the policy's provisions.

Co. v. Transamerica Ins. Co., 631 A.2d 907, 909-10 (N.H. 1993) (holding that the insured had entered into an indemnity agreement and hence was entitled to a defense by the insurer); *Aetna Cas. & Sur. Co. v. Spancrete of Ill., Inc.*, 726 F. Supp. 204, 207 (N.D. Ill. 1989) (applying Illinois law) (finding contractual liability coverage inapplicable since the third-party action was based on breach of contract); *Lafarge Corp. v. Hartford Cas. Ins. Co.*, 61 F.3d 389, 396-97 (5th Cir. 1995) (discussing contractual liability coverage under Texas law).

105 *Id.* at 477. The cases cited with approval for this proposition from Arkansas and Wisconsin were respectively, *Geurin Contractors, Inc. v. Bituminous Cas. Corp.*, 636 S.W.2d 638 (Ark. Ct. App. 1982) and *Sola Basic Indus., Inc. v. United States Fidelity & Guar. Co.*, 280 N.W.2d 211 (Wis. 1979).

106 *Id.* at 478; *see also Republic Nat'l Life Ins. Co. v. Heyward*, 536 S.W.2d 549 (Tex. 1976).

107 *Id.*; *see also Bituminous Cas. Corp. v. Vacuum Tanks, Inc.*, 75 F.3d 1048 (5th Cir. 1996).

In rejecting the insurer's claim that the contractual liability exclusion operated to preclude coverage the court noted:

> . . . Home argues that the City of Dallas never sought to recover indemnification for its own tort liability, but only for liability caused by Gibson's breach of contract and that the indemnity clause accordingly did not meet the requirements for an "insured contract."
>
> Such reasoning, however, seems to be inherently flawed. To be sure, the city does allege that its liability to the shop owners arose from Gibson's breach of the contractual agreement to conduct road construction in such a way as not to impede access to businesses along Main Street during business hours. The claims asserted against it by the shop owners, however, sought to hold the city liable not for breach of contract, but rather for negligence and unconstitutional deprivations of property. Clearly, the mere fact that the city's indemnification actions against Gibson arise from the parties contractual relationship does not convert the show owners' tort-based causes of action against the city into breach-of-contract claims.[108]

In view of the court's interpretation of the policy's contractual liability coverage, it granted the contractor's cross motion for summary judgment and found home had breached its duty to defend its insured against the city's claims.[109]

It is unusual for courts to conclude that a finding whether "property damage" and/or "occurrence" exist depends upon the nature of the underlying claim, would find much support in the jurisprudence of insurance law. The court's analysis, however, with regard to the contractual liability coverage issues is not all that unusual.[110]

1. Liability In Any Event

As discussed earlier, an exception to the contractual liability exclusion is for liability attaching to the ensured regardless of an indemnity obligation. This concept was discussed in *Reliance Insurance*

---

108 *Id.* at 479.

109 *Id.* at 479-480.

110 *See, e.g.*, Scott C. Turner, *Insurance Coverage of Construction Disputes* § 9.01, et seq. (1992).

v. Armstrong World Industries, Inc.,[111] which involved a sales contract which included a clause requiring the seller to indemnify the buyer for any damage or loss. The buyer sued the seller under several theories of liability including both the indemnity agreement in the sales contract and joint and several liability under CERCLA. The seller's CGL carrier pointed to the contractual liability exclusion to avoid liability. The court, however, held that the exclusion did not apply even though liability was assumed by the insured under a contract, because the insured would have been liable regardless of this contractual undertaking.[112] Because there was no evidence that the liability under the contract was any more extensive than the liability under CERCLA, the court concluded that the CGL policy covered the loss.[113] The seller had agreed contractually to provide a remedy to the buyer which was already provided under the CERCLA statutory scheme. As a result, the contractual liability exclusion did not apply.[114]

---

111 614 A.2d 642 (N.J. Super. Ct. Law Div. 1992), *opinion modified*, 625 A.2d 601 (N.J. Super. Ct. Law Div. 1993), *rev'd on other grounds*, 678 A.2d 1152 (N.J. Super. App. Div. 1152).

112 *Id.* at 658.

113 *Id.*; *see also Karadis Bros. Painting Co. v. Pennsylvania Nat'l Mut. Cas. Ins. Co.*, 292 A.2d 42 (N.J. Super 1972).

114 *See also Occidental Fire & Cas. Co. v. Lumberman's Mut. Cas. Co.*, 667 F. Supp. 679 (N.D. Cal. 1987); *Daily Express, Inc. v. Northern Neck Transfer Corp.*, 490 F. Supp. 1304 (M.D. Pa. 1980); *Trus Joist Corp. v. Safeco Ins. Co. of Am.*, 735 P.2d 125 (Ariz. Ct. App. 1986). The case of *Board of Trade Livery Co. v. Georgia Cas. Co.*, 200 N.W. 633 (Minn. 1924) is of interest on this subject. The insured contracted with the Northern Navigation Company which operated a line of passenger steam ships, to give steam ship passengers an automobile tour of Duluth, during a tour an accident occurred which injured two passengers, who successfully sued the Northern Navigation Company to recover damages for their injuries. The navigation company, in turn, sued the insured for indemnification. The insurer denied defense and indemnity on grounds of a policy provision excluding "liability to others assumed by the assured under any contractor agreement, oral or written." *Id.* at 634. The court determined that this exclusion did not relieve the insurer from its obligation to insure against the loss arising from the insured's own liability for personal injuries suffered by its passengers. The court emphasized that the insured had to reimburse the navigation company for the insured's own negligence, not for a secondary liability. For that

D. Coverage Afforded To Additional Insureds

It is common practice in the construction industry for owners to require contractors to name them as additional insureds under their CGL policies. Contractors, in turn, often require their subcontractors to do the same. The process to accomplish this is fairly routine. The obligation generally arises by way of a contract provision requiring one party to name the other as an additional insured under its liability policy. The specific mechanics for accomplishing coverage is typically through the issuance of an additional insured endorsement to the named insured. The additional insured often does not receive the actual endorsement but is notified by way of a certificate of insurance that it has been extended this coverage.

At the outset it is important to understand that there is no one standard additional insured endorsement form. There are many different such forms. The ISO publishes two standard forms for owners, lessees or contractors.[115] Common additional insured forms such as Form CG

---

reason, the exclusion did not apply. *See also American Cas. Co. v. Timmons*, 352 F.2d 563 (6th Cir. 1965) (contractor's negligence shoring of an electrical duct caused a collapse. Contractor had agreed in its contract with owner to be responsible for any damages that occurred due to the contractor's negligence. This agreement, however, did not trigger the contractual liability exclusion as the contractor would have been liable for negligence in the absence of the indemnity agreement.).

115 Form A or Form CG 20 09 10 93 (Form CG 20 09) contains a number of exclusions. *See, e.g.*, Susan J. Miller & Philip Lefebvre, *Miller's Standard Insurance Policies Annotated*, Vol. I at 444 (4th ed. 1995). Form B, also identified as CG 20 10 10 93 (Form CG 20 10), contains no express reference to exclusions. There are also additional ensured endorsements specifically for engineers, architects, or surveyors. *See, e.g.*, ISO Form CG 20 07 10 93 reprinted in Miller's *Standard Insurance Policies Annotated*, Vol. I at 444.1 (4th ed. 1995). There are other additional insureds covered by a type including construction managers, general contractor on subcontractor's policy, as well as subcontractor on general contractor's policy, contractual indemnities and sureties. *See, e.g*, Susan J. Miller & Philip Lefebvre, *Miller's Standard Insurance Policies Annotated*, Vol. I at 445.1 (4th ed. 1995). Beyond the insurance industry there are numerous additional insured endorsements ranging

20 09 (Form A), covering owners and contractors, have several edition dates currently circulating in the market place.[116] The coverages are not necessarily the same under even the various editions of the same form. For example, by its terms, the 1993 edition of form 20 09 provides coverage to additional insureds only for damages arising out of the named insured's "ongoing operations." In other words, the 1993 edition of this form, in its insuring language, extends coverage only to "ongoing operations."[117] There are also blanket additional insured endorsements. Where an insured faces repeated contractual demands to procure additional insurance coverage, it is more efficient for the insured to secure a blanket endorsement thereby making it unnecessary to issue separate endorsements every time it becomes necessary to add an additional insured to the policy.[118]

It is important to determine the specific endorsement language that is being offered or employed in order to understand the scope of coverage. In reviewing case law on this subject, it is often the case that courts will only briefly mention a party's status as an additional insured without indicating anything more. This can make the evaluation of the precedential value of such case law fairly challenging.

1. The Scope Of Additional Insured Coverage

Being named an additional insured under someone else's policy is not equivalent to purchasing your own coverage. Additional insured coverage is intended by the insurance industry to cover the vicarious liability of the additional insured arising out of the operations of the named insured. Thus, form GC 20 10 contains the operative language:

---

from governmental subdivisions (CG 24 09) to church members (CG 20 22).

116 Form CG 20 09 can be found in at least three different forms, *i.e.*, the 1973 edition, the 1985 edition and the 1993 edition.

117 All three editions of Form 20 09, however, contain a separate "completed operations" exclusion which is intended to limit coverage to additional insureds for liability arising once the named insured's work has been completed.

118 These blanket endorsements are sometimes referred to as "automatic additional insureds endorsements."

> Who is an Insured (Section II) is amended to include as an insured the person or organization shown in the schedule, but only with respect to liability arising out of your ongoing operations performed for that insured.[119]

The "arising out of" language contained in the additional insured endorsement has been interpreted to mean that coverage is limited to vicarious liability. In *Granite Construction Co. v. Bituminous Insurance Cos.,*[120] Granite, as general contractor, hired Joe Brown Company to remove asphalt from its construction site. Pursuant to the parties' subcontract, Joe Brown was required to name Granite as an additional insured under its liability policy, which it did. When one of Joe Brown's employees sued Granite claiming that the general contractor negligently loaded his truck causing it to overturn and injure him, the general contractor sought coverage under Joe Brown's policy. The carrier denied a tender of defense contending that the additional insured endorsement provided only coverage based upon the named insured's actions and not those of the additional insured. The court agreed with

---

119 One commentator has discussed the scope of coverage under this particular endorsement as follows:

> Based on the number and name of this endorsement, a person might surmise that it is very similar to CG 20 09 [Form A additional insured endorsement] however, CG 20 10 lists the scheduled person or organization as an additional insured only with respect to liability arising out of the named insured's ongoing operations performed for the additional insured; there is no mention of coverage for liability arising out of the acts or omissions of the additional insured in connection with the general supervision of the named insured's work. All exclusions on the CGL form apply under this endorsement and there are no additional exclusions to consider. Also, liability assumed under "insured contracts" is covered under CG 20 10.
>
> CG 20 10 would be used, for example, by an independent contractor hired to repair or work on property owned by Company A when Company A wants liability coverage for potential exposures but will not be involved in any supervision of the contractor's work.

FC&S Bulletins, Public Liability B. 1-4 (Sept. 1996).

120 832 S.W.2d 427 (Tex. Ct. App. 1992).

the insurer. The additional insured endorsement did not provide coverage to Granite for its own operations and because the underlying action alleged that Granite's liability arose out of its own loading operations, the additional ensured endorsement provided no coverage.[121]

Another common coverage afforded to additional insureds under certain endorsement types is the additional insured's liability arising out of its general supervisory operations over the named insured's activities. In this regard, coverage is extended somewhat beyond pure vicarious liability situations. As one might expect, there is not always uniform opinion over what is meant by "general supervision" liability. For example, in *Union Electric Co. v. Pacific Indemnity Co.,*[122] the term "general supervision" denoted the type of supervision that an owner-insured customarily undertakes with respect to work performed by an independent contractor. An owner supplying specifications and plats for work and reserving the right to inspect the work has been deemed to be "general supervision."[123] A construction manager's activities in coordinating, scheduling, and inspecting the work of contractors was deemed to be within "general supervision" activities.[124] With respect to a general contractor's activities, "general supervision" was construed to mean those oversight activities of a subcontractor's work necessary to see that the work is done in accordance with the plans and specifications.[125] The more control one exerts over the named insured's activities, however, the more likely it is that one's actions have gone beyond a general supervisory roll.[126]

As with respect to the named insured, an insurer's obligation to defend an additional insured may be broader than its indemnity

---

121 *Id.* at 429.

122 422 S.W.2d 87 (Mo. Ct. App. 1967).

123 *See, e.g, Western Cas. & Sur. Co. v. Southwestern Bell Telephone Co.*, 396 F.2d 351 (8th Cir. 1968).

124 *See, e.g,, Casualty Ins. Co. v. Northbrook Property & Cas. Co.*, 501 N.E.2d 812 (Ill. App. Ct. 1986).

125 *See, e.g., Ohio Cas. Ins. Co. v. Flanagin*, 210 A.2d 221 (N.J. 1965).

126 *Id.* At least one court has found the term "general supervision" to be ambiguous. In *City of Detroit Board of Water Commissioners v. Maryland Cas. Co.*, 1974 CCH (Fire & Casualty) 49 (Mich. App. 1974), the court found that the term had no customary usage or meaning and, therefore, construed it in favor of the insured.

obligations. As a general rule, if the pleadings do not clearly establish that the claims against the additional insured relate solely to the additional insured's own acts of negligence which do not fall within general supervision activities, then an obligation to defend will arise.[127] The existence of additional insureds creates defense cost allocation issues.[128] In *United States Fire Insurance Co. v. Aetna Life & Casualty,*[129] two insurance companies disputed which one had the duty to defend an underlying personal injury action. The court concluded that the subcontractor's CGL carrier, rather than the general contractor's insurer, was responsible for defending against the subcontractor's employee's personal injury suit. The subcontractor's policy contained an additional insured endorsement providing coverage to the general contractor and owner "but only with respect to acts or omissions of the named insured [subcontractor] in connection with the named insured's operations.[130] In this case, the worker was injured when he tripped on a conduit protruding from a concrete slab. While the subcontractor/employer was not named in the complaint, its liability to the worker potentially could have arisen from an act or omission, whether or not such act or omission rose to the level of negligence. Such a possibility was sufficient to trigger a duty to defend under the additional insured endorsement.[131]

The standard CGL policy contains an "other insurance" provision. This provision is intended to indicate when the CGL policies coverage is primary and when it is excess in the even other coverage exists. The standard "other insurance clause" does not specifically address the

---

127 See, e.g, *First Ins. Co. of Hawaii Ltd. v. State of Hawaii*, 665 P.2d 648 (Haw. 1983).

128 *See, e.g., Stonewall Ins. Co. v. City of Palo Verdes Estates*, 54 Cal. Tptr. 2d 176 (Cal. Ct. App. 1996); *Armstrong World Indus., Inc. v. Aetna Cas. & Sur. Co.*, 52 Cal. Rptr. 2d 690 (Cal. Ct. App. 1996); *Aerojet-General Corp. v. Transport Indem. Co.*, 53 Cal. Rptr. 2d 398 (Cal. Ct. App. 1996), *opinion modified*, 923 P.2d 76 (Cal. 1996), *aff'd in part, rev'd in part*, 948 P.2d 909 (Cal. 1997); *Buss v. Superior Court*, 50 Cal. Rptr. 2d 477 (Cal. Ct. App. 1996).

129 684 N.E.2d 956 (Ill. Ct. App. 1997).

130 *Id.* at 962.

131 *See also City of New York v. Consolidated Edison Co.* of N.Y., 655 N.Y.S.2d 496 (N.Y. App. Div. 1997).

"additional insured" endorsement. Instead, the standard clause makes the liability coverage excess to certain property and automobile coverages.[132] The additional insured that wishes to strengthen its position that its own coverage does not share in defense costs may consider adding language to its policy stating that it will be excess over any coverage available to it as an additional insured.[133]

2. *The Slippery Slope Of Additional Insured Coverage*

Confusion sometimes arises as to whether a party's status as an indemnitee for which contractual liability coverage is purchased by the indemnitor is equivalent to being an additional insured under the indemnitor's liability policy. The two are not equivalent. As an additional insured one has a direct contractual relationship with the insurer. One's status as a contract indemnitee, however, is different. In this case the insurer's contractual obligations run to its insured, the indemnitor, and not directly to the indemnitee.[134]

One practical effect of this difference is the application of anti-indemnity statutes to one status as a contract indemnitee as opposed to an additional insured. Where an indemnity agreement is in violation of state law the obligation to provide insurance to cover the void contractual obligation will not be enforceable.[135] The infirmity of the underlying contractual indemnity obligation should not, however, affect one's indemnity status as an additional insured. Sometimes this is not always the case.[136] The better reason decisions, unless the state's

---

132 *See, e.g.*, Susan J. Miller & Philip Lefebvre, *Miller's Standard Insurance Policies Annotated*, Vol. I at 416 (4th ed. 1995).

133 *See*, Wielinski, et al., *Contractual Risk Transfer* (International Risk Management Institute, 1995) at XI.C.12.

134 *See, e.g., Gotro v. The Town of Melvile*, 527 So. 2d 568 (La. Ct. App. 1988); *DiPietro v. City of Philadelphia*, 496 A.2d 407 (Pa. Super. Ct. 1985).

135 *See, e.g., Shaheed v. Chicago Transit Auth.*, 484 N.E.2d 542 (Ill. Ct. App. 1985).

136 *See, e.g, Posey v. Union Carbide Corp.*, 507 F. Supp. 39 (M.D. Tenn. 1980); *Transcontinental Ins. Co. v. National Union Fire Ins. Co. of Pittsburgh*, 662 N.E.2d 500 (Ill. Ct. App. 1996) (because subcontractor's agreement to indemnify and purchase insurance for general contractor violated Illinois anti-indemnity statute, contractor was not an additional

indemnity statute expressly disallows additional insured coverage, hold that the anti-indemnity bar does not affect additional insured coverage.[137]

The insurance industry believes that the additional insured endorsement should afford quite limited coverage. For the most part the industry believes that the endorsement limits coverage for vicarious liability. The plain language of many endorsements, however, does provide coverage for the additional insured's direct negligence, particularly for acts falling within the "general supervision" penumbra. Those securing additional insured status often claim that coverage is even broader. Sometimes the courts agree. For example in *Certainteed Corp. v. Employers Insurance of Wausau,*[138] a construction worker was injured while performing work on a manufacturing plant and sued the manufacturer. The manufacturer sought coverage as an additional insured under the general contractor's CGL policy. The insurer did not dispute that the manufacturer was an additional insured, but disagreed on the extent of coverage afforded by this policy endorsement. The carrier claimed that it was not responsible for insuring the manufacturer against acts of the manufacturer's own negligence. The court disagreed with the carrier. In construing the policy language of who is "an insured" with the scope of the insurance obligation placed on the general contractor under the Owner/Contractor agreement, the court concluded that the carrier owed an obligation to defend and indemnify the manufacturer as an additional insured.[139]

---

insured under CGL policy and carrier owed no obligation to defend or indemnify general contractor.).

137 *See, e.g.*, *County of Orange v. Hartford Accident & Indem. Corp.*, 641 N.Y.S.2d 118 (N.Y. App. Div. 1996); *Bosio v. Branigar Org., Inc.*, 506 N.E.2d 996 (Ill. App. Ct. 1987); *McAbee Constr. Co. v. Georgia Kraft Co.,* 343 S.E.2d 513 (Ga. Ct. App. 1986); *Shevron U.S.A., Inc. v. Bragg Crane & Rigging Co.,* 225 Cal. Rptr. 742 (Cal. Ct. App. 1986).

138 939 F. Supp. 826 (D. Kan. 1996).

139 There can be other surprises in policy interpretation. In *United States Fire Ins. Co. v. Aetna Life & Cas.*, 684 N.E.2d 956 (Ill. App. Ct. 1997), the endorsement provided coverage to additional insured "but only with respect to acts or omissions of the named insured in connection with the named insured's operations . . .". There was no suggestion that named insured was negligent. Coverage was still afforded to additional insured as language "acts and omissions" does not mean negligence, *i.e.*, "acts" do not equate with "negligent acts." *Id.* at 962; *see also, J.A. Jones*

3. *Conflict Issues Created By Additional Insureds*

Perhaps because there is not an ongoing business relationship between the insurer and an additional insured as there is with the named insured, it is not all that uncommon for insurers to respond more slowly to the demands made of them by additional insureds. This can result in problems for the insurers. The situation is compounded when the interests of the named insured potentially conflict with those of the additional insured. This most often arises where the limits of coverage may not be sufficient to satisfy all the claims. The California decision in *Shell Oil Co. v. National Union & Fire Insurance Co.*,[140] is instructive. National union provided coverage to both its named insured, a general contractor, and the project owner, Shell Oil, as an additional insured. An injured worker on the construction project sued both Shell and the contractor. The insurer settled the worker's claim against the contractor by paying the policy limits of $1,000,000. It paid no monies on behalf of Shell. When Shell sued the carrier argued that by paying its policy limits on behalf of the named insured, it was, in essence affording a $1,000,000 to Shell. The carrier argued that because the contractor and Shell were joint tort feasors, the worker's release of the contractor reduced his claim against Shell by $1,000,000. The court rejected these arguments, finding that Shell had received no "offset" for the insurer's payment because the settlement was simultaneous with and dependent upon Shell's own contribution of $2,000,000 with the worker. Therefore, the carrier's decision to pay $1,000,000 in settlement of the worker's claim against the contractor was a favoring of the interest of one insured over the other and thus amounted to bad faith. The trial

---

Constr. Co. v. Hartford Fire Ins. Co., 645 N.E.2d 980 (Ill. App. Ct. 1995) (endorsement language "but only with respect to your operations" did not require named insured to be negligent); *National Union Fire Ins. Co. v. Glenview Park Dist.*, 632 N.E.2d 1039 (Ill. 1994) (exclusion in endorsement that no insurance for "damages arising out of the negligence" of the additional insured did not preclude coverage for statutory liability); *Village of Hoffman Estates v. Cincinnati Ins. Co.*, 670 N.E.2d 874 (Ill. Ct. App. 1996), *appeal denied*, 671 N.E.2d 731 (Ill. 1996) (team "solely" in endorsement precluded coverage).

140 52 Cal. Rptr. 2d 580 (Cal. Ct. App. 1996).

court's ruling that Shell was entitled to $500,000, or half of the insurer's policy limit, was "conservative in view of Shell's sole negligence," in the cause of the worker's injury. In light of Shell's sole negligence in the cause of the worker's injuries, the general contractor owed no obligation to indemnify Shell because the indemnification provision was not a broad form clause. Nevertheless, this limitation on indemnity did not affect the insurance provisions which afforded coverage.

E. Coverage Trigger Issues: When Does Damage Occur

One of the more vexing issues arising under occurrence-based liability coverage is determining which policy or policies apply when damage occurs over long periods of time. This is particularly the case where the damage remains latent for much of the time. Additional challenges are presented if during some of this time there was no policy in place or if one of the carriers because insolvent. The coverage and allocation issues that these situations pose can be quite problematic. Unfortunately, the courts have taken widely divergent positions on many of the questions posed by these types of damages.

1. Coverage Trigger Theories Under Occurrence-Based Policies

The occurrence-based CGL policy provides coverage for bodily injury and property damage that occurs during the policy period. As such, the insurer's duty to indemnify is "triggered" by the existence of bodily injury or property damage during the policy period. Injury must take place during the policy period for coverage to exist.[141] Determining the date of occurrence is critical. Unfortunately, this can be frustratingly difficult in the case of certain latent yet progressive injuries. The courts have come up with at least four separate theories with respect to determining when an occurrence takes place for purposes of coverage: (1) the manifestation rule; (2) the exposure rule; (3) the injury-in-fact rule; and (4) the continuous injury (or multiple) rule.[142] Some

---

141 *See, e.g., Smith v. Hughes Aircraft Co.*, 22 F.3d 1432 (9th Cir. 1993); *Appalachian Ins. Co. v. Liberty Mut. Ins. Co.*, 676 F.2d 56 (3d Cir. 1982).

142 The California Supreme Court's *en banc* decision in *Montrose Chem. Corp. of Cal. v. Admiral Ins. Co.*, 913 P.2d 878 (Cal. 1995) provides a

jurisdictions apply different trigger theories, depending upon whether the policy at issue is first party coverage (*e.g.* property policy) or third-party coverage (*e.g.* liability policy).[143]

---

fairly exhaustive discussion on the various trigger theories and rationales underlying them.

143 California is such a jurisdiction. In *Prudential-LMI Comp. Ins. v. Superior Court (Lundberg)*, 798 P.2d 1230 (Cal. 1990), the Supreme Court of California applied a manifestation trigger theory in a first-party property insurance context. Five years later, in *Montrose Chem. Corp. v. Admiral Ins. Co.*, 913 P.2d 878 (Cal. 1995), the same court applied a "continuous injury" trigger for third-party policies. In discussing the difference between first-party and third-party coverage, the court noted:

> [A] first-party insurance policy provides coverage for loss or damage sustained directly by the insured (*e.g.* life, disability, health, fire, theft, and casualty insurance). A third-party liability policy, in contrast, provides coverage for liability of the insured to a "third party" (.*e.g.*, a CGL policy, a directors' and officers' liability policy, or an errors and omissions policy). In the usual first-party policy, the insurer promises to pay money to the insured upon the happening of an event, the risk of which has been insured against. In the typical third-party liability policy, the carrier assumes a contractual duty to pay judgments. The insured becomes legally obligated to pay his damages because of bodily injury or property damage caused by the insured.
>
> The difference in the nature of the risks insured against under first-party property policies and third-party liability policies is also reflected in the differing causation analyses that must be undertaken to determine coverage under each type of policy . . . [The] liability analysis [for third-party coverage] differs substantially from the coverage analysis in the property insurance context, which draws on the relationship between perils that either covered or excluded in the contract. *In liability insurance, by insuring for personal liability, and agreeing to cover the insured for his own negligence, the insurer agrees to cover the insured for a broader spectrum of risks.*
>
> The parties' expectations may also differ depending upon the type of coverage sought. First-party property coverage is typically purchased in an amount sufficient to cover the insured's maximum potential loss (*e.g.*, fire insurance typically covers the value of the property insured). Hence, there is no reason for a first-party insured to look to more than one policy

The particular coverage trigger employed can have significant ramifications on the obligations and responsibilities of the insurers involved as well as the insured. As a general rule, the adoption of a "continuous injury" trigger will broaden coverage for the insured.[144] A manifestation trigger analysis, on the other hand, can result in significantly less coverage, particularly if the injury manifests itself during a period of time that the insured's coverage is inadequate or nonexistent.[145]

2. *Stacking*

Those jurisdictions which have adopted a continuous trigger theory usually temper the holding with a prohibition against "stacking" or aggregate of consecutive policy limits. As discussed by the New Jersey Supreme Court in *Keene*:

---

in the event of loss (the policy in effect at the time of the fire). Third-party liability coverage differs substantially. As the Court of Appeal below observed, "At best, the insured makes an educated guess about its potential exposure to third parties. At worst, the insured's best guess falls far short of the mark."

Id. at 886-87.

144 *See, e.g., Keene Corp. v. Insurance Co. of N. Am.*, 667 F.2d 1034 (D.C. Cir. 1981) (1982). In this case, the court that asbestos injuries were progressive and constituted a single, continuous harm. The court reasoned that even during the period after exposure was ceased up to manifestation of an injury or disease, the claimant in the asbestos case was repeatedly re-exposed to the asbestos particles in his body. Consequently, coverage was triggered throughout this period. Thus, each insurer whose policy was in effect during the period of exposure, exposure-in-residence, or manifestation of injury would be jointly and severally liable up to its policy limits for the entire amount of the loss. Although, while the insured is free to choose which carrier will reimburse him, he is not entitled to stack or aggregate coverage limits.

145 *See, e.g., Commercial Union Ins. Co. v. Sepco Corp.*, 765 F.2d 1543 (11th Cir. 1985); *see also* William R. Hickman & Mary R. DeYoung, *Allocation of Environmental Cleanup Liability Between Successive Insurers*, 17 N. Ky. L. Rev. 291 (1990); Laura A. Foggan & John C. Yang, *Tort Feasors' Responsibility for Uninsured Periods: Allocation, Economics, and Market Demand*, 8 ENV. CL. J. (1996).

> [T]he principle of indemnity implicit in these policies requires that successive policies cover single asbestos-related injuries. That principle, however, does not require that *Keene* be entitled to "stack" applicable policies' limits of liability. To the extent, we have tried to construe the policies in such a way that the insurers' contractual obligations for asbestos-related diseases are the same as their obligations for other injuries . . . .
>
> [W]e hold that only one policy's limit can apply to each injury. Keene may select the policy under which it is to be indemnified.[146]

III.
Waiver Of Subrogation

Construction contracts often contain provisions which require the parties to waive their right to claim damages against one another up to the amount of coverage available for their losses. These "waiver of subrogation" provisions are intended to cut down the amount of litigation that might otherwise arise due to the existence of insured loss. Viewed globally these clauses make sense as an insurer on one project might very well be denied recovery by such a clause but benefit from the existence of such a waiver on another project.[147]

Most waiver of subrogation provisions limit the waiver to losses covered by property insurance. The most common waiver clause is the one contained in the American Institute of Architects General Conditions, which states in pertinent part:

> The owner and contractor waive all rights against (1) each other and any of their subcontractors, sub-subcontractors, agents and employees, each of the other, and (2) the architect, architect's consultants, separate

---

146 *Keene Corp. v. Insurance Co. of North Am.*, 667 F.2d 1034, 1049-50 (D.C. Cir. 1981). *But see, Insurance Co. of N. Am. v. Forty-Eight Insulations, Inc.*, 633 F.2d 1212 (6th Cir. 1980), *clarified and aff'd on reh'g*, 657 F.2d 814 (6th Cir. 1981) (joint and several paradigm rejected and allocation among policies and uninsured periods adopted).

147 From a global perspective, waiver of subrogation provisions merely eliminate the transaction costs involved in allocating insured losses to wrongdoers who are often themselves covered by insurance.

> contractors described in article 6 [separate contractors hired by owner], if any, and any of their subcontractors, sub-subcontractors, agents and employees, for damages caused by fire or other causes of loss to the extent covered by property insurance obtained pursuant to this paragraph 11.4 or other property insurance applicable to the work, except such rights as they have to proceeds of such insurance held by the owner as fiduciary . . . .[148]

Although exculpatory in nature, these waiver provisions are routinely upheld.[149] Due to the nexus with insurance proceeds, courts have been reluctant to void waiver of subrogation clauses merely because they are exculpatory in nature.[150]

---

148 AIA Document A201, General Conditions of the Contract for Construction, 15th ed. (1997).

149 *See, e.g., Burdue v. United States Steel Corp.*, 676 F.2d 1129, 1132 (6th Cir. 1982); *Willis Realty Assocs. v. Cimino Constr. Co.*, 623 A.2d 1287, 1288-89 (Me. 1993); *Chadwick v. CSI, Ltd.*, 629 A.2d 820, 826-27 (N.H. 1993); *Tokio Marine & Fire Ins. Co. Ltd. v. Employers Ins. of Wausau*, 786 F.2d 101, 104-05 (2d Cir. 1986); *Village of Rosemount v. Lentin Lumber Co.*, 494 N.E.2d 592, 600 (Ill. App. Ct. 1986); *Haemonetics Corp. v. Borphy & Phillips Co.*, 501 N.E.2d 524, 526 (Mass. Ct. App. 1986); *Home Ins. Co. v. Bauman*, 684 N.E.2d 828 (Ill. App. Ct. 1997).

150 The New Hampshire Supreme Court's decision in *Chadwick v. CSI, Ltd.*, 629 A.2d 820 (N.H. 1993), is a case in point. A school brought suit against a general contractor seeking recovery for losses incurred as a result of a fire in the school's auditorium. The New Hampshire Supreme Court denied recovery on the grounds that the parties' construction contract contained a waiver of subrogation provision, and the school had been paid for its losses by its property carrier. The court rejected the school's argument that the waiver of subrogation provision was a straight exculpatory provision prohibited under New Hampshire law. *See also Willis Realty Assocs. v. Cimino Constr. Co.*, 623 A.2d 1287 (Me. 1993)(waiver prohibited owner's property carrier from seeking recovery from general contractor for wall collapse even though builder's risk was purchased after the date of collapse.); *Len Immke Buick, Inc. v. Architectural Alliance* 611 N.E.2d 399 (Ohio App. Ct. 1992) (owner unsuccessfully argued that AIA waiver provision was unenforceable because it was inconsistent with the contract's indemnification provision. The court held that indemnity provisions simply required the contractor to protect the owner from liability for claims by third parties and was, therefore, not inconsistent with a waiver of the right to recover for

For example, a non-AIA waiver provision was construed in *Touchet Valley Grain Growers, Inc. v. Opp & Seibold Gen. Constr., Inc.*[151] The parties' agreement provided that:

> Subrogation rights, if any, are expressly waived by each party to the extent of insurance coverage afforded on any claim, loss or casualty arising from or in connection with the project.[152]

The contractor designed and constructed a steel framed flat house grain storage building. Several months after completion, the building's roof began to buckle and repairs were attempted. Eventually an exterior wall failed and the building was damaged beyond repair. The owner's property insurer "loaned" the owner monies to repair the structure. The owner brought suit against the contractor and its surety alleging design defects. The Washington Supreme Court held that the subrogation provision precluded suit against the contractor to the extent of insurance coverage, which was to be decided in a separate proceeding. The court also ruled that the owner was entitled to proceed against the subcontractor who supplied the design specifications because the subcontractor was not a party to the original contract containing the waiver provision.[153]

Once in a while the courts run into difficulties interpreting the provision. *In St. Paul Fire & Marine Ins. Co. v. Freeman-White Associates, Inc.*,[154] an architect defending a suit for property damages incurred by the owner of a project unsuccessfully argued that the owner had agreed to waive any property damage claims against the architect and look solely to the proceeds of builders risk insurance to cover the loss. The Supreme Court of North Dakota noted that various provisions of the owner/architect agreement stated that the owner and architect waive rights to claim damages against each other to the extent that these

---

damages for which compensation has already been received from another).

151 831 P.2d 724 (Wash. 1992).

152 *Id.* at 728.

153 *See also United States Fidelity & Guar. Co. v. Farrar's Plumbing & Heating, Co.,* 762 P.2d 641 (Ariz. Ct. App. 1988); *Hartford Ins. Co. v. CMC Bldrs, Inc.,* 752 P.2d 590 (Colo. Ct. App. 1988).

154 366 S.E.2d 480 (N.C. 1988).

damages were covered by property insurance. Nevertheless, the court refused to enforce the waiver as the agreement also required the architect to purchase professional liability insurance protecting the owner from direct and consequential damages as a result of errors or omissions committed by the designer. The court concluded that the requirement that the architect have such insurance was possibly an indication of an intention by the parties that the owner and its builder's risk insurer could pursue a subrogation claim against the architect of covered by errors and omissions insurance.

Most courts do not find the waiver provision to be invalid or made ambiguous by virtue of a contract's indemnity provisions or a state's anti-indemnification laws.[155]

As a general rule, if a claim is not barred by a waiver of subrogation clause the stated reason is that the provision does not apply to the particular facts of the case. Several scenarios occur. The clause may not prohibit suit because the party against whom recovery is sought is not covered by the provision. For example, in *Fortin v. Nebel Heating Corp.*,[156] the court held that the waiver in the prime contract did not constitute a waiver of the owner's right to recover from a subcontractor for fire damage otherwise covered by the owner's insurance.

Sometimes an issue will arise as to whether the particular claims being asserted fall within the scope of the waiver. The most common limitation along these lines focuses on the language "the Work." Losses to property which do not form part of the insured's "work," even thought covered by insurance, are often not affected by the waiver.[157] In *S.S.D.W. Co.* v. *Brisk Waterproofing Co.*,[158] the waiver provision was

---

155 *See, e.g.*, Annotation, *Insurance: Subrogation of Insurer Compensating Owner or Contractor for Loss Under "Builder's Risk" Policy Against Allegedly Negligent Contractor or Subcontractor*, 22 ALR 4th 701 (1983).

156 429 N.E.2d 363 (Mass. 1981).

157 *See, e.g., Fidelity & Guar. Ins. Co. v. Craig-Wilkinson, Inc.,* 948 F. Supp. 608 (S.D. Miss. 1996), *aff'd,* 101 F.3d 699 (5th Cir. 1996); *Employers Mut. Cas. Co. v. A.C.C.T., Inc.,* 568 N.W.2d 530 (Minn. Ct. App. 1997) (AIA waiver of subrogation clause does not apply to property damage caused by contractor to areas of the building not subject to the construction contract.).

158 556 N.E.2d 1097 (N.Y. 1990).

limited to "the Work" and did not extend to claims arising from damage to property outside the limits of "the Work." In so holding, the court noted:

> The waiver clause, if given its plain meaning, bars subrogation only for those damages covered by insurance which the owner has provided to meet the requirement of protecting the contractor's limited interest in the building—i.e., damages to the work itself.[159]

The theory behind limiting the scope of the waiver to the contractor's work is that the contractor should have liability coverage for losses incurred by property which is not the work of the contractor.[160]

IV.
Property Coverage

A. Introduction

The most prevalent form of property coverage pertaining to construction is the builder's risk policy. Because the risks to property are significantly different in nature when the property is undergoing construction, standard property coverages usually will not cover loss associated with buildings under construction.[161]

---

159 *Id.* at 1100; *see also Travelers Ins. Co. v. Dickey,* 799 P.2d 625 (Okla. 1990)(court held that the waiver provision did not exonerate the contractor from liability for damage to owner's property which was unrelated to contractor's scope of work.). *But see Public Employees Mut. Ins. Co. v. Sellen Constr. Co.,* 740 P.2d 913 (Wash. Ct. App. 1987) (fact issue created where microfiche records destroyed as the term "the Work" meant property under construction or property located at the construction site.).

160 *See, e.g., Public Employees Mut. Ins. Co. v. Sellen Constr. Co.,* 740 P.2d 913 (Wash. Ct. App. 1987). This theory, of course, assumes that poor workmanship and the damages resulting therefrom are an occurrence under a CGL policy.

161 The standard fire policy, which is a common property insurance form which includes perils beyond fire, contains the following language:

The insuring clause of the standard builder's risk coverage form is as follows:

a. Coverage

We will pay for direct physical loss of or damage to covered property at the premises described in the declarations caused by or resulting from any covered cause of loss.[162]

Conditions suspending or restricting insurance. Unless otherwise provided in writing added hereto this company shall not be liable for loss occurring

(a) While the hazard is increased by any means within the control or knowledge of the insured; or
(b) While a described building, whether intended for occupancy by owner or tenant, is vacant or unoccupied beyond a period of 60 consecutive days; . . .

See Susan J. Miller & Phillip Lefebvre, *Miller's Standard Insurance Policies Annotated*, Vol. I at 456 (4th ed. 1995). Structures under construction also present another challenged not faced by existing structures; namely, rapidly changing valuation issues. While existing structures can change in value over time this is through market forces and the changes are gradual. Construction, however, causes property to increase in value at a very rapid rate presenting peculiar pricing issues. Moreover, the parties who might conceivably have "an insurable interest" in a property under construction are generally much greater in number than the case with existing structures. Contractors, subcontractors, materialmen all potentially have an interest in a construction project. As a result, the standard property policy is not a particularly good vehicle for covering construction risks.

162 The language of property coverages has changed some what over the years. Property policies, including the builder's risk coverage form, developed before the mid-1980s used either the "all-risk" or "specified perils" terminology when setting forth the extent of coverage. The differences between the "all-risk" and the specified perils" policy language was significant beyond simply the scope of coverage afforded under either. Property coverage procured on a "all-risk" form creates a "special type of insurance" extending coverage to risks not usually anticipated, providing that such risks are "fortuitous." *See, e.g., Ariston Airline & Catering Supply Co. v. Forbes*, 511 A.2d 1278 (N.J. Super.

1986), citing Annotation, "Coverage Under 'All-Risk' Insurance," 88 A.L.R.2d 1122, 1125 (1963). The insured's burden of proof is less onerous under a "all-risk" policy form. The insured bears the burden of establishing a loss due to a fortuitous event. Once the insured establishes fortuitous loss, however, the burden shifts to the insurer to demonstrate that the loss is excluded under the terms of the policy. *See, e.g.*, *Texas Eastern Transmission Corp. v. Marine Office-Appleton & Cox Corp.*, 579 F.2d 561 (10th Cir. 1978). Where the cause of loss is difficult to ascertain, the "all-risk" form is of particular benefit to the insured. *See, e.g.*, *Underwriters Subscribing to Lloyd's Ins. v. Magi, Inc.*, 790 F. Supp. 1043 (E.D. Wash. 1991). *See also* Annotation, *Coverage Under "All-Risks" Insurance*, 88 A.L.R.2d 1122 (1963). Coverage secured under a "specified" or "named" peril form is more limited in scope. Whereas property insurance procured on a "all-risk" basis limits coverage through the use of exclusions, the "specified peril" form contains a more limited insurance grant. *See, e.g.*, *Avis v. Hartford Fire Ins. Co.*, 195 S.E.2d 545 (N.C. 1973). Furthermore, under a "specified peril" form, the insured bears the burden of proving that the loss resulted from the named or specified peril. *See, e.g.*, *Harbor House Condominium Ass'n v. Massachusetts Bay Ins. Co.*, 703 F. Supp. 1313 (N.D. Ill. 1988), *aff'd*, 915 F.2d 316 (7th Cir. 1990). New coverage forms began to emerge in the 1980s. The "all risk" and "specified peril" language has been replaced with coverage forms setting forth "causes of loss." There are three "causes of loss" forms in general use. The basic form provides coverage for losses resulting from fire, lightening, explosion, windstorm or hail, smoke, non-owned aircraft or vehicles, riot or civil commotion, vandalism, sprinkler leakage, sink hole collapse, and volcanic action. The "broad form" provides additional coverages for breakage of glass, falling objects, weight of snow, ice or sleet, and water damage from systems or appliances. There is also a "special form" which provides coverage for all risks of direct physical loss subject to various exclusions. Because the exclusions contained in the "special form" are also contained in the basic and broad forms, this form tends to provide greater coverage. *See, e.g.*, James T. Hendrick & James A. Riddle, *Commercial General Liability and Commercial Property Insurance for Contractors*, 7 CONSTR. LAW. 5 (April 1987); Lauren E. Roberts, *All-Risk Property Insurance: Problems in Determining the Scope of Coverage*, 53 INS. COUNS. J. 88 (1986).

1. Covered Property

Covered property, as used in this coverage part, means the following type of property for which a limit of insurance is shown in the declarations:

> Building under construction, meaning the building or structure described in the declarations while in the course of construction, including:
>
> a. Foundations;
>
> b. If intended to become a permanent part of the building or structure described in the declarations, the following property located in or on the building or structure or within 100 feet of its premises:
>
> (1) Fixtures, machinery and equipment used to service the building; and
>
> (2) your building materials and supplies used for construction;
>
> c. If not covered by other insurance, temporary structures built or assembled on site, including cribbing, scaffolding and construction forms.[163]

While the builder's risk policy is the main vehicle for providing property coverage for structures under construction, it is not the only property insurance available to contractors and owners. The AIA General Conditions requires the owner to purchase boiler and machinery insurance.[164] This insurance is sometimes also referred to as some form

---

163 *See* Susan J. Miller & Phillip Lefebvre, *Miller's Standard Insurance Policies Annotated,* Vol. I at 468 (4th ed. 1995).

164 AIA Document A201, General Conditions of the Contract For Construction, ¶ 11.4.2 (1997). This provision requires that:

> The owner shall purchase and maintain boiler and machinery insurance required by the contract documents or by law, which shall specifically cover such insured objects during instillation and until final acceptance by the owner; this insurance shall include interests of the owner,

of "breakdown" insurance. This coverage is necessitated because of standard exclusions contained in most property policies.[165] Boiler and machinery policies cover loss resulting from "sudden and accidental" breakdown of insured equipment.[166] This insurance covers the costs of repairing the insured equipment and any damage to adjacent property or equipment and resulting business interruption.[167]

---

contractor, subcontractors and sub-subcontractors in the work, and the owner and contractor shall be named insureds.

165 ISO Form CP 10 20 10 91, "causes of loss-broad form" contains the following exclusion:

> 2. We will not pay for loss or damage caused by or resulting from:
>
> a Artificially generated electrical current, including electric arcing, that disturbs electrical devices, appliances or wires. But if loss or damage by fire results, we will pay for that resulting loss or damage.
>
> b. Explosion of steam boilers, steam pipes, steam engines or steam turbines owned or leased by you, or operated under your control . . . .
>
> c. Mechanical breakdown, including rupture or bursting caused by centrifugal force.
>
> d. Rupture or bursting of water pipes (other than automatic sprinkler systems if sprinkler leakage is a covered cause of loss) unless caused by a covered cause of loss.

See Susan J. Miller & Phillip Lefebvre, *Miller's Standard Insurance Policies Annotated*, Vol. I at 475 (4th ed. 1995).

166 *See, e.g.*, *Cafritz Co. v. Employers Commercial Union Ins. Co. of Am.*, 309 A.2d 302 (D.C. 1973) (terminal board not directly damaged by accident was not compensable under boiler and machinery coverage); *Central La. Elec. Co. v. Westinghouse Elec. Corp.*, 579 So. 2d 981 (La. 1991) (corrosion which was a direct result of accident and resulting damage covered under boiler and machinery coverage even where corrosion damage itself outside of policy coverage).

167 *See, e.g.*, *Cyclops Corp. v. Home Ins. Co.*, 352 F. Supp. 931 (W.D. Pa. 1973).

Under the AIA insurance scheme, the owner has the option of purchasing and maintaining loss of use insurance.[168] Builder's risk policies typically contain an exclusion for any type of consequential or indirect loss. This would include losses due to delays in construction resulting from the structure being damaged by an insured "cause of loss." Certain commercial owners can incur significant damages if their projects are delayed. These "time element" or "soft costs" damages are covered under "loss of use" insurance. This coverage can be purchased by endorsement to the builder's risk policy.[169]

Another common property coverage is known as the "instillation floater." This policy is in reality a specialized form of builder's risk coverage. It is usually purchased by equipment suppliers or

---

168 AIA Document A201, General Conditions of the Contract for Construction ¶ 11.4.3 (1997). This provision reads as follows:

> The owner, at the owner's option, may purchase and maintain such insurance as will insure the owner against loss of use of the owner's property due to fire or other hazards, however caused. The owner waives all rights of action against the contractor for loss of use of the owner's property, including consequential losses due to fire or other hazards however caused.

169 A standard vehicle for securing this coverage is ISO form CP 00 30 10 91, the "business income coverage form (an extra expense)." The coverage grant states in part:

> We will pay for the actual loss of business income you sustain due to the necessary suspension of your "operations" during the "period of restoration." The suspension must be caused by direct physical loss of or damage to property at the premises described in the declarations, including personal property in the open (or at least a vehicle) within 100 feet caused by or resulting from any covered cause of loss.

See Susan J. Miller & Phillip Lefebvre, *Miller's Standard Insurance Policies Annotated*, Vol. I at 482 (4th ed. 1995). In addition to reimbursement for business income lost as a result of property loss, the policy also provides "extra expense" coverage which entails reimbursement for expenses incurred during the "period of restoration" that the insured would not have incurred if there had been no direct physical loss or damage to property caused by or resulting from a covered cause of loss.

subcontractors responsible for supplying and/or installing expensive equipment or materials that are not otherwise covered by the standard builder's risk policy. Typical items covered by an instillation floater are compressors, generators or other sizable equipment.[170]

A number of issues recur with regard to property coverage for construction projects. Perhaps the most significant is whether a builder's risk policy or other property coverage is afforded for damages caused, in whole or in part, by poor workmanship. An examination of this issue inevitably leads one to standard property coverage exclusions. The analysis also, however, requires an examination of concurrent causation issues. As with the situation involving liability coverage, the question of poor workmanship often raises the issue of whether the loss is "fortuitous."[171]

Another issue that arises with respect to a builder's risk policy is whether the parties seeking coverage has an "insurable interest." The requirement that a policy beneficiary have an "insurable interest" is by no means unique to builder's risk coverage. It is a crucial element in all property insurance. Having an insurable interest is the first prerequisite for obtaining coverage under a property policy.[172] An insurable interest exists where the insured can suffer a loss if the subject property is

---

170 *See, e.g.*, International Risk Management Institute, Inc., Construction Risk Management, Vol. II § VII.A.2 (Second Reprint, March 1995).

171 *See, e.g.*, *Compagnie des Bauxites de Guinee v. Ins. Co. of North Am.*, 724 F.2d 369 (3d Cir. 1983) (collapse of building was fortuitous and therefore insurable); *Kilroy Indus. v. United Pacific Ins. Co.*, 608 F. Supp. 847 (C.D. Cal. 1985) (insured's loss of income following government order to vacate building was fortuitous as neither insurer nor insured was aware of defects in building as a result of faulty workmanship at time they entered into contract); *Essex House v. St. Paul Fire & Marine Ins. Co.*, 404 F. Supp. 978 (S.D. Ohio 1975) (failure of brick facade on new apartment building was covered loss under a "all-risk" policy even though loss was due to negligent design and construction). *See also* Cozen & Bennett, *Fortuity: The Unnamed Exclusion*, 20 FORUM 222 (1985).

172 *See, e.g.*, *Stebane Nash Co. v. Campellsport Mut. Ins. Co.*, 133 N.W.2d 737 (Wis. 1965); *Phalen Park State Bank v. Reeves*, 251 N.W.2d 135 (Minn. 1977).

damaged.[173] A party does not need to have an absolute right to the subject property. It may belong to the insured in whole or in part. Thus, an insurable interest can come about if the insured is an owner, lessee, or bailee (*i.e.*, holding the property for someone else). The insured must suffer some loss if the property is destroyed, regardless of whether the insured has title to, lien upon, or possession of the property itself.[174] For example, the loss of future profits derived from property caused when rain fell during an outdoor concert was held an insurable interest.[175]

The "insurable interest" issue most often arises in the construction context where the loss occurs after the contractor or subcontractor has completed its work. Courts will sometimes conclude that once a contractor's work is complete it no longer has an insurable interest in the property. For example, in *Baker v. Aetna Ins. Co.,*[176] fire damage to structure after the contractor had completed construction although it had not yet received final payment. The court held that the contractor had no insurable interest in the structure at the time it was damaged and, therefore, it was not protected under the policy.[177]

---

173 *See Northwestern Nat'l Bank of Minneapolis v. Maher*, 258 N.W.2d 623 (Minn. 1977).

174 *See, e.g., Webb v. M.F.A. Mut. Ins. Co.*, 620 P.2d 38 (Colo. Ct. App. 1980); *Antell v. Pearl Assurance Co.*, 89 N.W.2d 726 (Minn. 1958).

175 *See, Casablanca Concerts, Inc. v. American Nat'l Gen. Agencies, Inc.*, 407 N.W.2d 440 (Minn. App. 1987).

176 262 S.E.2d 417 (S.C. 1980).

177 *But see General Elec. Co. v. Zurich-Am. Ins. Co.*, 952 F. Supp. 18 (D. Me. 1996) (supplier had insurable interest even though it had finished its turbine-supply contract as it had an interest in being held free from liability arising from its involvement in the construction project.); *Dyson & Co. v. Flood Eng'rs, Archs, Planners, Inc.*, 523 So. 2d 756 (Fla. Dist. Ct. App. 1988) (court adopted majority view that builder's risk policy included liability as well as property interests and found engineer had "insurable interest" in builder's risk coverage because it had an interest in being free from liability with respect to the project.); *Automobile Ins. of Hartford Co. v. United H.R.B. Gen. Contractors*, 876 S.W.2d 791 (Mo. Ct. App. 1994)(contractor that had finished work no longer had an "insurable interest" in the work so that owner/contractor provision that required waiver of claims to the extent of coverage applied to claims arising prior to final payment and the waiver provision with respect to

1. *Coverage for Defective Work*

a. The Fortuity Requirement

As was the case with liability insurance, property coverage does not insure against nonfortuitous loss. As the Sixth Circuit discussed in connection with an owner's claim under an all-risks property policy for the cost of removing asbestos-containing materials from one of its buildings:

> The application of the implied requirement of fortuity is "universally recognized." . . . [T]he Ohio Court of Appeals adopted the definition of a fortuitous event found in the Restatement of Contracts § 291, cmt. a (1932), which defines a fortuitous event as one "which so far as the parties to the contract are aware, is dependent on chance. It may be beyond the power of any human being to bring the event to pass; it may be within the control of third persons; it may be a past event, as the loss of a vessel, provided that the fact is unknown to the parties". . . . A fortuitous event is an event "occurring by chance without evident causal need or relation or without deliberate intention." In addition, the design of all-risk insurance to "extend protection against the kind of 'fortuitous loss' which is not usually covered under other insurance," demonstrates that insurers are offering protection against unexpected, accidental occurrences that are difficult for persons seeking insurance to designate in advance, as is required under traditional named-peril insurance policies. This view of fortuity is consistent with the idea that "[i]nsurance should only cover losses resulting from a *casualty. There is no casualty unless some risk is involved.*"[178]

---

waiving all claims except for defective work upon final payment applied when the contractor had an "insurable interest" in the work.)

178 *University of Cincinnati v. Arkwright Mut. Ins. Co.*, 51 F.3d 1277, 1281 (6th Cir. 1995) (citations omitted); *see also Adams-Arapaho Joint Sch. Dist. No. 28-J v. Continental Ins. Co.*, 891 F.2d 772 (10th Cir. 1989); *Derby v. Westminster Found. of Ohio*, 103 N.E.2d 10 (Ohio Ct. App. 1951); *Fidelity and Guar. Ins. Underwriters, Inc. v. Allied Realty Co.*, 384 S.E.2d 613 (Va. 1989); *HRG Dev. Corp. v. Graphic Arts Mut. Ins. Co.*, 527 N.E.2d 1179 (Mass. Ct. App. 1988); *Standard Structural Steel Co. v. Bethlehem Steel Corp.*, 597 F. Supp. 164 (D. Conn. 1984).

The Sixth Circuit found that the owner's removal of asbestos was a business decision and not a fortuitous event. To hold otherwise would "convert the plaintiff's all-risk insurance policy into a cash fund for plaintiff's business plans."[179]

As a practical matter, the fortuity requirement has not been a particularly onerous obstacle to overcome for insureds seeking property coverage for defective construction. For example, in *Diana v. Western National Assurance Co.*,[180] an owner of a all-risk homeowners' insurance policy successfully recovered for structural damages incurred to her home when a contractor removed some interior and exterior walls. The renovation contractor removed the walls in order to achieve "a feeling of special openness." Unfortunately, this feeling came with a price. The contractor removed structural walls, thereby causing the house to incur structural damage. The contractor then walked off the job without replacing the ceiling and wall supports.[181] In rejecting the insurer's contention that the damage was not caused by an "accident," but was due to deliberate choices made by the owner, the court noted that the owner, as a matter of law, would not be held to have known that the removal of the walls would cause the damage to the house. The court also rejected the insurer's argument that remodeling materially altered the insurer's risk and voided the property policy. The policy did

---

179 *Id.* at 1282. The Sixth Circuit's analogy to other products is reminiscent of the judicial fondness of likening insurance policies to performance bonds if coverage is afforded for defective work. The Sixth Circuit's analogy is not particularly more persuasive. Insurance is, after all, simply a form of financing and, as such, viewing insurance as a substitute for a "cash fund" for certain emergencies is not all that unusual. *See, e.g.*, International Risk Mgmt. Inst., Inc., *Construction Risk Management*, Vol. I at V.A.1 (3d reprint 1996) (discussing insurance as a form of risk financing; *i.e.*, a method of funding losses). Property owners have not been terribly successful in securing insurance coverage for asbestos removal costs. These claims have been denied not only on fortuity grounds, but also because the presence of asbestos is not a direct physical loss and, furthermore, fell within the pollution exclusion of most property policies. *See, e.g.*, *Great Northern Ins. Co. v. Benjamin Franklin Fed. Sav. & Loan Ass'n*, 793 F. Supp. 259 (D. Or. 1990), *aff'd*, 953 F.2d 1387 (9th Cir. 1992).

180 785 P.2d 479 (Wash. Ct. App. 1990).

181 *Id.* at 481.

not require that the insured notify the insurer of any remodeling to the property and there was no implied duty of such notification. The policy did not expressly exclude renovation from coverage.[182]

Fortuity in the property coverage arena generally involves a subjective analysis similar to that encountered with liability insurance. In *Baugh-Belarde Construction Co. v. College Utilities Corp.*,[183] a question arose as to whether a subcontractor's negligence in causing a fire was a fortuitous loss. The court found that it was, noting that one looks at what the parties to the contract knew or should have known at the time the policy was written. This inquiry was not made with the benefit of hindsight but, rather, by looking at what the parties knew or reasonably should have known at the time coverage was secured.[184]

The limitations of the fortuity requirement are apparent in the New York decision of *A & B Enterprises v. Hartford Insurance Co.*[185] A general contractor sought recovery from its property insurer for the loss of equipment taken by one of its subcontractors in a payment dispute. The jury found for the insurer and the trial court denied the contractor's motion for judgment notwithstanding the verdict. The Appellate Court, however, reversed, finding that the contractor's loss was fortuitous because the loss was "to a substantial extent beyond the contractor's control."[186] The fact that the contractor and subcontractor had a payment dispute was irrelevant to the issue of fortuity because there was no evidence that the subcontractor had a legal claim to the equipment.

2. *The Requirement That There Be Physical Loss*

The coverage grant of the builder's risk coverage form states, in pertinent part:

---

182 *Id.* at 483.

183 561 P.2d 1211 (Alaska 1977).

184 *Id.* at 1215; *see also C.H. Leavell & Co. v. Fireman's Fund Ins. Co.*, 372 F.2d 784 (9th Cir. 1967); *General Am. Transp. Corp. v. Sun Ins. Office, Ltd.*, 239 F. Supp. 844 (E.D. Tenn. 1965), *aff'd,* 369 F.2d 906 (6th Cir. 1966); *Associated Eng'rs., Inc. v. American Nat'l Fire Ins. Co.*, 175 F. Supp. 352 (D.C. Cal. 1959); *Federal Ins. Co. v. Tamiami Trail Tours*, 117 F.2d 794 (5th Cir. 1941); *Redna Marine Corp. v. Poland*, 46 F.R.D. 81 (S.D.N.Y. 1969).

185 604 N.Y.S.2d 166 (N.Y. App. Div. 1993).

186 *Id.* at 168.

> We will pay for direct physical loss of or damage to covered property at the premises described in the declarations caused by or resulting from any covered cause of loss.[187]

As with the case with liability coverage, the term "physical loss" is intended to eliminate coverage for economic losses or diminution in value. From the insurance industry's perspective, "physical loss" requires some physical change in the condition of the covered property. The mere fact that defects exist in one's property is generally insufficient, in and of itself, to trigger coverage.[188] Thus, where an insured really discovers a construction defect sometime during the term of the policy has not established that "physical loss" has occurred to the covered property. What has changed is the insured's state of knowledge, rather than the condition of the covered property.[189] Where the defect is a progressively-deteriorating condition, the analysis grows more complicated. For example, in *St. Michael's Orthodox Catholic Church v. Preferred Risk Mutual Insurance Co.*,[190] a church sought coverage for roof leakage which occurred before and after the issuance of the policy. The church had the burden of proof that its loss resulted from a covered peril. The church claimed that the water damage had worsened after the policy commenced, although it failed to distinguish between the portion

---

187 *See* Susan J. Miller & Phillip Lefebvre, *Miller's Standard Insurance Policies Annotated*, Vol. I at 468 (4th ed. 1995).

188 *See, e.g., State Farm Fire & Cas. Co. v. Superior Court (Aegea Homeowners Ass'n, Inc.)*, 264 Cal. Rptr. 269 (Cal. Ct. App. 1989) (defect that caused loss of value to building was not covered as "[n]either diminution in value nor the cost of repair or replacement are active physical forces–they are not the cause of the damage to the structures, they are the measure of the loss or damage.").

189 *Id.* at 1445; *see also Mellon v. Federal Ins. Co.*, 14 F.2d 997 (2d Cir. 1926) (property insurer is not a guarantor of the good quality of the property insured); *Green v. Cheetham*, 293 F.2d 933, 937 (2d Cir. 1961) (carrier not responsible for defect in property existing prior to the commencement of coverage); *Acme Galvanizing Co., Inc. v. Fireman's Fund Ins. Co.*, 270 Cal. Rptr. 405, 412 (1990) (to require a property insurer to pay for undisclosed defects would turn insurance policy into a warranty of fitness).

190 496 N.E.2d 1176 (1986).

of the loss that occurred after it was insureds and the portion that predated coverage.[191]

Progressive loss situations often give rise to "trigger" issues. As was the case the liability coverage, if physical loss extends over a long period of time, then there is a question as to which property policies answer to what losses. While the outcomes may be somewhat different from the case law discussing the same subject with respect to third-party liability coverage, the analysis is essentially the same. As with the liability cases, most decisions discuss and elect one of four trigger theories: (1) the exposure trigger, (2) the manifestation trigger, (3) the injury-in-fact trigger, and (4) the continuous trigger.[192]

3. *Is Poor Workmanship or Defective Construction a Covered Peril/Cause of Loss*

The type of policy at issue is a critical factor in determining whether faulty workmanship is a covered peril or cause of loss. As a general rule, an "all-risk" policy will be interpreted to include faulty workmanship or negligent conduct as a covered peril or cause of loss

---

191 *Id.* at 1179. Because the requirement of establishing direct physical loss is a condition to coverage, it is generally held that the burden of proof falls upon the insured. *See, e.g., Morrison Grain Co. v. Utica Mut. Ins. Co.*, 632 F.2d 424 (5th Cir. 1980).

192 *See, e.g., Continental Ins. Co. v. Northeastern Pharmaceutical & Chem. Co.*, 11 F.2d 1180 (8th Cir. 1987), *aff'd on reh'g*, 842 F.2d 977 (8th Cir. 1988) (exposure theory); *Fireman's Fund Ins. Co. v. Ex-Cell-O*, 662 F. Supp. 71 (E.D. Mich. 1987) (exposure theory); *U.S. Gypsum Co. v. Admiral Ins. Co.*, 643 N.E.2d 1226 (Ill. App. Ct. 1994) (criticizes exposure theory); *Prudential-LMI Commercial Ins. v. Superior Court Lundberg*, 798 P.2d 1230 (Cal. 1990) (manifestation trigger); *Home Ins. Co. v. Landmark Ins. Co.*, 253 Cal. Rptr. 277 (Cal. Ct. App. 1988) (manifestation trigger); *Maryland Cas. Co. v. W.R. Grace & Co.*, 4 F.3d 155 (2d Cir. 1993), *amended superseding opinion*, 23 F.3d 617 (injury-in-fact trigger); *Triangle Pubs., Inc. v. Liberty Mut. Ins. Co.*, 703 F. Supp. 367 (E.D. Pa. 1989) (injury-in-fact trigger); *Harleyville Mut. Ins. Co. v. Sussex, County Del.*, 831 F. Supp. 1111 (Del. 1993), *order aff'd*, 46 F.3d 1116 (3d Cir. 1994) (continuous trigger); *United States Fidelity & Guar. Co v. Thomas Solvent Co.*, 683 F. Supp. 1139 (W.D. Mich. 1988) (continuous trigger theory).

subject to elimination or restriction by exclusion.[193] An all-risks property insurance policy, as opposed to a named-perils policy insures against all risks of direct physical loss or damage to property unless a peril is specifically excluded.[194] Thus, for example, if a trier of fact determined that negligent repair was the proximate cause of damage to a vessel, then coverage under an all-risk policy was afforded.[195]

Coverage was afforded in *National Fire Insurance Co. of Pittsburgh v. Valero Energy Corp.*,[196] where an owner sustained substantial damages as a result of the faulty design of an addition to one of its plants. As a result of design error, substantial corrosion was caused. The builder's risk policy specifically excluded coverage for loss caused by naturally-occurring corrosion. Yet, as the corrosion was caused by the faulty design rather than by a natural process, the damage fell within coverage.

4. *Concurrent Causation*

What are the coverage ramifications where an excluded cause of loss is intertwined with a cause that is not excluded under the policy? In the construction context, this issue often presents itself as a juxtaposition between external causes of loss, such as fire or windstorm, with such internal causes, as latent defect and inherent vice. The ISO causes of Loss–Broad Form and Special Form attempt to address this situation through the following language:

B. Exclusions.

---

193 *See, e.g., Essex House v. St. Paul Fire & Marine Ins. Co.*, 404 F. Supp. 978 (S.D. Ohio 1975); *Mattis v. State Farm Fire & Cas. Co.*, 454 N.E.2d 1156 (Ill. App. Ct. 1983); *General Am. Transp. Corp. v. Sun Ins. Office, Ltd.*, 239 F. Supp. 844 (E.D. Tenn. 1965), *aff'd* 369 F.2d 906 (6th Cir. 1966); *American Motors Ins. Co. v. R & S Meats, Inc.*, 526 N.W.2d 791 (Wis. Ct. App. 1994); *Standard Structural Steel v. Bethlehem Steel Corp.*, 597 F. Supp. 164 (D. Conn. 1984); *Adams-Arapaho Joint Sch. Dist. v. Continental Ins. Co.*, 891 F.2d 772 (10th Cir. 1989).

194 *See, e.g., Hilt Truck Lines, Inc. v. Riggins*, 756 F.2d 676, 679 (8th Cir. 1985).

195 *Egan v. Washington Gen. Ins. Corp.*, 240 So. 2d 875 (Fla. Dist. Ct. App. 1970).

196 777 S.W.2d 501 (Tex. Ct. App. 1989).

1. We will not pay for loss or damage caused directly or indirectly by any of the following. Such loss or damage is excluded regardless of any other cause or event that contributes concurrently or in any sequence to the loss.[197]

The case law on this subject is somewhat mixed. California is a leading jurisdiction, advancing the "efficient proximate cause" test. Under this test, where loss is caused by a combination or covered and specifically-excluded risks, the loss is covered if the covered risk was the efficient proximate cause of the loss.[198] The "efficient proximate" cause test must be distinguished from the "concurrent" cause analysis employed in third-party liability coverage disputes.[199] In *Garvey*, it was

---

197 *See* Susan J. Miller & Phillip Lefebvre, *Miller's Standard Insurance Policies Annotated*, Vol. I at 475 (4th ed. 1995).

198 *See, e.g., Garvey v. State Farm Fire & Cas. Co.*, 770 P.2d 704, 706-7 (Cal. 1989).

199 The *Garvey* decision contains a rather lengthy discussion about the differences between liability and property insurance. The concurrent causation test, when applied to a property coverage dispute, allows coverage whenever a covered peril is a concurrent proximate cause of the loss, without regard to the application of specific policy exclusion clauses. This approach effectively nullifies many policy exclusions in "all-risk" policies. *See, e.g., Garvey v. State Farm Fire & Cas. Co.*, 770 P.2d 704 (Cal. 1989). In adopting the more stringent, "efficient proximate cause" test, the court made the following distinctions between property and liability coverage.

Property insurance . . . is an agreement, a contract, in which the insurer agrees to indemnify the insured in the event that the insured property suffers a covered loss. Coverage, in turn, is commonly provided by reference to causation, *e.g.*, "loss caused by . . . " certain enumerated perils. The term "perils" in traditional property insurance parlance refers to fortuitous, active, physical forces, such as lightening, wind, and explosion, which bring about the loss. *Thus, the "cause" of loss in the context of a property insurance contract is totally different from that in a liability policy.* This distinction is critical to the resolution of losses involving multiple causes. Frequently, property losses occur which involve more than one peril that might be considered legally sufficient. If one of the causes (perils) arguably falls within the coverage grant–

a question for the jury to decide whether damage to plaintiff's house addition was caused by an excluded peril (earth movement) or a nonexcluded peril (negligent construction).[200]

A similar fact pattern arose in *State Farm Fire & Casualty Co. v. Von Der Lieth*,[201] where the insured's home sustained damage due to massive landslides. The policy contained an "earth movement" exclusion.[202] The insurer denied coverage, because the losses were caused by "earth movement or natural groundwater."[203] The insureds countered by arguing that the efficient proximate cause of their loss was the negligence of several entities, including the state of California and a local homeowners' organization. They complained that the State was negligent in removing a portion of the Big Rock Mesa mountain slope to construct the Pacific Coast Highway in the 1930s. They also claimed that the homeowners' association failed to properly maintain its drain system. The jury sided with the homeowners.[204]

---

commonly either because it is specifically insured (as in a named peril policy) or not specifically accepted or excluded (as in an "all-risks" policy)–disputes over coverage can arise. The task becomes one of identifying the most important cause of the loss and attributing the loss to that cause.

Id. at 710 (quoting Bragg, *Concurrent Causation and the Art of Policy Drafting: New Perils for Property Insurers*, 24 FORUM 385, 386-87 (1985)). Liability insurance draws on traditional tort concepts of faults, proximate cause, and duty. The liability analysis differs substantially from the coverage analysis in property insurance, which draws on the relationship between perils that either covered or excluded in the contract. In liability insurance, by insuring for personal liability, and agreeing to cover the insured for his own negligence, the insurer agrees to cover the insured for a broader spectrum of risks.

200 *Id.* at 302.

201 820 P.2d 285 (Cal. 1991).

202 *Id.* at 287.

203 *Id.* at 288.

204 *Id.* at 289. This was a bad faith claim, and the insureds were awarded $1,100 for reasonable costs incurred to repair their property to protect it from further damage and compensatory damages of $55,000 for the insurer's breach of contract, breach of the covenant of good faith and fair dealing, and unfair claims practices.

In upholding the jury verdict, the California Supreme Court remarked:

> [I]f third-party negligence is not excluded under such a policy [all-risks property policy], it is a covered peril. As we stated, third party negligence under a homeowner's policy is a "risk of physical loss" under the policy.
>
> Moreover, the third party negligence that occurred here cannot be considered a remote cause of loss under wither *Garvey* or *Sabella*. The expert testimony overwhelmingly supported the jury's determination that the predominating cause of the loss at Big Rock Mesa was third-party negligence. By developing the hillside with septic tanks instead of sewers and failing to properly dewater the hillside, it was inevitable the ancient landslide would be reactivated, causing damage to a substantial number of properties on the Mesa.[205]

An insured must, however, do more than simply characterize damage as resulting from "negligent construction." If the evidence establishes that the loss was, in fact, occasioned by only a single cause, albeit one susceptible to various characterizations, the efficient, proximate cause analysis has no application. In *Chadwick v. Fire Insurance Exchange*,[206] homeowners sought coverage for structural damage to their home due to defective framing. The insurer denied coverage, citing the "latent defect or inherent vice" exclusion.[207] The homeowners countered by claiming that their loss was caused by builder negligence. The court rejected the homeowners' reasoning:

> An insured may not avoid a contractual exclusion merely by affixing an additional label or separate characterization to the act or event causing the loss. Thus, in *Finn v. Continental Ins. Co., supra*, a loss was caused by leakage from a broken sewer pipe and coverage had been denied under an exclusion for seepage for leakage. The insured

---

205 *Id.* at 292 (citations omitted).

206 21 Cal. Rptr. 2d 871 (Cal. Ct. App. 1993).

207 Inherent vice has been defined as "a loss entirely from internal decomposition or some quality which brings about its own injury or destruction. The vice must be inherent in the property for which recovery is sought." *State Farm Fire & Cas. Co. v. Volding*, 426 S.W.2d 907 (Tex. App. 1968).

> did not show some additional force, act, or event had broken the pipe, but simply argued the "break," rather than the "leak," was the efficient proximate cause. The Appellate Court rejected that argument, holding the hypothesized "break" was not conceptually distinct from the observed "leak"; rather, "leakage" and "seepage" necessarily imply some break of gap in the thing leaking. Consequently, the court held, the case "involved not multiple causes, but only one, a leaking pipe."[208]

The "efficient proximate cause" test is not the only causation doctrine in use. Some courts employ what appear to be more relaxed tests, such as a concurrent causation analysis.[209] In *Safeco Insurance Co. of America v. Guyton*,[210] damage caused by flood waters was covered by an all-risks homeowner's policy despite a flood exclusion because the flooding was caused by a third-party's negligent maintenance of flood control structures.[211]

---

208 *Id.* at 874. *But see Essex House v. St. Paul Fire & Marine Ins. Co.*, 404 F. Supp. 978 (S.D. Ohio 1975) (falling brick facade not excluded under latent defect/inherent vice exclusion as cause was negligent construction resulting in greatly over-stressed header. The "latent defect" exclusion has also been limited by a line of authority which holds that "latent" means that it could not have been discovered by "known scientific testing." *Mattis v. State Farm Fire & Cas. Co.*, 454 N.E.2d 1156 (1983). In other words, latent doesn't just simply mean "unknown," but "could not have been reasonably discovered by experts." *See, e.g., General Am. Transp. Corp. v. Sun Ins. Office, Ltd.*, 239 F. Supp. 844 (E.D. Tenn. 1965), *aff'd* 369 F.2d 906 (6th Cir. 1966); *American Home Assur. co. v. Shea*, 445 F. Supp. 365 (D.C. 1978).

209 *See, e.g., Mattis v. Sate Farm Fire & Cas. Co.*, 118 Ill. App. 3d 612, 454 N.E.2d 1156 (1983); *see also* Houser, *The Rise and Fall of Concurrent Causation: Background in Current Trends Affecting Property Insurance Coverage*, 44 FICC QTRLY. 10 (1993).

210 692 F.2d 551 (9th Cir. 1982).

211 *See also Norfolk & Dedham Mut. Fire Ins. Co. v. DeMarta*, 799 F. Supp. 33 (E.D. Pa. 1993), *aff'd*, 993 F.2d 225 (3d Cir. 1993) (ordinance exclusion did not apply where City required demolition of house as demolition was ordered due to collapse due to hidden decay, a covered peril).

5. *The Workmanship Exclusion*

The standard form builder's risk policy contains the following workmanship exclusion:

> We will not pay for loss or damage caused by or resulting from any of the following. But if loss or damage by a covered cause of loss results, we will pay for that resulting loss or damage.
>
> c. Faulty, inadequate, or defective: (1) planning, zoning, developing, survey, siting; (2) design, specifications, workmanship, repair, construction, renovation, remodeling, grading, compaction; (3) materials used in repair, construction, renovation, or remodeling, or (4) maintenance;
>
> (4) of part or any of any property on or off the described premises.[212]

Because the language of property policies can differ, it is important to refer to the specific policy language. The Delaware Supreme Court's decision in *Phillips Home Builders, Inc. v. Travelers Insurance Co.*,[213] is instructive. Phillips, a home builder, sought recovery under its builder's risk policy for damages incurred when a concrete slab it poured settled and caused damage to surrounding walls. The policy contained the following exclusionary language:

> We will not pay for a "loss" caused by or resulting from any of the following. But if "loss" by a covered cause of loss results, we will pay for that "resulting loss" . . .
>
> c. Faulty, inadequate or defective:
>
> (2) Workmanship, repair, construction, renovation, remodeling, grading, compaction; . . .
>
> (e) setting; cracking, shrinking, bulging, or expansion.[214]

---

212 *See* Susan J. Miller & Phillip Lefebvre, *Miller's Standard Insurance Policies Annotated*, Vol. I at 476 (4th ed. 1995).

213 700 A.2d 127 (Del. 1997).

214 *Id.* at 128.

The contractor argued that the policy should provide coverage where the settling and cracking was caused by a covered cause of loss, such as water seepage. Travelers' interpretation was that losses were caused by settling and not by whatever caused the slab to settle and because settling was an excluded cause of loss, there was no coverage.

The Delaware Supreme Court found both analyses lacking:

> We find problems with both sides' interpretations. Neither one gives full effect to all of the contract language and both could be applied in ways that a reasonable person probably would not have intended. The critical language is the second sentence of the quoted exclusion paragraph. The first sentence says that Travelers will not pay for losses caused by or resulting from the events and conditions listed in the subsections. Those include "settling, cracking, shrinking, bulging, or expansion." The next sentence provides an exception to the exclusion. It says that Travelers will pay, "if 'loss' by a covered cause of loss results . . ." What does that mean?" Phillips argues that the exception to the exclusion grants coverage for an excluded condition, such as settling, if the settling is caused by a covered cause of loss. Thus, for example, settling would not be covered if the settling were caused by poor workmanship, another excluded cause of loss. On the other hand, if the settling were caused by a fire, which is a covered cause of loss, then the settling losses would be covered.[215] . . . Travelers takes a different approach. It contends that the exception to the exclusion applies only if a covered cause of loss occurs as a result of an excluded loss. Travelers suggests that if settling caused a fire, the losses associated with the fire would be covered under the exception to the exclusion.[216]

---

215 The court took issue with Phillips' interpretation because it made the exception significantly broader than the exclusion. By the terms of the policy, every risk of direct physical damage is a covered cause of loss unless a particular cause of loss is listed in the exclusions. A relatively-small number of events and conditions, such as earthquakes, volcanic eruptions, and nuclear hazards, are listed as exclusions. Thus, most ordinary risks of loss are covered causes of loss and, under Phillips' interpretation, the settling exclusion would be operative only in very limited circumstances.

216 *Id.* at 129. The court took exception with this analysis as it ignored the fact that settling could be both a loss and a cause of loss. Therefore, under Travelers' interpretation, losses caused by settling are not covered,

Under the circumstances, the court found the language to be ambiguous and construed the policy against Travelers.[217]

A significant amount of case law pertaining to the various versions of the "faulty workmanship" exclusionary language focuses upon exceptions to the exclusion. In *Rosenberg v. First State Ins. Co.*,[218] completion of the insured's commercial structure was delayed as a result of problem arising during the course of construction. In particular, the shoring used when pouring the concrete in the subterranean garage was removed prematurely which resulted in the deflection of the first floor and parking level, the development of cracks, and the ponding of water.[219] The builder's risk policy provided for coverage of $3,600,000 for buildings and $800,00 for loss of rents, and it contained a faulty workmanship exclusion as follows:

> This policy does not cover: . . . cost of making good faulty or defective workmanship, material, construction, or design, but this exclusion shall not apply to damage resulting from such faulty or defective workmanship, material, construction, or design.[220]

---

and losses caused by a fire resulting from settling are covered. Travelers does not explain how its policy language operates in a situation such as the one under consideration, where the settling is not the cause of the damage; settling is the damage.

217 The court remanded the case for a determination of whether the cause of the damage to the concrete slab was a covered loss, such as water seepage and frost following heavy rain, or an excluded cause, such as faulty workmanship. *See also Allstate Ins. Co. v. Smith*, 929 F.2d 447 (9th Cir. 1991) (term "faulty workmanship" is ambiguous as it may be interpreted to describe a flawed process or a flawed product.); *Nationwide Mut. Fire Ins. Co. v. Tomlin*, 352 S.E.2d 612 (Ga. Ct. App. 1986) (the term "collapse" ambiguous and insured's interpretation that it did not require a "suddenness" element adopted); *but see State Farm Fire & Cas. Co. v. Superior Court (Aegea Homeowners Ass'n, Inc.)*, 264 Cal. Rptr. 269 (Cal. Ct. App. 1989) (insurance policy unambiguously excluded coverage for deficiencies caused by faulty construction and latent defects).

218 280 Cal. Rptr. 388 (1991).

219 *Id.* at 390.

220 *Id.* at 389.

Relying on the policy's faulty workmanship exclusion, the insurer denied the claim for economic damages, including lost rents, which were incurred as a result of the need to repair the structure. The trial court granted judgment in favor of the insurer, but the Court of Appeals reversed. The Appellate Court analyzed the policy's insuring clause, which required the existence of "physical loss." The court found that the complaint contained allegations of cracking and unlevel concrete floors which provided a record of physical damage sustained by the structure.[221]

The court then proceeded to analyze the faulty workmanship exclusion:

> Defendant urges that "faulty workmanship is . . . not a covered peril" and, accordingly, the policy provides no coverage for loss of rents resulting from such faulty workmanship. We disagree. The policy's exclusion of coverage for the cost of "making good" faulty workmanship does not establish that faulty workmanship is not a covered peril. To the contrary, the phrase stating this exclusion goes to clarify that "this exclusion does not apply to damage resulting from such faulty or defective workmanship, material, construction, or design." Because damage resulting from defective workmanship is expressly covered by the policy, the conclusion is inescapable that faulty workmanship is a covered peril.[222]

---

221 *Id.* at 392.

222 *Id.* at 392. *But see United States Indus., Inc. v. Aetna Cas. & Sur. Co.*, 690 F.2d 459 (5th Cir. 1982) (damages resulting from steel tower that became deformed while a heat treatment was improperly performed by insured contractor were excluded under "faulty workmanship" exclusion as "defects in the product (*i.e.*, the tower) caused by faults in the construction process."); *Dow Chem. Co. v. Royal Indem. Co.*, 635 F.2d 379 (5th Cir. 1981) (collapse of concrete dome excluded due to faulty workmanship exclusion); *Derenzo v. State Farm Mut. Ins. Co.*, 533 N.Y.S.2d 195 (N.Y. Sup. Ct. 1988) (settlement of footings and cracking of basement walls due to improper workmanship were "latent defects" excluded by the policy); *Merz v. Allstate*, 677 F. Supp. 388 (W.D. Penn. 1988) (collapse caused by improper construction excluded as latent defect).

The court, furthermore, found coverage for the consequential damages, *i.e.*, the lost rents, in addition to the costs associated with the repair of the physical damage:

> The rule emerging from the cases is that the insurer, while not liable for repair or replacement of the contractor's defective work or product, is nonetheless liable for damages to other property caused by the defects . . . We conclude that the phrase in the contract of insurance, "this exclusion shall not apply to damage resulting from such faulty or defective workmanship, material, construction, or design," encompasses economic damages resulting from defective workmanship as well as physical damage.[223]

A closely-related exclusion to "faulty workmanship" is that for "defective design." Just what constitutes defective design can be problematic. For example, in *American Home Assur. Co. v. J.F. Shea Co., Inc.*,[224] an excavation-support wall on a subway project buckled as it was insufficient to stand actual lateral pressures. The insurer denied coverage on the ground that the failure to accurately predict the lateral pressures was a design error; the court disagreed, finding that the failure to predict actual pressures was not a design defect where the designer was not negligent in estimating the anticipated pressures.[225]

---

223 *Id.* at 392-93.

224 445 F. Supp. 365 (D.D.C. 1978).

225 *Id.* at 367-68; *see also Henning Nelson Constr. Co. v. Fireman's Fund & Life Ins. Co.*, 383 N.W.2d 645 (Minn. 1986) (concrete block wall collapse due to insufficient strength was not a design error, even though a poured wall would have been stronger).

CHAPTER 7

DECIDING TO LITIGATE: IN THE BANKRUPTCY COURT

Jay M. Mann
Robert J. Berens

I. Introduction

Every bond default is a potential bankruptcy filing, and every claims decision of the surety should be influenced by the knowledge that a bankruptcy filing is a distinct possibility. It is not enough for the surety to understand its rights in bankruptcy. To protect itself from the bankruptcy to the greatest extent possible, the surety must understand how to use bankruptcy law and procedure to its greatest advantage. The surety should be an active and aggressive player in the principal's, the obligee's, and even the payment bond claimant's bankruptcy proceeding. Otherwise, the surety risks becoming a victim of the bankruptcy process.

This chapter is not intended to parallel in scope any of the recognized bankruptcy treatises. Rather, this chapter's limited scope is to provide instruction to the surety on how to best manage a large and complex case after a bankruptcy proceeding has been commenced. Discussion in this chapter is restricted to postpetition litigation strategies that may be available to the surety in the bankruptcy proceeding, with a special emphasis upon selected topics which either frequently arise or are emerging areas of concern to the surety. The scope of this chapter is further limited because the various bankruptcy issues are not addressed objectively, but rather from the prospective of the surety, consistent with the authors' purpose to instruct surety practitioners on how to better present their positions in complex bankruptcy cases.

II.
Implications Of The Principal's Bankruptcy

A. First Days After The Bankruptcy Filing

Upon the surety's claims representative and/or counsel becoming aware that a principal under a bond has filed for bankruptcy, certain actions must be taken immediately. The first is to file a Notice of Appearance and Request for Notice in the newly-filed bankruptcy proceeding with copies being sent to the debtor's counsel and the United States Trustee's Office (who supervises Chapter 11 cases). This notice will notify debtor's counsel as to who represents the surety. Additionally, the filing of the notice will result in surety's counsel being added to the master mailing list for this bankruptcy case. Many notices in the bankruptcy proceeding, including the date, time, and place of the meeting of creditors, the deadline to file a proof of claim, and the deadline to object to a plan of reorganization, will be sent to the parties listed in the master mailing list.[1]

Another early to-do item is to file a Notice of Non-Consent to Use of Cash Collateral, which is a formal document that affirmatively puts the debtor on notice that the surety does not consent to the use of its bonded contract proceeds or "cash collateral." The term "cash collateral" means cash or other cash equivalents in which the debtor's bankruptcy estate and an entity other than the estate have an interest.[2] Under Section 363(c)(2)[3] of the Bankruptcy Code, the debtor is only permitted to use cash collateral if all parties with an interest therein consents *or* if the court authorizes such use. This formal document will clear any uncertainty regarding the surety's position on the debtor's use of bonded contract proceeds. Thereafter, the debtor should not use the surety's cash collateral without a court order authorizing such use.

The formal document also serves the purpose of notifying debtor's counsel that the surety claims an interest in bonded contract proceeds. In many cases a cash collateral stipulation is entered into between the

---

1 Many notice requirements are provided for in Rule 2002 of the Federal Rules of Bankruptcy Procedure.

2 *See* 11 U.S.C. § 363(a).

3 Unless otherwise indicated, all references to the "Code," "Bankruptcy Code," or "Section" throughout this chapter are to Title 11 of the United States Code, as amended through December 31, 1997.

debtor and the secured lender without any consideration of the surety's rights of subrogation and equitable lien on contract proceeds. The authors have seen stipulations of this type providing that the secured lender's lien on all contract proceeds is superior to all other third parties. It is imperative that the surety not let stipulations on cash collateral become final orders without properly providing for the surety's equitable lien on bonded contract proceeds.[4] A cash collateral order that is adverse to the surety's interests will become binding if it is not appealed even if the order runs contrary to established controlling precedent.[5]

As with all bond defaults, the surety must analyze the status of all bonded projects. From a bankruptcy perspective, the surety may attempt to wrestle control of the bonded projects from the debtor so that the surety may determine for itself how to best fulfill its performance bond obligations. As discussed below, the surety should file a motion to compel the debtor to decide whether to assume or reject executory contracts. A key issue concerning the motion to compel assumption or rejection is whether the costs to complete exceeds the remaining contract balances on a project-by-project basis. Since the surety will probably want to raise this issue early in the bankruptcy proceeding, analysis of the costs to complete bonded projects should be a high priority on the surety's to-do list.

B. The Need For Information: Rule 2004 Examination

In the early days and weeks of any bond default, the need for information is critical. Information concerning bonded contract status, unpaid suppliers and subcontractors, job progress, etc. may be forthcoming from the bond principal. However, when a principal files bankruptcy with little or no advance warning to the surety, it may be

---

4 Pursuant to Rule 4001(b) of the Federal Rules of Bankruptcy Procedure, the bankruptcy court may not "commence a final hearing on a motion for authorization to use cash collateral" any sooner than fifteen days after the service of the cash collateral motion. FED. R. BANKR. P. 4001(b). After a final hearing on use of cash collateral, the bankruptcy court's order will become a final, non-appealable order.

5 *See*, *e.g.*, *In re Ivory*, 70 F.3d 73 (9th Cir. 1995) (holding that county was bound by Chapter 13 plan even if the bankruptcy court did not have jurisdiction over the county).

difficult to obtain that information, or an update of previously provided information, notwithstanding the surety's contractual right to records and information under the terms of a general indemnity agreement (GIA). Without some of this information, strategic decisions as to how to react to the principal's bankruptcy cannot be made intelligently.

A party that has just filed bankruptcy, particularly a Chapter 11 bankruptcy, is faced with a myriad of conflicting demands and administrative obligations. Confronted with these burdens and deadlines, it is not unusual for the debtor to view the surety's demand for information as being of secondary importance. This is particularly true if the debtor views the surety as an adversary who wishes to "steal" the bonded projects.

Fortunately, the Federal Rules of Bankruptcy Procedure provide an avenue of relief to the surety under these circumstances. Rule 2004 authorizes an examination, similar to a deposition, where the debtor or its agents may be compelled to testify and to produce documentation. A major advantage of Rule 2004 is that it does not require the surety to commence litigation against the debtor before it may be used. Additionally, it also does not mandate an extended waiting period before the examination is allowed or the documents are produced, unlike ordinary discovery procedures under the Federal Rules of Civil Procedure.

Rule 2004 of the Federal Rules of Bankruptcy Procedure requires only that the surety, requesting the issuance of an order for examination, file a motion or application.[6] The order of the court under Rule 2004 is sufficient to compel the attendance of the debtor at the examination. The scope of Rule 2004 examination is very broad, as the language of the rule would suggest.[7] Thus, the surety can use this rule to examine the debtor with respect to all aspects of bonded job performance, subrogation issues, and any other particular matters of interest to the surety.

---

6 FED. R. BANKR. P. 2004.

7 Courts have stated the scope of the Rule 2004 examination to be "so all encompassing as semantically to include and encourage harassment on every human subject." *In re Georgetown of Kettering*, 17 B.R. 73, 75 (Bankr. S.D. Ohio 1981).

C. The Surety's Right To Bonded Contract Funds

A performing surety's right to contract proceeds under the doctrine of equitable subrogation developed through a line of early United States Supreme Court cases.[8] In *Pearlman v. Reliance Insurance Co.*,[9] the United States Supreme Court unequivocally held that as between a payment bond surety and a trustee in bankruptcy, the surety is entitled to receive the contract proceeds to the extent necessary to reimburse it for its losses. Although the ultimate holding of *Pearlman* remains unshaken, there are a number of reported bankruptcy cases in which the courts have struggled to create exceptions to the surety's paramount priority to contract funds under the doctrine of equitable subrogation.

1. Are Bonded Contract Funds "Property of the Estate"

The early Supreme Court decisions establishing the surety's priority interest in bonded contract proceeds seem at odds with the broad definition of "property of the estate" contained in Section 541 of the Bankruptcy Code. The statute's legislative history supports a broad interpretation of its scope.[10] It is important to understand that most of the "equitable subrogation" case law relied upon by the surety to demonstrate its superior claim to bonded contract funds was decided prior to the 1978 enactment of the Bankruptcy Code. The definition of "property of the estate" was substantially expanded under the 1978 Bankruptcy Code from the definition in Section 70a of the predecessor Bankruptcy Act. Accordingly, some doubt has been cast upon the continued validity of the older case law holding that the bankrupt principal had no property interest in bonded contract funds.

Some courts have concluded that the debtor has no interest in contract proceeds where the debtor's failure to pay for labor and materials would excuse payment under the construction contract. For example, in *In re Modular Structures, Inc.*,[11] the debtor/contractor's

---

8 *United States v. Munsey Trust Co. of Wash., D.C.*, 332 U.S. 234 (1947); *Henningsen v. United States Fidelity & Guaranty Co.*, 208 U.S. 404 (1908); *Prairie State Nat'l Bank v. United States*, 164 U.S. 227 (1896).

9 371 U.S. 132 (1962).

10 H.R. REP. NO. 95-595, 95th Cong., 1st Sess. 367-68 (1977); S. REP. NO. 95-989, 95th Cong., 2d Sess. 82-83 (1978).

11 27 F.3d 72 (3rd Cir. 1994); *see also Merchants Bonding Co. v. Pima*

secured creditor sought to obtain the unearned contract progress payments and retention held by the owner. The construction contract required the debtor to pay its subcontractors, laborers and material suppliers. The contract also contained provisions providing that the debtor would not be entitled to be paid if these lower tiers were not paid. The *Modular Structures* court held that as a consequence of the debtor's failure to pay subcontractors, the owner owed the debtor no monies. Accordingly, the court held that the contract balances held by the owner were not property of the debtor's estate.

In *In re Pacific Marine Dredging & Construction*,[12] a surety was required to pay the debtor/contractor's laborers and material suppliers as a result of having issued a payment bond on the debtor's behalf. The issue was whether the surety or the debtor-in-possession had the superior claim to the retained contract funds. The court held that the funds were not property of the estate: "This court agrees with other courts that have found that the contractor's failure to pay for labor and materials is just as much a failure to perform and carry out the terms of the contract as an abandonment of work. *In short, plaintiff is not contractually obligated to pay the fund to debtor. Due to debtor's breach of contract, the debtor does not have any legal or equitable interest in the fund. Accordingly, the fund is not property of the estate.*"[13]

Similarly, in *In re Massart Co.*[14] the court required an unpaid progress payment to be turned over to the surety:

> However, United Pacific [the surety] clearly stepped in for Massart [the debtor] and performed Massart's obligations, albeit financial rather than structural. United Pacific deserves compensation for this

---

County, 860 P.2d 510, 511 (Az. Ct. App. 1993) (court cited *Pearlman* and held that the principal's failure to provide lien waivers required by the contract documents caused the contract payment not to be due: "[p]ayment was not due and payable and it is therefore arguable whether the payment was part of the bankruptcy estate . . ."). *But see In re Construction Alternatives, Inc.*, 2 F.3d 670 (6th Cir. 1993) (although the surety argued that the principal was contractually obligated to ensure that that subcontractors and material suppliers were paid, the court found that there was no express condition precedent and refused to conclude that the funds were not due and payable).

12 79 B.R. 924 (Bankr. D. Or. 1987).

13 *Id.* at 929 (emphasis added).

14 105 B.R. 610 (W.D. Wash. 1989).

> no less than if United Pacific had actually performed the contract for Massart.
>
> ***
>
> It makes no sense to punish a surety for paying claims after a contractor declares bankruptcy, and the equitable rights of subrogation exist to protect a surety in just these circumstances. Therefore, the court finds that United Pacific's equitable lien arose when it executed its bonds in favor of Massart, November 9, 1983.[15]

The court concluded that the debtor did not have a legal or equitable interest in the progress payment at the time it filed for bankruptcy. The holding, citing *Pearlman* and Section 541 of the Bankruptcy Code, was that the progress payment was not property of the estate.

Other courts have determined that contract proceeds were not property of the estate upon different grounds. Some state laws impose a trust on money owed to a contractor where subcontractors and material suppliers remain unpaid. It is well settled that property held in trust by the debtor does not become property of the bankruptcy estate.[16] In *Universal Bonding Insurance Co. v. Gittens & Sprinkle Enterprises, Inc.*,[17] the Chapter 11 debtor-in-possession asserted that it had a superior right to the bonded contract funds that were collected and that were still due from various governmental agencies. The surety argued, among other things, that the contract funds paid and still due were held in trust for the benefit of laborers, and material suppliers and were, therefore, not property of the debtor's bankruptcy estate. Pursuant to a New Jersey statute, funds received by a general contractor for any public improvement are held in trust for the benefit of subcontractors, laborers,

---

15 *Id.* at 613.

16 *See Begier v. Internal Revenue Serv.*, 496 U.S. 53 (1990) ("A debtor does not own an equitable interest in property he holds in trust for another, that interest is not 'property of the estate . . . "); *United States v. Whiting Pools, Inc.*, 462 U.S. 198, 205 n.10 (1983) ("Congress plainly excluded property of others held by the debtor in trust at the time of the filing of the petition."); *In re California Trade Technical Sch., Inc.*, 923 F.2d 641 (9th Cir. 1991) (funds held "in trust" were not property of the debtor for preferential transfer purposes).

17 960 F.2d 366 (3rd Cir. 1992).

and material suppliers.[18] The *Universal Bonding* court held that the contract funds paid to the debtor were subject to the statutory trust, and that the contract funds due the debtor would constitute an equitable trust for the benefit of laborers and material suppliers once the debtor received them. The court further held that the surety would become the beneficiary of this trust after it fulfilled its obligation to compensate the laborers and material suppliers.[19]

Similarly, courts have held that bonded contract funds are not property of the bankruptcy estate based on contract provisions impressing a trust on contract proceeds.[20] Generally, the GIA will contain a pledge by the principal to hold bonded contract funds in trust for the benefit of subcontractors and material suppliers. Other agreements used by the surety, such as "fund control" agreements, may contain a similar trust provision. However, blind reliance on a contract trust provision may not be sufficient for the creation of a trust. The parties must have acted in conformance with state law requirements for the funds to be held in trust prior to bankruptcy.[21] In other words, failure to observe trust formalities as required by applicable state law may defeat the surety's claim that the bonded contract proceeds are held in trust.

---

18 N.J. STAT. ANN. § 2A:44-148 (West 1995). Other states have similar trust fund statutes, including Kentucky, Michigan, New York, and Oklahoma. These trust fund statutes are not limited to public improvements.

19 For cases with holdings similar to the holding in *Gittens* see *Parker v. Klochko Equip. Rental Co., Inc.,* 590 F.2d 649 (6th Cir. 1979); *In re Dunwell Heating & Air Conditioning Contractors Corp.*, 78 B.R. 667 (Bankr. E.D.N.Y. 1987); *In re D&B Elec., Inc.*, 4 B.R. 263 (Bankr. W.D. Ky. 1980).

20 *See, e.g., Federal Ins. Co. v. Fifth Third Bank*, 867 F.2d 330 (6th Cir. 1989) (court held funds were held in trust based upon provision in bonded construction contract); *In re Gonzales*, 22 B.R. 58 (B.A.P. 9th Cir. 1982) (same).

21 *See, e.g., In re Construction Alternatives, Inc.*, 2 F.3d 670, 677 (6th Cir. 1993) (court held that the indemnity agreement's trust provision was insufficient to find contract funds held in trust where the principal did not segregate funds); *In re H. & A. Constr. Co., Inc.*, 65 B.R. 213 (Bankr. D. Mass. 1986).

2. *Surety's Priority To Bonded Contract Funds Held To Be "Property Of The Estate"*

A number of bankruptcy courts have concluded that the bonded contract proceeds are "property of the estate."[22] Nonetheless, many of these courts have gone on to reaffirm the surety's priority to the contract funds over the debtor. In *In re Alliance Properties, Inc.*,[23] the court concluded that the contract proceeds were property of the estate, noting the expansive definition of this phrase in Section 541 of the Bankruptcy Code. The court also concluded that once the surety paid various subcontractors' claims the surety obtained an equitable lien on the contract proceeds superior to any rights of the defaulting contractor and/or the estate. Thus, even though the *Alliance Properties* court concluded that the bonded contract funds were property of the estate, the court awarded the funds to the surety because the funds were subject to the surety's equitable lien, relating back to the time the bonds were executed.

Other bankruptcy courts have found that the surety's equitable lien does not relate back to the date that the bonds were executed.[24] These cases conclude that the surety only obtains the priority to contract proceeds held by the party that it paid. These decisions seemingly ignore the legion of authority which holds that the surety is also subrogated to the principal's and owner's rights to bonded contract proceeds.[25] In *In re Universal Builders, Inc.*,[26] the court correctly concluded that the surety was subrogated to the rights of material suppliers and subcontractors whose claims it had paid. However, the court improperly held that the laborers and material suppliers had only unsecured claims against the estate and, therefore, the surety held only

---

22 *See, e.g.*, *In re Glover Constr. Co., Inc.*, 30 B.R. 873, 878 (Bankr. W.D. Ky. 1983) (court concluded that progress payments were "property of the estate" noting the "remarkably broadened definition" of this phrase in Section 541 of the Bankruptcy Code).

23 104 B.R. 306 (Bankr. S.D. Cal. 1989).

24 *In re Levitz Elec., Inc.*, 100 B.R. 602 (Bankr. S.D. Fla. 1989); *In re V. Pangori & Sons, Inc.*, 53 B.R. 711 (Bankr. E.D. Mich. 1985).

25 *See, e.g.*, *In re Padula Constr. Co., Inc.*, 118 B.R. 143 (Bankr. S.D. Fla. 1990).

26 53 B.R. 183 (Bankr. M.D. Tenn. 1985).

an unsecured claim. The *Universal Builders* court failed to recognize that the surety was also subrogated to the obligee's rights.

3. Earned But Unpaid Progress Payments: A Problem Area

A number of troublesome bankruptcy court decisions come into play when the principal files under Chapter 11 of the Bankruptcy Code and makes claim to unpaid progress payments on ongoing projects. *In re Glover Construction Co., Inc.*,[27] involved a surety's and Chapter 11 debtor's conflicting claims to unpaid progress payments. The *Glover* court distinguished *Pearlman*, noting that *Pearlman* involved claims to retainage funds while *Glover* involved claims to unpaid progress payments. The court's basis for distinguishing *Pearlman* was tenuous. The result in *Glover* seems significantly influenced by the court's displeasure with the surety's lack of performance:

> [The surety] seeks unlimited authority over the construction funds but treats the duty of performance with patrician disdain. At no point has it sought to exercise its contractual option to step in and take over the jobs in question, although we have suggested in court that the surety's most practical method of analyzing its potential exposure would be to periodically determine from one progress payment to the next, whether Glover can complete the work cheaper than a replacement contractor.[28]

Similarly, in *In re Ram Constr. Co., Inc.*,[29] the surety's lack of involvement in completing the construction projects was pivotal in the court's determination relating to unpaid progress payments. The court stated:

> Under the present circumstances, the Court believes that RAM is acting as a contractor for [the surety] in order to complete these contracts. The surety conducts an interesting dance to avoid this explicit statement. The Court has no inhibitions about stating it explicitly. The Court believes that this arrangement permits the Debtor to operate and to complete these contracts at a lower cost than

---

27 30 B.R. 873 (Bankr. W.D. Ky. 1983).
28 *Id.* at 882.
29 32 B.R. 758 (Bankr. W.D. Pa. 1983).

> the introduction of a new contractor. If that is not the case, [the surety] would or could easily replace RAM, in order to reduce their losses.[30]

Consequently, the debtor in *RAM* case was permitted to use the progress payments with limitations designed to protect the surety's interests.

While *Glover* and *RAM* recognize the surety's right of equitable subrogation, they regrettably represent a chink in the surety's armor by creating a distinction between earned but unpaid contract proceeds and retainages. A surety's superior right to retainages, even in a Chapter 11 case, is well settled. The more difficult situation relates to unpaid progress payments in a Chapter 11 case with ongoing projects. Thus, it may be important, in a particular case, for the surety to take some of the actions suggested in the following section to fully preserve the surety's priority claim to bonded contract funds.

D. Actions To Protect Bonded Contract Funds

Upon learning that its principal has filed for bankruptcy, a surety should act promptly to protect its interests in bonded contract proceeds and to limit the surety's liability under its bonds, by sending carefully worded notices of its priority claim to contract proceeds to bond obligees. The surety should also consider filing motions: (1) for sequestration of its cash collateral (the bonded contract proceeds); (2) for adequate protection, if the court permits the debtor to use the surety's cash collateral; and (3) for relief from the automatic stay.

1. Notify The Obligee To Withhold Bonded Contract Funds

When a bond default has occurred, and the principal has not filed bankruptcy, one of the surety's first actions will be to send a letter to bond obligees. This letter will typically advise obligees of the default, but its fundamental purpose will be to place a demand upon obligees that no further contract funds be disbursed to the principal. This letter is such a basic part of the surety's response to a bond default that the letter now appears in the published literature as a practice form.[31]

The principal may have filed for bankruptcy prior to the surety having an opportunity to send such notices to bond obligees. Can the

---

30 *Id.* at 761.

31 BOND DEFAULT MANUAL 428, Exhibit 2.3 (D. Clore ed., 2d ed. 1995).

surety still send the notice to an obligee demanding that further contract proceeds be withheld from the principal? As discussed below, there is authority to support the argument that a carefully worded notice may be sent by the surety to the obligee without violating the automatic stay that arose upon the principal's bankruptcy filing. However, the surety must avoid declaring a default or demanding that the obligee withhold contract proceeds. Such a declaration or demand will probably violate the automatic stay if the bankruptcy court determines that the contract funds are "property of the estate." Fortunately, something less than a declaration or demand will likely accomplish the surety's desired result.

Section 362(a) of the Bankruptcy Code is aptly titled the "automatic stay" provision of the Code because the stay arises automatically upon the debtor filing its bankruptcy petition. Section 362(a) imposes a stay, applicable to all entities, against virtually all collection activities or other acts that would improve a creditor's position against the debtor or the debtor's property. The focus under Section 362(a)(3) is whether the sending of a notice to the obligee demanding that the funds be withheld is an action to exercise control over "property of the estate," namely the bonded contract proceeds.

Great care should be taken when preparing the notices to be sent to the obligee post-petition. If the bankruptcy court determines that the automatic stay was applicable, then the surety's notice to the bond obligee will be void and of no force and effect.[32] A bond obligee paying funds to the principal over the surety's objection might normally be subject to a set off or even an affirmative claim by the surety for the prejudice caused to the surety. However, if the condition precedent of notice to the bond obligee by the surety is held by the bankruptcy court to have been "void" because it violated the automatic stay of Section 362(a), then the surety will lose this valuable right if the demand is ignored.[33] Moreover, the bankruptcy court may award damages to the debtor against the surety if it finds the stay violation was willful.[34]

Two reported decisions from the United States Bankruptcy Court for the Southern District of Ohio offer advice on how the surety should notify the obligee to withhold funds without violating the automatic stay. These decisions, issued by two different judges from that court,

---

32 *See In re Schwartz*, 954 F.2d 569 (9th Cir. 1992); 3 COLLIER ON BANKRUPTCY, ¶ 362.11[1] at 362-115 (15th ed. rev. 1997).

33 *See, e.g., In re Carroll*, 903 F.2d 1266 (9th Cir. 1990).

34 11 U.S.C. § 362(h).

both hold that it is not a violation of the automatic stay to send a notice to a third party advising them of their rights to withhold funds from the debtor.[35]

In *In re Hughes-Bechtol, Inc.*,[36] for example, a surety sent letters to bond obligees. The letters advised the obligees that the bond principal had failed to pay certain claims and that those claimants had asserted claims against the bond issued by the surety. The surety advised that it had paid claims and that it believed it had a direct right to the contract proceeds held by obligees. In non-demanding language, the surety went on to state that it believed its claim to the contract funds was superior to that of the principal, and that the surety might have a claim against an obligee if the contract funds were disbursed or any other action were taken by the obligee that prejudiced the surety's position.[37] The debtor complained that the surety had violated the automatic stay. The court found that the surety's correspondence was predominantly informational in character, and represented an isolated contact which was not a violation of the automatic stay.[38]

What distinguishes the letter in *Hughes-Bechtol* from the normal surety form "notice to obligee" is that the *Hughes-Bechtol* letter contained no demand for the funds, nor did it direct how the funds should be paid. Instead, the *Hughes-Bechtol* letter advised the bond obligee of certain facts and stated the surety's position with respect to the bonded contract funds. The warning included in the letter of the consequences that may accompany any payment of the funds by the bond obligee to the principal was clearly a veiled threat, but one found non-objectionable by the court. As a practical matter, such a letter will be sufficient in most cases to cause bonded contract funds to be withheld until the priorities of competing claims are determined. The *Hughes-*

---

35 *In re Hughes-Bechtol, Inc.*, 117 B.R. 890 (Bankr. S.D. Ohio 1990); *In re United States Elec., Inc.*, 123 B.R. 262 (Bankr. S.D. Ohio 1990); *see also Merchants Bonding Co. v. Pima County*, 860 P.2d 510 (Ariz. Ct. App. 1993) (surety's postpetition request that the owner make no further payments to the debtor was held not to violate the automatic stay).

36 117 B.R. 890 (Bankr. S.D. Ohio 1990).

37 *Id.*

38 *Id.* It appears the *Hughes-Bechtol* court knew the importance that would be placed upon its holding. The court attached copies of the surety's letters as an appendix to the reported decision, removing any doubt as to the exact phraseology it found did not violate the automatic stay.

Bechtol letter also avoids any statement that there is a legal default, or that the letter constitutes a declaration of default by the surety against the principal under the GIA or otherwise.

Thus, the surety's preferred practice is to send letters to bond obligees which closely track the language approved in *Hughes-Bechtol*. If desired, it is acceptable to add case law citations and other references to legal authorities, but demands, directions, and default declarations must be avoided. Use of the *Hughes-Bechtol* language does not guarantee that the principal will refrain from complaining loudly to the bankruptcy court that the surety has violated the automatic stay, nor does it guarantee that the bankruptcy court will find that the letter has not violated the stay. However it can be persuasively argued that an "informational notice," previously approved in a reported bankruptcy court decision, is not the type of conduct the automatic stay was meant to prohibit and, furthermore, should not be the basis for a "willful" violation of the automatic stay provision by the surety.

2. *Move To Sequester Cash Collateral And For Adequate Protection*

When the surety has an interest in bonded contract funds that are deemed to be property of the bankrupt estate, those funds are referred to as the surety's "cash collateral."[39] The trustee or debtor-in-possession is required to segregate and account for any cash collateral in its possession, custody, or control.[40] The Chapter 11 debtor is not permitted to use any cash collateral unless (a) the surety consents; or (b) the court, after notice and a hearing, authorizes such use.[41]

Although the provisions of the Bankruptcy Code severely restrict a trustee's or debtor-in-possession's use of cash collateral, it is not uncommon for the surety to discover that cash collateral has been dissipated. The surety should promptly put the debtor on notice, by a brief court filing, that it does not consent to the use of bonded contract funds. Additionally, the surety should file a motion pursuant to Section 363(e) of the Code to prohibit or condition the debtor's use of cash

---

39 11 U.S.C. § 363(a) provides in pertinent part: "In this section, 'cash collateral' means cash, negotiable instruments, documents of title, securities, deposit accounts, or other cash equivalents whenever acquired in which the estate and an entity other than the estate have an interest."

40 11 U.S.C. § 363(c)(4).

41 11 U.S.C. § 363(c)(2).

collateral. The motion should request that the debtor be prohibited from using the surety's cash collateral. As an alternative position, the motion should request that the cash collateral be segregated and the surety be provided "adequate protection" for any use of the funds by the debtor.

When earned but unpaid progress payments are at issue, the surety should emphasize to the bankruptcy court that it intends to take over and complete the projects and that the surety has or will be filing a motion for relief from the automatic stay. The surety's intention to perform will distinguish the cases that have allowed a debtor to use the surety's cash collateral.

If the court does not prohibit the debtor's use of cash collateral, then the surety should suggest the form of adequate protection.[42] A debtor-in-possession completing a bonded contract can be required to: (1) limit its disbursement of funds to only those expenses required to keep the bonded contracts going; (2) permit the surety to control the debtor's payment of expenditures to assure the bonded monies are dedicated to the ongoing expenses of the bonded contracts; (3) provide an accounting for each project, so that income and expenses can be reviewed; (4) limit any salary payments to the debtor's stockholders, officers, or their relatives; and (5) not permit payment for equipment or expenses not necessary on the ongoing projects.[43]

---

42 *See In re Ram Constr. Co., Inc.*, 32 B.R. 758, 759 (Bankr. W.D. Pa. 1983). The court in *In re Glover Constr. Co., Inc.*, 30 B.R. 873, 882 (Bankr. W.D. Ky. 1983), required progress payments to be used first to pay all bona fide claims against the bonded projects that the surety may become liable, and any surpluses could be used for the debtor's job-related operating and overhead expenses "most stringently viewed."

43 *But see In re Universal Builders, Inc.*, 53 B.R. 183 (Bankr. M.D. Tenn. 1985) (court denied surety's action for adequate protection as concerns unpaid progress payments finding that the surety had not perfected its interest in the funds; further, the court improperly stated that even if the surety had perfected its interest, it would have been avoided under Section 552); *In re Kuhn Constr. Co., Inc.*, 11 B.R. 746 (Bankr. S.D. W. Va. 1981) (court concluded that surety's equitable right of subrogation was subject to requirements of chapter 9 of the Uniform Commercial Code, held that the surety was not entitled to adequate protection, and allowed the debtor to use the cash collateral without the surety's consent). However, the surety's equitable right of subrogation to bonded contract proceeds exists and arises independently of the Uniform Commercial Code, and the surety is not required to observe Uniform Commercial

3. *Seek Relief From The Automatic Stay*

If the surety's argument that the bonded contract funds are not "property of the estate" is sustained, then the automatic stay preventing any action to obtain "property of the estate" is inapplicable. However, the surety should not attempt to obtain the contract funds without first seeking the bankruptcy court's determination on whether the bonded contract funds are or are not property of the estate.

In opposition to the surety's stay relief motion, the debtor will argue that the contract funds are essential to its effective reorganization. Especially at the inception of a case, a court may be persuaded to permit the debtor to use progress payments. The real risk for the surety is that the bonded contract funds may be used to complete non-bonded projects or for expenses besides the completion of bonded projects. Hopefully, the surety can obtain the order discussed above, which restricts the debtor's ability to use progress payments, thereby adequately protecting the surety's interest in the bonded contract funds.

The surety may also seek, under Section 362(d)(1) of the Bankruptcy Code,[44] to have the debtor-in-possession terminated from

---

Code filing and perfection requirements to enforce those rights. *See, e.g.*, *National Shawmut Bank of Boston v. New Amsterdam Cas. Co.*, 411 F.2d 843 (1st Cir. 1969); *Transamerica Ins. Co. v. Barnett Bank of Marion County, N.A.*, 540 So. 2d 113 (Fla 1989).

44 11 U.S.C. § 362(d) provides in pertinent part:

> On request of a party in interest . . . the court shall grant relief from the stay provided under subsection (a) of this section, such as by terminating, annulling, modifying, or conditioning such stay—
>
> (1) for cause, including the lack of adequate protection of an interest in property of such party in interest;
>
> (2) with respect to a stay of an act against property under subsection (a) of this section, if—
>
> (A) the debtor does not have an equity in such property; and
>
> (B) such property is not necessary to an effective reorganization . . . .

The two subsections in Section 362(d) are in the disjunctive. Therefore, Section 362(d)(1) and (d)(2) provide separate bases for relief from stay. *In re Sun Valley Ranches, Inc.*, 823 F.2d 1373, 1376 (9th Cir. 1987).

controlling the work on the bonded projects. The "cause" as required under Section 362(d)(1) for the automatic stay to be lifted, with regard to a particular bonded contract, is that the contract is an executory contract which is not assumable by the principal. The contract may not be assumable because the debtor cannot cure the existing defaults or because the debtor cannot provide adequate assurance of its future performance of the contract. Ordinarily, the debtor will not have paid numerous subcontractors, laborers, and material suppliers. This failure will be a default under most construction contracts. The surety (and possibly the obligee if it has joined in the motion) should be prepared to show at a hearing that the contract is not assumable due to the debtor's financial inability to cure defaults or adequately assure future performance. These factors, plus the surety's prior right to the contract proceeds under the doctrine of equitable subrogation, should establish the "cause" required under Section 362(d)(1) to obtain stay relief to complete the bonded projects.

E. Executory Contracts And Completion Of Bonded Projects

The Bankruptcy Code does not contain a definition of the term "executory contract." The legislative history defines it as a contract "on which performance remains due to some extent on both sides."[45] Under this definition, most uncompleted construction contracts will be executory contracts. The same is true for a bonded principal's contracts with its suppliers and subcontractors. However, if the bonded contract is

---

45 "Though there is no precise definition of what contracts are executory, it generally includes contracts in which performance remains due to some extent on both sides. A note is not usually an executory contract if the only performance that remains is repayment. Performance on one side of the contract would have been completed and the contract is no longer executory." H.R. REP. NO. 95-595, 95th Cong., 1st Sess. 347-48 (1977); S. REP. NO. 95-989, 95th Cong., 2d Sess. (1978), *reprinted in* 1978 U.S.C.C.A.N. 5787, 5844. A substantial number of courts utilize the following definition: "[A] contract under which the obligation of both the bankrupt and the other party to the contract are so far unperformed that the failure of either to complete the performance would constitute a material breach excusing the performance of the other." Countryman, *Executory Contracts in Bankruptcy, Part I*, 57 MINN. L. REV. 439, 460 (1973); *see also* Countryman, *Executory Contracts in Bankruptcy, Part II*, 58 MINN. L. REV. 479 (1974).

fully terminated and all cure periods have lapsed prior to bankruptcy, it will not be an executory contract in a bankruptcy proceeding.[46] For example, a bonded contract is not executory if the project owner has sent a prepetition notice to the contractor, pursuant to the construction contract, that the contract would terminate if various defaults were not cured within a specified period. If the cure period passed with no cure of the defaults prior to the contractor filing for bankruptcy, then the contract will have terminated prior to the bankruptcy filing.

The debtor's interest in an executory contract is property of the estate under the Bankruptcy Code, and is protected by the automatic stay.[47] For this reason, if the owner has not terminated a bonded contract prepetition, the bankruptcy filing of the contractor will preclude the owner or the surety from immediately sending a default notice. The owner or surety will have to seek relief from the automatic stay before terminating the contract, or before attempting to take over and relet the contract.

Under Section 365 of the Bankruptcy Code 365, the trustee or debtor-in-possession can assume, assign, or reject an executory contract if certain requirements are met. If the Chapter 7 petition has been filed, the Chapter 7 trustee has sixty days from the date of the bankruptcy filing to assume (*i.e.*, accept) or reject the contract, or it is deemed rejected after the expiration of this sixty-day period.[48] If the debtor has filed a Chapter 11 proceeding, the debtor-in-possession can assume or reject executory contracts at any time before confirmation of a plan of reorganization.[49] Since the debtor in a Chapter 7 case does not have standing to assume or reject an executory contract,[50] most assumption

---

46 *See, e.g., In re Munple, Ltd.*, 868 F.2d 1129 (9th Cir. 1989); *Moody v. Amoco Oil Co.*, 734 F.2d 1200 (7th Cir. 1984); *In re Butchman*, 4 B.R. 379, 381 (Bankr. S.D.N.Y. 1980) ("When a debtor's legal and equitable interests in property are terminated prior to the filing of the petition with Bankruptcy Court . . . , the Bankruptcy Court cannot then cultivate rights where none can grow."); *see generally*, Griffin, *Post-Termination Bankruptcy Considerations for the Defaulted Contractor*, 17 CONSTR. LAW. 24 (Jan. 1997).

47 *In re Computer Communications, Inc.*, 824 F.2d 725 (9th Cir. 1987).

48 11 U.S.C. § 365(d)(1).

49 11 U.S.C. § 362(d)(2).

50 *See In re Tompkins*, 95 B.R. 722 (B.A.P. 9th Cir. 1989).

and rejection issues involving the surety arise in the course of Chapter 11 cases.

Section 365(d)(2) of the Code provides that "any party to such contract" with the debtor can move the court to require the debtor "to determine within a specified period of time whether to assume or reject such contract." Because the surety is not a direct party to the contract that it bonded, the strict language of the Code would not seem to permit the surety to file a motion to compel assumption or rejection. The surety can argue that its bonding of the contract should give it standing to compel assumption or rejection. But it may be necessary for the surety to persuade the bond obligee to join with the surety in bringing the motion to compel assumption or rejection.[51]

At the hearing on a motion to compel assumption or rejection, the surety's goal will be to prove that the debtor cannot meet the requirements to assume an executory contract. The surety should request the court to set an immediate deadline for the assumption or rejection. A good example of how this process should work can be found in *In re C.M. Systems, Inc.*[52] In that case, the debtor-in-possession had not been declared in default under the subject contract when the Chapter 11 petition was filed. The United States Bankruptcy Court for the Middle District of Florida analyzed the information about the project presented by the surety and the obligee, and determined that the debtor was behind schedule and that the project would lose money. Based upon this analysis, the court denied the debtor's motion to assume the contract, notwithstanding the fact it was not technically in default. The court noted that the rejection would benefit the estate by reducing claims. The court further noted that "in the case of a large construction project where there is a payment and performance bond posted, an early takeover of the project by the bonding company may minimize, if not completely eliminate, damage claims."[53] In *CM Systems*, the surety had the good fortune to have the debtor promptly file a motion to assume the contract, thereby placing the issue in front of the court on a prompt basis. A more savvy debtor might simply have chosen to defer the

51 The bond obligee may be a public entity who will be resistant to, or incapable of, prompt and decisive action. The surety should accompany the request with an offer to bear the cost of preparing the necessary court filings.

52 64 B.R. 363 (Bankr. M.D. Fla. 1986).

53 *Id.* at 365.

assumption/rejection issue and to continue to perform, thereby depriving the surety of this quick and easy method to place the performance issues in front of the court.

To assume an executory contract, the debtor must show that the contract is executory in nature, *i.e.*, that it has not been completely terminated prior to the date of the bankruptcy filing. If the contract is executory and there have been defaults, the debtor must promptly cure the defaults,[54] and must demonstrate the ability to provide adequate assurances of future performance under the executory contract.[55] Where there has been no default by the debtor, the debtor may assume an executory contract without satisfying the above tests.[56] The debtor involved with a project with no defaults must still establish that, in the exercise of its reasonable business judgment, assumption of the executory contract will benefit the estate.[57]

The executory contract doctrine presents several problems to the surety. First, if a bonded construction contract has not been terminated by the obligee before the principal files bankruptcy, the obligee will not be able to immediately terminate the contract or to terminate the principal's rights therein. Also, the surety may not displace the principal from the construction contract site, since to do so would be to "exercise control over property of the estate," an action prohibited by Section 362(a)(3) of the Bankruptcy Code. Thus, the automatic stay may delay the surety's post-petition efforts, significantly increasing completion costs.

The surety has two options to oppose assumption of a bonded executory construction contract (where there are multiple executory contracts, each contract must be considered independently to determine whether it is or is not assumable). First, the surety can take the position that debtor is unable to meet the requirements of Section 365(b)(1) for assumption because the debtor cannot promptly cure its defaults and/or cannot provide adequate assurance of the future performance.[58] Section

54 11 U.S.C. § 365(b)(1)(A).

55 11 U.S.C. § 365(b)(1)(B), (C).

56 *See In re Great Northwest Recreation Ctr., Inc.*, 74 B.R. 846, 856 (Bankr. D. Mont. 1987); *In re Westview 74th St. Drug Corp.*, 59 B.R. 747, 754 (Bankr. S.D.N.Y. 1986).

57 *In re Chi-Feng Huang*, 23 B.R. 798 (B.A.P. 9th Cir. 1982); *In re Huff*, 81 B.R. 531 (Bankr. D. Minn. 1988).

58 Section 365(b)(1) provides:

365(b)(1)(A) requires a "prompt" cure of any defaults prior to the assumption of a defaulted executory contract. To be "prompt," the cure must occur very shortly after assumption.[59] Certain defaults, such as the payment of penalties, may be excepted from the payments necessary to cure a default under an executory contract.[60] As to the requirement of adequate assurance of future performance, this element of Section 365(b)(1) of the Code frequently requires a great deal of conjecture and expert testimony regarding whether the debtor has the resources to properly complete the contract.

Second, the surety can oppose debtor's assumption of a bonded contract by filing a motion for relief from the automatic stay pursuant to Section 362(d) of the Code. As discussed above, the stay relief motion should request the court to lift the automatic stay to permit the surety to obtain and control the bonded contract proceeds. The surety should persuade the bond obligee to join in the stay relief motion, so that the obligee may obtain court authorization to default the debtor and to terminate the debtor's rights in the bonded contract. Finally, the stay relief motion should request court authorization for the surety to discharge its performance bond obligations, including reletting the contract if necessary.

If there has been a default in an executory contract or unexpired lease of the debtor, the trustee may not assume such contract or lease unless, at the time of assumption of such contract or lease, the trustee—

(A) cures, or provides adequate assurance that the trustee will promptly cure, such default;

(B) compensates, or provides adequate assurance that the trustee will promptly compensate, a party other than the debtor to such contract or lease, for any actual pecuniary loss to such party resulting from such default; and

(C) provides adequate assurance of future performance under such contract or lease.

11 U.S.C. § 365(b)(1).

59 *See In re Berkshire Chem. Haulers, Inc.*, 20 B.R. 454 (Bankr. D. Mass. 1982) (proposal to cure defaults over eighteen-month period is not a "prompt" cure).

60 The Bankruptcy Reform Act of 1994 amended Section 365(d)(2)(D) of the Bankruptcy Code to provide that the satisfaction of "any penalty rate" under an executory contract was added as a contract provision that does not need to be cured to assume the contract.

Under Section 362(f) of the Code, the stay relief motion may be heard and granted without hearing, where relief from the stay is necessary to prevent irreparable damage. It may be possible to meet this elevated standard where stay relief is needed to protect valuable work in progress or to assure the timely completion of certain types of projects.[61] However, stay relief motions can generally be heard thirty to sixty days after the stay relief motion is filed, especially if it is accompanied by a request for an expedited hearing. Although this time frame is generally quicker than waiting for the debtor's election to assume or reject, the result may still be a substantial delay in the completion of the executory contract which is extremely costly to the surety.

There is no question that the executory contract problem is a difficult one for the surety. The debtor can help the surety considerably by acknowledging its inability to complete and turning over the bonded contracts to the surety voluntarily. The surety should nonetheless encourage the debtor to file a stipulation to reject the executory contract, with a proposed order that formally rejects the executory contract. To ignore the Code's requirements is to take the chance that the debtor will reconsider its decision and reassert its claim to possession and control of the project. The debtor's original acquiescence in the surety's takeover may not be given great weight by the court, since the executory contract and stay relief issues were never presented to and decided by the court.

Where the debtor is cooperative after its bankruptcy filing and agrees to turn over the bonded contracts, or where the bonded contracts have been terminated prepetition, the debtor may further assist the surety by stipulating to assume and assign executory subcontracts and supplier contracts on the bonded projects. If these subcontracts and material supply contracts have favorable terms, their retention will benefit both the surety and the debtor by minimizing the bond losses. The debtor should be encouraged to file a stipulation to assume these executory contracts, and to assign them to the surety or the surety's completing contractor under Section 365(f) of the Code. The defaults may be promptly cured by the payment of outstanding balances. Such payments do not represent an added cost to the surety, since the claims are typically covered by the surety's payment bond. The surety's

---

61 Examples would be construction of prison beds required by court order or emergency repairs to public improvements to protect health, safety, and welfare of citizens.

acceptance of an assignment of these contracts, or the completion contractor's surety bonds, will provide adequate assurance of the future performance of the debtor's obligations under the subcontracts and supply contracts. In cooperation with the debtor, these subcontracts and supply contracts may be assigned directly to the surety's completing contractor. Thus, the use of assumption and assignment procedures under Section 365 can retain these favorable contracts, assuming they have not been conclusively terminated prior to the debtor's bankruptcy filing.

F. Can The Surety Be Held Liable For The Prepetition Payments Made By Its Principal

If certain characteristics are present in a transaction, it may be "avoided" under bankruptcy preference law. The "avoidance" of a transfer means that the trustee or debtor-in-possession can force the creditor who received a payment from the debtor before bankruptcy to repay the money to the bankruptcy estate. This repayment may be compelled notwithstanding the fact that there was nothing fraudulent about the payment, and the creditor was entitled to have its debt paid at the time the payment was made.

This section discusses an emerging issue in the law of suretyship and bankruptcy. The issue is whether a preference action may be brought against a surety to recover pre-bankruptcy payments made by the debtor to laborers, subcontractors, and material suppliers on bonded projects. The majority of preference actions are brought against the party who actually received funds or other property from the debtor. However, this preference action against the surety does not involve any property received directly by the surety. Rather, the preference action is based on the theory that the payments to project laborers, subcontractors, and material suppliers indirectly benefited the surety because the surety was relieved from paying debts guaranteed under the surety's bonds.[62]

---

62 *See, e.g.*, *Newbery Corp. v. Fireman's Fund Ins. Co.*, 106 B.R. 186, 187 (D. Ariz. 1989) ("[The principal's] theory is based on a bankruptcy maxim that the estate may proceed directly against a surety, instead of circuitously recouping the preference from the creditor, forcing the creditor to proceed against the surety and then waiting for the surety to assert a claim against the estate for reimbursement.").

This preference claim against the surety is sometimes referred to as an "indirect preference action."[63] This is based on established case law that a guarantor of a debtor's debts may have preference liability if the debtor pays the guaranteed debts ahead of other non-guaranteed debts.[64]

The trustee or debtor-in-possession has the burden of proof in an indirect preference action against the surety under Section 547(b) of the Bankruptcy Code. The elements to be proven with respect to each alleged preferential transfer are: (1) that the transfer took place; (2) that the transfer was made with the debtor's property; (3) that the transfer was for the benefit of the surety; (4) that the transfer was for or on account of an antecedent debt; (5) that each transfer was made while the debtor was insolvent; (6) that the transfer occurred during the preference period; and (7) that the surety received more than it would have received in a Chapter 7 liquidation if the transfer had not been made. If the trustee or debtor fails to prove any of these elements of preference, then the debtor's preference action must fail.[65]

One element of the debtor's prima facie case is that the transfers were for the benefit of the surety.[66] For the debtor's payments to benefit the surety, the payments must be of debts covered under the surety's payment bonds. However, the debtor may not confine its indirect preference action to obligations covered under the surety's payment bonds. Obviously, payments made by the debtor on unbonded projects, or payments on bonded projects for expenses not covered under the bonds, would not benefit the surety. Therefore, these types of payments are not preferential to the surety.

The debtor may allege that it can recover as preferential transfers all payments that were made on bonded projects because the surety guaranteed performance of these projects in its performance bonds.

---

63 For a more in-depth and complete discussion of the indirect preference action and how a surety can defend this type of preference action see Berens, *Bankruptcy: Can A Surety Be Held Liable For The Prepetition Payments Made By Its Principal?,* NORTON'S ANNUAL SURVEY OF BANKRUPTCY LAW 209 (1995-96 ed.).

64 *See generally* 5 COLLIER ON BANKRUPTCY ¶ 547.03[3][a], at 547-30, 31 (15th ed. rev. 1997); *see also In re Robinson Bros. Drilling, Inc.*, 97 B.R. 77 (W.D. Okla. 1988), *aff'd,* 892 F.2d 850 (10th Cir. 1989).

65 11 U.S.C. § 547(g); *In re Bullion Reserve of N. Am.*, 836 F.2d 1214, 1217 (9th Cir. 1988).

66 11 U.S.C. § 547(b)(1).

However, this argument is not consistent with the rationale of an indirect preference action. The debtor can proceed directly against the surety as to all transfers in which transferees receiving payments could have recovered against the surety. The transferees, absent unusual circumstances, would have no right to recover under the performance bonds because these bonds are solely for the benefit of the named project owner/obligee.[67] A surety is, in the vast majority of cases, not liable for a project expense incurred by its principal unless that expense is covered under a payment bond.[68] The reason for this is payment bonds are retrospective from the day of default and guarantee laborers and suppliers that the surety will make good on the debtor's unpaid obligations.[69] For the debtor to sustain its burden to prove that its payments benefited the surety, as required under Code 547(b)(1), it is essential that the debtor prove that each payment that it made was a legitimate bond obligation.

The debtor also has the burden to demonstrate that any transfer be of an "interest of the debtor in property." The Bankruptcy Code does not define the phrase "property of the debtor" as it is used in Section 547(b). However, property generally "belongs to the debtor for purposes of § 547 if its transfer will deprive the bankruptcy estate of something which could otherwise be used to satisfy the claims of creditors."[70]

The surety may utilize a trust theory to demonstrate that the debtor's payments to trade creditors were not transfers of the debtor's property. Under this theory, the contract funds assume a "trust nature" until the job is complete and laborers, material suppliers, subcontractors, and other trade creditors are paid. In *In re Building Dynamics, Inc.*,[71] the United States Bankruptcy Court for the Western District of New York

---

67 *See United States Fidelity & Guar. Co. v. A & A Mach. Shop, Inc.*, 330 F. Supp. 1403 (S.D. Tex. 1971); *Norton v. First Fed. Sav.*, 624 P.2d 854 (Ariz. 1981).

68 *See Pioneer Concrete & Fuel v. Apex Constr., Inc.*, 664 P.2d 938, 941 (Mont. 1983); *Hewson Constr., Inc. v. Reintree Corp.*, 685 P.2d 1062, 1066 (Wash. 1984).

69 *See Rhea v. Meadowview Elderly Apartments, Ltd.*, 676 S.W.2d 94, 97 (Tenn. Ct. App. 1984). Conversely, performance bonds are prospective from the date of default, guaranteeing to the owner or bond obligee that the surety will perform the debtor's unfinished contract.

70 *In re Bullion Reserve of North America*, 836 F.2d at 1217.

71 134 B.R. 715 (Bankr. W.D.N.Y. 1992).

addressed the issue of a debtor/contractor's claim that a payment to its subcontractor to pay in full all persons who furnished material or performed labor on the property was a preference. Pursuant to a New York lien law, funds received by a general contractor for the improvement of real property were held in trust prior to bankruptcy.[72] Based upon this New York lien law the subcontractor contended that the payment to it was from trust funds that were not the debtor's property. The *Building Dynamics* court agreed and held that the prepetition payment to the subcontractor was not preferential because the funds paid to the subcontractor were not property of the debtor. In states with a trust fund statute, the surety may defend a preference action because the monies used to pay those debts were held in trust. Since the trust funds would not be property of the debtor, the payments to various trade creditors would not be preferential.

Other elements of the debtor's prima facie preference action can also be challenged by the surety. To the extent the surety is successful in challenging any element in regards to the debtor's prima facie case, the transfer cannot be avoided.

Even if the debtor has met its burden of proving the elements of a preference action, under Section 547(b) of the Code, several affirmative defenses are available to the Surety under Sections 547(c) and 546(a) which may reduce in part, or eliminate entirely, the indirect preference liability. The applicable affirmative defenses are: (1) that in exchange for the debtor's preferential payments the surety contemporaneously provided a benefit to the debtor by releasing its equitable lien on bonded contract proceeds;[73] (2) that the debtor's payments were made in the ordinary course of business;[74] (3) that the debtor's project creditors extended new unsecured credit (*i.e.*, "new value") subsequent to the time they received payments from the debtor;[75] (4) that payments to project creditors were in satisfaction of perfected statutory liens that were fully secured;[76] and (5) that the limitations period to bring a preference action, as provided for in Section 546(a), has expired.

---

72 N.Y. LIEN LAW § 71(2) (McKinney 1987).

73 11 U.S.C. § 547(c)(1).

74 11 U.S.C. § 547(c)(2); *In re Tolona Pizza Prods. Corp.*, 3 F.3d 1029 (7th Cir. 1993); *In re Yurika Foods Corp.*, 888 F.2d 42 (6th Cir. 1989).

75 11 U.S.C. § 547(c)(4).

76 11 U.S.C. § 547(c)(6).

One affirmative defense that the surety will typically be able to assert, under Section 547(c)(1),[77] is referred to as the "contemporaneous exchange of new value" preference exception. The principal's indirect preference action against the surety is based upon the theory that, when the principal paid creditors on bonded jobs, the payments released the surety from the contingent liability to pay those bills under its bonds. In effect, the principal's payment to creditors was an indirect benefit to the surety. This analysis, however, cuts both ways. When the principal made payments to its laborers, material suppliers, subcontractors, and other trade creditors on bonded projects (the payments which the principal may argue were preferences) the trade creditors released their claims for payment against the surety. Contemporaneously therewith, the surety released its equitable claim to the contract funds to the extent of the payments made and the principal was then entitled to receive the contract funds from the owner. The surety is entitled to assert its release of its equitable claim to contract funds as a defense to the principal's indirect preference action under Section 547(c)(1).

In *In re E.R. Fegert, Inc.*,[78] two subcontractors each brought suit against their general contractor and the general contractor's surety. The general contractor eventually paid the two subcontractors and their suits were dismissed and all claims against the surety were released. Subsequently, the general contractor filed for bankruptcy. The Chapter 7 trustee of the general contractor's bankruptcy claimed that the prepetition payments to the subcontractors were preferences to the surety because the surety was indirectly benefited. The *Fegert* court rejected the trustee's indirect preference claim against the surety, as follows:

---

77 The "contemporaneous exchange of new value" preference exception is provided for in Section 547(c)(1):

> (c) The trustee may not avoid under this section a transfer—
> (1) to the extent that such transfer was—
> (A) intended by the debtor and the creditor to or for whose benefit such transfer was made to be a contemporaneous exchange for new value given to the debtor; and
> (B) in fact a substantially contemporaneous exchange . . . .

11 U.S.C. § 547(c)(1).

78 88 B.R. 258 (B.A.P. 9th Cir. 1988), *aff'd*, 887 F.2d 955 (9th Cir. 1989).

In this case the surety did not make the payments in question to the subcontractors. Instead, the payments were made pre-petition by the Debtor as part of a tripartite agreement under which the subcontractors released their claims against [the surety] in exchange for payments directly from the Debtor. Under these circumstances, the [Bankruptcy Appellate] Panel does not find that the fact that the surety did not actually make the payments to the subcontractors requires the application of a different equitable rule. If the Debtor had not made the payments to [the subcontractor] then [the surety] would have been called upon to advance the funds and then exercise its lien rights against payments due or to become due to the Debtor. The Trustee does not dispute that the estate received post-petition more than the total paid to both subcontractors from the White River Road Contract. Since the Debtor's payments to the subcontractors avoided the imposition of an equitable lien by the surety on future payments under the contract, there was no diminution of the estate. Under these facts, the release of the subcontractors' rights against the surety, which in turn could have exercised its lien rights, constituted "new value" being given in a substantially contemporaneous exchange.[79]

Based upon the analysis established in *Fegert*, the surety can move to dismiss the debtor's indirect preference action based upon the contemporaneous exchange of new value defense. The debtor may oppose the surety's motion on the basis that the debtor's payments were greater than the surety's release of its equitable lien on bonded contract proceeds.[80] In other words, the court must measure the amount of the "new value" the surety gives to the debtor to determine whether it is equivalent to the alleged preferential transfers.

---

79 *Id.* at 260.

80 *See, e.g., In re Nucorp Energy, Inc.*, 902 F.2d 729 (9th Cir. 1990) (to the extent that a preferential payment exceeds the amount of new value given by the creditor the exception to preference, under section 547(c)(1), provides no defense); *In re Robinson Bros. Drilling*, 877 F.2d 32 (10th Cir. 1989) (court imposed same valuation requirement, and stated that the time to value the release of a lien is the time when the lien is released); *cf.*, *In re George Rodman, Inc.*, 792 F.2d 125, 128 (10th Cir. 1986) (court stated that "[t]he plain language of the definition [of Section 547(c)(1)] does not require the valuation of the property transferred.").

In *Newbery Corp. v. Fireman's Fund Ins. Co.*,[81] the district court held that the surety was required to show the amount of new value actually given to the debtor. It is important to recognize, however, that the court agreed with the *Fegert* court's analysis and adopted its holding; the court simply found that the amount of new value must be proven.[82]

Another affirmative defense typically available to the surety, under Section 547(c)(4) of the Code,[83] is referred to as the "subsequent advance of new value" preference exception. The purpose of this preference exception is to protect creditors who, after receiving a preferential payment, extend new unsecured credit to the debtor. Trade creditors may have extended unsecured credit to the debtor within the ninety days before bankruptcy by supplying materials or services to the debtor. If a trade creditor extended unsecured credit after a preferential payment was made to the trade creditor, then the subsequent advance of new value defense may be applicable.

The court in *In re IRFM, Inc.*[84] applied Section 547(c)(4) to a series

---

81 106 B.R. 186 (D. Ariz. 1989).

82 In *Newbery*, Fireman's Fund Insurance Company's motion to dismiss was granted by the bankruptcy court. The district court reversed the bankruptcy court's dismissal of Newbery's indirect preference action, and remanded the action for the bankruptcy court to determine whether the new value given by Fireman's Fund Insurance Company as a result of its release of its equitable liens was equivalent to the amount of the alleged preferential payments.

83 The "subsequent advance of new value" preference exception is provided for in Section § 547(c):

> (c) The trustee may not avoid under this section a transfer . . . .
>
> (4) To or for the benefit of a creditor, to the extent that, after such transfer, such creditor gave new value to or for the benefit of the debtor—
>
> (A) not secured by an otherwise unavoidable security interest; and
>
> (B) on account of which new value the debtor did not make an otherwise unavoidable transfer to or for the benefit of such creditor . . . .

11 U.S.C. § 547(c)(4).

84 52 F.3d 228 (9th Cir. 1995).

of alternating preferential payments and subsequent advances of new credit by the payee. In this case, the debtor made a number of preferential payments to a single vendor and the vendor periodically extended unsecured credit to the debtor. The court examined the legislative history of Section 547(c)(4) as it relates to an alternating series of preferential payments and extensions of new value, and concluded that the vendor had a valid subsequent advance of new value defense even though its earlier extensions of new credit were repaid by the debtor.[85]

The surety should prepare a spreadsheet, similar to the one utilized by the *IRFM* court, that separately analyzes the advances made by each trade creditor that allegedly received preferential payments. The reason the spreadsheet must be prepared on a vendor-by-vendor basis is because the surety is subrogated to each trade creditor's subsequent advance of new value defense. If the debtor could not avoid its payments to the trade creditors due to the subsequent advance of new value defense available to these transferees, then the debtor cannot recover the alleged preferential payments from the surety.[86] Thus, the trade creditors' subsequent advances of new value to the debtor may be asserted by the surety as an affirmative defense to the debtor's preference action.

The discussion above involved an analysis of Section 547 of the Bankruptcy Code to determine whether the debtor's prepetition payments on bonded projects were avoidable as preferences to the surety. However, even if the payments to project creditors are determined to be avoidable preferences under the above Section 547 analysis, the surety may still not be held liable due to Section 550(a). Various cases have held that there may be no recovery from a party that did not receive the transfer of preferential or fraudulently conveyed property, if there was no intent by the debtor to benefit that party at the time the transfer was made. For example, in *In re Bullion Reserve of North America*,[87] the parties stipulated that a $1.5 million transfer from the debtor to its president was a fraudulent conveyance under Section 548 of the Code. The sole issue was whether the fraudulent conveyance could be recovered from the defendant, a party involved with the

85 *Id.*

86 *In re H&S Transp. Co., Inc.*, 939 F.2d 355 (6th Cir. 1991); *Newbery Corp.*, 106 B.R. at 187 n.2.

87 922 F.2d 544 (9th Cir. 1991).

transaction. The court held that the defendant was not an initial or subsequent transferee because he never had dominion over the fraudulently conveyed funds. The only other type of liability would be if the defendant was an "entity for whose benefit such transfer was made," under Section 550(a)(1). The *Bullion Reserve* court held that for a party to come within this portion of Section 550(a)(1), the initial transfer must have been made with an intent to benefit that party. Since the initial transfer was not made for the defendant's benefit, the court held that the Chapter 7 trustee could not recover the fraudulent conveyance from the defendant. Regarding the indirect preference case against the surety, the surety may be able to prove that the debtor's intent when paying its project creditors was to keep the business "afloat" or otherwise to benefit itself rather than to benefit the surety. In these situations, Section 550 should provide the surety with a defense to the indirect preference action.

G. Recoupment And Setoff Defenses

Recoupment and setoff are distinct but related doctrines that may enable an unsecured or undersecured creditor to significantly reduce its loss in a bankruptcy case, and may have surprising applicability to the surety in bankruptcy situations. The distinction between recoupment and setoff defenses, for bankruptcy purposes, is whether the debt and claim arose out of the same transaction or out of different transactions.[88] In an adverse proceeding or contested matter in bankruptcy, the advantage to a creditor asserting a recoupment defense, as opposed to the setoff defense, is the ability to recoup the debt against the debtor without first obtaining relief from the automatic stay[89] and without establishing that the opposing obligations both "arose before the

---

88 *Newbery Corp. v. Fireman's Fund Ins. Co.,* 95 F.3rd 1392 (9th Cir. 1996).

89 *See* 11 U.S.C. § 362(a)(7), which enjoins "the setoff of any debt owing to the debtor that arose before the commencement of the case . . . against any claim against the debtor."

commencement of the case."[90] These prerequisites to asserting a setoff defense are significant procedural and substantive hurdles.[91]

An example of the distinction between recoupment and setoff is demonstrated in *Newbery Corp. v. Fireman's Fund Insurance Co.*[92] In *Newbery,* the principal, Newbery Electric, Inc. (Newbery), defaulted and walked off all of the projects bonded by Fireman's Fund Insurance Company (Fireman's Fund). Subsequently, but prior to Newbery filing for bankruptcy, Fireman's Fund, Newbery, and Newbery's secured lender, United Bank of Arizona (the Bank), entered into an agreement which allowed Fireman's Fund and its completing contractors to use Newbery's equipment in consideration of Fireman's Fund paying reasonable rent to the Bank. Fireman's Fund attached a copy of the GIA executed by Newbery, and incorporated the document into the agreement by an appropriate reference. Postpetition, the Bank assigned its contractual right to receive rent to Newbery. Newbery brought suit against Fireman's Fund seeking rent for the equipment Fireman's Fund and its completing contractors had used to complete the bonded projects. Fireman's Fund asserted a recoupment defense, utilizing its multimillion dollar indemnity claim against Newbery, to abate Newbery's smaller rent claim against Fireman's Fund. Newbery argued that Fireman's Fund was not entitled to assert recoupment as a defense because Fireman's Fund's indemnity claim and Newbery's equipment rental claim did not arise out of the same transaction. The *Newbery* court held that the written agreement between the parties, which incorporated the GIA, was a single transaction out of which both the rent claim and indemnity claim arose. Therefore Fireman's Fund's larger indemnity claim recouped and reduced Newbery's rent claim to zero, and Fireman's Fund had no liability to Newbery.

In *Newbery,* Fireman's Fund alternatively asserted a setoff defense, and Newbery contended that the requirements for setoff, under Section 553(a) of the Bankruptcy Code, were not met. In this regard Newbery argued that its rent claim was a postpetition claim, and that Fireman's Fund's indemnity claim related back to the prepetition indemnity agreement. If Newbery's position had been accepted, then setoff would

---

90 11 U.S.C. § 553(a).

91 *See generally* Hill, *An Overview of Setoff and Recoupment Issues in Bankruptcy Cases*, NORTON'S ANNUAL SURVEY OF BANKRUPTCY LAW 243 (1995-96 ed.).

92 95 F.3d 1392 (9th Cir. 1996).

not have been permitted because Newbery's postpetition rent claim could not be offset by Fireman's Fund's prepetition indemnity claim (*i.e.*, "mutuality" would be lacking). The lower court rejected Newbery's argument, and held that the rent claim and the indemnity claim were both prepetition because they both related back to the prepetition agreements from which they arose. The Ninth Circuit did not rule upon this setoff issue because its ruling on the recoupment defense was dispositive of the case, and so the validity of a surety's setoff defense in this situation is still an open question.

Newbery is an example of a principal making a claim against the surety. In such a situation, the surety should look to the *Newbery* decision to determine whether it is entitled to assert its bond losses, on the basis of either recoupment or setoff, as a defense to the principal's claim.

Additionally, on the basis of recoupment and setoff, the surety may be able to assert its principal's defenses and affirmative claims against payment bond claimants after the principal has filed for bankruptcy. For example, a supplier provides defective goods or causes some delay to the project. The general contractor and its surety may be able to assert the defects or delay as part of a recoupment defense to the payment bond claim of the supplier.[93] There is, however, case law that provides that a surety may not assert these defenses.[94] The principal's claims against trade creditors may arise out of a different transaction than the payment bond claimant's claim. In this situation, setoff rather than recoupment would apply. The question then becomes whether the surety asserting

---

93 *See United Structures of Am., Inc. v. G.R.G. Eng'g, S.E.*, 9 F.3d 996 (1st Cir. 1993) (extensive discussion of distinctions between recoupment and setoff, and rejects Ninth Circuit authority that refuses to allow setoff in the context of the Miller Act by one not in privity); *C.A. Oakes Constr. Co., Inc. v. Ajax Paving Indus., Inc.*, 652 So. 2d 914 (Fla. Dist. Ct. App. 1995) (recognizing the doctrine of subrogation, *i.e.*, surety to its general contractor/principal, general contractor to rights of subcontractors against suppler/claimant).

94 *See United States. ex rel Bartec Indus., Inc. v. United Pac. Co.*, 976 F.2d 1274 (9th Cir. 1992), *opinion revised*, 15 F.3d 855 (9th Cir. 1994); *United State ex rel. Martin Steel Constructors, Inc. v. Avanti Constructors, Inc.*, 750 F.2d 759 (9th Cir. 1984) (cases rejecting assertion of "setoff" in the context of the Miller Act by one not in privity).

its principal's setoff defense violates the automatic stay by exercising control over property of the estate?

The principal's defenses and/or affirmative claims against trade creditors are clearly within the very broad definition of "property of the estate" as defined in Section 541(a) of the Code. If the surety attempts to utilize the principal's defenses and/or affirmative claims as part of the surety's defense of a payment bond claim, is the surety violating the automatic stay by "exercis[ing] control over property of the estate" as provided in Section 362(a)(3)? An argument can be made that the surety has the right to assert the principal's defenses postpetition as part of the surety's recoupment defense against a payment bond claimant's claim. The rationale is the surety's assertion of recoupment is not the improper exercise of control over the principal/debtor's property but rather an act to determine the appropriate amount owed under the contract upon which suit is brought.[95] The same argument cannot be made if the opposing claims do not arise out of the same transaction, and a setoff defense is asserted. If the principal has a defense or affirmative damage claim against a trade creditor, then that defense or claim is "property of the estate." Additionally, if that defense or damage claim could be used as a setoff by the principal, the setoff would be property of the estate.[96] Therefore, the surety's assertion of the principal's setoff defense is an exercise of control over that claim, which would be a violation of the automatic stay.[97]

---

95 "In proper form, a right of recoupment is merely a defensive right that assists in the just and proper determination of the defendant's liability." 5 COLLIER ON BANKRUPTCY, ¶ 553.10 at 553-102 (15th ed. rev. 1997).

96 *Id.* at ¶ 553.03[7][b].

97 *But see Merritt Commercial Sav. & Loan, Inc. v. Guinee*, 766 F.2d 850 (4th Cir. 1985) (court held that surety may assert bankrupt principal's setoff defenses, citing *In re Yale Express System, Inc.*, 362 F.2d 111 (2d Cir. 1966), and *United States ex rel. Johnson v. Morley Constr. Co.*, 98 F.2d 781 (2d Cir. 1938)). However, the *Merritt Commercial Savings* case contains no analysis of whether the principal's setoff rights might be property of the estate under Section 541(a), and no analysis of how the automatic stay precludes the surety from asserting its principal's setoff defense.

H. The Principal's Bankruptcy Filing Does Not Toll The Bond Limitations Period

After the principal files for files for bankruptcy, Section 362(a) of the Bankruptcy Code, the automatic stay provision, prevents the commencement or continuation of lawsuits against the debtor. To prevent a creditor from losing its claim due to expiration of a limitations period, whether statutory or contractual, the Bankruptcy Code contains tolling statutes that extend various limitations periods if the applicable time period did not expire prior to the bankruptcy filing.[98] One tolling provision is Section 108(c) which extends the time "for commencing or continuing a civil action . . . *against* the debtor" until the *later* of: (1) the end of such period and any suspensions thereof or (2) "30 days after notice of termination or expiration of the stay . . . with respect to such claim." In other words, Section 108(c) provides that when applicable non-bankruptcy law requires that a lawsuit be commenced within a certain time period, and the bankruptcy petition is filed prior to the time period running, then the time period to file suit is extended until thirty days after the stay is lifted.

The Bankruptcy Code's tolling statutes do not, however, toll the period for institution of suit against the surety. Several reported cases have held that the limitations period for bringing a bond claim against the surety is not extended by the bankruptcy filing of the bond principal.[99]

---

98 11 U.S.C. § 108.

99 *See United States ex rel. American Bank v. C.I.T. Constr., Inc. of Tex.*, 944 F.2d 253 (5th Cir. 1991) (court rejects bond claimant's argument that Section 108(a) extended the statute of limitations); *Cumberland Metals, Inc. v. Kentucky Ins. Guaranty Ass'n*, 801 S.W.2d 339 (Ky. Ct. App. 1990) (statute of limitations as to claim against surety not extended by principal/contractor's bankruptcy filing).

III.
Implications Of The Owner/ Obligee's And Payment Bond Claimant's Bankruptcies

A. First Days After The Bankruptcy Filing And The Need For Information

As in a bankruptcy filed by a principal/contractor, the surety should file a Notice of Appearance and Request for Notice at the earliest opportunity in the owner/obligee's or payment bond claimant's newly-filed bankruptcy case. This will lead to surety's counsel being added to the master mailing list, and thus receiving all important notices sent in the case.

The surety's need for information is as great in an owner/obligee's or payment bond claimant's bankruptcy as in a principal/contractor's bankruptcy. If the debtor is not immediately forthcoming with the information required by the surety, then the surety should promptly utilize Rule 2004 of the Federal Rules of Bankruptcy Procedure. As discussed in Section II(B) above, Rule 2004 is an extremely broad discovery device that allows examination of any person under oath on all matters that relate to the "acts, conduct or property or to the liabilities and financial condition of the debtor, or to any matter which may affect the administration of the debtor's estate, or to the debtor's right to discharge." In Chapter 11 cases, the scope is even broader, with examination allowed to matters that "relate to the operation of any business and the desirability of its continuance, the source of any money or property acquired or to be acquired by the debtor for purposes of consummating a plan and any other matter relevant to the case or to the formation of a plan."[100] The bankruptcy court may order not only the debtor to appear at a Rule 2004 examination, but also "any person" who might have knowledge of the debtor's affairs.[101] A creditor can use a Rule 2004 examination to review books and records of a debtor held by the debtor's accounting firm.[102]

In the owner's bankruptcy proceeding, pursuant to Rule 2004, the surety can compel production of documents and testimony on the

---

100 *In re Continental Forge Co., Inc.*, 73 B.R. 1005 (Bankr. W.D. Pa. 1987).

101 *In re Nixon Elec. Supply, Inc.*, 85 B.R. 988 (Bankr. W.D. Tex. 1988) (Rule 2004 examination of former officer and shareholder permitted).

102 *In re Mittco, Inc.*, 44 B.R. 35 (Bankr. E.D. Wis. 1984).

owner's ability to pay remaining contract balances, the status of the bonded projects, and any information relating to the owner's ability to reorganize under Chapter 11. The surety can also compel production of documents and testimony from various third parties, including the owner's accountants and banks. Similarly, the surety can compel production of documents and testimony on the payment bond claimant's ability to complete the contract with the bond principal, whether there are any defaults in relation to the relevant contract as well as the payment bond's intentions to complete any executory contracts with the principal. Thus, the surety can use Rule 2004 to examine the owner, the payment bond claimant, and various third parties concerning all aspects of the bonded projects, ability to perform, claims, and any other matters of interest to the surety.

B. Motion To Compel Assumption Or Rejection Of Executory Contracts

As discussed above in Section II(E) of this chapter, the Bankruptcy Code allows a debtor in a Chapter 11 case until confirmation of a plan of reorganization to decide whether it will assume or reject its executory contracts.[103] For the surety involved with an owner/obligee's bankruptcy, this excessive amount of time for the owner to decide to assume or reject an executory construction contract can lead to substantially increased losses. For example, the contractor/principal may continue working on the project for months only to have the owner decide that it will reject the executory contract. Although the owner will not be able to make a performance bond claim on the rejected contract, the lower tier project creditors will still be able to assert claims against the surety's payment bond. Therefore, either by itself or in conjunction with its principal, the surety should move, under Section 365(d)(2) of the Code, for the bankruptcy court to require the owner to decide within a short time period whether to assume or reject the bonded executory contracts. If the bankruptcy court sets a short time period for the owner

---

103 However, if the owner or payment bond claimant filed a Chapter 7 bankruptcy petition, the Chapter 7 trustee has sixty days from the date of the bankruptcy filing to assume or reject the contract, or it is deemed rejected after the expiration of the sixty day period. 11 U.S.C. § 365(d)(1).

to decide to assume or reject, the surety's liability exposure under its bonds will be reduced considerably.

Although not typical, it is conceivable that the surety may also be in a position to move to compel assumption or rejection of an executory contract in a payment bond claimant's bankruptcy. For example, the surety may be completing a bonded project on behalf of a defaulted principal, and thus may be assigned or subrogated to a favorable subcontract entered into by the principal. If the payment bond claimant files a Chapter 11 bankruptcy petition, the surety, either by itself or in conjunction with its principal, should move the bankruptcy court to require assumption or rejection of the executory subcontract as soon as possible.

The principles of assumption and rejection of executory contracts, as discussed in Section II(E) with respect to the debtor/principal, are equally applicable to obligee and payment bond claimant bankruptcy proceedings. The debtor must demonstrate that the contract is executory in nature, namely that the contract has not been terminated prior to the bankruptcy filing. If there has been a default under the executory contract, the debtor must meet the requirements of Section 365(b)(1) of the Bankruptcy Code, namely that it can cure defaults promptly and can adequately assure future performance of its contractual obligations. If there are no defaults under the executory contract, the debtor must still establish that, in the exercise of its reasonable business judgment, assumption is beneficial to the bankrupt estate.

C. Recoupment And Setoff Defenses

As discussed in Section II.G. of this chapter, it is to the surety's advantage to assert a recoupment defense, as opposed to a setoff defense. The benefits to asserting a recoupment defense are this defense can be asserted without obtaining relief from the automatic stay, and the opposing obligations do not have to arise before the bankruptcy filing.

During the owner's bankruptcy proceeding the owner may have a claim to assert against the surety's performance bond. The surety, however, may have a countervailing defense to assert against the owner's claim. For example, the surety may have a valid defense due to the owner having improperly paid contract funds for work that the owner knew or should have known was defective.[104] If the surety's overpayment defense arises out of the same construction contract that the surety bonded, then the surety has a valid recoupment defense because the opposing obligations "arise out of the same transaction."

If the scenario were revised slightly, however, the surety may be required to assert a setoff defense. Assume that the surety has the same overpayment defense, but it relates to Project A. The owner is asserting a claim against the surety's performance bond in relation to Project B. Under these circumstances the opposing claims arise out of different transactions, namely different construction contracts. Thus, the surety would be required to seek relief from the automatic stay prior to asserting a setoff defense.[105] Additionally, the surety would have to prove that both of the opposing claims "arose before the commencement of the case."[106] If the owner made the improper overpayment after the owner's bankruptcy filing, then the surety's setoff defense may be invalid because the surety's defense may not arise prepetition. However, the surety could argue that its overpayment defense relates back to the execution of the construction contract for Project A. If a court were to conclude that the surety's argument is correct, then the surety would be able to successfully assert the setoff defense in the multi-project example.

Similarly, in the payment bond claimant's bankruptcy the claimant (a subcontractor or supplier) and the surety may have opposing claims. As an example, the surety may be able to assert its principal's claim that the claimant supplied defective materials. The claimant may argue that the recoupment defense is not applicable because the surety's claim arises from the supply contract between the principal and the claimant, but the claimant's claim arises from the payment bond. The surety can argue that the term "transaction" comprehends "a series of many occurrences, depending not so much upon the immediateness of their connection as upon their logical relationship."[107] In other words, the supply contract is logically related to the surety's payment bond, and so

---

104 Philip L. Bruner, *et al.*, *The Surety's Analysis of Investigative Results: "To Perform or Not to Perform—That is the Question," in* BOND DEFAULT MANUAL 57, 107-109 (D. Clore ed., 2d ed. 1995).

105 11 U.S.C. § 362(a)(7).

106 11 U.S.C. § 553(a).

107 *Newbery Corp. v. Fireman's Fund Ins. Co.*, 95 F.3d at 1392, 1402 (9th Cir. 1996).

the opposing claims arise out of the same transaction. If a court accepts this argument, then the surety's recoupment defense should be granted. Otherwise, the surety will have to meet the stringent requirements under Section 553(a) of the Code for it to successfully assert a setoff defense.

D. The Bankruptcy Filing May Toll The Bond Limitations Period

As discussed in Section II(H) of this chapter, the Bankruptcy Code contains certain tolling provisions that extend various limitations periods if the applicable time period did not expire prior to the bankruptcy filing. Two tolling provisions are Section 108(a), which extends the time "within which the debtor may commence an action," and Section 108(b), which extends the time within which the debtor "may file any pleading, demand, notice or proof of claim or loss, cure a default, or perform any other similar act," where such time period has not expired prior to the bankruptcy filing. If applicable, Section 108(a) tolls (or extends) a limitations period, whether statutory or contractual, until the *later* of: (1) the end of such period and any suspensions thereof; or (2) "two years after the order for relief." If applicable, Section 108(b) tolls (or extends) a limitations period, whether statutory or contractual, until the *later* of: (1) the end of such period any suspensions thereof; or (2) "60 days after the order for relief." The term "order for relief" for a voluntary bankruptcy means the date the bankruptcy petition is filed, and for an involuntary bankruptcy means the date the party against whom the involuntary petition was filed is adjudicated to be a "debtor." It is possible for litigation over the propriety of the involuntary bankruptcy filing to go on for many months, which will mean extension of the time periods provided in Section 108(a) and (b) for an equivalent time period. Thus, an owner/obligee in bankruptcy may have additional months, and perhaps even an additional two years, in which to bring suit against a surety's performance bond.

Regarding the payment bond claimant in bankruptcy, one bond notice that is clearly affected by the Code's tolling provisions is the ninety day notice requirement under the Miller Act and many states' Little Miller Acts. Notice requirements set forth in non-statutory bonds are equally subject to the Bankruptcy Code's tolling provisions. It is very possible that the ninety day notice period in a payment bond may be extended to as much as a 150-day period, due to a bankruptcy, and even longer if the payment bond claimant was involuntarily placed into bankruptcy.

The important point for the surety is that there is the potential for dormant claims "to come out of the woodwork" in both obligee and payment bond claimant's bankruptcies. A surety involved in either of these bankruptcies is well advised to investigate the possibility of "extended claims" prior to consenting to final disbursements of bonded contract funds or to release of any collateral held by the surety.

IV.
Surety's Strategies In A Complex Bankruptcy Case

A. Withdrawal Of Bankruptcy Court Litigation To The District Court (Sometimes Referred To As "Withdrawal Of The Reference")

Section 112 of the Bankruptcy Reform Act of 1994 clarified that bankruptcy judges may conduct jury trials if the following three conditions are satisfied: (1) the right to a jury trial applies; (2) the judge is specifically designated to conduct jury trials by the district court; and (3) all parties consent. Based on this provision, if a party has a right to a jury trial and does not want the bankruptcy court to conduct the jury trial, the party may have the proceeding transferred to the district court pursuant to Section 157(d) of Title 28 of the United States Code. Thus, in a situation in which the trustee or debtor-in-possession has brought suit against the surety in bankruptcy court, the surety has the advantage of "forum shopping" (assuming it is entitled to a jury trial). If the surety decides that the district court would be a better forum to conduct the jury trial, the case should be transferred to the district court merely by the surety not consenting to the bankruptcy court's conduct of the trial.

It must be noted, however, that if a creditor files a Proof of Claim in the bankruptcy court, then the creditor waives its right to a jury trial.[108] Therefore, if the surety wants to retain its right to a jury trial (and, thereby, retain its right to have the litigation transferred to the district

---

108 *Langenkamp v. Culp*, 498 U.S. 42 (1990) (court held that a creditor filing a claim against a bankruptcy estate triggers the process of "allowance and disallowance of claims," thereby subjecting themself to the bankruptcy court's equitable power); *In re Locke*, 205 B.R. 592, 599-600 (B.A.P. 9th Cir. 1996) (same).

court), the surety should wait as long as possible to file its Proof of Claim. Of course, the surety should not miss the deadline for filing the Proof of Claim, because to do so would risk any pro rata distributions from the bankruptcy estate. The point is that by delaying the filing of a Proof of Claim, the surety may be able to have its jury trial held in the district court and still participate as a creditor in any distributions from the bankruptcy proceedings.

B. Motion To Appoint A Trustee Or To Convert The Proceeding To Chapter 7

Ordinarily, a Chapter 11 debtor-in-possession will remain in possession of the bankruptcy estate throughout the Chapter 11 proceedings.[109] But in certain cases, the surety may wish to have the debtor's management displaced in favor of an independent trustee. This may be accomplished by the surety making a motion to have an independent trustee appointed to manage the Chapter 11 case and/or to have the Chapter 11 bankruptcy case converted to a Chapter 7 proceeding.

Both appointment of a trustee and conversion are drastic actions by the bankruptcy court, and both require a substantial showing of "cause" by the movant. Section 1104(a) of the Bankruptcy Code allows for appointment of a trustee where there is some form of fraud or mismanagement, or when the overall interests of the creditors will be served thereby.[110] Evidence of embezzlement or intentional fraud will

---

109 *See, e.g.*, *In re TS Indus., Inc.*, 125 B.R. 638 (Bankr. D. Utah 1991) (same); *In re Colorado—Ute Elec. Ass'n, Inc.*, 120 B.R. 164 (Bankr. D. Colo. 1990) (appointment of a Chapter 11 trustee is an extraordinary remedy because there is a strong presumption that the debtor should remain in possession of the estate).

110 Section 1104(a) provides in pertinent part:

> at any time after the commencement of the case but before confirmation of a plan, on request of a party in interest or the United States trustee . . . the court shall order the appointment of a trustee—
>
> (1) for cause, including fraud, dishonesty, incompetence, or gross mismanagement of the affairs of the debtor by current management, either before or after the commencement of the case, or similar cause . . .

clearly support the appointment of a trustee. Other behavior that will permit the appointment of a trustee include failure to pay post-petition taxes, failure to maintain insurance, commingling of personal and business assets, gross mismanagement, or an inability to reorganize the debtor.

Conversion to a Chapter 7 proceeding, addressed in Section 1112(b) of the Code, will also result in the appointment of an independent trustee. The standard for cause that must be shown for conversion is slightly different than the standard for appointment of a trustee. Section 1112(b) sets forth certain non-exclusive examples of the cause that must be shown, including "continuing loss to or diminution of the estate" coupled with "absence of a reasonable likelihood of rehabilitation."

A surprisingly small percentage of cases commenced under Chapter 11 actually result in confirmed plans of reorganization. In most cases where the surety is confronted with a principal/debtor who is in bankruptcy and merely uncooperative, the time and expense attendant to a motion to appoint a trustee or to convert will not be justified. But, in an appropriate situation, these remedies under the Code can be successfully used by the surety to supplant the principal's management and terminate unreasonable and economically wasteful behavior of the principal/debtor.

C. Motion To Appoint Examiner

Moving for the appointment of an examinor is an infrequently used remedy by the surety and other creditors in a Chapter 11 proceeding. But an examiner, if appointed, may be quite helpful to the surety dealing with an uncooperative principal/debtor.

Unlike a trustee, the examiner will not displace the debtor's management of the bankruptcy estate. Rather, the examiner's responsibility is to investigate and report to the bankruptcy court as to selected areas of inquiry. The examiner can review and advise on the debtor's financial condition and controls (where the debtor has incompetent management, the examiner's report may result in the appointment of a trustee). The examiner may review, and his report may

---

(2) if such appointment is in the interests of creditors, any equity security holders, and other interests of the estate . . . .

11 U.S.C. § 1104(a).

ultimately reduce inflated management salaries or other overhead costs of the debtor. Examiners have even been appointed for the purpose of reviewing litigation the debtor is prosecuting or proposing to file.[111] An unfavorable examiner's report may lead the bankruptcy court to restrict the debtor from filing meritless litigation against third parties (possibly including the surety), when such litigation will dissipate the assets of the bankruptcy estate.

Because the examiner is principally an investigator, the bankruptcy court may be more willing to appoint an examiner than a trustee in the early stages of the bankruptcy proceeding. Section 1104(c) of the Bankruptcy Code permits appointment of an examiner where it is either "in the interests of the creditors, any equity security holders, and other interests in the estate," or where "the debtor's fixed, liquidated, or unsecured debts . . . exceed $5,000,000."[112] Thus, a lesser showing of "cause" is required for appointment of an examiner than for appointment of a trustee. Particularly where fraud, dishonesty, or gross mismanagement by the debtor is suspected, a motion for examiner should be considered by the surety.

D. Determination Of Dischargeability Of Debt

The Bankruptcy Code provides for discharge of debts in Chapter 7 and Chapter 11 cases.[113] In appropriate circumstances, the surety may wish to contest the dischargeability of the principal/debtor's indemnity obligation to the surety.

Two grounds for nondischargeability of debt are set forth in Section 523(a)(2) of the Code. Generally, there are two different subsections of

---

111 *Williamson v. Roppollo*, 114 B.R. 127 (W.D. La. 1990); *In re Jartran, Inc.*, 78 B.R. 524 (Bankr. N.D. Ill. 1987); *In re Carnegie Int'l Corp.*, 51 B.R. 252 (Bankr. S.D. Ind. 1984).

112 Section 1104(c)(2) seems to suggest that upon request of a party in interest, an examiner is mandatory where non-trade debt of the debtor exceeds $5,000,000. However, the "mandatory" nature of this section has been questioned. *In re Bradlees Stores, Inc.*, 209 B.R. 36, 38-39 (Bankr. S.D.N.Y. 1997).

113 Section 727 grants a discharge to debtors filing for relief under Chapter 7, and Section 1141 grants a similar discharge to debtors filing for relief under Chapter 11. The discharge under either case will dispose of any claim by the surety for indemnity against the debtor.

fraudulent conduct by the debtor that will render its liability to the surety nondischargeable. Both subsections of conduct relate to the giving of false information to the creditor (*i.e.*, the surety) by the debtor to induce the issuance of credit (*i.e.*, the surety's bonds). Under Section 523(a)(2)(A), it must be shown that debtor employed "false pretences, a false representation, or actual fraud" to obtain bonding from the surety.[114] Recently, the United States Supreme Court held that a debt will be nondischargeable under Section 523(a)(2)(A) if the creditor "justifiably relies," rather than "reasonably relies," on the debtor's fraudulent misrepresentations.[115] Under Section 523(a)(2)(B), it must be shown that debtor provided a false financial statement to the surety, with the intent to deceive the surety. The surety, as the creditor objecting to dischargeability of debt, has the burden of proof under both subsections and that burden is preponderance of the evidence.[116]

If the surety has a legal basis for nondischargeability, then it may contest discharge by filing a complaint to determine nondischargeability of the debt. The time limit for bringing a complaint of this type is relatively brief. The complaint must be filed within sixty days following the date set for the first meeting of creditors under Section 341(a).[117]

Frequently confused with the complaint to determine nondischargeability of debt is the complaint objecting to the debtor's discharge under Section 727. The successful objection to discharge under Section 727 will cause all debts, owed to all creditors, to survive the bankruptcy. In some cases, the showing of nondischargeability under Section 523 will also support the denial of discharge under Section 727. From the surety's standpoint, relief under Section 523 is preferable because it results in only the debt to the surety being excepted from discharge (under Section 727, no debts are discharged). This makes debtor's postpetition assets, if any, subject to collection only by the surety rather than the entire group of creditors of the bankruptcy estate.

---

114 A detailed explanation of the elements to be proven for nondischargeability under this subsection is found in 4 COLLIER ON BANKRUPTCY ¶ 523.08 (15th ed. rev. 1997).

115 *Field v. Mans*, 516 U.S. 59 (1995).

116 *Grogan v. Garner*, 498 U.S. 279, (1991).

117 FED. R. BANKR. P. 4007(c). A creditor can seek to extend this deadline to file a nondischargeability complaint "for cause," but the motion must be filed prior to the deadline passing. *Id.*

Where individual indemnitors have filed bankruptcies concurrently with the debtor/principal, the surety may also file complaints to determine nondischargeability or objection to discharge in those individual bankruptcies. Even when there is little hope of immediate recovery from the principal or indemnitors, there may be a strategic reason for filing the nondischargeability litigation. If the principal and indemnitors have not been cooperative in the surety's completion of bonded projects, the possibility of survival of the surety's indemnity claim may cause the principal and indemnitors to be more helpful and, thus, minimize the bond losses incurred by the surety.

E. Consideration Of The Surety's Use Of Involuntary Bankruptcy

Sureties generally decry the bankruptcy filing of their principal. However, under particular circumstances it may be to the surety's advantage to commence an involuntary bankruptcy petition against the principal. Generally, the benefit to be derived from an involuntary bankruptcy is the prevention or "recapture" of fraudulent conveyances and preferential transfers of the principal's property.[118] An involuntary bankruptcy proceeding may be justified where the surety becomes aware that the proceeds from bonded projects are substantially funding the completion of unbonded work. It is not unusual for a cash-starved principal to "rob Peter to pay Paul." If the amounts diverted were significant, or if the practice is ongoing, the surety may wish to consider recourse to the bankruptcy court.

---

118 To illustrate, when a principal will imminently default on its bonded projects, the principal may transfer assets without consideration, or may repay loans to insiders (*i.e.,* officers, directors, and related family members) or other favored creditors. The surety will desire to stop this hemorrhaging of assets. The involuntary bankruptcy proceeding can quickly divest the principal of control of its assets. Pursuant to the avoiding powers of the Bankruptcy Code, it will also be possible to file actions against those who have received fraudulent conveyances or preferential transfers and to recover those funds for distribution pro rata to the principal's creditors. The surety's claim may be a substantial component of the total pool of unsecured claims, thereby making the pro rata distribution to the surety substantial enough to justify the expense of an involuntary bankruptcy proceeding.

The filing of an involuntary petition is authorized by Section 303 of the Code against an entity that is generally not paying its debts as they become due.[119] A petitioning creditor must hold a claim that is not contingent as to liability and is not the subject of a bona fide dispute. Generally, the involuntary petition must be commenced by three creditors whose unsecured claims aggregate over $10,000. However, pursuant to Section 303(b)(2) a single petitioning creditor is permissible, if the debtor has less than twelve creditors. A single petitioning creditor

---

119 Section 303 provides:

(b) An involuntary case against a person is commenced by the filing with the bankruptcy court of a petition under Chapter 7 or 11 of this title—

(1) by three or more entities, each of which is either a holder of a claim against such person that is not contingent as to liability or the subject of a bona fide dispute, . . . if such claims aggregate at least $10,000 more than the value of any lien on property of the debtor securing such claims held by the holders of such claims

(2) if there are fewer than 12 such holders, excluding . . . any transferee of a transfer that is voidable under [the Bankruptcy Code] by one or more of such holders that hold in the aggregate at least $10,000 of such claims; . . . .

(h) If the petition is not timely controverted, the court shall order relief against the debtor in an involuntary case under the chapter under which the petition was filed. Otherwise, after trial, the court shall order relief against the debtor in an involuntary case under the chapter under which the petition was filed, only if—

(1) the debtor is generally not paying such debtor's debts as such debts become due unless such debts are the subject of a bona fide dispute; or

(2) within 120 days before the date of the filing of the petition, a custodian, other than a trustee, receiver, or agent appointed or authorized to take charge of less than substantially all of the property of the debtor for the purpose of enforcing a lien against such property, was appointed or took possession.

11 U.S.C. § 303(b) and (h).

must be able to demonstrate that it made some reasonable inquiry to determine that the debtor had less than twelve creditors.[120]

If the petition is challenged, it will be the petitioning creditor's burden to prove that the debtor is failing to pay its debts as they become due.[121] Among the factors that will be considered in making this determination are: (1) the number of unpaid claims; (2) the amount of such; (3) the materiality of the non-payments; and (4) the debtor's overall conduct of its financial affairs.[122]

The determination whether a creditor holds a claim that is not subject to a "bona fide dispute" requires the consideration of factors in a case-by-case analysis. The court will generally look to the nature of the dispute, the extent of evidence supporting a creditor's claim and a debtor's objection, whether the creditor's claim and the debtor's objection appear to have been made in good faith and without fraud or deceit, and whether, on balance, the interests of the creditor outweigh those of the debtor.[123]

The principal disadvantage to the surety who files an involuntary petition is the risk that the court will find the petition was filed in bad faith.[124] If the bankruptcy court dismisses the involuntary bankruptcy proceeding, then the parties that filed the involuntary petition may be subject to an award of attorneys' fees and costs. Additionally, if the bankruptcy court concludes that the involuntary bankruptcy petition was filed in bad faith, it may assess punitive damages against the petitioning creditors.[125]

An illustration of how a surety's petition can result in bad faith finding is found in *In re Vincent J. Fasano, Inc.*[126] In *In re Vincent J.*

---

120 *In re Race Horses, Inc.*, 207 B.R. 229, 232-3 (Bankr. E.D. Okla. 1997).

121 *See* 11 U.S.C. § 303(h)(1); *In re Fischer*, 202 B.R. 341, 350-51 (E.D.N.Y. 1996) ("There is substantial authority for the proposition that even though an alleged debtor may owe only one debt, or very few debts, and order for relief may be granted where such debt or debts are sufficiently substantial to establish the generality of the alleged debtor's default").

122 *Id.*

123 *In re Audio Visual Workshop, Inc.*, 211 B.R. 154 (Bankr. S.D.N.Y. 1997).

124 11 U.S.C. § 303(i).

125 11 U.S.C. § 303(i)(2).

126 55 B.R. 409 (Bankr. N.D.N.Y. 1985), *rev'd,* 58 B.R. 1008 (N.D.N.Y. 1986).

Fasano, Inc., the surety filed an involuntary bankruptcy petition. The court found the surety acted in bad faith, due to the manner in which it solicited other parties to be co-petitioners in the involuntary bankruptcy. The court's concern related to how the surety solicited an unpaid subcontractor to join in the involuntary petition. The surety made this solicitation, but withheld information from the subcontractor regarding his bond rights. The debtor argued that the surety purposely delayed payment of the subcontractor's bond claim until it was time-barred, so as to qualify the subcontractor to be a petitioning creditor. The bankruptcy court accepted the debtor's version of the facts and found bad faith on the part of the surety, although the district court later reversed due to the bankruptcy court's failure to consider the surety's subjective motivation for filing the petition.

Without question, the involuntary bankruptcy remedy is one that should be used by the surety sparingly and only in appropriate circumstances. However, the pre-bankruptcy dissipation of assets by a principal may warrant this drastic remedy. Furthermore, the gross misuse of bonded contract funds may also justify an involuntary filing by the surety. Indeed, where the surety is suddenly confronted with mounting potential losses due to a principal's decision to repay bank loans or make other conveyances that undermine the principal's financial position, the involuntary bankruptcy may be the only way for the surety to preserve assets for later distribution to itself and other creditors of the bankrupt estate.

V.
CONCLUSION

All creditors experience frustration because of the slow and uncertain pace of Chapter 11 proceedings. Contract sureties face especially acute problems because of the potential for bonded project deterioration, and assessment of liquidated damages, which may result from the delays attendant to the bankruptcy process.

The surety, aware of its rights and potential pitfalls, should aggressively act in the principal's, the owner's, or the payment bond claimant's bankruptcy proceeding to limit its liability. If the party that is in financial turmoil is cooperative with the surety, a joint effort can lead to a substantial decrease in the surety's ultimate liability. If the party is antagonistic to the surety, a strategy based on knowledge and resolve is preferred to merely reacting to the initial flurry of motions filed by the debtor (and frequently joined by the debtor's secured creditor). The bottom line is that a proactive strategy by the surety should limit liability exposure and enhance salvage, and ultimately reduce the surety's losses as a result of the bankruptcy filing.

CHAPTER 8

ORGANIZING AND CONTROLLING THE MASSIVE DOCUMENT/DEPOSITION CASE

Deborah S. Griffin
Robert A. McCall[1]

I. Introduction

The most revolutionary developments in managing and litigating large cases are (1) reforms in the judicial administration intended to streamline discovery; and (2) advances in the availability of computer software and hardware that permits lawyers to have at their finger tips case documents, deposition transcripts, and videotapes of deposition testimony. Without efficient use of current technologies, the lawyers will not be able to deal cost-effectively with such reforms as automatic disclosure, discovery event limitations, and accelerated dockets in the context of a large case. Technological advances rather than well-meaning judicial case management reform efforts will in the long run bring greater cost savings in the large case. Linked with powerful relational databases, these tools will both make the large complex case more manageable for lead counsel and bring clients significant savings in the sometimes astronomical paralegal, clerical, and document retention costs.

---

1 The authors are grateful for the assistance on various technical issues from Katherine Pike and Barbara Mende, both of Peabody & Arnold.

II.
Casing The Case

The key to organizing and controlling high volume documents and depositions is early identification of important factual issues, legal issues, sources of document collections, types of documents, and individuals and companies with knowledge of facts. In the first chapter of this book, various strategies were discussed by which the surety can identify and analyze the key factual issues surrounding claims and liabilities of the parties. This process will be refined further if there has been a period of negotiation or mediation in an effort to reach a settlement of the disputes.

If the surety has had the opportunity through these mechanisms, prior to litigation, to become familiar with the factual and legal issues and the parties to the dispute, the initial steps to organizing and controlling the massive document and deposition case will have been taken. Staying on top of these issues is also essential to meeting the types of automatic disclosure requirements required by Rule 26(a)(1) of the Federal Rules of Civil Procedure. These initial steps entail identifying the individuals and companies in possession of facts and articulating the disputed issues of fact and law most likely to be the subject of a trial.[2] The pleadings served by the parties at the commencement of the litigation will also help to define the issues. The process of identifying individuals, companies, and issues should result in the generation of two lists or databases which will serve as the basic organizational framework in discovery, and eventually at trial: a names directory and an issues directory.

A. The Names Directory

In identifying individuals and companies with information, one must accumulate more than their names. A data entry form should be designed onto which key information can be recorded.

Such key information should include the: (1) individual's first name and middle initials; (2) individual's last name (3) individual's address; (4) individual's telephone number; (5) individual's employer; (6) employer's address; (7) employer's telephone number; (8) individual's

---

2 FED. R. CIV. P. 26(a)(1).

position with employer; (9) start date of individual's involvement with project; (10) end date of individual's involvement with project; (11) individual's immediate superior in company; (12) issues about which individual has knowledge (using issue codes or uniform terminology developed for the Issues Directory, discussed in Section II.B); (13) location of individual's working files; (14) custodian of individual's working files; (15) counsel; (16) individual's ID code (assigned); and (17) attorney comments. A sample data entry form is included in the Appendix as Exhibit 8.1.

Not all of this information is likely to be available immediately, but holes will be filled as the case progresses. Indeed, interrogatories, document requests, and deposition questions can be framed for the specific purpose of obtaining the missing information. Records also will be added as the case progresses as additional individuals are identified through discovery.

In a case in which the volume of information is expected to be more than can be managed manually, the data collected with regard to each individual should be entered into a computerized database, with each of the items listed above constituting a separate data field. Numerous off-the-shelf database products can be adapted to this task, but a product should be selected that has the capacity and flexibility to accommodate the additional fields that will be added as the litigation progresses,[3] as well as electronic deposition and trial transcripts.[4] The product should

---

3 Additional fields likely to be added can include those tracking whether, on what dates, and by whom the individual was interviewed or deposed; the specific attorney having responsibility for examining or preparing the individual to testify; and whether the individual will be available for trial or must be deposed for trial purposes.

4 Products especially designed for use in litigation support include Summation-Blaze and Discovery ZX. Software and hardware products change so rapidly that literature on such products becomes dated rapidly. Some of the more recent articles regarding hardware and software for litigation support include *Litigation & Administrative Practice Course Handbook Series*, Practicing Law Institute, Nos. H4-6253, H4-6264 (1997); Daniel B. Evans & Storm M. Evans, *New Products*, 23 NO. 3 LAW PRAC. MGMT. 6 (1997); Todd H. Flaming, *Litigation Support Part I: Deposition Transcript Software*, 85 ILL. B.J. 383 (1997); Thomas C. Seawell, *Introduction to Automated Litigation Support—Part II*, 26 COLO. LAW 61 (1997).

also be able to connect electronically to document images in the event imaging is to be used.[5]

In order to maximize the usefulness of the database, information entered into each field should be entered in a uniform format, so that word searches and alphabetical or numerical sortings can be performed reliably. For example, fields calling for entry of a date should be formatted so that the month, day, and year are always entered: MM-DD-YYYY, or 10-03-1997, so that entries can be date-sorted without missing entries incorrectly formatted DD/MM/YYYY, or, for single digit months or days, M-D-YYYY.[6]

A unique identification code should be coined for each individual, preferably using the individual's initials so that the codes will be recognizable.

B. The Issues Directory

The issues directory should list and assign a three- or four-letter code to each factual and legal issue in dispute. In addition, important undisputed factual issues should be listed. The initial list of issues should be teased from the pleadings, settlement communications, and consultants' reports. The list should include each element of a *prima facie* case on each legal theory of recovery, as well as each affirmative defense to those theories. If the pleadings include counterclaims, the theories of recovery and affirmative defenses pertaining to those counterclaims should be included as well. For example, in a case involving claims under the performance bond and for bad faith arising from the principal's failure to complete work within the time required, and counterclaims for delay damages, the issues list might include:

1.	contract formation	KF
2.	plans and specifications	PS
3.	bidding and estimating	BID

---

5 For additional information on imaging, *see infra* Section V.

6 In the past, dates were formatted with only the last two digits of the year, *e.g.* 10/03/97. With the approach of the millennium, however, the practice increasingly is to use all four digits. Indeed, some software products now require use of four digits. Some software products recognize dashes and slashes interchangeably, but others may not. The requirements of the particular software should be checked.

4.	issuance of bond	ISSB
5.	original completion date	OCD
6.	time extensions approved	TXA
7.	time extensions denied	TXD
8.	change order 1	CO1
9.	change order 2	CO2
10.	notice	NOT
11.	critical path	CPTH
12.	acceleration	ACC
13.	home office overhead	HOO
14.	owner's damages	DAMO
15.	principal's damages	DAMP
16.	surety's response to claim	RESP
17.	failure to mitigate	MIT
18.	payments	PAY
19.	surety's financial condition	SFIN
20.	principal's financial condition	PFIN
21.	attorneys' fees	ATTF
22.	architect's fees	ARCF

The list can and should be expanded as issues evolve. In some cases, it may be appropriate to make use of standard Construction Industry Codes for the various trades involved in the project. If issues center on particular sections of the specifications, those section numbers can be incorporated into the issue codes. Whatever coding system is adopted should have a common sense basis. That is, the letters and numbers of the codes should bear a logical relationship to the words describing the issues, so that they can be learned and recognized readily by the attorneys, paralegals, and technical support staff who will be using them.

The surety's consultants should have a major role in compiling the list of issues involving technical disputes. For the most important issues, subissues may be identified. However, if the list of issues is too long or too refined, the matching of codes to individuals and documents, which comes later in the process, will be less reliable. The more judgment that has to be brought to bear on whether a given document relates to a given issue, the less likely it is that the coding of documents will be performed in a consistent manner. Inconsistency in coding will arise when more than one person is doing the coding work as well as when only one person is doing the coding over an extended period of time.

III.
Collecting The Paper

A. Obtaining All Important Document Collections

It is likely that, during the course of conducting its investigation, the surety will have accumulated documents from various sources. These documents will have been useful in creating the names directory and issues directory. However, the accumulation of documents likely will have to be more thorough and systematic than the accumulation of documents in investigation. Consideration should be given to making a formal, comprehensive document request even from individuals and entities from whom the surety obtained documents during investigation, to ensure the completeness of the surety's document collection.

The names directory should be used to generate a list of potential document sources by generating a report of all individuals' names and the names of their employers. This report can be cross-checked against the information contained in the directory about the locations and custodians of the individual's working files to ensure that no document collections are overlooked.

The report of document collections and locations can serve multiple functions. It can be used as a checklist for the parties and non-parties upon whom requests for production of documents or subpoenas *duces tecum* will be served. It can be used in the drafting of the description of documents requested to be produced in document requests and subpoenas.[7] It can also be used to generate the list of documents which must be included in an automatic disclosure statement under Rule 26(a)(1)(B) of the Federal Rules of Civil Procedure as a precondition to initiating discovery.[8]

---

7 For suggestions concerning the types of documents to request, see John R. Henderson, *Paper Discovery in Construction Litigation: Finding the Forest Through the Trees*, unpublished paper presented at MANAGING AND LITIGATION THE COMPLEX CONSTRUCTION CASE, American Bar Association, Forum on the Construction Industry (Oct. 1998).

8 Rule 26(a)(1)(B) requires each party to provide "a copy of, or a description by category and location of, all documents, data compilations,

Once responses to document requests and subpoenas begin to come in, or the surety is given the opportunity to inspect and copy document collections in the locations where they are maintained, the surety will be faced with questions about whether to be selective or all-inclusive in copying documents produced. In cases involving a large volume of documents, counterintuitive though it may seem, the best course is to obtain hard copy of *all* documents tendered in document production, even at the risk of copying multiple copies of the same document, either from within a given party's production or from multiple parties. There are both economic and strategic reasons for this. The cost of photocopying, numbering, bar-coding, indexing, and storing multiple copies of a document is likely to be far less than the cost of having a person with an hourly billing rate of tens or hundreds of dollars ponder or research whether a copy has already been obtained. In addition, selectivity in copying entails the risk that a document believed to be a duplicate is not and thus is missed. Further, in some instances it can be important to know that a particular individual *had* a copy of a given document in his possession, a fact which will be lost if documents are copied selectively. To that end, it is advisable to copy file folders in which the documents are maintained, or their labels, and to replicate the filing system employed by the person whose records are produced by clipping, banding, or placing the documents in folders at the time they are copied.

Whenever possible, documents and other project-related materials should be obtained in electronic format as well as in hard copy. Materials most likely to have been created or maintained electronically are accounting records, requisitions, project schedules, and critical path analyses. If CAD technology was used in preparation of design and construction drawings, such materials may also be available electronically. In cases involving allegations of design defects or delay claims, having drawings, schedules, and critical path analyses in electronic format will save the expense of having consultants and testifying experts create electronic files from scratch, and will provide a baseline from which consultants and experts can create electronic

---

and tangible things in the possession, custody, or control of the party that are relevant to disputed facts alleged with particularity in the pleadings." FED. R. CIV. P. 26(a)(1)(B).

simulations and analyses of causations, alternative "what if" scenarios, time lines, and other visual aids.

B. Documenting The Collection Process

As documents are collected, basic information about the collection process should be recorded. Such information should include, as to each document or collection: (1) the source of the document (utilizing the identification code assigned in the names directory to the individuals from whose files the document is obtained); (2) the location from which the document is obtained; (3) the date on which the production occurs or begins; and (4) the name or initials of the person obtaining the document on behalf of the surety.

Information concerning the location from which the document is obtained should be more specific than the street address. If materials are being produced from multiple offices at a given street address, or if multiple file drawers of materials are being produced from a given office, the person obtaining production should diagram the filing locations and label discrete offices, cabinets, and drawers on the diagram. Each file or sequence or pages should then be identified to its filing location by reference to the office, cabinet, and drawer label. As previously noted, it can be important to know that a particular individual *had* a copy of a given document in his possession, and recording the precise filing location of a particular copy of a produced document can establish such possession. When preparing for that person's deposition, one can readily retrieve that person's entire file, folder labels and all, and lessen the chance that important areas of questioning will be overlooked as to documents not specifically addressed to or authored by that person.

If documents are obtained through the deposition of a keeper of the records of a business, questions should be asked in the deposition to secure the same information. In addition, the examiner should try to establish, through the keeper of the records, a foundation for the authentication of the documents and for the qualification of the documents as business records under applicable rules of evidence. If responses to such foundational questions are unsatisfactory, they should be addressed again to a witness designated to testify on those subjects pursuant to Rule 30(b)(6) of the Federal Rules of Civil Procedure or a corresponding state rule of procedure.

C. Numbering And Bar-Coding

Just as important as obtaining complete copies of all important document collections is separately numbering each page of each collection. Assigning a unique number to each page provides several advantages. It provides an easy way to refer to specific pages in framing questions in interrogatories, in deposition, or at trial. It enables someone reviewing a document to see that the document is complete if there are no gaps in its page numbers. It provides an economical way of referencing a document when identifying it as an identical copy, a draft, or a copy with handwritten variations of another document. Depending upon the numbering system used, it can also provide information as to the source of the document that will be apparent from the face of the document. If all pages in a given collection are numbered in the order in which they were maintained, and if the files were maintained chronologically, then the proximity of pages in a numerical sequence can provide information about when a document came into possession of the person in whose collection it was found.

The numbering system used can be as simple as beginning with the number one and increasing sequentially from there. However, in complex cases with a high volume of documents, an electronic database likely will be used to index and retrieve documents. In such cases, an estimate of the total volume of documents should be made at the outset, and the number of digits to be used in numbering selected accordingly. For example, if it is anticipated that the document collection ultimately will have over one million pages but less than ten million, all page numbers should have at least seven digits. Thus, the first page will be numbered "0000001" rather than "1."

In addition, the numbering system can be designed to include a numerical or alphabetic code as to the source of the document. This code can be the same as the individual identification code in the names directory. Another option is to create a code that merely identifies the party from whose collection to the document was produced (*e.g.*, SU for surety, GC for general contractor, AR for architect, OW for owner, etc.). A variation on this system would be to subdivide each party's collection by location, so that documents from the surety's home office might be coded SH, while documents from the surety's local office would be coded SL. These codes would be used as prefixes to the numbers themselves. Accordingly, if the first 7,500 pages of documents were copied from the general contractor's field office, they would bear numbers GL0000001 through GL0007500. If the next 4,000 pages were copied from the architect's project manager's files, they would bear numbers AM0007501 through AM0011500. If another group of general contractor's documents were numbered next, they would begin with page number GL0011501.

There are several techniques for applying page numbers to documents:

1. *Stamping*. Numbers can be stamped onto the page with a Bates numbering stamp, which can be set so that the numbers automatically advance after each stamping. This method can be performed by almost anyone, but suffers from irregularity in the clarity of the ink impression it leaves on the page. With each successive generation of photocopying, the numbers become less legible. If a mistake is make in numbering, it is difficult to correct. There is a risk, if more than one person is numbering at the same time, that the same number will be applied inadvertently to two documents.
2. *Stickering*. Numbers can be applied by self-adhesive stickers which have been pre-printed with number sequences in the desire format. Once a run of numbers has been printed, several people can apply the stickers simultaneously without risk of duplication. With practice, one can apply stickers much more rapidly than stamping. Some types of stickers, although self-adhesive, can be removed and replaced for a short time after initial use, so that misplacements can be corrected easily.[9]
3. *Laser numbering*. Pages can be numbered by a specially equipped copier which numbers and copies pages simultaneously. Numbers can be applied in any sequence and additional information or labels such as "confidential" can be included. This type of numbering is useful in a situation where the original document is not to be numbered and copies

---

9 When a prefix is used in the numbering system, stickers must be numbered as the numbering process progresses, rather than pre-printed in full, so that the appropriate prefixes are printed with the applicable sequence of page numbers.

need to be made. The greatest risk in using this numbering system is that if the copy machine jams, as often occurs, the numbering sequence can be disrupted with loss of efficiency and quality control. A variation on laser numbering is a machine which prints numbers on pages without copying them. One such machine is called Videojet, which prints in red so as to differentiate originals from photocopies. Such a machine can number as many as 10,000 pages per hour if the originals are clean and able to be fed through the machine mechanically.

4. *Bar coding*. Bar codes can be applied by any of the page-numbering methods described above. The page number is simply supplemented to show both arabic numerals and corresponding bar codes. Bar codes are useful mainly when a document is going to be imaged. When the image is scanned, the scanner reads the bar code on the same pass, thus automatically creating a skeleton database to which detail can be added later. The page number is put into the database electronically, rather than by error-prone fingers. Bar coding saves time and accuracy when documents are being used in deposition or at trial with electronic presentation systems. Instead of the presentation system operator typing in a number, one can simply waive the bar code reader wand at a bar code (on a hard copy or on a list in the presenter's notebook), thereby calling up the image of the document on the electronic presentation screen. In addition to numbers, bar codes can be used to correspond to the parties and individuals in the names directory and to issues in the issues directory.

IV.
Indexing And Coding Documents

A. Database Creation

The only practical way to control a massive document collection is to create a computerized database in which information on or about documents is recorded in standardized categories or fields, using uniform terminology and formatting. Once the database is created, the fields in the database can be searched for selected word, number, or letter combinations, and the document summaries containing the selected search terms can be sorted in a desired order. The sorted search results can be viewed on line or printed on paper.

The fields in a database are not predetermined by the database software. Rather, most databases allow the user to define the fields and to arrange the display on the screen or on paper. Faced with an almost limitless range of choices about the design of a document database, the litigator must specify fields to be established which are most likely to serve the ultimate purposes of retrieving needed information in usable form. Thus, the database should be designed, and document records created, with the ultimate uses in mind.

The questions a litigator asks most frequently about documents collected in discovery or about information in those documents are:

(1) What documents did a particular person author or receive?
(2) In what sequence was a particular issued addressed?
(3) About what issues has a particular person written?
(4) What did a particular person know and when did she/he know it?
(5) Where can I find a particular document I remember seeing three years ago?

The database should be designed so that searches for documents in the database will provide answers to these kinds of questions.

With these objective in mind, most document databases contain the following fields:

1. Start Number (the number assigned to the first page of the document);
2. End Number (the number assigned to the last page of the document);
3. Document Date (using a uniform format);
4. From (code of the person who authored the document);
5. To (code for the primary addressee of the document);
6. Copies To (codes for all individuals to whom the document is shown as having been circulated);
7. Mentions (codes for all individuals mentioned in the document other than as an addressee, recipient or author);

8. Document Type (*e.g.*, letter, memo, notes, report, invoice, agreement, file);
9. Subject (a brief description of the subject matter and content of the document);
10. Topic (issue codes applicable);
11. Attorney Comments (a place to record attorneys' mental impressions about the document);
12. Priority (a code indicating the relative importance of the document);
13. Cross Reference (a place to record information such as the page numbers of duplicate copies of the document; this field can be supplemented with another field designating one such copy as the "master" and others as "duplicates," which can be used to ensure uniform treatment of all copies);
14. Deponent (the name of a deponent in whose deposition the document has been marked as an exhibit);
15. Exhibit Number (the exhibit number assigned in deposition);
16. Production Status (codes or words indicating when and by whom or even whether the document has been produced; this information can be tied to a separate database or listing of production events); and
17. Privilege (codes to indicate, with respect to documents withheld for privilege, the grounds for the claim of privilege).

Once records have been created for the collected documents, and applicable information about those documents recorded in each of the above-listed fields, the database can be searched for answers to the commonly asked questions. For example, in answering the question "What documents did the architect's project manager receive?" one would perform a search for: (1) all documents numbered with the prefix AM; and (2) all documents with the individual identification code of the architect's project manager in either the "To" field (field 5) or the "Copies To" field (field 6). In answering the question "In what sequence was the issue of change order 2 addressed?" one would perform a search for (1) all documents with an issue code CO2 in the "Topic" field (field 10); and (2) sorted in chronological order in the

"Document Date" field (field 3). Other combinations of fields and directions for the order in which records meeting the search criteria are sorted should provide reasonably responsive search results.

B. Prioritizing

The two primary mechanisms for designating the relative importance of documents in the database are propriety coding in the "Priority" field (field 12) and the level of indexing. Database records can be created for a single document or for a group of documents maintained in a file or box. The most important or unusual documents will be indexed on a document-by-document basis and each document assigned a priority code in the "Priority" field. For less important documents, a single entry covering the entire file or box can be created. Occasionally, an entire file or box is important enough to warrant its own priority code.

Coding documents for priority will always be a matter of judgment. The fewer priority categories used, the greater the likelihood that judgment will be applied consistently across a collection. A coding scale of three to five levels of importance is workable for most cases. If a priority of "1" signifies the most important documents, it should be assigned only to those documents which contain information or admissions and are certain to be used as trial exhibits. This category should be reserved for "smoking gun" documents. The number signifying the least important documents in a collection should be assigned to documents that the review of which can be foregone in all but the most thorough searches. In other words, assigning the lowest priority code should be consistent with the decision that, unless one needs to do a search leaving no stone unturned, the attorney need not even look at a summary of these lowest priority documents. Between the high and the low are any number of graduations of importance.

In a case involving tens of thousands of pages of documents, it is unlikely to be practical, economical, or necessary to index each document individually with the level of detail required to complete all fields in the database. More likely, less than a quarter of any collection merits such detailed indexing. For the remainder of the documents, indexing on a file or box level suffices. For documents not important enough to need a complete database record for each document, a single entry can be created to track the source, page numbers, and general contents of the file or box. Such database records will utilize the "Start Number" (field 1), "End Number" (field 2), "Mentions" (field 7), "Document Type" (field 8), "Subject" (field 9), "Topic" (field 10), and "Production Status" (field 16) fields. Use of these fields will record the page numbers of the first and last pages in the file or box and the source of it. In the "Mentions" field, the initials of all individuals mentioned in the page sequence, *including* as an addressee, recipient, or author, can be listed. The "Topic" field can include all issue codes applicable to the contents of the file or box. The "Subject" field can be as general or as detailed as the file or box seems to merit. For example, a file containing a dozen requisitions could be described in the "Subject" field as "Requisitions 1-12." A file of correspondence spanning a six month period could be described as "General Contractor's Home Office Correspondence File, 7/1/95-12/31/95. Significant letters indexed separately." In order to designate such a database entry as relating to a file rather than an individual document, the "Document Type" field would contain the word "file."

Indexing on the file or box level makes it possible to locate at a later date documents whose significance was not recognized in the early stages of indexing. If an exhaustive search for documents relating to a particular topic or individual is needed, database records by box or file level will appear in the search results. The user of the search results can then decide whether to examine the file or box.

The key to creating a useful database that does not index every document on a document level, but utilizes as appropriate proportion of file and box level indexing, is to have the person or people doing the indexing be sufficiently knowledgeable about the important issues in the case so that they can distinguish documents that merit document level indexing from those that do not. However, it is not practical to have the trial attorneys to do the indexing. Rather, paralegal or other specialized support staff can perform the indexing functions, provided they are given a thorough briefing by counsel as to the important legal and factual issues, including important terms of art which may serve as red flags for important documents.

After the initial orientation, the staff doing the indexing should begin by indexing several thousand pages and submitting the documents, along with the database records as indexed, to counsel for review. Counsel should be able to determine on the basis of such a review whether important documents are being identified as such and whether documents warranting only file or box level treatment are being indexed individually. Such quality control early in the process should make for efficient and reliable indexing. Occasional spot-checking throughout the process is also desirable. As the database is used, errors invariably will be found and should be corrected promptly.

C. Privileged Documents

It is assumed that, before producing documents in its own collection, the surety will screen them for privilege or other immunity from production. As decisions are made regarding whether a particular document is privileged or otherwise can be withheld from production, those decisions should be recorded in the document database in the "Privilege" field (field 17). One-word labels or codes for each ground for withholding a document should be created and applied consistently, the most common being attorney-client privilege, work product protection, and confidentiality obligations pursuant to agreement or protective order.[10] The "Privilege" field should also be able to accommodate annotations with regard to challenges to a claim of privilege, and the resolution of those challenges. For example, if a particular document is the subject of a motion to compel production and the motion is allowed, the "Privilege" field for that document should be updated in the database to indicate that the motion to compel was allowed. In addition the "Production Status" field (field 16) should be updated to indicate when a document previously withheld for privilege was produced.

Maintenance of the privilege field on a current basis enables one to use the database to create a privilege log which is often required to be furnished to the party requesting production of documents. The database fields one might want to include in a privilege log would be the page number fields (fields 1 and 2); the "From," "To," and "Copies To" fields (fields 4, 5, and 6); the "Document Type" field (field 8); and the "Privilege" field (field 17). The privilege log should also include a broad description of the subject matter, sufficient to enable the recipient of the privilege log or a court to understand the basis for the claim of privilege. Although the database contains a subject field, in order to be

10 It is beyond the scope of this chapter to articulate all possible grounds for a claim of privilege or other immunity from production.

as useful as possible to the party creating the database, the information in the database subject field may be more revealing as to the contents of the documents than one would want to include in a privilege log. Thus, although the subject field already in the database may be used as a starting point for the subject field in a privilege log, it should be reviewed carefully by counsel and edited as appropriate to make it suitable for inclusion in a privilege log. A sample privilege log is included in the Appendix as Exhibit 8.2.

D. Privileged Fields

One tends to assume that the work product of counsel which leads to the creation of a document database is protected from discovery in litigation as work product. However, the work product protection of Rule 26(b)(3) of the Federal Rules of Civil Procedure is not absolute.

The courts recognize two types of work product: "ordinary" or fact-based work product and "opinion" work product, consisting of the "mental impressions, conclusions, opinions, or legal theories of the attorney or other representative of a party concerning the litigation."[11] Opinion work product is accorded a higher degree of protection than ordinary work product. In light of this distinction, one should not assume that all fields of a document database will be entitled to the maximum protection. Rather, it is likely that only the field for attorney comments, and possibly the fields for priority and topic, are likely to warrant the level of protection accorded to opinion work product. Thus, guidelines for completion of the other fields should take into consideration the possibility that the contents of the database, other than the fields mentioned above, are at risk of ultimately becoming available to the other parties to the litigation.

V.
Imaging

Prior to the widespread availability of imaging technology, deposition preparation in the large document case was frequently

---

11 *Data Gen. Corp. v. Grumman Sys. Support Corp.*, 139 F.R.D. 556, 557 (D. Mass. 1991) (quoting FED. R. CIV. P. 26(b)(3)); *see also Upjohn Co. v. United States*, 449 U.S. 383, 400-401 (1981).

cumbersome. After searching for a document on a database, a lawyer would request that a particular document be retrieved by a clerk, either from off-site or on-site storage, and then copied. For example, to prepare a witness notebook (or boxes) for a deposition, the lawyer would need to (1) direct searches; (2) request retrieval of selected documents; (3) review each document; and (4) make a leap of faith that he or she accurately predicted what the lawyer on the other side would inquire about at deposition. The obvious problems are: (1) the repeated photocopying of the same documents is costly; (2) storing documents relatively nearby is costly, if available at all: (3) retrieving documents is time-consuming; (4) documents are at risk of being misfiled or lost; and (5) relevant documents needed for use at a deposition are bulky and difficult to transport.

Imaging permits the lawyer to carry with his or her laptop computer the ability to retrieve documents both before and during a deposition. In fact, a search through an imaging database can save twenty-three minutes per search over traditional document storage and retrieval methods.[12]

Document imaging (Tagged Image File Format) is a process by which a digital copy of an image of a document is made (usually on a CD-ROM). The initial advantage of imaging is that a single CD-ROM can hold nearly 15,000 pages of documents, the equivalent of five or six boxes of documents. The cost of scanning documents can range from $.10 to $1.00 a page, for most purposes ranging from $.20 to $.30 per page. These costs alone are much less than the costs of traditional paper document management such as costs of on site storage and labor costs of document retrieval. In addition, paper copies of imaged documents can print on an office laser copier.

---

12 *See generally* Michael O. Miller & Brenden J. Griffin, *Computer Database: Forced Production Versus Shared Enterprise*, LITIG. 40 (1997); Clifford Franklin Shnier, *Decoding Imaging & Document Management*, 4 LAW TECH. PRODUCT NEWS (Jan. 1997).

One important cautionary note about imaging is the cost of creation of a database from which a user can retrieve the images on line.[13] This cost cannot be avoided. The ability to retrieve specific documents from images requires a database. Otherwise, the lawyer will set at his or her computer scanning over each image page by page. For most people, it is quicker to flip through a hard copy of the documents than to scroll through documents page by page on a computer monitor. The costs of creating a separate database for imaged documents can range from $.05 to $.75 per page. The good news is that the same document database discussed in Section IV.A can serve as the search and retrieval tool for imaged documents.

Combined with the cost of imaging, the creation of a computer retrievable, readable, and printable database (for discovery purposes) can reasonably be expected to cost between $.50 and $1.00 per page. Therefore, in the truly massive case of over three thousand boxes (or 7.5 million pages), imaging every document can cost between $3.8 and $7.5 million. These cost considerations dictate careful review of documents. Thus, the question whether to image documents in a case turns on the expected quantity of key documents relative to the stakes. For example, at the other extreme, in a case with hundreds of boxes of raw documents where only two or three thousand pages (a box full) are key, imaging, although not costly, may be unnecessary; indeed, flipping through a box of key documents marked with tabs may be cheaper and more efficient than computerization.

VI.
Depositions

A. Deposition Transcripts

The traditional approach to deposition transcripts is either (1) for an attending attorney to draft a summary memorandum after the deposition but before receiving the transcript; or (2) for a paralegal or associate to

---

13 Remember that a document image is not a computer searchable text file. Depending on the quantity and quality of the imaged documents, one can convert an image into a text file through Optical Character Recognition. This type of work can be extremely labor intensive. In addition, many law firm text file database and retrieval systems, such as Soft Solutions, require a form of document coding.

draft a deposition digest after receipt of the transcript (usually two weeks later). Neither process is perfect. The first approach is subject to the fallibility of the attorney's notes and memory of what was said. The second approach is flawed because it does not capture a contemporaneous, first hand assessment of the witness. Where multiple lawyers within the same office are working on a large case, technology presents a number of alternatives to make deposition transcripts manageable.

1. Live Transcripts

A number of software programs such as Case View permit lawyers to "hook up" to the court reporter to obtain live or simultaneous transcription of the deposition. Live transcription costs about twice the cost of traditional court reporting but offers a number of advantages both during and after the deposition.

During the deposition an attorney can highlight portions of the transcript on his or her own laptop for a variety of purposes, such as further questioning by the examiner or clarifying testimony with a client deponent. Also during the deposition, the lawyer can make notes on specific portions of the transcripts. These notes and marketings are later useful for drafting a summary memo, for planning further discovery, and for trial preparation.

2. Deposition Control

After the deposition, an electronic copy of the final transcript should be loaded onto the lawyer's litigation database along with an electronic copy of any notes and markings. The document database should be updated to include (or note on previously coded document entries) any newly marked deposition exhibits and to cross-reference use of any previously marked exhibits at a later deposition. Any collection of images should be updated to include new deposition exhibits.

In addition to the transcript, the attorney should complete and load onto the database a summary memorandum. A sample is included in the Appendix as Exhibit 8.3. The ability to review notes, markings, summary memoranda, and results of transcript searches renders obsolete the traditional deposition digest. Computerized annotations to transcripts can and should use the same codes for referring to people and issues as in the document database so that a common search of both databases will yield complete results. For attorneys who still prefer paper transcripts, current software permits one to print a full size or compressed transcript with a word index. An electronic transcript should be updated to reflect any errata noted by the witness.

B. Videotape

Videotape at pure discovery depositions, although gaining use, should be employed only as a necessary evil during fact depositions. The use of videotaping at all depositions adopted by some counsel is both wasteful and counterproductive. Unless there is reason to believe a witness either may not be available for trial or may exhibit particularly noteworthy behavior or may be more disciplined because of the presence of the camera, the camera will have the effect of causing the witness to be more guarded and less forthcoming. The simple fact is that, because of the high cost to integrate video into a litigation defense, videotaped depositions in the massive case are largely left to accumulate dust and take up space.

The lawyer conducting a video deposition should request that the court reporter "time track" the transcript to the video. Second, the parties should establish a protocol for all depositions where no local rules apply. For example, state courts may have a rule governing protocol for videotaping to which the parties can agree to adhere in a federal court action.

C. Conclusion

In the case with a large number of depositions, failure to plan can be costly to your case strategy and to the client in terms of fees. Even if the same lawyer attends every deposition, there is no realistic expectation of having an ability to prepare for and use effectively deposition transcripts.

VII.
Anticipation Of Trial: Case Strategy

In the massive case, discovery can take on a life of its own seemingly divorced from the issues of the case and the anticipated use of discovery materials at trial. Frequently, the client and counsel will have only the most general idea what facts each side will need to prove at trial. Seldom does the plaintiff file a complaint which alleges the facts of the claims with sufficient specificity to provide a road map. The first step a lawyer should consider is whether contention interrogatories are necessary before any substantive work begins. At this earliest stage, it is imperative that experts become involved in mapping out what needs to be learned from discovery. The second step should be for the lawyers to maintain and update a work plan and schedule throughout discovery with input from the experts.

Several months before the end of discovery trial counsel should begin preparation of the pretrial memorandum. This process (including summarization of expected evidence) will highlight any gaps in the discovery in sufficient time to serve additional document requests and complete any further depositions. Throughout discovery, counsel should review and update the names and issues directories regularly.

VIII.
Conclusion

Use of the tools discussed in this chapter is not calculated to transform a complex case into a simple case. Rather, they have the potential, if used effectively and diligently, to make the collection, assimilation, and control of a high volume of documents and depositions manageable. The highest degree of control will be achieved if the management task is confronted very early in the life of the complex case, in collaboration with the surety's consultants, and reviewed and maintained through the life of the case.

CHAPTER 9

PROVING (AND DEFENDING) QUALITY OF WORK ISSUES

J. William Ernstrom
John W. Dreste

I. Introduction

The proof or defense of a principal's quality of work may be required in a variety of circumstances. The most common situation arises during the course of a bonded project where the owner/obligee has declared the principal in default as a result of claimed work-related defects. A second typical situation may arise as a result of assertion of warranty claims for defective work discovered after the completion of a bonded project, but prior to the expiration of the surety's liability under the particular bond terms. A third situation may arise following the discharge by a surety of its performance obligations to an owner/obligee when it seeks to obtain indemnification from the principal in the face of a principal's allegations that the surety proceeded as a volunteer in correcting work that was not defective. A final situation may arise when surety or principal pursues claims against the obligee for extra work attributable to having been directed by the obligee or its agents to perform work of a higher quality than required by the plans and specifications.

This chapter cannot exhaustively treat or consider every possible circumstance in which a surety may be faced with a need to prove or defend the quality (or lack thereof) of its principal's work product. However, the intent of this chapter is to present and outline the essential legal considerations, together with a framework for investigating and presenting at trial the facts and circumstances underlying the ultimate determination of work quality.

II.
Legal Implications Of Quality Of Work Issues

A. Performance Bond Language

The need to prove or defend the quality of a principal's work hinges in part on the language of both the performance bond utilized and the underlying bonded contract. The most widely utilized bond form is probably the American Institute of Architects (AIA) A311.[1] The language impacting quality of work issues in that bond provides: "if Contractor shall promptly and faithfully perform said Contract, then this obligation shall be null and void." The language further provides: "Whenever Contractor shall be, and declared by Owner to be in default under the Contract, the Owner having performed Owner's obligations thereunder, the Surety may promptly remedy the default."

Simply stated, if the bond obligee/owner claims that the principal/contractor has defaulted due to performance of defective work, then the bond obligations may be triggered. Conversely, if alleged defects are, in fact, not present, not material if present, or not the responsibility of the principal/contractor, then a defense under the bond may exist.

Because a wide variety of other performance bond forms are utilized throughout the country, including those drafted by municipalities, the bond language must be considered carefully. The City of New York has historically[2] utilized a form which provides:

> [I]f the Principal, his or its representative or assigns, shall well and faithfully perform the said Contract . . . according to its terms and its true intent and meaning, including repair and/or replacement of

---

1 American Institute of Architects, Document A311 (1970 ed.).

2 The city of New York has presently unified its construction and design services through the Department of Design and Construction. Currently, the Department of Design and Construction is in the process of promulgating new bond forms which will likely be the source of vigorous dispute and debate due to the radical departure of the proposed forms from what has been a traditional surety/obligee/principal relationship.

defective work . . . then this obligation shall be null and void . . .

The Surety . . . , for value received, hereby stipulates and agrees, if requested to do so by the City, to fully perform and complete the Work to be performed under the Contract, pursuant to the terms, conditions and covenants thereof, if for any cause, the Principal fails or neglects to so fully perform and complete such Work . . . .

As with the AIA A311, the city of New York's quoted language permits the surety to implicitly assert contract defenses based upon a showing that the principal/contractor has, in fact, performed the contract work as required. The surety merely proves that the quality and adequacy of the principal's work conforms to the requirements of the bonded contract. A determination that the principal/contractor has adequately performed will exhonerate the surety from liability.

In any subsequent effort by the surety to obtain repayment/reimbursement from the principal or its indemnitors, the surety also may be met by a defense that the surety essentially acted as a volunteer in unnecessarily performing or correcting work. This is the dilemma presented by the double edged sword of suretyship.

B. Impact Of Quality Of Work Issues Upon The Surety's Right Of Indemnification

A classic example is *Arntz Contracting Co. v. St. Paul Fire & Marine Insurance Co.*[3] Although the facts in *Arntz* are complex, the case grew out of a termination of the bond principal (Arntz) based upon complaints by the owner/obligee that Arntz had performed deficient work. The owner demanded that St. Paul, as surety, complete the project. All of the parties then became immersed in various disputes, one of which was over the measures needed to correct alleged deficient stucco and other work. Although St. Paul and Arntz initially had a friendly relationship under which St. Paul allegedly agreed to continue bonding Arntz in exchange for $1,000,000 in collateral, Arntz ultimately sued St. Paul and the owner seeking a declaration that it was unnecessary to completely remove and replace the disputed stucco. Needless to say, the relationship between Arntz and St. Paul quickly deteriorated and bonding was cut off. After the dust had settled, Arntz

3 54 Cal. Rptr. 2d 888 (Cal. Ct. App. 1996).

recovered $16.5 million in compensatory damages, due to St. Paul's refusal to further bond Arntz and its performance of unnecessary corrective work on the disputed stucco. The jury also awarded $100 million dollars in punitive damages that subsequently was vacated at the trial level. For its part, St. Paul's award on its indemnification claims, after substantial off-sets, was only about $813,000.00, due in part to its recognition as a "volunteer" in repairing the disputed stucco.[4] That, of course, only served to reduce the amount due to Arntz. What is often lost in the shuffle of this case is that the relationship between St. Paul and Arntz deteriorated based upon St. Paul's insistence that the project stucco be replaced as demanded by the owner/obligee. That quality of work issue snowballed into a surety's nightmare.

The operative language within the Arntz General Indemnity Agreement provided that St. Paul had the right "at its option and in its sole discretion to take possession of all or part of the work . . . whether, in its sole opinion, such action is desirable or necessary, and at the expense of [Arntz] to complete . . ."[5] The trial court found that St. Paul was indeed entitled to take over the project and recover the expense of completing the work, except for those expenses incurred without a good faith belief that they were "desirable or necessary."[6] The appellate court rationalized the trial court's ruling by noting that a surety, in such a situation, must weigh the viability of competing claims and the consequences of its possible responses to those claims so it can, in good faith, decide if it is "desirable or necessary" to take over the project to protect its interests.[7] In applying that standard, it was determined that St. Paul did not have a good faith belief that it was desirable or necessary to follow the directive of the owner/obligee to completely remove and replace the stucco, sheet metal, and roofing on the project. In fact, the evidence showed that St. Paul incurred the expenses for the stucco, sheet metal, and roofing work despite being presented with overwhelming proof that those expenses were "unnecessary and unwarranted" and that St. Paul's authorization of those expenses was without "rational justification" and constituted "unreasonable economic waste."[8] Thus, St. Paul was not entitled to indemnification for those

---

4 *Id.* at 892.
5 *Id.* at 898.
6 *Id.*
7 *Id.*
8 *Id.* at 899.

expenses, but the entire dispute appears to have arisen from those very issues.

When seeking indemnification from indemnitors who challenge expenses as unnecessary and voluntary, the surety may show that its expenditures under the bond were both obligatory and reasonable. In *General Insurance Co. of America v. K. Capolino Construction Corp.*,[9] the United States District Court for the Southern District of New York denied a surety indemnification for expenses incurred in discharging a performance obligation where the court was unable to determine, as a matter of law, that the surety acted in good faith in deciding to perform remaining work. The court permitted indemnification for payment bond losses, but observed that the performance bond language only required the surety to comply with the owner's demand for completion if the owner itself was not, in fact, actually in default. The court distinguished the issue of whether the principal was actually in default from whether the owner was actually in default, based upon the performance bond terms which were prefaced by the clause: "If there is no Owner Default, the Surety's obligation under this Bond shall arise . . . ."[10]

As in *Arntz*, the principal/contractor, Capolino, contested the default termination by the owner/obligee. The surety, based upon its own investigation, concluded that Capolino was in default and elected to perform. The court observed that New York law provides that a party who pays a claim it is not obligated to pay is a volunteer and may not recover such expenses. Thus, the surety could not recover from Capolino unless, under the terms of the performance bonds issued on Capolino's behalf, the surety was obligated to complete the projects.[11] In addition to the above-quoted language, the bond form (AIA A312) defined owner default to include: "[failure] of the Owner . . . to pay the contractor as required by the Construction Contract or to perform and complete or comply with the others terms thereof . . . ."[12]

---

9 903 F. Supp. 623 (S.D.N.Y. 1995).

10 *Id.* at 626. The court also observed that it is well-settled under New York law that a party that pays a claim it is not obligated to pay is a volunteer and may not recover those expenses. *Id.* (citing *National Union Fire Ins. Co. v. Ranger Ins. Co.*, 599 N.Y.S.2d 347 (N.Y. App. Div. 1993)) (other citations omitted).

11 *Id.* at 627.

12 *Id.* at 626.

In the context of denying summary judgment to the surety, the court ruled that it was necessary for the surety to factually demonstrate that the owner was not in default itself. The surety was basically faced with a need to prove the reasonableness of the obligee/owner's action. These cases illustrate the need to quickly, effectively, and accurately determine whether a quality of work issue exists.

C. Quality Of Work Issues

As a general rule, a contractor extends an implied warranty that its performance will be in a "good and workmanlike manner,"[13] and frequently agrees to provide an express warranty in the bonded contract.[14] For its part, the owner extends an implied warranty of the adequacy of its detailed design plans and specifications[15] and has an implied duty of full disclosure of information material to the contractor's performance.[16] Quality of work issues most frequently arise out of: (1) conflicts or ambiguities in the design documents; (2) incompatibilities between construction materials; or (3) imposition of hypertechnical test or inspection standards. The implied warranties may be altered or disclaimed by the express provisions of specifications. The owner may expressly specify stricter standards than required by custom or applicable codes.[17] Where the contract documents affirmatively

---

13 *St. Paul Fire & Marine Ins. Co. v. Pearson Constr. Co.*, 547 N.E.2d 853 (Ind. Ct. App. 1989); *Korte Constr. Co. v Deaconess Manor Ass'n*, 927 S.W.2d 395 (Mo. Ct. App. 1996); *Middleton v. Benne*, 849 S.W.2d 265 (Mo. Ct. App. 1993); *Dixon v. Mountain City Constr. Co.*, 632 S.W.2d 538 (Tenn. 1982). What is a "good and workmanlike manner" may be interpreted according to technical standards developed over many years, such as those promulgated by the American Society of Testing and Materials (ASTM) or the American Concrete Institute (ACI).

14 *See, e.g.*, Article 3.5, AIA A201-1997 General Conditions of Contract ("the work will be free from defects not inherent in the quality required or permitted and . . . will conform to the requirements of the Contract Documents"); *see also* Article 12.2, AIA A201-1997 (addressing correction of defective work before and after substantial completion).

15 *United States v. Spearin*, 248 U.S. 132 (1918).

16 *City of Indianapolis v. Twin Lakes Enters., Inc.*, 568 N.E.2d 1073 (Ind. Ct. App. 1997).

17 *M.A. Mortenson Co*, ASBCA No. 49,917, 97-1 B.C.A. ¶ 28,682 (1996). In *Mortenson*, the specifications referenced a masonry standard which

specify a method of performance not in compliance with applicable state or local law, the contractor may be entitled to recover its increased costs of compliance, notwithstanding a general contract requirement that the contractor "comply with all applicable state and local laws and ordinances."[18] A contract calling for work in violation of applicable regulations may even be deemed void and unenforceable.[19]

The contractor's general duty is to perform in a workmanlike manner, and in conformance with specifications or applicable codes and regulations.[20] The contractor is not responsible for verifying that the design prepared by the owner's design professionals are in compliance with code and, unless any non-conformance with code is patently obvious, the non-conformance may entitle a contractor to additional compensation incurred in attempting to comply with contract documents in conflict with code.[21] The contractor ordinarily is responsible for determining applicable codes or regulations applicable to the installation of its work and for adhering to them in the course of performance.[22]

---

was more stringent than the ASTM standard that was incorporated by reference as well. The Board agreed that there was a conflict between the ASTM and the specification, but that the government was able to impose a higher standard. *Id.*

18 *Hemphill Contracting Co., Inc.*, ENGBCA No. 5698, 94-1 BCA ¶ 26,491 (1993). Even though the contract included a standard permits and responsibilities clause requiring contractor compliance with all applicable state and local laws and ordinances, the contract affirmatively specified a work method contrary to local laws and ordinances. The contractor incurred extra expenses to comply with the local laws and ordinances which were subsequently enforced, and was entitled to recover. *Id.*

19 *Hieb v. Opp*, 458 N.W.2d 797 (S.D. 1990).

20 *See e.g.*, Article 3.7, AIA A201-1997 (contractor must comply with "all laws, ordinances, rules, regulations and lawful orders of public authorities applicable to performance of the work," but is not obliged to ascertain if the contract documents conform to code).

21 *Castle Constr. Co., Inc.*, ASBCA No. 28,509, 84-1 B.C.A. ¶ 17,045 (1993); *Green v. City of N.Y.*, 128 N.Y.S.2d 715 (N.Y. App. Div. 1954) (where city prepared plans and specifications setting forth certain construction requirements contrary to an applicable local law, the contractor was entitled to recover the additional costs of compliance even though contract provided that contractor would comply with rules and regulations).

22 *D.E.W., Inc.*, ASBCA No. 35,759, 92-1 B.C.A. ¶ 24,520 (1991) (National Plumbing Code incorporated into contract by reference was binding on

This does make sense as it is the contractor who holds itself out to be an expert in construction. A truly qualified contractor should be well aware of all pertinent industry practices, codes, and regulations applicable to performance of its work.

Delegation of design responsibility to the contractor by the owner's performance specifications may extend significantly the contractor's liability for design adequacy. This evolving area of conflict relates to the distinction between design and performance specifications. Under a design specification, the owner/obligee essentially warrants the sufficiency of the design so that the contractor, if it performs according to that design, is entitled to expect that the resulting work product will function as intended. This rule is the well-known *Spearin* doctrine which derives its title from United States Supreme Court case of that name.[23] The opposite of a design specification is the performance specification, under which the contractor is provided only with the intended end result or performance characteristics for the project.[24] It is then up to the contractor to prepare a design and construct the work so as to achieve the desired end result. In such circumstances, the contractor is obligated to produce an end product that performs as specified. A pure performance specification may run afoul of certain states' prohibitions concerning the delegation of design or the practice of design by a contractor, which is discussed in some detail below. However, it is more common to encounter a hybrid situation where the

---

contractor and government's approval of non-conforming shop drawings did not waive requirement); *Gardenvillage Realty Corp. v. Russo*, 366 A.2d 101 (Md. App. 1976) (violation of a building code can be evidence of breach of a duty owed); *Barry v. Saratoga Homes, Ltd.*, 524 N.Y.S.2d 869 (N.Y. App. Div. 1988) (contractor obligated to construct in accord with uniform fire prevention and building code where contract provided that contractor would comply with all relevant laws and applicable codes and regulations); *Spotsylvania County Sch. Bd. v. Seaboard Sur. Co.*, 415 S.E.2d 120 (Va. 1992) (contractor was required to comply with building code requirements during construction, as well as upon contract's completion, but violation of building code must constitute a substantial violation of contract in order to justify termination).

23 *United States v. Spearin*, 248 U.S. 132 (1918).

24 *See Stuyvesant Dredging Co. v. United States*, 834 F.2d 1576 (Fed. Cir. 1987); *J.L. Simmons Co. v. United States*, 412 F.2d 1360 (Ct. Cl. 1969); *Fruin-Colnon Corp v. Niagara Frontier Transp. Auth.*, 585 N.Y.S.2d 248 (N.Y. App. Div. 1992).

specifications are a combination of design and performance requirements which must be viewed as a whole and interpreted in order to determine the contractor's responsibility.

In many states, it is illegal for those who are not professionally licensed to practice engineering or architecture.[25] In these states, a contractor who is obligated by the contract documents to design portions of a structure or project may actually be committing a felony under state law.[26] These legal prohibitions run afoul of the expanding attempt by some design professionals to foist design responsibilities on contractors. Although such efforts are far-ranging, they are typically more common in areas involving structural steel connections or similar specialized or detailed areas.[27] From the surety's standpoint, a broad performance bond obligation could be interpreted to impose responsibility for a contractor's assumed design responsibilities, which may come as a surprise to a performing surety upon a declaration of default.

The initial analysis of quality of work issues must hinge on an interpretation of the underlying contract documents, and the performance bond itself. Simply because a structure or project fails or the owner is dissatisfied with the work product does not mean that the contractor/principal is at fault. Construction contracts, like all contracts, should be initially read in an effort to determine the intent of the parties by giving effect to all provisions. However, if an ambiguity exists, then the issue may become whether or not the ambiguity is latent or patent. If a latent ambiguity exists, then the contractor may escape responsibility, if it acted as a reasonable and prudent contractor in performing.[28] If a patent ambiguity exists, then the contractor may be

---

25 States in which such practice may be illegal include New York and Missouri, among others.

26 *See Charlebois v. J.M. Weller Assocs.*, 531 N.E.2d 1288 (N.Y. 1988) (recognizing that "grave contractual and criminal consequences" can result if an unlicensed contractor engages in the practice of engineering).

27 *See Duncan v. Missouri Board*, 744 S.W.2d 524 (Mo. Ct. App. 1988)(structural engineer's license revoked due to gross negligence in failing to review structural steel connections designed by steel fabricator for Hyatt Hotel walkway that collapsed).

28 *See Stroh Corp. v. General Servs. Admin.*, GSBCA No. 11,029, 93-2 BCA ¶ 25,841 (1993); *Maguire Co., Inc. v. Herbert Constr. Co., Inc.*, 945 F. Supp. 72 (S.D.N.Y. 1996); *Pennsylvania Dep't. of Transp. v. Bracken Constr. Co.*, 457 A.2d 995 (Pa. 1983).

held liable for failing to draw the ambiguity to the owner's attention, prior to performing work.[29]

The 1997 edition of AIA Document A201 became effective in late October 1997. The revised A201 General Conditions deals with the contractor's responsibilities concerning ambiguities.[30]

---

29 *See Blinderman Constr. Co., Inc. v. United States*, 17 Cl. Ct. 860 (1989); *H.V. Allen Co., Inc.*, ASBCA No. 38,377, 90-2 BCA ¶ 22,933 (1990); *Green Island Constr. Co., Inc. v. County of Chenango*, 622 N.Y.S.2d 132 (N.Y. App. Div. 1995) (contractor required under contract notice provisions to report inaccuracies it discovered in bid specifications and is precluded from recovering payment for work performed under inaccurate specification).

30 The 1997 revision provides:

3.2 REVIEW OF CONTRACT DOCUMENTS AND FIELD CONDITIONS BY CONTRACTOR

3.2.1 Since the Contract Documents are complementary, before starting each portion of the Work, the Contractor shall carefully study and compare the various Drawings and other Contract Documents relative to that portion of the Work, as well as the information furnished by the Owner pursuant to Subparagraph 2.2.3, shall take field measurements of any existing conditions related to that portion of the Work and shall observe any conditions at the site affecting it. These obligations are for the purpose of facilitating construction by the Contractor and are not for the purpose of discovering errors, omissions or inconsistencies in the Contract Documents; however, any errors, inconsistencies or omissions discovered by the Contractor shall be reported promptly to the Architect as a request for information in such form as the Architect may require.

3.2.2 Any design errors or omissions noted by the Contractor during this review shall be reported promptly to the Architect, but it is recognized that the Contractor's review is made in the Contractor's capacity as a contractor and not as a licensed design professional unless otherwise specifically provided in the Contract Documents. The Contractor is not required to ascertain that the Contract Documents are in accordance with applicable laws, statutes, ordinances, building codes, and rules and regulations, but any nonconformity discovered by or made known to the

The new provisions follow the time-honored custom of requiring the contractor to notify the architect of any design errors discovered in reviewing the contract documents. If the contractor becomes aware of design defects and fails to notify the architect, the contractor may be liable to the owner for all costs and damages that could have been avoided.[31]

The contractor's design clarification obligation clearly extends to patent errors, inconsistencies, or omissions discovered by or made known to the contractor. Expert input or testimony may become necessary at this point as to just what a reasonable contractor objectively should have discovered.

A conflict within the contract documents ordinarily is resolved based upon a standard of reasonableness.[32] Many contracts also contain

---

Contractor shall be reported promptly to the Architect.

3.2.3 If the Contractor believes that additional cost or time is involved because of clarifications or instructions issued by the Architect in response to the Contractor's notices or requests for information pursuant to Subparagraphs 3.2.1 and 3.2.2, the Contractor shall make Claims as provided in Subparagraphs 4.3.6 and 4.3.7. If the Contractor fails to perform the obligations of Subparagraphs 3.2.1 and 3.2.2, the Contractor shall pay such costs and damages to the Owner as would have been avoided if the Contractor had performed such obligations. The Contractor shall not be liable to the Owner or Architect for damages resulting from errors, inconsistencies or omissions in the Contract Documents or for differences between field measurements or conditions and the Contract Documents unless the Contractor recognized such error, inconsistency, omission or difference and knowingly failed to report it to the Architect.

31 Paragraph 3.2.3 contains language providing that:

If the Contractor fails to perform the obligations of Subparagraphs 3.2.1 and 3.2.2, the Contractor shall pay such costs and damages to the Owner as would have been avoided if the Contractor has performed such obligations.

32 An ambiguity generally concerns an omission or inconsistency which permits two or more reasonable interpretations. A conflict, however, references instances where different portions of a contract contain provisions which are irreconcilable. *See* 19 CONSTR. CLAIMS MONTHLY,

"order of precedence" clauses which provide for the controlling priority of the various sections of the contract documents. For instance, it may be provided within the contract that the specifications shall govern over the drawings, in which event the contractor may reasonably perform according to the specifications and may not be held responsible if a conflicting notation on a drawing provides otherwise. If a contractor can reasonably resolve a conflict between provisions of the contract by reference to an "order of precedence" clause, it is entitled to proceed with the work accordingly. In the absence of an "order of precedence" clause, basic rules of interpretation will be applied in an effort to resolve the ambiguity. These rules construe the whole agreement and grant priority to specification provisions over conflicting provisions in plans or in industry codes incorporated by reference.[33] If the conflict cannot otherwise be resolved, then the ambiguity may simply be construed against the drafter of the ambiguous provisions.[34]

Whether or not an ambiguity exists within the contract documents will likely require technical input from an appropriate expert. The resulting mix of legal and factual questions may ultimately result in a battle of experts, but the prudent surety will, at a minimum, conduct its initial investigation at least so that it can determine how to proceed and preserve its rights under the applicable indemnification agreement with the principal.

Incompatibility between construction materials is a significant quality of work issue. Manufacturers of construction products, intent on capturing expanded market shares, have been known to promise more for less cost and have pushed new products into the market with insufficient laboratory research and almost total reliance on "field testing" (*i.e.* "try it in the field and see what happens"). Not unexpectantly, quality of work issues have arisen out of incompatible

---

Business Publishers, Inc. (July 1997) for a discussion of this distinction.

33 *See Tarlton Constr. Co. v. General Servs. Admin.*, GSBCA No. 10,528, 94-1 BCA ¶ 26,279 (1993).

34 The rule of *contra proferentum* provides that as a last resort, ambiguities or conflicts are to be resolved as against the drafter of the document. *Blount Bros. Constr. Co. v. United States,* 346 F.2d 962 (Ct. Cl. 1965); *P.R. Post Corp. v. Maryland Cas. Co.*, 242 N.W.2d 62 (Mich. Ct. App. 1976), *aff'd in part, rev'd in part,* 271 N.W.2d 521 (Mich. 1978); *Pennsylvania Dept. of Transp. v. Mosites Constr. Co.*, 494 A.2d. 41 (Pa. 1985).

materials used in roof insulation and decks under roof membranes;[35] adhesives;[36] windows, glass and frames;[37] and even out of brick materials incompatible to the environment in which they were used.[38]

Imposition of new higher standards of quality also has raised quality of work issues. For instance, subjection of work to tests not spelled out in the specifications and unreasonable under the circumstances may constitute a material breach of contract.[39]

III.
Implications Of Building Codes And Industry Standards On Quality Of Work Issues

A. The Contractor's "Good Workmanship" Standard

Construction contracts generally require the contractor to perform in accordance with a "good" or "reasonable workmanship" standard, and frequently give more specific meaning to this standard by the express provisions of the construction contract documents.[40] Contract documents routinely incorporate by reference codes and industry standards. When not expressly incorporated by reference into the plans and specifications, the standard of workmanship prevailing in the area in conformity with

---

35 *See Certain-Teed Products Corp. v. Goslee Roofing & S.M., Inc.*, 339 A.2d 302 (Md. App. 1975); *State v. GAF Corp.*, 747 P.2d 1326 (Kan. 1987).

36 *See Floorcraft v. Parma Community General Hospital Ass'n*, 560 N.E.2d 206 (Ohio 1990).

37 *See R.W. Murray Co. v. Shatterproof Glass Corp.*, 697 F.2d 818 (8th Cir. 1993).

38 *See Trustees of Indiana Univ. v. Ætna Cas. & Surety Co.*, 920 F.2d 429 (7th Cir. 1990).

39 *See State v. Buekner Constr. Co.*, 704 S.W.2d 837 (Tex. App. 1985).

40 *See M.A. Mortenson Co.*, ASBCA No. 49,917, 97-1 BCA ¶ 28,682 (1996) (specifications can impose stricter requirements than custom or standards); *St. Paul Fire & Marine Ins. Co. v Pearson Constr. Co.*, 547 N.E.2d 853 (Ind. Ct. App. 1989) (concerning implied warranty that performance will be in a workmanlike manner); *American Towers Owners Ass'n, Inc. v. CCI Mechanical, Inc.*, 930 P.2d 1182 (Utah 1996) (owner can contract for low-grade materials that meet only requirements of a building code and contractor should be shielded from subsequent action for poor quality of construction).

codes will be the basis for judging the adequacy of the contractor's work.[41]

An owner, of course, may contract for materials that meet only minimum requirements of a building code or may specify materials substantially in excess of code requirements. When minimum quality materials are specified, a contractor is not liable for the resulting poor general quality.[42] Furthermore, a finding that a building code has been violated may not be sufficient to constitute a material breach of the construction contract, depending upon the nature of the violation.[43] If the owner's design is in violation of code, the contractor should be absolved of liability. If the contractor's workmanship violates code, the contractor usually must be given the opportunity to "cure" the deficiency.[44] If the deficiency is substantial and remains uncured so as to constitute a material breach of contract, termination may be warranted. Materiality is an issue of fact for the jury.[45] If the code violations are minor, the work may be determined to substantially conform to the contract and immaterial deviations will preclude termination, although the owner may recover the cost of correcting or completing the work.[46]

---

41 *See Carter v. Krueger*, 916 S.W.2d 932 (Tenn. Ct. App. 1995).

42 *See American Towers Owners Ass'n*, 930 P.2d at 1182; *United States v. Spearin*, 248 U.S. 132 (1918).

43 *See Spotsylvania County Sch. Bd. v. Seaboard Sur. Co.*, 415 S.E.2d 120 (Va. 1992).

44 *See McClain v. Kimbrough Constr. Co., Inc.*, 806 S.W.2d 194 (Tenn. Ct. App. 1990).

45 *Spotsylvania*, 415 S.E.2d at 120.

46 *Granite Constr. Co. v. United States*, 962 F.2d 998 (Fed. Cir. 1992) (owner may not require replacement of work that would amount to economic waste and owner's remedies to take a credit for non-conforming work left in place); *Behling v. Ruckstaetter*, 394 S.E.2d 539 (Ga. Ct. App. 1990) (owner's measure of damages for defective work was diminution in fair market value of building); *M.W. Goodell Constr. Co. v. Monadnock Skating Club, Inc.*, 429 A.2d 329 (N.H. 1981) (costs of repairs to work not done according to code was less than five percent of contract price and were not economically wasteful); *Mort Wallin of Lake Tahoe, Inc. v. Commercial Cabinet Co.*, 784 P.2d 954 (Nev. 1989) (measure of damages for defective work is diminution in fair market value of property); *Ferris v. Mann*, 210 A.2d 121 (R.I. 1965) (where builder has not willfully deviated from specifications and all that remains

Contractors may not simply ignore what they know to be a conflict between the owner's design documents and an important code or standard. Contractors have an implied duty to draw the owner's attention to conflicts within the contract documents themselves.[47] The new A201 goes one step further and expressly imposes liability on the contractor for knowingly performing work contrary to codes, rules, or regulations.[48]

B. Primary Work Areas Of Concern

Issues relating to workmanship and construction defects can range from the merely irksome to the tragic. The latter extreme was most vividly illustrated by what is commonly referred to the as the Hyatt disaster which occurred at the Hyatt Regency Hotel in Kansas City, Missouri in 1981.[49] In an incident which was characterized by the National Bureau of Standards in terms of loss of life and injuries as the "most devastating structural collapse ever to take place in this country," the second and fourth floor walkways of that hotel collapsed and fell to the floor, resulting in 114 deaths.[50] The walkways at issue had been designed by a structural engineering firm, but during construction a change to the original design was made at the suggestion of the steel fabricator. The structural engineer reviewed and approved the drawings that contained the change, but there was no contractual relationship between that engineer and the steel fabricator and no evidence of coordination of design responsibilities between those two parties.[51] The proof also showed that no one checked for compliance with the city's building code. The National Bureau of Standards issued a formal report

---

to be done is of trivial or minor nature, builder is entitled to recover contract price with adjustments made by court to compensate defendant for unfinished or unsatisfactory work).

47 *Blount Bros. Constr. Co. v. United States*, 346 F.2d 962 (Ct. Cl. 1965).

48 *See* Article 3.7.4 of the A201-1997 General Conditions.

49 *See Duncan v. Missouri Bd. Architects, Prof. Eng'rs & Land Surveyors*, 744 S.W.2d 524 (Mo. Ct. App. 1988).

50 *Id.*

51 *Id.* It is notable that the engineer, who was retained by the project architect, was found guilty of gross negligence for failing to review the shop drawings and failing to perform engineering tests and calculations to determine the safety of certain connections.

in which it determined that the collapse was caused by the design change.[52] In this particular case, it was ultimately determined that the structural engineering firm bore the primary responsibility for the disaster by improperly delegating or otherwise failing to discharge its professional responsibilities to determine the engineered soundness of the resulting structure.

In most instances confronted by sureties, the consequences and damages being faced are solely economic in nature. However, similar determinations are required with respect to the applicability of codes and standards in relation to the express provisions of the contract documents. Apart from structural steel, areas typically creating problems include concrete construction, geotechnical investigations, and roofing standards and warranty issues.

C. Determining Compliance Or Default In Fact: The Need For An Independent Investigation

The surety's "do nothing" option has resulted in vigorous attacks.[53] The surety is well advised to immediately commence a prompt, independent, and aggressive investigation of the competing claims of the principal and obligee.[54] The failure to adequately investigate and remedy an evident default may subject a surety to a claim of bad faith.[55]

Sureties, or their counsel, should have working relationships with the necessary consultants or other experts as may be required in default circumstances. In most cases, consulting forensic engineers should review all technical construction issues bearing upon the question of

---

52 *Id.*

53 *See Cates Constr., Inc. v. Talbot Partners*, 62 Cal. Rptr. 2d 548 (Cal. Ct. App. 1997); *George A. Fuller Co. v. Kennsington-Johnson Corp.*, unpublished memorandum decision (Dec. 22, 1993), *modified on appeal*, 651 N.Y.S.2d 58 (N.Y. App. Div. 1996); *see also Pivotal Legal Precedent Affecting Underwriting and Claim Handling By a Surety,* unpublished paper presented to the 1995 Mid-Winter Meeting of the Fidelity & Surety Law Committee, American Bar Association (discussing *Fuller*). The trial court in *Fuller* observed that the surety's obligations arose from AIA form A311 (1970 ed.) and set forth three options or duties for the surety, none of which entitled the surety to do nothing.

54 *See generally Pivotal Legal Precedent Affecting Underwriting and Claim Handling By a Surety, supra* note 54.

55 *Cates Constr.*, 62 Cal. Rptr. 2d at 548.

whether or not the principal is actually in default of its contract obligations. A detailed physical inspection of the work and all available job site records is critical. Depending upon the nature of the project and the claimed default, it is always prudent for the surety's expert to photograph and videotape site conditions. That videotaping could be conducted by a retained consultant that the surety would consider utilizing at trial.

The surety should take care to keep the principal and obligee apprised of the steps the surety is taking and the anticipated time period in which a preliminary investigation will be completed. If additional time is required, the surety should keep all parties informed. In all events, an adequate documentation trail is crucial.

This initial investigative use of experts should be commenced with a knowledge of applicable disclosure or evidentiary rules in the jurisdiction where the case may be tried.[56]

---

56 *See, e.g.,* Rule 26 of the Federal Rules of Civil Procedure, provides in relevant part:

> Except to the extent otherwise stipulated or directed by order or local rule, a party shall, without awaiting a discovery request, provide to the other parties:
>
> * * *
>
> (2) Disclosure of Expert Testimony
>
> (A) . . . a party shall disclose to other parties the identity of any person who may be used at trial to present evidence under Rules 702, 703, or 705 of the Federal Rules of Evidence.
>
> (B) Except as otherwise stipulated or directed by the Court, this disclosure shall, with respect to a witness who is retained or specially employed to provide expert testimony in the case or whose duties as an employee of the party regularly involve giving expert testimony, be accompanied by a written report prepared and signed by the witness. The report shall contain a complete statement of all opinions to be expressed and the basis and reason therefor; the data or other information considered by the witness in forming the opinion; any exhibits to be used as a summary of or support for the opinions; the qualifications of the witness, including a list of all publications authored by the witness within the

Ordinarily, a party may obtain discovery from or depose only persons who have been identified as experts to be called to provide expert testimony at trial. Facts known or opinions held by an expert who has been retained or specially employed by a party in anticipation of litigation or preparation for trial but who is not expected to be called as a witness at trial, are discoverable only upon a showing of exceptional circumstances under which it is impracticable for the party seeking discovery to obtain such facts or opinions on the same subject by other means.

Upon a declaration of default, the surety's first role is to act impartially to determine, from a "construction" standpoint, which of the parties is correct in their factual and legal positions. Regardless of the ultimate conclusion reached by the surety, what is vital is that a showing can later be made that a thorough and competent investigation and review was made which supports the surety's conclusions. It may later be determined that the surety's conclusion was wrong, but if a surety's procedures in arriving at the conclusion were sound, a principal or obligee may find it difficult to claim that the surety acted in bad faith or in some manner contrary to sound practice.

IV.
Preservation And Presentation Of Issues At Trial

If the surety has determined that the principal is in default, or that there is a probability that the principal is in default, immediate steps

---

preceding ten years; the compensation to be paid for the study and testimony; and a listing of any other cases in which the witness has testified as an expert at trial or by deposition within the preceding four years.

(C) These disclosures shall be made at the times and in the sequence directed by the court. In the absence of other directions from the court of stipulation by the parties, the disclosure shall be made at least 90 days before the trial date or the date the case is to be ready for trial or, if the evidence is intended solely to contradict or rebut evidence on the same subject matter identified by another party . . . , within 30 days after the disclosure made by the other party . . . .

FED. R.CIV.P. 26.

must be taken to further trigger rights afforded under the applicable indemnification agreement. Under most circumstances, the surety can demand that the principal immediately post collateral security sufficient to hold the surety harmless if the principal maintains that it has a defense to the declared default. This contractual right available to the surety is not as important from a standpoint of providing actual collateral as it is important from the standpoint that if the principal fails to do so, then the surety may consider proceeding based upon the principal's default under its contractual indemnification obligations.[57] Regardless of which way the surety proceeds, it must continue to take care to document its steps so that adequate facts are preserved for later presentation, if necessary, at trial.

A. Utilization Of The Engineering Expert

There are five important elements in any case: the facts, the facts, the facts, the facts, and the law. The marshaling and presentation of the facts supporting the client's position is the critical difference between an outstanding success and a less than satisfactory result. If the facts and circumstances strongly support the client's position and dictate that the only fair result is a result consistent with the client's position, legal precedent can generally be found to support the desired result.

The facts and circumstances surrounding quality of work issues, as well as their relationship to the contractor's costs of performance, can be confusing and complex. It should be the attorney's goal in representing the client's interest in any construction matter to make the complex simple and persuasive. When married to the applicable law, the facts must persuasively demonstrate the desired result by virtue of the justice, equity, and fairness of that position.

The attorney must be able to recognize the value of the appropriately qualified expert to assist and provide guidance as a consultant during performance of the completion work, during the preparation of the case, and as a witness during litigation. The expert is the key to the achievement of the goal of persuasion. The goal: to marshall and present the facts establishing the client's position which when married to

---

57 *See General Accident Ins. Co. of Am. v. Merritt-Meridian Constr. Corp.*, 975 F. Supp. 511 (S.D.N.Y. 1997) (observing that the exercise of the surety's rights which arose after the principal declined to post collateral is not evidence of bad faith).

the law persuasively demonstrates the client's entitlement to the requested relief. Success depends primarily upon detailed preparation and persuasive presentation of the facts.

While it may be necessary to review literally thousands of documents including years of job site records, the key is to present only the critical portion of those documents in a fashion which focuses the attention of everyone on the key facts and questions. In other words, try to use a rifle shot (the critical facts and materials) rather than a shotgun blast (a mass of materials, facts, and dates that becomes a jumble of confusion). The services of a skilled expert with broad construction knowledge and experience can be invaluable in assembling and analyzing the relevant facts, as well as in providing opinions based upon ether actual knowledge of the facts or assumed hypothetical facts.

There are two general prerequisites to the introduction of an expert's testimony. First, the subject of the testimony must involve questions beyond ordinary experience and knowledge. Second, the expert must be appropriately qualified. Under the Federal Rules of Evidence opinions and expert testimony are admissible at trial under certain circumstances:

> If scientific, technical or other specialized knowledge will assist the trier of fact to understand the evidence or to determine a fact in issue, a witness qualified as an expert by knowledge, skill, experience, training or education, may testify thereto in the form of an opinion or otherwise.[58]

The expert utilized by the surety may be the individual retained to assist in the initial determination of whether or not a default occurred, or may be a subsequently retained engineer who has reviewed the course of the project. In either circumstance:

> The facts or data in the particular case upon which an expert bases an opinion or inference may be those perceived by or made known to the expert at or before the hearing. If of a type reasonably relied upon by experts in the particular field in forming opinions or inferences upon the subject, the facts or data need not be admissible in evidence.[59]

---

58 FED. R. EVID. 702.
59 FED. R. EVID. 703.

Of course, the expert opinion offered may embrace the ultimate issue to be determined by the trier of fact.[60]

If the expert who is testifying assisted the surety in the initial determination relating to the default declaration, then that witness may actually be testifying as both a fact witness and an expert. If a case has reached the point where this testimony is being presented to a jury, it is vital that the witness be an individual who is capable of conveying to the lay person the facts which support the surety's position. It is a well-known fact of life to the surety and construction practitioner that both judges and juries tend to get glassy eyed during the trial of the cases on which we make our living. We will reserve for later comment what it means that we find these cases fascinating, but it must be remembered that the goal of the trial is to put a compelling and understandable case before the jury. Toward that goal, it is imperative that a theme be developed that the jurors can identify with and around which testimony revolves.

The nature of the expertise needed in a construction dispute requires consideration of a broad spectrum of specialized knowledge. Although it may be possible to utilize the services of an expert whose expertise is limited to understanding only one component of the construction project, it is far better to have the advantages inherent in the expert who has: (1) an understanding of the management of construction projects and the traditional owner-architect-consultant-contractor-subcontractor relationships as they exist under various types of contractual relationships utilized; and (2) a basic working knowledge of the major trade activities on a construction project, *i.e.*, excavation, concrete, structural steel, masonry, carpentry, mechanical, and electrical trades.

Of course, it is not necessary for the expert to be an "expert" in all aspects of construction. It is important, however, for the expert to understand and comprehend the practical issues related to the construction of a complex project. The most important qualification for the expert is his or her ability to convincingly communicate and explain his or her advice and conclusions. One cannot do that without a broad understanding of the project components.

At the same time, any applicable sub-specialty or focus of expertise which may be applicable must be identified. Typical sub-specialties include structures (the design of bridges, dams, tunnels, and buildings);

---

60 FED. R. EVID. 704.

sanitary (the design of sewage treatment processes); transportation (the design of transportation facilities such as railroads, highways, and airports); and hydraulics (the design of water distribution systems and sewage systems).[61] Further, it is recognized that civil engineers may further specialize and it is likely that an expert can be found who devotes much of his or her professional practice to topics or areas presented in any given case. Thus, it is imperative that an early recognition be obtained of any sub-specialty or issues which will be involved. Technical experts can be located through universities (this may be one of the best spots), professional societies, or private consulting companies.

In obtaining an expert, carefully evaluate the consultant's educational background and experience. This should include a review of all articles and books the consultant has written on the subject matter as well as a review of the consultant's prior litigation experience. If the consultant has gone on record with opinions contrary to the client's position, this should be known before he or she is retained.

Analyze the expert's ability to persuade the finder of fact. The outcome of the case may well hang on the impression the expert witness will leave with the jury. Self confidence, credibility, clarity, and the ability to communicate easily with the jury in understandable lay terms are all important considerations. You will not be well served by a highly qualified expert who lacks the basic oral skills to convey thoughts and impressions to the finder of fact. It is often preferable to retain an expert with less impressive credentials than an academician with impeccable professional qualifications, but lacking in self confidence, enthusiasm, and the ability to speak and convincingly convey ideas in ordinary language and practical terms.

The physical appearance and the demeanor of an expert witness is also important. How will a judge or jury react to the witness? Does the witness appear honest and credible or does he or she seem sly and perhaps devious? Is the witness pleasant and courteous or does he or she convey pompousness and arrogance? You must weigh carefully not only what the expert is saying, but how it is being said. A conservative image is most impressive to a jury to instill the idea that your expert is a person of integrity, sincerity, and competence. The expert's appearance should be restrained. His or her deportment should be quiet, yet self

---

61 Practising Law Institute, *Using Experts in Civil Cases*, CONSTR. CASES (1977).

assured, courteous, and poised even in the face of possible sarcastic comments or hostile challenges during cross examination.

Effective use of the engineering expert virtually mandates that the expert be utilized during the early parts of the case, including the formulation of discovery and case strategy. However, the practitioner must take care not to compromise the expert's position by involving the expert in legal discussions. Rather, the expert could and should participate in factual review and strategy only. The practitioner should also consider limiting any discussions with the client beyond purely factual review in the presence of the expert as the attorney-client privilege relating to discussions with the client could be compromised by the presence of the non-client expert. Also from a practice standpoint, any written reports or materials prepared by the engineer/expert should be submitted directly to counsel, as opposed to the client.

B. The Role Of The Expert In Litigation

There is little to be gained by the surety in keeping the information imparted by the expert "secret" for too long. As soon as the surety has determined that it is necessary to hire outside counsel, costs and damages are being incurred by the surety for the principal's account. If the expert can quickly recognize compelling points or arguments, counsel should determine whether or not it makes sense to attempt to quickly resolve and/or settle the matter through the use of such information. While not necessarily disclosing that the expert has been retained, the information and arguments developed by the expert certainly provide worthwhile ammunition.

Once the expert is chosen, it is generally best to be over inclusive as opposed to under inclusive with respect to the review of documentation held by the principal. In addition, it is good practice to always encourage, if not require, the expert to "get his hands dirty" and physically visit the project site. In some circumstances, the construction defect alleged to exist cannot be seen and the site visit will essentially be useless. However, it is better to have the expert testify to that effect at trial than to provide an opportunity for the opposition to repeatedly point out to the jurors that the surety's expert never even visited the site. Next, the practitioner should consider issues relating to the discoverability of the expert's work product and whether it may make sense to meet in person to obtain an oral report regarding the expert's conclusions. If it appears that the expert will be helpful, a written report can be considered.

When obtaining the oral report from the expert, it is usually a good idea to adopt somewhat of a cross-examination approach. In particular, it is prudent to review all materials that the expert has used in arriving at his or her conclusions. If there are any holes or gaps in the expert's explanation as to how he or she arrived at a given conclusion, it is better to learn that sooner rather than later. An expert's testimony is less then compelling when he or she is reduced at trial to asking a jury to believe his or her conclusion simply because the expert "says so."

From a surety's perspective, it is also vital that all pertinent records of the principal be preserved. It is presumed that the principal is being cooperative, at least at the beginning of the case, in order for an expert to have ready access to documentation. However, the surety may find itself in a situation, depending upon the length of a case, where the principal has effectively ceased to exist and documentation has either been scattered or, in some instances, put into the hands of a bankruptcy trustee. If the surety determines from its preliminary review that the principal's substantive defense is meritorious, it is vital that evidence necessary at trial be preserved.

The expert witness takes as much preparation as any other witness and maybe more. The attorney must: (1) make sure all raw data on which the expert's opinion is based has been reviewed by the expert and that such data is available for presentation in evidence; (2) review the expert's background and writings on the particular subject to determine his or her vulnerability on cross examination; (3) make sure the expert understands the whole picture so he or she will know where and how the testimony fits into the case; this will enable the expert to anticipate problem areas on cross examination; (4) be certain that the expert knows all the facts—good and bad; (5) prepare the expert for testifying utilizing essentially the same guidelines as for a lay witness; and (6) caution the expert to be straightforward and avoid the role of advocate.

To persuasively prove the client's case, it is necessary that the position be understood by the trier of fact. Therefore, "keep it simple stupid" (KISS). The expert's primary value in major litigation is his or her ability to distill an understandable opinion from a confusing, complex factual situation with a seemingly myriad number of potential problems. In addition, the expert must be able to explain the process in a convincing fashion in terms which are readily understood by the lay person and, if challenged, furnish sufficient detail and reflect a comprehensive understanding of the facts to buttress the basis of his conclusions.

1. Qualifying The Expert

The qualifications of an expert must be presented to the court before he or she will be allowed to testify as an expert. There are no fixed rules regarding the prerequisites which must be met and it is within the discretion of the court to determine if appropriate qualifications are shown. The court's ruling will not be reversed in the absence of an abuse of that discretion.

Frequently, the court or opposing attorney will attempt to obtain a stipulation as to the qualifications of a well known expert. Such an offer should be rejected, especially in jury trials. Explaining the qualifications of your expert is an opportunity for the expert to sell him or herself to the trier of fact.

2. The Nature Of The Expert's Testimony

Opinion testimony of an expert witness may be based upon facts which are within the expert's own knowledge, or upon hypothetical facts assumed to be true. In a completion situation, it is desirable for the expert to testify and to render his or her opinions on the basis of facts which are within his or her own knowledge as a result of having been involved in the construction process. For an expert who has not been so involved, a major pitfall is the failure to provide the expert with reliable source material upon which to base his or her opinion. Frequently, an expert's testimony is rebutted by attacking the source material upon which the expert based his or her opinions rather than the expert's qualification or the opinion itself. Thus, extreme care should be exercised to ensure that the opinions stated by an expert witness are based upon reliable factual information and data if the underlying facts and data are not within the experts own knowledge.

The bottom line is that an expert is the key to "putting the puzzle together." He or she must be familiar with the basic factual data so as to be thorough, logical, and persuasive, but not so complex as to confuse.

If possible, the use of what amounts to visual aids should also be considered, if they can be reasonably calculated to lead to a greater understanding on the part of the jurors. Experience shows that jurors pay close attention to photographs and other types of demonstrative evidence. Such items also provide the jurors with something to actually grasp and take with them into the deliberation room, as opposed to being generally relegated to having testimony read upon request. It is vital that any items prepared by the expert be checked and double-checked for accuracy. In one anecdote we recall, a client spent tens of thousands of dollars on the creation of a scale model of an HVAC system which was at issue in a case. However, the "expert" firm which prepared the model inexplicably inserted a small wooden figure to roughly depict a person among the mechanical system. On voir dire, the opposing attorney noted that the wooden figure was grossly out of scale and that any person matching that scale would be something on the order of ten feet tall. As a result of that inaccuracy (and we assume some heated debate and argument) the court excluded the entire exhibit.

The same generally holds true for summaries prepared by or utilized by the expert witness. For instance, Rule 1006 of the Federal Rules of Evidence permits a presentation at trial of a chart, summary, or calculation relating to voluminous writings or similar documentation which cannot conveniently be examined in court.[62] As with the example of the model, it is imperative that the summaries be accurate. A minor inconsistency or inaccuracy may result in the entire summary being deemed inadmissible with the result that the voluminous documents which were sought to be avoided must be resorted to. It may also be a good idea, depending on what the summary imparts, to utilize a "blow-up" of such an exhibit. However, be careful to ensure that such an exhibit actually shows the jury something, as opposed to simply reciting numbers or figures. Similarly, photographs of construction elements are extremely helpful as it is difficult for the average juror to mentally envision what is being testified to in most construction cases.

Apart from seeking to educate the jurors, it is also advisable to try to educate the judge as much as possible. The end result may be the ability to gain some beneficial pre-trial instructions to the jury relating to the technical aspects of the case as well as the nature of suretyship, if applicable. If such is obtained, the opening statement should allude to the favorable aspects of any such charge and also to at least a generic

---

62 FED. R. EVID. 1006. For further discussion of Rule 1006 summaries, see *infra* Chapter 12.

description of the conclusions anticipated to be reached by the scheduled expert witness.

C. Other "Experts" Employed By Principal Or Obligee

It is not the intent of this chapter to provide an exhaustive trial practice manual. As a result, this chapter will not cover any detailed cross-examination techniques or similar topics. However, suffice it to say that, because construction and suretyship are not generally well understood, it is perhaps more crucial that sound cross-examination be prepared and conducted in a construction/surety context than in most cases.

In many cases, the best that can be done in cross-examining an opponent's expert is to establish what that expert does not know. The goal is to not hurt your own case any further than the expert's testimony may already have accomplished. In a construction or surety case, however, there should always be some area that all experts will agree on and there will usually be areas where one or more experts simply cannot make conclusions. Further, if the surety is defending a bad faith claim, it may be vital to simply establish that opinions may differ on a certain point. If the expert retained by a party seeking bad faith damages agrees that opinions may differ, then the surety that has relied upon such a differing opinion may escape a finding that it failed to act in good faith.

You need an expert during the discovery process to begin planning an attack on the opposing expert. It is necessary to evaluate the information the expert is utilizing in preparing his or her evaluation of the issue and the methodology and logic the expert intends to use. In addition, during discovery you need to learn: (1) if there are any writings by the expert which may provide sources of information to contradict his or her testimony in the case; and (2) the names of individuals or other sources in the field upon whom the expert indicates he or she relies, which may contain theories or methodologies that contradict those utilized by the expert in the case being tried; and (3) the factual and logical basis for the expert's conclusions and objectivity and the extent and thoroughness of the expert's understanding of the project and the project's problems and other related questions. We are taught to never ask a question of an expert unless we know the answer. That axiom does not apply in the discovery setting.

In the cross-examination setting, counsel is permitted broad discretion with the opposing expert. The weight to be given to an expert's testimony by the trier of fact in its determination of the ultimate issues is exclusively for the trier of fact's determination. With that in mind, do not simply attack the expert's ultimate conclusions. If you do you simply give the expert an opportunity to reinforce his or her own conclusion through repetition. Probe the underlying facts and assumptions for weaknesses, inaccuracies, or partial consideration of the facts. Challenge any faulty logic or methods, all so that you can point out to the jury at closing why the opposing expert's conclusions should not be given any weight.

D. Avoiding The "Catch-22"

Traditionally, a surety that was required to go to trial on a quality of work issue was almost always defending the principal's work against a default claim by the obligee. However, recent case developments have raised the specter that the surety may also finding itself in the position of defending the obligee's position in the context of defending an affirmative claim by the principal that the surety acted in bad faith. A detailed and effective early use of an appropriate engineer/expert will not only permit the surety to intelligently determine whether the principal or obligee is more likely correct, but can also serve as a later shield against allegations that the surety made its determination without due regard to the rights of the principal or obligee. Such an investigation will also serve as a basis for the surety to assert its rights under any applicable indemnification agreement, such as may require the principal to post collateral security in the face of any alleged defense by the principal to a default declaration.

At minimum, the surety should position itself adequately, should it be required to adopt the position of the obligee against the principal or vice versa. The dual obligations of the surety running to the contractor and obligee can place the surety in a precarious position. The best defense is to adequately investigate the competing positions and to proceed on a reasonable and informed basis, while preserving and asserting all rights under the bonds and any indemnity agreement.

CHAPTER 10

PROVING (AND DEFENDING) TIME IMPACT CLAIMS

Jon M. Wickwire
Peter A. "Tony" Warner

I.
Scheduling Basics

A. Why Parties Need Schedules

1. The Owner

Schedules provide several advantages to the owner. First, schedules provide the owner with an early indication of when the contractor plans to complete the project. Further, schedules provide the owner with an indication of what the general contractor believes will be the critical path of the project. In turn, the owner can help ensure that these activities are kept on schedule (*i.e.*, expediting inspections, approvals, vacating areas of the project, delivery of the owner FFE, etc.). Better cash flow forecast also can be generated through the dollar loading of a detailed schedule. Consequently, funding can be structured in a way to reduce interest costs. The owner can react to the contractor's schedule and problems more effectively by being aware of the current posture of the project and the critical areas thereof. The payment process can be accomplished more effectively and accurately through the use of a dollar loaded schedule. The schedule can help avoid "front-end loading" and create a process by which payment is rendered for actual work accomplished at its reasonable value or cost. Through the update process, the owner is constantly advised of the status of the project, schedule trends, what is critical to the completion of the project, etc., and can re-evaluate its plan accordingly, including: (1) re-evaluate its move-in plans; (2) re-schedule OFE/FFE deliveries and interfaces; (3) re-evaluate funding requirements; and (4) re-schedule separate contractors. Finally, schedules permit all parties to more accurately evaluate the impact of delays on the project. In the case of the owner caused delays (*i.e.*, changes), the owner and/or architect are provided a current schedule with which to evaluate the impact of the change rather than relying solely on the contractor to offer its opinion in this regard—an opinion at times structured to support a claim for added costs.

2. The General Contractor

Schedules also provide several advantages to the general contractor. First, a schedule satisfies, in part, the general contractor's obligation to schedule and coordinate the project. The scheduling process also provides the general contractor the opportunity to discuss, in detail, the various subcontractors' and suppliers' plans for performance and the compatibility of those plans within the overall project objectives. During the planning process, alternative schedules can be evaluated quickly and conveniently. A schedule also provides the general contractor a dynamic tool with which to monitor the progress of the project and all the parties thereto. A schedule can give the general contractor, its subcontractors, and material suppliers a focus on critical long lead items, and provides an agreed upon bench mark with which all parties can analyze the influence of delays and evaluate time extension requests. Finally, effective schedules, and those projects completed on schedule, provide the general contractor, its staff, and its subcontractors and suppliers a track record of on-time performance, which is one of the single most important factors in the evaluation process for future work.

3. The Subcontractor And Material/Equipment Suppliers

Schedules provide several advantages to subcontractors and material and equipment suppliers as well. By participating in the preparation of the initial schedule, subcontractors and suppliers can inform the general contractor and its project management staff of their performance requirements. Accurate, detailed schedules provide these parties advance notification as to when they are required to perform their work and/or deliver necessary materials or equipment. A detailed schedule should give a subcontractor notice of the environment in which it will be required to perform its work. A critical path schedule will advise these parties as to the relative importance of their performance and whether their work is critical, near critical, or has ample float associated with it. Manpower and equipment resources can be evaluated and allocated more precisely utilizing detailed schedules. Through the update process, the schedule is revised and the subcontractors' and material suppliers' portion of the project is re-evaluated as well. Through the update process, subcontractors and suppliers are provided an early warning system as to the influence of delays and changes on their work. Early notice can be given protecting their rights in regard to delay damages.

B. Advantages Of CPM Schedules

Once one acknowledges the justification for advanced scheduling techniques on today's projects, the advantages of CPM or network schedules become obvious. First, a CPM schedule tells accurately how long a project will take and what task must be kept on time to meet the schedule. A CPM schedule makes personnel think about a project in more than usual detail and tends to prevent omission of important actions in the project plan. It also shows the proper relationships between all tasks typically not easily recognizable on a bar chart. A CPM schedule simplifies advance work assignments and helps improve communications among those responsible for the project, and can assist by simulation the evaluation and forecast of the outcome of alternate plans of action. It focuses attention on potentially troublesome tasks and allows pinpointing of effective action rather than wasteful scatter shooting, and fixes responsibility and provides a permanent record of assignments. A CPM schedule can insure continuity of action even with changes of personnel or organization and provides a measure of performance. It encourages individuals at all management levels to contribute to effective job planning and simplifies periodic evaluation and re-scheduling of the project. A CPM schedule measures proposed changes against time, money, manpower, and equipment demands and limitations. It also provides a graphic picture of the work to be done and encourages making prompt, accurate field decisions where necessary. A

CPM schedule encourages accurate and continuous control of the procurement process, and provides all persons involved with a common language with which to communicate about a job.

C. Basics Of CPM

1. Definitions: Words And Phrases

This chapter utilizes terminology and definitions that are part of the body of knowledge required in construction scheduling. Below is a list of terms with the appropriate definition or explanation.

Activity: Any portion of a project which consumes time or resources and has a definable beginning and ending.

Node: Used to represent the beginning and ending points of activities in arrow diagramming. Synonymous are "event" and "connector."

Event: See Node.

Link: The path between two nodes. Network terminology synonymous with arrows in arrow diagramming.

Dependency: The relationship between two activities with regard to beginning and/or completion of the two activities relative to each other.

Dummy Activity: Arrows used to show dependencies only, in arrow diagrams. (Dummy activities do not consume resources.)

Predecessor: Relative to the activity under consideration, the predecessor activity restrains the activity under consideration.

Successor: Relative to the activity under consideration, the successor activity is restrained by the activity under consideration.

Activity-on-Arrow: A type of network diagramming that identifies activities as arrows.

Activity-on-Node: A type of network diagramming that identifies activities as geometric shapes, nodes, and uses arrows to show dependencies only.

Gnatt Chart: Early type of bar chart developed by Gnatt (WWI era).

Bar Chart: A graphic portrayal of activities plotted against a time scale. Activities are shown as bars from the start to the completion of the activity.

Harmonygraph: A bar-chart with time plotted vertically which recognized simple precedence relationships.

Network: A graphical representation of a project plan, showing the interrelationships of the various activities.

Schedule: The addition of activity durations to a network and the computation of start and finish times results in a schedule

As-Planned Schedule: The initial schedule for a project, representing the contractors original intentions with regard to the progression of the work.

Resources: A commodity consumed by an activity.

Milestone: Used to indicate the start or end of a series of related activities. Milestones indicate a significant point in time.

PERT: Project Evaluation and Review Technique. A network based scheduling method which uses a statistical approach to activity durations.

CPM: Critical Path Method. A network based scheduling method which uses "fixed" time durations.

2. *Forward Pass / Early Start And Finish Dates*

The main objective of forward pass computations is to determine the duration of the project. This time span is found by searching through the network for a continuous chain of activities beginning at the initial node, ending at the terminal node, and having the greatest total time duration. To compute the earliest start and finish dates for each activity in the network diagram, the forward pass computation is initiated by assigning an arbitrary earliest start time to the initial node. A value of zero is usually used for this start time since subsequent earliest times can then be interpreted as the project duration up to the point in question. The computation then proceeds by assuming that each activity starts as soon as possible, *i.e.*, as soon as all of its predecessor activities are completed.

The most common procedure for finding the project duration is to successively add activity durations along chains of activities until a merge node is found. At the merge node the largest value of the sum of the activity times from each path entering the node is taken as the start of the succeeding activities. Addition continues to the next merge point, and these steps are repeated until the terminal node is reached.

Forward Pass Rules: Computation of Early Start and Finish Times:

RULE 1. The initial project activity is assumed to occur at time zero. Letting the initial activity be denoted by 1, this can be written as:

$\text{Activity}_1 = 0$

RULE 2. All activities are assumed to start as soon as possible, that is, as soon as all of their predecessor activities are completed. For an arbitrary activity (i-j) this can be written as:

Es_{ij} = Maximum of EF's of activities immediately preceding activity (i-j), *i.e.*, all activities ending at node "I"

RULE 3. The early finish time of an activity is merely the sum of its early start time and the estimated activity duration. For an arbitrary activity (i-j) this can be written as:

$$Ef_{ij} = ES_{ij} + D_{ij}$$

3. Backward Pass / Late Start And Finish

The backward pass computation provides the latest possible times for the start and finish of each activity. These values are derived by finding the path with the greatest time duration extending from the terminal node, back to the initial node of the project network.

The calculations are started by assigning a late finish date to all the activities merging at the last node in the project and then successively subtracting activity durations along chains of activities until a burst node is found. At the burst node the smallest value of late starting dates calculated for each path leaving that node is taken as the late finish date for all activities that enter the burst node. Once this decision is made, subtractions continue until the next burst point is met. A new decision is then made, and the process continues in a like manner until the initial node is reached. As in the forward pass computations, this decision-making process does not reveal the activities that lie on the longest time path.

Thus, the latest allowable finish time for an activity is the smallest, or earliest, of the latest allowable start times of its successor activities. Finally, the latest allowable start time for an activity is merely its latest allowable finish time minus its duration time.

Backward Pass Rules: Computations of Latest Allowable Start and Finish Times:

RULE 1. The latest allowable finish time for the project terminal activity (t) is set either equal to an arbitrary scheduled completion time for the project, Ts, or equal to its earliest occurrence time computed in the forward pass computations.

Lt = Ts or Et

RULE 2. The latest allowable finish time for an arbitrary activity (i-j) is equal to the smallest, or earliest, of the latest allowable start times of its successor activities.

Lfij = Minimum of LS's of activities directly following activity (i-j)

RULE 3. The latest allowable start time for an arbitrary activity (i-j) is merely its latest allowable finish time minus the estimated activity duration time.

LSij = LFij - Dij

4. *Calculation Of Float*

All basic scheduling computations first involve a forward and then a backward pass through the network. Based on a specified project start time, the forward pass computations proceed sequentially from the beginning to the end of the project giving the earliest (expected) start and finish times for each activity and also the total time necessary for the completion of the project. Backward pass calculations give the latest times for every activity's beginning and ending.

Whether calendar days or work days are used to prepare the arrow diagram, the methods for calculating the forward and backward pass remains the same.

In addition to a unique i and j node, activity description and duration, each activity in a network diagram has associated with it four values that can be identified as follows:

ES: Early Start Date
EF: Early Finish Date
LS: Late Start Date
LF: Late Finish Date

The early start date, ES, for an activity is the earliest point in time that any activity bursting from its i node can start. Similarly, the late finish date, LF, is the latest point in time that any activity which merges at the activity's ending or j node can finish. The early finish date, EF, is dependent upon the determination of the early start date, ES, and the late start date, LS, is related to the late finish date, LF. See Appendix at Exhibit 10.1. Note that there is no relationship directly linking the LS and the EF of the activity. The schedule that results when all activities of the network are positioned in the early start location is termed an early start schedule. Similarly, when every activity is in the late position, the resulting schedule is a late finish schedule.

D. The Preliminary Schedule

Some experienced network users will say that all one needs to begin networking is a large piece of paper, several sharp pencils, and a large eraser. Actually, there is a bit more to it than that. Several general questions need to be raised and answered before detailed project planning should begin. These questions include:

(1) What are the project objectives?
(2) Who will be charged with the various responsibilities for accomplishing the project objectives?
(3) What organization of resources is available or required?
(4) What are the likely information requirements of the various levels of management to be involved in the project?

Of course, these questions are fundamental to project management and should not be passed over lightly. Discussions of responsibilities and resources can bring to light erroneous assumptions or misunderstandings in these areas. Naturally, it is well to resolve these matters before proceeding with networking (and certainly before beginning the project).

Several of the most common terms in networking are defined below. Terms associated with scheduling computations are explained in later sections.

An *activity* is any portion of a project which consumes time or other resources and has a definable beginning and ending. Activities may involve labor, paper work, contractual negotiations, equipment operations, etc. Commonly used terms synonymous with "activity" are "task" and "job." In the arrow scheme of networking, activities are graphically represented by arrows, usually with descriptions and time estimates written along the arrow.

An arrow representing merely a dependency of one activity upon another is called a *dummy* activity. A dummy carries a zero time estimate. It is also called a "dependency arrow." Dummies are often represented by dashed-line arrows or solid arrows with zero time estimates.

The beginning and ending points of activities are called *events*. Theoretically, an event is an instantaneous point in time. Synonymous are "node" and "connector." If an event represents the joint completion of more than one activity, it is called a "merge" event. If an event represents the joint initiation of more than one activity, it is called a "burst" event. An event is often represented graphically by a numbered circle, although any geometric figure will serve the purpose.

A *network* is a graphical representation of a project plan, showing the interrelationships of the various activities. Networks are also called "arrow diagrams." When the results of time estimates and computations have been added to a network, it may be used as a project schedule.

The few rules of networking based on the arrow scheme may be classified as those basic to all arrow networking systems, and as those imposed by the use of computers or tabular methods of critical path computation.

Basic Rules of Network Logic:

RULE 1. Before an activity may begin, all activities preceding it must be completed. (Activities with no predecessors are self-actuating when the project begins.)

RULE 2. Arrows imply logical precedence only. Neither the length of the arrow nor its "compass" direction on the drawing have any significance. (The exception to this rule are "Time-scaled Networks")

Additional Rules Imposed by some Computer or Tabular Methods:

RULE 3. Event numbers must not be duplicated in a network.

RULE 4. Any two events may be directly connected by no more than one activity.

RULE 5. Networks may have only one initial event (with no predecessor) and only one terminal event (with no successor).

Rules 4 and 5 are not required by all computer programs for network analysis. Hand computation on the network does not require Rules 3,4, or 5, but the network must not have loops.

E. The Initial As-Planned Schedule

As one embarks on a construction project sufficiently large enough and/or complex enough to require the use of a CPM or comparably detailed/sophisticated scheduling technique, there are a wide range of considerations that must be addressed in regards to the preparation, submittal, and maintenance of that schedule. The following offers an outline and check list of those factors that will have a particular bearing on your efforts in this regard. A recognition of these factors and the appropriate evaluation of them will make the planning and scheduling process more comfortable and result in a product that is more meaningful as a source of accurate and current information. As such, you can avoid some of the pitfalls and liabilities of a project that is poorly planned, inadequately scheduled, and as a result thereof, at least in part, poorly managed.

First, recognize that there is a difference between planning the project and scheduling the project. There are many different ways of building most construction projects. In fact, it is highly unlikely that any two project managers would plan and schedule the same project identically.

The planning process involves the development, review, and evaluation of a variety of different approaches, each of which would attempt to satisfy the stated objectives such as budget, time of performance, etc. of the project.

As a background to the entire process of project planning, there is the necessity that the "scheduler" (hopefully the superintendent and/or the project manager) know the project well. To know the project well, they must study all the contract documents and understand the full and complete scope of the work. Sources of background information would include, for example, the estimate, pre-bid subcontractor input, or a bid schedule. With this foundation, establish your objectives. Your goals may be as simple as performing the project in the stated, contractual term and within the budget determined during the preparation of your bid. On the other hand, those objectives may be far more complex.

Additionally, within the framework of the established objectives, realize that there are certain schedule considerations over which you will have little or no influence which will restrain your ability to achieve those objectives including:

(1) When do you expect to receive a contract and notice to proceed?
(2) Are there any delayed start considerations (delayed access of a part of the project)?
(3) Are there interim completion dates established by the contract? (very typical of mass transit projects)
(4) What is the required contract term of performance and do you anticipate taking the full term or have you based your estimate on an early completion?
(5) What are the realistic delivery dates for schedule sensitive materials and/or equipment (structural steel, hollow metal, hardware, elevators, curtain wall, etc.)?
(6) Are there other contractors performing within the limits of the project and to which you must interface?

(7) Recognize that you and your subcontractors will be restrained in some way by the availability of resources.
(8) Weather.
(9) Access to and within the site.
(10) Necessary actions on the part of the owner, architect, and/or construction manager (*i.e.*, inspections, approvals, etc.)

Having established a sound background and knowledge of the project, approach the subcontractors that will (or may) perform work on the project. Typically the subcontractors perform between sixty and ninety-five percent of the work. Consequently their input and impact on the project is both important and vital to the planning process. As specialists, they can offer suggestions in regard to the performance of their work that will have a favorable influence on the overall project. Of particular importance, establish the anticipated delivery of their schedule sensitive materials and/or equipment, examine the availability of the resources necessary to perform their work, break their portion of the work into components of reasonable detail, and establish realistic durations to perform these activities. Allow the subcontractor the opportunity to offer its suggestions in regard to the direction, timing, and priority of the work.

With the combination of knowledge gained through this research and investigation, explore an appropriate number of plans or options and evaluate these as they relate to and satisfy your objectives. The trade-off between time of performance and anticipated cost is always an interesting process. Once a plan has been established, that plan is then refined into a schedule, generally referred to as the "as-planned" schedule. The "as-planned schedule" should be in a form compatible with the contract requirements (IJ/ADM versus PDM or bar-chart), submitted in a timely manner, reviewed with all the parties, adopted as the schedule for the project, and maintained and updated throughout the project term.

As you start the scheduling process, make sure you recognize and conform to the scheduling requirements of your contract. Some specifications require a statement of experience in the preparation of schedules, on the part of the general contractor's staff and/or the outside consultant to be utilized.

Over and above the particular requirements related to the schedule itself, many of the more complex scheduling specifications require additional material to be submitted with the initial schedule including: (1) a time-scaled summary diagram; (2) a cash flow forecast (normally a Corps of Engineers and GSA requirement); (3) a manpower and/or equipment utilization forecast; (4) a narrative describing the schedule, resources to be utilized, etc.; and (5) customized accounting reports.

Satisfy these obligations and keep the owner/architect informed of the progress in this regard. Often, the specification requires the submittal of the initial schedule in a short and unreasonable time frame, *i.e.*, within ten or fifteen days after Notice to Proceed. The transmittal of a preliminary or draft schedule will normally satisfy this requirement for an interim period.

The "as-planned" schedule should serve two primary purposes. First, it should project and provide a basis with which to monitor job progress. The "as-planned" schedule also should satisfy the general contractor's obligation to schedule and coordinate the project.

In this regard, the schedule should be as accurate as possible and not unrealistically optimistic nor inflated to constantly reflect the project being on schedule. It should satisfy all contractual obligations in regard to performance, including overall term, delayed start, and interim completion requirements. And, as an initial schedule, it should not include any delays experienced during the early part of the project even though you are aware of these delays during the development of the "as-planned" schedule.

Submit the "as-planned" schedule to the owner, architect or construction manager in the form required. Normally, the specifications require a formal review and evaluation process. If at all possible, get the "as-planned" schedule approved by the owner or its agent. This eliminates the contention later that the original schedule was an unreasonable one. This is particularly true if the schedule is an early finish schedule. Once the schedule is approved, distribute copies to your subcontractors and material suppliers. In doing so insure that you do not guarantee performance as portrayed in the schedule. Instead, make those parties aware that the schedule is an estimate and that they should be prepared to perform their work in a time frame compatible with the schedule or updates thereto as their subcontracts require.

To implement the schedule effectively you should make sure the various parties involved understand the schedule. Offer to have a meeting with the owner and the A/E to review the schedule with them. You should also designate someone to be responsible for the maintenance and updating of the schedule and establish regular update and monitoring procedures so that the schedule is an accurate measure of the status of the job. Make sure your entire project management staff understands the contract language in regards to term of performance, delays, notice of delays, time-extensions, etc. It is important to use and review the schedule during progress meetings. Finally, post and distribute the schedule.

A positive attitude toward the planning, scheduling, and updating of the schedule normally results in an accurate, effective, and meaningful schedule which should serve all parties and contribute to the best interest of the construction efforts.

F. How To Properly Update A Schedule

The primary purpose of updating the project schedule is to evaluate the current status of the work, based on progress to date, and to forecast a realistic project completion date given that progress.

In those instances where there is a CPM scheduling specification, that requirement will normally dictate that the schedule be updated monthly. Likewise, the specification will probably describe the process to develop the data to be included in the update itself. Often the specification will require a meeting between the resident engineer and contractor's representative. The information normally discussed in those meetings or the data that must be considered to update any detailed schedule, includes the following:

1. Data Date

The data date is the date (or the point in time) for which progress is reflected in that update. For example, a data date of June 30 means progress through that date is delineated in the update by incorporating percent complete for activities underway, remaining durations, and/or actual start and/or finish dates for those activities that have been advanced during the reporting period, *i.e.*, through June 30.

Report dates, however, are not typically the same as the data date. Report dates usually represent the date that the update was run on the computer (calculated) and not the date for which progress is reported.

Updates are typically required monthly. Consequently, contractors routinely use the 25th or the end of the month as a data date for the update. While the update progress is typically determined by inspecting the project on the data date, occasionally an owner will allow the contractor to forecast progress through some future date. For example, the contractor and owner may walk the project on June 25 and forecast the progress of work through June 30. In this circumstance, June 30 would be the data date. Seldom will an owner allow a contractor to forecast more than a week of progress.

2. *Construction Activities On Which Work Was Performed During The Reporting Period And The Amount Of Progress Realized*

The typical process used to update activities in a schedule is the percent complete. By updating all the activities within the schedule on which progress was realized during the reporting period and then processing (recalculating) the update, the schedule is updated as of its data date.

The determination of the status of an activity (its percent complete) is typically made by either the general contractor, often with input from its subcontractors, or in a meeting between the contractor and owner. In either case, a walk through of the project is appropriate to observe status and to make appropriately accurate evaluations of progress. Using a copy of the preceding month's update, percent complete notations are entered on an activity by activity basis. The percent complete is simply an estimate of the amount of work completed on an activity at the data date. That percentage will be incorporated into the update when it is processed in two ways. First, the percent complete will be the basis on which a remaining duration will be calculated which, in turn, will be used in the forward pass, *i.e.*, if an activity whose duration is ten days is thirty percent complete, the remaining duration for that activity will be seven days (30% x 10 = 3 days "complete" or 7 days remaining). Second, if the schedule is dollar loaded, the percent complete will be used to determine the contractor's earnings. In the above case, assume the activity carries a value of $10,000. The thirty percent complete will generate an earnings of $3,000 (30% x $10,000) at the data date and an accumulation of similar activity earnings as will produce an overall gross earnings or payment due the contractor prior to any calculated retainage. It should be noted that most CPM programs allow the user to input separate time and cost percentages where such differences exist, *i.e.*, the time percentage complete could be ten percent while the cost percentage could be fifty percent.

3. Procurement And/Or Administrative Activities On Which There Has Been Progress During The Reporting Period And The Amount Of Progress Realized

Much like the process used to update the construction activities, the activities delineating the procurement of materials or equipment must also be updated to reflect their status or progress at the data date. These activities can be updated using either a percent complete or remaining duration. If done correctly, the early finish of the procurement or fabrication and delivery activity should accurately project the earliest potential delivery date for that material or equipment.

Of the two alternatives, most people utilize the remaining duration as the more accurate and convenient update mechanism. For example, if at data date 200 (day 200 could equal June 30), the submittal and approval of the material X is complete, obviously those activities should be updated as one hundred percent complete. Assume that the approval took place on day 175 and that the fabrication and delivery activity of fifty days has advanced during the update period to a point where material X could be delivered within twenty-five days of the data date. This activity can be updated by either using a fifty percent completion, generating a twenty-five day remaining duration and early finish for the material X delivery of day 225 (data date 200 plus 25 days) or by simply using the twenty-five day remaining duration directly.

4. Anticipated Changes In Estimated Durations Of Activities Not Yet Completed

As the contractor updates the schedule by incorporating percent complete data, one must constantly reevaluate the durations themselves to make certain that their durations are still accurate. If not, they should be revised.

Consider the possibility that the contractor has worked continuously, with a full crew, on an activity whose estimated duration is ten days and is fifty percent complete at the data date. Unchanged, the fifty percent complete will generate a five day remaining duration, suggesting that the activity can be completed in five additional days. However, the accuracy of that forecast is challenged if the contractor worked on the activity ten days (not five) to achieve the fifty percent complete status. In this case a more realistic remaining duration is ten days. To achieve the refinement of the schedule necessitated by this example, the contractor can either change the estimated duration to twenty days (holding the fifty percent complete) or input a remaining duration of ten days (again holding the fifty percent complete). In both cases an early finish of the data date plus ten days will be generated assuming all predecessor activities have been completed.

5. Changes In The Logic Of the Network To Reflect The Current "Game Plan"

Whether dealing with an initial "as-planned" schedule or an updated schedule, the schedule must reflect the contractor's plan (the combination of logic and durations) to accomplish the work if the schedule is expected to be accurate. Often a contractor changes/revises its plan or schedule during the course of the work as a result of changes, delays, or a variety of other possible influences. Once that revised plan is adopted, the schedule should be revised to reflect that new plan. Typically this will require that the logic be revised, *i.e.*, the original schedule may depict a bottom to top sequence of finishes while a change order may require that the changes be performed from the top down. In turn, the schedule should be revised from a 2-3-4-5 floor series of finishes to a 5-4-3-2 series of finish activities. This will require a number of logic revisions.

The update of an outdated schedule will serve little or no purpose. Revise the schedule, keep it current, and make it reflect the current plan/schedule for accomplishing the work, and, in turn, the update should generate accurate information.

6. *Duration And Logic Revisions Necessitated By Changes Instituted By The Owner And/Or Architect*

To the extent that the project has been changed (the scope revised) via a change order or contract modification, the schedule should be revised to incorporate that change. The revisions to the schedule can take two different forms. The first form is to add an activity or "fragnet" to reflect the added work/scope with appropriate durations and dollar values (if the schedule is dollar loaded). The new activities' description should include the change order or modification number and a short description of the change, *i.e.*, "CO 1 - Furnish and Install Flag Pole." If the materials or equipment included in the change are readily available, no activity need be added to reflect that procurement effort. On the other hand, if the procurement is significant, a separate activity can be added to the network to reflect it, *i.e.*, "CO 1 Furnish (Submit/Approve/Fabricate and Deliver) Flag Pole" and a separate successor activity used to identify the installation of the flag pole (CO 1 Set and Correct Flag Pole).

As an alternative to the first revision form, existing activities in the network can be altered/revised to reflect the change. Assume that there is an activity in the schedule to erect site light standards or fixtures and that the flag pole, added by CO 1, will be set at the same time but take an additional day for the crew setting the lights. The existing activity "Set Site Light Fixtures" (2 days) could be revised to incorporate the flag pole added by CO 1 ("Set Site Light Fixtures + CO 1/Flag Pole" 3 days). Or assume that the millwork in a courtroom is increased by a change order (CO 2). The existing activity describing the base contract work could be altered/revised as follows: "Install Base Contract Millwork + CO 2 - Additional Millwork" with the duration changed from three to five days (and the value revised from $3,000 to $6,000 to incorporate the value of the change).

Obviously, it is important to determine if the added work embodied in the change order or modification is critical to project completion. By correctly incorporating the change into the network, that determination can be made. This topic will be dealt with separately.

7. Other Delays Or Disruptions Experienced

During the update process, the contractor can incorporate activities into the schedule to reflect delays or disruptions experienced. If the contractor chooses or is required to do so by the contract, this process must be executed very carefully. The corresponding data and projections used may play a large part in a time extension request and/or subsequent claim.

To incorporate delays and/or disruptions into the schedule, the contractor can insert this data into the schedule much the same way as it does change orders or contract modifications, either by adding activities or revising existing activities to reflect the delay or disruption. The following may serve as examples of the two options in a case where rock, a differing site condition, was encountered during excavation of the footings, defined in the network with an activity "Excavate Footings - 6 days."

As the first option, the following activity could be added to depict the unanticipated rock which was encountered after two days of earth excavation:

> Excavate Footings - 2 days
> (This is the base contract part of the original activity)
>
> Excavate Rock - 10 days
> (This is the added activity describing the removal of the rock)

Absent any other information one would assume that four days (original duration for excavate footings of six days less two days of progress on actual earth excavation) of the ten days is the equivalent of base contract work. Therefore, since the excavation of footings took a total of twelve days and the estimate was six days, the rock excavation took an additional, unanticipated six days. This extra time may or may not be critical to contract completion.

As an alternative option, the contractor could incorporate the rock excavation into the existing activity for footing excavation as follows:

> Excavate Footing (Base Contract) + Rock Excavation (Change Order X) 12 days (Actual)

If a change order is applicable, its number should be incorporated into the activity description.

In either case, the actual start and finish dates for the footing excavation and rock excavation should be recorded carefully and accurately in contemporaneous documents and the update, if required.

8. *Actual Start And Finish Dates*

Most government scheduling specifications require the contractor to document and enter into its schedule updates actual start and finish dates. Absent such a requirement, a contractor may find it beneficial or informative to record these dates as a routine part of its update procedure. In either case, the contractor and/or owner need only record and enter these dates into the update much the same as the percent completion data is entered. The software program will provide a position in the report format for this as-built data, typically in the early start and finish fields so that the data is maintained from update to update. Also, if an actual date is entered most systems place an "A" next to the date to signify that its an actual date.

If the above procedure is followed throughout the project term, at the end of the project the contractor will have a complete record, on an activity by activity basis, of actual performance expressed as actual start and finish dates for all the activities in the schedule.

9. *Time Extensions Granted*

If a time extension is granted by the owner or architect (typically requiring a formal change order or contract modification), the schedule should be revised to reflect that time extension. Typically, the contract completion date is plugged or fixed to the last activity in the network (*i.e.*, Project Completion) as its late finish date. In turn, float or slack is generated by comparing projected early dates (starts and finishes) to calculated late dates (starts and finishes). If the status of the project via the update is expected to be accurate, as it should be, the then current contract completion date must be the date fixed in the schedule for project completion and the basis for the backward pass generating late dates.

The following may serve as an appropriate example: the original contract required completion within three hundred calendar days or by June 15. The initial schedule correctly incorporates June 15 as the late finish date for Project Completion. A subsequent update run at day 200 (data date) generates an early finish for Project Completion of June 20 and a status of -5 (calendar) days. Consider the following possibility: if, in fact, a time extension of ten days had been granted prior to the update but not incorporated into the schedule. Therefore, the then current (at data date 200) contract completion date was June 25 (June 15 + 10 days). Had the time extension been incorporated into the schedule as it should have been, the project status would have been +5 (5 calendar days ahead of schedule) rather than -5 (5 calendar days behind schedule).

Based on all of the above considerations, the schedule should be updated and distributed to the owner and subcontractors. Often the contract will require the contractor to prepare and submit a monthly progress narrative to accompany the update itself. That narrative will normally state the current status of the project, the forecasted project completion date, and the critical path of the remaining work as identified in the update. This narrative should also identify the factors that have delayed or disrupted progress during the reporting period and further identify any factors currently influencing the work.

In the absence of a particular requirement, a schedule should be updated when: (1) progress varies from the "as-planned" schedule measurably; (2) at a point in time at the initiation and termination of a significant delay; (3) at a point in time necessary to support a payment request; (4) after a time extension has been granted; (5) any time that the contractor revises its plan (schedule); or (6) any time that there is the realization that portions of the schedule are inaccurate for any reason. A record of the update should be set aside by all parties for future reference.

The owner and architect should review the update, verify the data contained therein, particularly if the updated schedule serves as the basis of the monthly requisition, and react to it accordingly.

G. How To Properly Revise A Schedule

A revised schedule is a schedule in which there has been significant revisions from the logic and durations comprising the original "as-planned" schedule. Typically this revision is prompted by the contractor adopting a new "game plan" for any number of reasons or an owner demanding a revised schedule to demonstrate timely performance (also known as a recovery schedule).

If the contractor adopts a revised "game plan," it must incorporate those revisions into the schedule if the updates thereafter are expected to be accurate. Most CPM scheduling specifications require the general contractor to advise the owner when making significant changes to the schedule (a revised schedule) so that the new schedule can be reviewed in depth and approved.

If the project is inexcusably behind schedule (delays that are the responsibility of the contractor or subcontractors and therefore not a basis for a time extension in the contract), an owner may demand that the contractor prepare a revised or recovery schedule demonstrating timely (by the contract completion date) completion. To shorten the term of performance as demonstrated in the schedule, the general contractor will normally revise logic to show a more aggressive approach to the work and shorten durations creating a revised or recovery schedule. This new schedule may demonstrate an accelerated approach to the work.

To the extent that the contractor believes that the project was excusably delayed, notice of those delays should have been given earlier. That notice should inform the owner that the acceleration may result in additional, unanticipated costs for which the contractor will hold the owner responsible (a constructive acceleration).

Once the revised schedule is refined and its underlying game plan adopted, the new revised schedule should be distributed to the owner, its agent, and the subcontractors for review and approval. Thereafter, the revised schedule should become the basis for future updates.

II.
Use Of CPM Techniques On Contract Claims

1. Status Of Acceptance Of Courts And Boards Of CPM Techniques On Contract Claims

a. CPM Techniques Accepted And Preferred As Device To Prove Contract Claims Which Relate To Issues Of Delay And Disruption

Network analysis techniques were first introduced into the construction field in the early 1960s. With the specification by the Government for the use of network analysis techniques on major projects and the perception by contractors (after a period of initial resistance) that network analysis techniques could be extremely important tools for project management, the use of CPM techniques to plan and schedule the work has become the accepted standard in the construction industry.[1] Further, the boards and courts have shown their willingness to utilize the network analysis techniques to identify delays and disruptions on projects, as well as the causes of delays and disruptions.

The basic technique utilized in the evaluation of the use of CPM on contract claims is to compare the as-planned CPM with the as-built CPM. To state it more simply, several questions should be asked:

(1) How was it planned that this project would be constructed?
(2) How did the project construction actually occur on the project?
(3) What are the variances, or differences, between the plan for performance and the actual occurrence of performance with

---

1 *See, e.g., H. W. Detwiler Co., Inc.*, ASBCA No. 35,327, 89-2 BCA ¶ 21,612 (1989); *Al Johnson Constr. Co. v. United States*, 854 F.2d 467 (Fed. Cir. 1988); *Umpqua Marine Ways, Inc.*, ASBCA Nos. 27,790, 29,532, 89-3 BCA ¶ 22,099 (1989); *Tyger Constr. Co. v. United States,* 31 Fed. Cl. 177, 256-59 (1994); *Wilner v. United States*, 26 Cl. Ct. 260 (1991), *aff'd*, 994 F.2d 783 (Fed. Cir. 1993); *vacated and remanded on other grounds*, 24 F.3d 1397 (Fed. Cir. 1994).

respect to activities, sequences, durations, manpower, and other resources?

(4) What are the causes of the differences or variances between the plan for performance and the actual performance of the project?

(5) What are the effects of the variances and sequence, duration, manpower, etc., as it relates to the costs experienced, both by the contractor and the owner for the project?

The decision by the United States Court of Claims in *Haney v. United States*[2] provides a good basic example of the court's acceptance of the CPM process:

> Essentially, the critical path method is an efficient way of organizing and scheduling a complex project which consists of numerous interrelated separate small projects. Each sub-project is identified and classified as to the duration and precedence of the work . . . The data is then analyzed, usually by computer, to determine the most efficient schedule for the entire project. Many sub-projects may be performed at any time within a given period without any effect on the completion of the entire project. However, some items of work are given no leeway and must be performed on schedule; otherwise, the entire project will be delayed. These latter items of work are on the "critical path." A delay, or acceleration, of work along the critical path will affect the entire project.[3]

In allowing the plaintiff contractor to recover for delay, the court noted that the CPM analysis "took into account, and gave appropriate credit for all of the delays which were alleged to have occurred."[4]

2. *Level Of Knowledge On The Part Of Courts And Boards Has Risen To High Degree Of Sophistication*

A few courts have not yet totally embraced CPM scheduling concepts and may not be completely comfortable with the use of such

---

2 676 F.2d 584 (Ct. Cl. 1982).

3 *Id.* at 595.

4 *Id.* at 596.

aids to assist the judge or fact finder.[5] However, more often than not the courts and boards have shown an increasing level of knowledge and sophistication in using this tool to assist their decisions. This significant level of understanding is represented by the *Utley-James* and *Weaver-Bailey* decisions.

In *Utley-James, Inc.*,[6] the Board of Contract Appeals provided a lengthy essay on CPM scheduling which includes consideration of such important concepts as the critical path, resource leveling, acceleration, and buy back time. In *Weaver-Bailey Contractors, Inc. v. United States*,[7] the Claims Court provided a detailed exposition on float which implicitly recognized the availability of float to the parties to the project for use as an expiring resource. Theses decisions confirm the increasing sophistication of the courts in working with CPM techniques.

B. Current Recognition By Courts And Boards Of Dynamic Nature Of CPM Process

1. What Exactly Do We Mean When We Indicate That The CPM Process Is Dynamic

The critical path through any CPM network is the longest chain or chains of connected activities through the project in terms of time. The great advantage to the CPM planning techniques is that it is not static. The plan reflected by the critical path network (and the critical path itself) will change as work is performed and events occur which are later or earlier that than originally anticipated. The beauty of the CPM process is that it is dynamic and allows the executor of the schedule at any given point in time to react to events as they change so that resources (labor, equipment, time, etc.) can be applied in a different fashion and still achieve the planned project completion or minimize the effect of delays.

If we take away the ability of the executor of the project schedule to look forward and adjust his plan at any given point along the time line

---

5 *See, e.g., Indiana & Mich. Elec. Co. v. Terre Haute Indus., Inc.*, 507 N.E.2d 588.

6 GSBCA No. 5370, 85-1 BCA ¶ 17,816 (1984).

7 19 Cl. Ct. 474 (1990).

represented by the project, we take away his control of the project, indeed, the life blood of his ability to control the project.

The boards and courts are also fully aware of the dynamic nature of the CPM process. The United States Claims Court, in *Fortec Constructors v. United States,*[8] recognized that the control of the project, as well as the time extension process, is lost if the parties do properly update the critical path diagram to properly reflect delays and time extensions. The court stated: "If the CPM is to be used to evaluate delay on the project, it must be kept current and must reflect delays as they occur . . . . Reliance upon an incomplete and inaccurate CPM to substantiate denial of time extensions is clearly improper."[9]

2. *Recognition By Major Public Owners Of Dynamic Nature Of CPM Process And Necessity To Properly Update Schedules And Incorporate Changes Into The CPM As They Occur*

The standard operating manuals for the major government contracting agencies also recognize the dynamic nature of the CPM

---

8 8 Cl. Ct. 490 (1982), *aff'd*, 804 F.2d 141 (Fed. Cir. 1986).

9 *Id.*; *see also Continental Consolidated Corp.*, ENGBCA Nos. 2743, 2766, 67-2 BCA ¶ 6624 (1967) (rejecting reliance on the CPM, stating, "It is essential that any changes in the work and time extensions due to the contractor be incorporated into the progress analysis concurrently with the performance of the changes or immediately after the delay and thus integrated into the periodic computer runs to reflect the effect on the critical path."); *J. A. Jones Constr. Co.*, ENGBCA Nos. 3035, 3222, 72-1 BCA ¶ 9261 (1972) (rejecting reliance on the original, unadjusted CPM, stating, "The value and usefulness of the CPM . . . is dependent upon . . . the contractor promptly revising and updating the CPM chart to incorporate time extensions, whether they be tentative or finally determined, within a short time after occurrence of the delay."); *Ballenger Corp.*, DOTBCA Nos. 74-32, 74-32A, 74-32H, 84-1 BCA ¶16,973 (1984) (noting that the CPM's "usefulness as a barometer for measuring time extensions and delay damages is necessarily circumscribed by the extent to which it is employed in an accurate and consistent manner to comport with the events actually occurring on the job . . . This is the single most important factor in determining the acceptability of the analysis.").

process, the necessity for updating schedules in a timely fashion, and the comparative process to be utilized in analyzing delays. For example, the October 15, 1985 version of ER1-1-11, the Department of Army Corps of Engineers regulation governing network analysis systems, states: "Changes to the work and occurrences which impact progress must be entered in the schedule logic in order to keep the schedule up-to-date and reflect actual job progress conditions."[10]

Further, the extremely detailed and valuable Modification Impact Evaluation Guide of the Department of the Army Corps of Engineers, EP 415-1-3, provides a detailed summary of the procedures to be utilized in determining time extensions based upon up-to-date information as the job progresses. The Manual further states: "The process of accurately identifying and evaluating impact depends largely on an up-to-date CPM progress schedule."[11]

C. Major Developments In CPM Delay Analysis

The 1974 Wickwire and Smith article, *The Use of Critical Path Method Techniques on Contract Claims*,[12] (the 1974 Article) suggested approaches for the proof of delay and acceleration claims. These approaches, which have been recognized in court and board decisions, include a number of elements. For delay claims the 1974 Article

---

10 ERI-I-II, at 2.

11 Manual, at 3-7 to 3-8. The Manual provides:

> (3) Procedures for developing a revised schedule (para 3-3)
> (a) Revise the schedule to show actual job status;
> (b) Insert the directly changed work;
> (c) Recalculate affected unchanged work (retaining presently assigned durations);
> (d) Reestablish critical path, and note time extension justified by direct changes;
> (e) Analyze schedule for impacted unchanged activities; assign new durations to these activities as appropriate; and
> (f) Reestablish critical path, and note any slippage of final completion date indicated in (d), above. Difference is amount of time extension justified because of impact.

12 7 PUB. CONT. L. J. 1 (1974)

included: (1) reasonable as-planned CPM; (2) as-built CPM; (3) as-built CPM including recognition of all owner, contractor, excusable delays; and (4) adjusted CPM establishing completion absent owner delays (the date when the contractor would have finished but for owner delays). For acceleration claims, the 1974 Article suggested the use of: (1) reasonable as-planned CPM; (2) as-built diagram and analysis; (3) as-built diagram including recognition of all types of delays; and (4) adjusted CPM (to demonstrate that the contractor was making adequate progress toward completion at time of the act triggering acceleration and was entitled to finish at a later date than actual completion).

Since 1974, courts and boards have addressed a variety issues that significantly advanced the state of the law with respect to a number of additional critical questions. In addition, the period since 1974 has stressed significant advancement with respect to the use of computers and computer programs directly on the jobsite of construction projects. This development has greatly advanced the utility and access of CPM programs for smaller construction projects, but has also created certain problems related to the dispute resolution process when it is related to CPM issues. This section reviews the key findings by the courts and boards with respect to these advancements since 1974, and also reviews potential abuses of computer programs available.

1. CPM Acknowledged As The Preferred Method Of Proof Of Delays

The fact that CPM techniques have come to be acknowledged as the preferred method of the proof of delays is reflected in the increasing number of court and board decisions reflecting the use of CPM techniques to assist the finder of fact.[13] Three recent decisions clearly

---

13 *See, e.g.*, *Hensel Phelps Constr. Co. v. King County*, 787 P.2d 58 (Wash. Ct. App. 1990); *Attlin Constr., Inc. v. Muncie Community Schools,* 413 N.E.2d 281 (Ind. Ct. App. 1980); *Boston Edison Co. v. Department of Pub. Utilities*, 471 N.E.2d 54 (Mass. 1984); *In re Yeager Bridge & Culvert Co.*, 389 N.W.2d 99 (Mich. Ct. App. 1986); *Walter Toebe & Co. v. Department of State Highways,* 373 N.W.2d 233 (Mich. Ct. App. 1985); *McDevitt & St. Co. v. Marriott Corp.*, 911 F.2d 723 (4th Cir. 1990).

demonstrate the preference of the judiciary that CPM proof be provided to prove cause and effect on construction claims. For example, in *Al Johnson Construction Co. v. United States,*[14] the Federal Circuit fully supported the lower Board of Contract Appeals decision which had criticized "the 'bar chart' appellant provided which lacked a 'critical path,' a favorite device with present day fact finders in contract disputes."[15]

Similarly, in *H.W. Detwiler Co., Inc.*,[16] the Board of Contract Appeals found that the selection by the contractor of a bar chart to schedule its work rather than a CPM meant that the contractor was unable to satisfy the standard of proof the Board of Contract Appeals was looking for in establishing delays to the project. The Board stated that because the contractor: "chose to use a bar graph to assess its progress, its proposed schedule was necessarily general and somewhat vague with respect to planning discrete construction activities." The Board, quoting *Wilner Construction Co.*,[17] noted that "'in order to recover for alleged compensable delay a contractor must demonstrate that delay was caused by the Government and, with a reasonable degree of accuracy, the extent of such compensable delay.'"[18] Because the contractor only "provided very general, almost vague, evidence to the effect that the parties' failure to resolve the problem with the power lines delayed the masonry work," the contractor failed to meet its burden of proof.

In another decision, *Umpqua Marine Ways, Inc.*,[19] the Board, finding that the contractor had not provided proper proof of Government delays to convert a termination for default to a termination for convenience, stated that the contractor had "failed to prove" the delays "by critical path analysis or otherwise."[20]

---

14 854 F.2d 467 (Fed. Cir. 1988).

15 *Id.* at 470.

16 ASBCA No. 35,327, 89-2 BCA ¶ 21,612 (1989).

17 ASBCA No. 26,621, 84-2 BCA ¶ 17,411 (1984).

18 *Id.*

19 ASBCA Nos. 27,790, 29,532, 89-3 BCA ¶ 22,099 (1989).

20 *Id.; see also Industrial Constructors Corp.*, AGBCA No. 84-248-1, 90-2 BCA ¶ 22,767 (the Board refused to apportion responsibility for concurrent delays between the parties in the absence of proper CPM

2. *Necessity To Establish Cause And Effect Relationship*

Recent board decisions, in *Preston-Brady Co., Inc.,*[21] and *Titan Mountain States Construction Corp.*,[22] confirm the necessity to fully prove the specific cause and effect relationship between the plan for performance and the variances in that performance. In *Titan Mountain States*, the Board, in circumstances where the contractor's CPM expert did not personally examine any CPM print-outs, quality control reports, quality assurance reports, or payrolls, and only cursorily reviewed project correspondence or other documents, held that the "contractor was not entitled to time extensions for delay and impact allegedly resulting from modifications, because his critical path method analysis did not establish a causal relationship between the modifications and the alleged delays attributable to them."

The *Preston-Brady* decision demonstrates the necessity that the CPM schedule must be kept current, and that the comparative presentation of the plan versus as-built condition on the project must be grounded in the actual project records. The Board specifically stated that "[a] CPM schedule, in order to properly demonstrate delay on a project, must be kept current to reflect those delays as they occur."

Recent state court decisions also recognize the necessity for the contractor to explain not just the variances between the planned and actual CPMs, but also to prove the specific and effect relationship

---

proof); *Freeman-Darling, Inc.,* GSBCA No. 7112, 89-2 BCA ¶ 21,882 (1989) (the Board did not allow the contractor to recover for delays where the contractor's presentation of an as-built CPM contained numerous and serious inconsistencies and errors).

21 VABCA Nos. 1892, 1991, 2555, 87-1 BCA ¶ 19,649 (1987).

22 ASBCA Nos. 22,617, 22,930, 23,095, 23,188, 85-1 BCA ¶ 17, 931 (1985); *see also C & D Lumber, Inc.*, VABCA Nos. 2877, 3204, 91-1 BCA ¶ 23,544 (1990) (holding that the contractor must establish that the delay was caused by the Government; the key to establishing the necessary causation is showing "the absence of concurrent delay which would have legally delayed the contract."); *Woodington Corp.*, ASBCA no. 37,885, 90-1 BCA ¶23,579 (1990) (holding that the delayed work must be on the critical path for performance to recover for delay); *Jordan & Nobles Constr. Co.,* GSBCA No. 8349, et al., 91-1 BCA ¶23,659 (1990) (same); *Wilner V. United States*, 23 Cl. Ct. 241 (1991) (same).

between the plan for performance and the variances in that performance (*e.g.*, owner, contractor, or excusable delays).[23]

3. *Confirmation Of Requirement For Contemporaneous Baseline For Measuring Quantum Of Delay*

One issue related to the use of critical path method techniques on contract claims has been the most significant since the mid-1970s: the particular reference point to be utilized to determine if the critical path has been affected by a delay and to determine the quantum of delay. The questions have been presented as follows:

(1) Do we value the delay at the inception of the delay? (In other words, do we forward price the delay?)

(2) Do we price the delay as the delay is occurring or immediately after the delay? (We might describe this method as the real time, or contemporaneous delay determination and calculation)

(3) Or, do we determine and value the delay after the fact, after the project is over? (In other words, do we utilize hindsight pricing of delays?)

The problem in question may not always arise since the contemporaneous determination and pricing of delays may coincide with pure hindsight determination in valuation of delays to the critical path.

The whole concept of critical path method techniques is a dynamic one where the executor of the schedule analyzes the network planning techniques during the life of the project to identify and react to problems through the updating process to apply appropriate resources to the critical path and other work activities to achieve the best timing for completion considering the goals of both cost and time.[24] Absent proper use of the updating process, the utility of the CPM is lost. On the other

---

23 *See, e.g., Indiana & Michigan Elec. Co., v. Terre Haute Indus., Inc., 507 N.E.2d 588 (Ind. Ct. App. 1987)* (the court, while accepting the critical path concept for establishing the extent of project delays, made clear the necessity for specific proof of the actual cause of project delays).

24 GSBCA No. 2429, 68-2 BCA ¶ 7377 (1968).

hand, the contractor only receives time extensions to the extent that the project is *actually* delayed by a particular delay.[25]

The Board's decision in *Blackhawk Heating & Plumbing Co., Inc.*[26] poses the problem quite clearly. The Board noted that "the amount of delay granted can well depend on the point in time at which the delay claim is analyzed and acted upon," and that a contractor "could be granted a time extension because of delay in an apparently critical activity when later evidence might show the activity noncritical and the time extension therefore unwarranted."[27]

What, then, does the evolving body of contract clauses, manuals, regulations, and case authority tell us with respect to this critical issue?

The clear weight of authority (case authority, contract clauses, and agency manuals) since 1974 has given credence to the dynamic nature of the CPM process, and requires that the determination of delays affecting the critical path, as well as the development of quantum of such delays, should be developed contemporaneously as the project history unfolds utilizing the updating process as the point of reference.

Procedures to "contemporaneously" evaluate delays requires starting the analysis with the beginning of the project and then stopping at various reference points monthly or quarterly during the life of the project to determine the location of the critical path, along with any delays or positive time gains to the project (by evaluating positive or negative float generated on the critical path during the period in question) until the completion of the project is reached.

Contemporaneous determinations and pricing of delays on the critical path can also be developed after the fact by reviewing project records starting from the beginning of the project. The occurrences during the project and delay impacts are then reconstructed by moving through the various time reference points until the completion of the

---

25 *Montgomery-Macri Co.,* IBCA Nos. 59, 72, 1963 BCA ¶ 3819 (1963).

26 GSBCA No. 2432, 75-1 BCA ¶ 11,261 (1975).

27 *Id.; see also Fletcher & Sons, Inc.* VABCA No. 2502, 88-2 BCA ¶ 20,677 (1988).

project is reached. Historical records are used in this manner to perform the contemporaneous analysis.[28]

Clear case authority supporting the use of contemporaneous calculations of delay is represented by a number of cases decided beginning in the 1980s.[29] These decisions emphasize the use of current updates of the CPM, specific reference points during the life of the CPM project to assess the location of the critical path as well as the pluses and minuses of slack on the critical path as it accumulates through the project. These decisions also provide support for the contractor's ability to collect on delay claims which represent more time than the period for which the project is actually delayed past the contract completion date. This principle appears to apply even where the original CPM only showed completing on the contract completion date.

As a hypothetical, we can consider a project with an eighteen-month contract completion date and with an as-planned CPM submitted by the contractor showing the full eighteen months to complete the project. With twelve months of the construction expired, the update shows the contractor to be four months ahead on its critical path schedule with a new completion date at only fourteen months rather than eighteen months. Starting with the next month, the project is delayed by a further ten months, pushing the project out to a total twenty-four months to completion as a result of a major owner modification in the project. Obviously, under the use of the current CPM updates to evaluate delays to contractors, delay entitlement would be the total of the twenty-four months actual completion, less the fourteen months that it would have

---

28 Example of contract clauses utilizing such contemporaneous or current valuations of delays are represented by the Veterans Administration and Corps of Engineers specifications.

29 *See, e.g., Titan Mountain States Constr. Corp.*, ASBCA Nos. 22,617, et al., 85-1 BCA ¶ 17,931 (1985); *Fred A. Arnold, Inc.*, ASBCA Nos. 27,151, et al., 84-3 BCA ¶ 17,517 (1984); *Santa Fe, Inc.*, VABCA Nos. 1943, 84-2 BCA ¶ 17,341 (1984); *Santa Fe, Inc.*, VABCA No. 2168, 87-3 BCA ¶ 20,104 (1987); *Williams Enters. v. Straight Mfg. & Welding*, 728 F. Supp. 12 (D.D.C. 1990), *aff'd in part, remanded in part. sub nom., Williams Enters., Inc. v. Sherman R. Smoot Co.*, 938 F.2d 230 (D.C. Cir. 1991); *Hull-Hazard, Inc.*, ASBCA No. 34,645, 90-3 BCA ¶ 23,173 (1990); *Gulf Contracting, Inc.*, ASBCA Nos. 30,195, 32,839, 33,867, 89-2 BCA ¶ 21,812 (1989), *aff'd*, 23 Cl. Ct. 525 (1991).

taken to complete the project, even though the original CPM provided for an eighteen-month completion. Thus, the contractor would receive ten months' compensable delay, even though the project completion date only was extended by six months, and even though the original CPM planned on utilizing the full eighteen-month contract period.

The use of the contemporaneous schedule information to evaluate delays and the recognition that CPM is a dynamic process has continued in the 1990s.[30]

4. Float: Recognition Of Float As An Expiring Resource Available To All Parties To The Project

One of the most significant developments during the period represented by the 1980s and 1990s has been the clear recognition that float is an expiring resource available to all parties working on the project.

"Float" as described previously, is the excess time contained on side path activities not on the critical path. Float is important because it allows a contractor to shift resources from an activity with float to apply pressure to an activity on the critical path that is behind schedule and delaying the project. The float time permits the contractor to subsequently return to the non-critical activity and complete it on time.

Government delays to a non-critical activity that expend all of the float time and impact the critical path entitle a contractor to increased costs for extended duration. The matter becomes more complex when different types of delay to non-critical activities expend all of the float and subsequently delay the entire project. For example, if an activity that has twenty days of float is delayed thirty days, the entire project will be delayed by ten days. However, where the thirty day delay includes ten days of contractor delay, ten days of excusable delay, and ten days of owner delay, it is unclear whether the contractor is entitled to recover for the ten-day delay to the project. Where the contractor delay occurred first, excusable delay second, and the owner delay third, recovery for the ten days delay is likely, since the owner delay forced the non-critical

---

30 *See, e.g., American Int'l Contractors, Inc./Capitol Indus. Constr. G Groups, Inc.*, ASBCA Nos. 29,544, *et al.*, 95-2 BCA ¶ 27,920 (1995); *J.A. Jones Constr. Co.*, ENGBCA No. 6252, 97-1 BCA ¶ 28,918 (1997).

activity onto the critical path. Where the contractor delay occurs last and pushes the activity onto the critical path, the opposite result is likely.

The contractor should be awarded its increased costs even where, as in the example above, the contractor delay occurred last. In *Joseph E. Bennett*,[31] the Board noted that float provides the project manager with latitude in scheduling the non-critical activities to effect trade-offs of resources, to decrease costs, or shorten the length of the project. As such, float time is a resource used in scheduling the work.

Today, managers or executors of projects, usually the contractors, no longer have exclusive use of the float resource. Almost all significant public procurements include contract clauses that provide that the float is not for the exclusive benefit of any party to the project. Typically, no time extensions are allowed until the float is exhausted on the activity in question. These clauses permit the individual who "gets to" the float first to gain the benefit of the float. This also makes good sense when considering the causation principle of proximate cause.

If the owner delays a particular activity during the period when float is available until only five days of float remain and thereafter the contractor delays that activity for sixty days, the contractor is responsible for fifty-five days of delay to the project. Of course, if the contractor uses the float first and the owner then delays the activity to the point that it impacts the critical path and the project is delayed, the owner becomes responsible for the delays to the project and for damages. Since 1974, decisions by the courts and boards have confirmed that the party who gets to the float first receives the benefit of the float (until the float is actually used up on the project and the project is delayed).[32]

---

31 GSBCA No. 2362, 72-1 BCA ¶ 9364 (1972).

32 *See e.g., Dawson Constr. Co., Inc.*, GSBCA No. 3998, 75-2 BCA ¶ 11,563 (1975); *Titan Pac. Constr. Corp.*, ASBCA No. 24,148, 87-1 BCA ¶ 19,626 (1987); *Williams Enters. v. Straight Mfg. & Welding*, 728 F. Supp. 12 (D.D.C. 1990), *aff'd in part, remanded in part sub nom, Williams Enters., Inc. v. Sherman R. Smoot Co.*, 938 F.2d 230 (D.C. Cir. 1991); *Ealahan Elec. Co.*, DOTCAB No. 1959, 90-3 BCA ¶ 23,177 (1990); *Weaver-Bailey Contractors, Inc. v. United States*, 19 Cl. Ct. 474 (1990); *Robglo, Inc.*, VABCA Nos. 2879, 2884, 91-1 BCA ¶ 23,357 (1990).

All of the court and board decisions regarding float tell us the CPM analyses for delays must take the project as it finds it at the time that the delay is occurring. Thus, if a delay occurs when the project still has significant float available on the activities which have been delayed, no delay to the entire project is encountered and the party is not going to be charged for extending the project duration. However, if the delay pushes the activity onto the critical path and the entire project is extended, then the party responsible for that particular delay will be held to have proximately caused delays to the project.

E. The Demise Of The "Impacted As Planned" As A Means Of Proving Delays

One favorite device of CPM experts presented throughout the 1970s and the 1980s for asserting delay claims against owners was to take the as-planned CPM or an adjusted as-planned CPM for the project and then to impact that diagram or network solely by the acts and claims asserted to be the owner's responsibility. The "great lie" with such an analysis is that it fails to require the contractor to accept responsibility for its own delays and to give the owner credit for delays which might not be the responsibility of either the contractor or the owner such as weather or strikes. Recent landmark decisions, make it clear that such CPM analyses will not be acceptable in the 1990s.[33]

The objection to the use of impacted as-planned CPMs as a basis for measuring delay is its failure to answer the question (at least on each monthly update), "When would the contractor have finished absent thee owner delays?" As detailed later, the "But For" standard is alive and well.

---

33 *See e.g., Gulf Contracting, Inc.*, ASBCA Nos. 30,195, 32,839, 33,867, 89-2 BCA ¶ 22,812 (1989), *aff'd*, 23 Cl. Ct. 525 (1991); *Titan Pacific Constr. Corp. v. United States,* 17 Cl. Ct. 630 (1989), *aff'd*, 899 F.2d 1228 (Fed. Cir. 1990); *Ealahan Electric Elec. Co., Inc.*, DOTCAB No. 1959, 90-3 BCA ¶ 23,177 (1990).

6. *Distinctions In Proof Required For Liquidated Damages And Time Extensions Versus Proof Required For Extended Duration Claims And Concurrent Delay And Apportionment*

The section concerns two basic questions. First, must the contractor satisfy the same standard to establish the right to time extensions (thereby avoiding liquidated damages) as that required to prove an extended duration claim? "Extended duration claim," as used here refers to a contractor's affirmative claim for delay costs. The second question concerns the evaluation of concurrent delays. What developments have occurred with respect to the willingness of courts and boards to apportion concurrent delays between the parties utilizing CPM principles?

Decisions in the 1980s and the 1990s address the differences in proof between time extension and extended duration claims. These decisions indicate there is a difference in the proof required. In *Cline Construction Co.*,[34] the Board in discussing the effect of concurrent delays noted:

> Concurrent delay does not bar extensions of time, but it does bar monetary compensation for daily fixed overhead costs of the type claimed by Cline because such costs would be incurred on account of the concurrent delay even if the Government-responsible delay had not occurred. *Commerce International Co. v. United States* [9 CCF ¶ 72,781], 167 Ct. Cl. 529, 338 F.2d 81 (1964). . . . Cline has presented no evidence that those segments of the work which were delayed by the Government-responsible causes cost more to perform in a later period than they would have cost if performed at the time originally schedule . . . .

In *Titan Pacific Construction Corp.*,[35] the Board in evaluating the record regarding various delays stated:

> Although our findings establish that appellant incurred many delays through its own fault and that of its subcontractors which prevented it from completing Phase II until 30 October 1978 . . . they also establish

34 ASBCA No. 28,600, 84-3 BCA ¶ 17,594.

35 ASBCA Nos. 24,148, et al., 87-1 BCA ¶ 19,626.

> that the Government contributed to the delays by issuing change orders, modifying Phase II requirements, as late as 20 July 1978, . . . 31 August, 1978, . . . 20 September 1978, . . . and 19 October 1978. . . .
>
> Under those circumstances, the delays are not compensable so as to entitled appellant to delay damages therefore, but the Government's actions relieve appellant from liability for liquidated damages. *Myers-Laine Corporation*, ASBCA No. 18234, 74-1 BCA ¶ 10,467. Consequently, the Government's assessment of liquidated damages amounting to $8,787 for Phase II was improper and must be set aside.

Similarly, in *Utley-James*,[36] the Board noted:

> A delay for which the Government is responsible is excusable by definition, and it may also be compensable. The rule is that for a delay to be compensable under either the Changes clause or the Suspension of Work clause, it must result solely from the Government's action. . . . If a period of delay can be attributed simultaneously to the actions of both the Government and the contractor, there are said to be concurrent delays, and the result is an excusable but not a compensable delay. . . .

Consequently, for purposes of establishing entitlement to a time extension, a contractor need only demonstrate that concurrent causes of delay resulted in a specific amount of delays to project completion. The contractor need not demonstrate that project completion would have occurred earlier than actual completion but for the government's actions.

The distinction in the differences in proof required between time extensions and extended duration claims was again confirmed by a decision in 1989, the decision of the Board of Contract Appeals in *Freeman-Darling, Inc.*[37] This case which concerned a remodeling contract for a United States Customs Service Facility addressed the issues of an extended duration claim on the part of the contractor as well as the assessment of liquidated damages on the part of the owner. In finding that the contractor was not entitled to recovery for extended

---

36 GSBCA No. 5370, 85-1 BCA ¶ 17,816, *aff'd*, *Utley-James Inc. v. United States*, 14 Cl. Ct. 804 (1988),

37 GSBCA No. 7112, 89-2 BCA ¶ 21,882.

duration expense, but was still entitled to remission of liquidated damages, the Board stated:

> That delay was concurrent with delays due to changes and strikes. The law is well settled that where both parties contribute to the delay neither can recover damages, unless there is clear evidence by which we can apportion the delay and the expense attributable to each party. *Blinderman Construction Co. v. United States*, [30 CCF ¶ 70,619], 695 F.2d 552, 559 (Fed. Cir. 1982); *Active Fire Sprinkler Corp.*, GSBCA No. 5461, 85-1 BCA ¶ 17,868, at 89,484. Since no method is apparent for apportioning the delays, appellant may not recover increased costs for the period of June 25 to August 2, 1982. Correspondingly, for purposes of liquidated damages, appellant must be credited with an extension equal to the delay that occurred during that period.

A number of cases provide guidance on how courts and boards treat the use of CPM techniques to prove apportionment or non-apportionment of concurrent delays. *John Murphy Construction Co.*,[38] reflects the traditional unwillingness of the courts and boards to apportion concurrent delays. However, this decision involved a project where CPM was not specified. Further, the issue presented to the Board was whether the contractor would be entitled to recovery from the government. In denying recovery, the Board refused to apportion concurrent delays where the contractor was responsible for concurrent delays affecting work in the orderly sequence of events necessary to timely completion of the contract. In another 1990 decision, *Industrial Contractors Corp.*,[39] the Board made it clear that the lack of proper CPM evidence caused its inability to apportion delays.

In addition, we have a number of examples where CPM network analysis systems have been used in analyzing concurrent delays. In *Santa Fe, Inc.*, [40] the Board advised that to excuse the contractor from the assessment of liquidated damages, the contractor must clearly establish that the concurrent owner delay affected the critical path. After noting that the contractor accepted responsibility for a major portion of the delay in completion, the Board questioned whether the government

38 AGBCA No. 418, 79-1 BCA ¶ 13,836.

39 AGBCA No. 84-348-1, 90-2 BCA ¶ 22,767.

40 VABCA Nos. 1943, 84-2 BCA ¶ 17,341.

contributed to the contractor's delayed performance by using change orders, thereby forfeiting all or part of its right to assess liquidated damages. The Board upheld the assessment when it concluded that the contractor had failed to prove any impact on the contract completion date since the CPM presentation showed that there was sufficient float time available to absorb the delays. In doing so, the Board relied primarily on the following quote from the decision in *Blackhawk Heating and Plumbing Co., Inc.*,[41] where the GSA Board upheld an assessment of liquidated damages after concluding that the late contract completion was attributable to contractor delays:

> The rule has been applied by this Board in situations where it was difficult to apportion responsibilities for delay as between the parties; and, too, the record did not disclose whether or not the delays were critical to project completion. Since liquidated damages are only imposed for delays in project completion, it is manifest that only those delays should be considered which actually affect project completion. By their nature the delayed activities involved must necessarily lie on the critical path as it was actually completed. In terms of the concurrent delay rule then, the concurrent delay must pertain to activities whose completion was critical to completion of the project itself. Appellant cannot successfully urge, as it apparently seeks to do, that because critical contractor-caused delays . . . were concurrent with noncritical Government delays . . . the imposition of liquidated damages may be avoided. Relief from the imposition of liquidated damages must depend upon showing concurrent delay in respect to activities on the critical path.

In *Williams Enterprises, Inc. v. Strait Mfg. & Welding Inc.*,[42] the court considered strenuous claims by the structural steel subcontractor who had been responsible for a collapse in the steel erection process, that a concurrent delay had been presented by the delay in the approval of shop drawings for the structural steel and precast fabrication. Here the District Court, in a clear decision apportioning the delay responsibility between the parties, found that the delays in the approval of the shop drawings for the structural steel had occurred in the first few

41 GSBCA No. 2432, 76-1 BCA ¶ 11,649.

42 728 F. Supp. 12 (D.D.C. 1990).

months of the project and that these delays had not affected the project critical path during the later period when the critical path of the project was actually delayed by the steel collapse.

In *H & S Corp.*,[43] the Board of Contract Appeals apportioned alleged concurrent delays between the parties arising out of the construction of a Naval Container Operation Facility. Like the decision in *Williams*, the Board found that one of party's allegation of concurrent delay was no defense to a claim by the other party, since the alleged concurrent delay did not indeed affect the critical path of the project. In *H & S Corp.*, the government was the party asserting the concurrent delay as a defense. In denying this defense, the Board stated:

> Turning first to the concurrent delay issue, we agree with the Government that a contractor-caused delay affecting the construction of the footings would have a significant effect on this contract's critical path. The overwhelming flaw in the Government's concurrent delay argument, however, resides in the premise that appellant's failure to gain approval of the building system caused the delay.
>
> * * *
>
> The Government cannot, from these facts, construct a scenario in which the very real delay caused by the unsuitable soil and additional earth work is nullified by the time it took to resolve the building load error. This failure proves fatal to the Government's defense. "The defense of concurrent delay is valid only when applied to an actually established delay, not merely an alleged delay." *Essential Construction Company, Inc. and Himount Constructors, Ltd., A Joint Venture,* ASBCA No. 18706, 32-2 BCA ¶ 16,906 at 84,114.

The 1990, *Sierra Blanca, Inc.*,[44] also provides us with a clear example of apportionment between the contractor and an owner of concurrent delays using critical path principles. The Board stated:

> The issue before us is whether or not appellant is entitled to the day-for-day delay . . . days it contends it experienced or to some lesser

---

43 ASBCA No. 29,688, 89-3 BCA ¶ 22,209.

44 ASBCA 32,161, 90-2 BCA ¶ 22,846.

amount as a result of concurrent delays attributable to other non-Government responsible causes.

* * *

Since appellant began work on the critical metal deck forty eight days late, any extension of the contract completion date must take this into account. Since the overall project was impacted 152 days as a result of the delays to the masonry, appellant is not entitled to an extension for more than one hundred and four days (152-48) and the costs attributable to that delay.

In *Utley-James, Inc.*,[45] the GSA Board demonstrated a willingness to apportion or discriminate between concurrent delays to determine the actual delays to the critical path. The Board specified the finder of fact should not assess the contractor with responsibility for delays which would not have affected the critical path. Since delays fall within the category of "why hurry up and wait." Consider the following passage:

> [S]trictly speaking, there can be but a single delay over a given period of time, and when that delay has multiple, indivisible causes, it is attributable not to either party but to both. Hence it would probably be more accurate to speak not of concurrent delays but of a single delay with concurrent causes. We note this even though, for convenience, we will use the standard terminology for the most part.
>
> When venturing into this area, we must be wary of deciding too readily that there was a concurrent delay. We considered this issue in *Warwick Construction*, . . . and concluded that, at the very least, we would not require a contractor claiming a compensable delay to prove that in the absence of the Government's delaying actions, it would have completed the job on schedule. However, we also adverted to *Warwick, id.*, to the basic principle of *Wunderlich Contracting Co.* . . . which requires that a contractor seeking compensation establish "the fundamental facts of liability, causation, and resulting injury." That, we said, "has always been the law," and we adhere to it in this appeal as we have in the past.

---

45 GSBCA No. 5370, 85-1 BCA ¶ 17,816.

> The lesson of *Warwick* is that certain kinds of second-guessing are proscribed. To take an easy example, if the job schedule was originally such that the contractor needed certain widgets on hand by January 1, but because of a six-month delay attributable to the Government, the contractor rescheduled the delivery for July 1, the Government cannot be heard to say the delays were concurrent because the contractor would have had to wait six months for the widgets anyway. In such a situation there is no reason to doubt that the contractor could have had the widgets on January 1 and proceeded on schedule absent the Government-caused delay. Such a simplistic example poses no problem at all. The problem lies not in reaching the right conclusion, given such an example, but in determining whether a given fact situation is an example of such an occurrence or is instead an example of a true concurrent delay.

Based on *Utley-James*, where the owner causes delays to the critical path, it is permissible for the contractor to relax its performance of non-critical work to the extent that it does not impact project completion (and still avoid the defense of concurrent delay). This concept is the same basic principle as that represented in *Weaver-Bailey v. United States*,[46] in the Claims Court landmark decision on float. There the court found that the contractor had the right to plan its work in accordance with its own plans or reasonable resource usage (regardless of the fact that this utilized float).

In *MCI Constructors, Inc.*,[47] the Board determined that the District of Columbia had wrongfully terminated MCI and converted the default termination into a termination for convenience. This case represented a reaffirmation of the principle that parties are entitled to use additional float that is created in the schedule as a result of another party's delays, the "why hurry up and wait concept." MCI sought delay damages consisting of extended labor supervision and administration, unabsorbed or extended home office overhead, and costs for extended equipment and general conditions. The District sought to diminish MCI's recovery for critical path delays by the government owner on the basis that MCI was also responsible for concurrent critical path delays. However, the Board found that the delays in question did not affect the critical path for

---

46 19 Cl. Ct. 471 (1990).

47 DCCAB No. D-924, 1996 WL 331212 (1996).

the project given the earlier delays by the government owner to the critical path, which had pushed out the completion date for the work. The Board held, "Although it is clear that MCI completed roofing later than the date called for it the as-planned schedule, and it is true that MCI was having difficulty prodding Fischer & Porter in early 1988 to complete its delivery of instrumentation, we conclude that those delays simply did not affect project completion in view of the overriding District-caused critical path delays."

In *Federal Insurance Co.*,[48] the contractor, Rogers Constr. Co. (who assigned its claim to Federal Insurance Co.), submitted an equitable adjustment claim for delay against the Bureau of Reclamation (BOR) on a contract for a pumping plant, alleging defective design and constructive changes. The Board looked to the contractor's responsibility to meet its burden through the use of CPM analyses. In its consideration of the issue of concurrent delays, the Board confirmed the contractor's "burden to establish the proportion of delay and expense attributable to the [owner], or recovery will be denied." The Board then found that the "BOR's delay, attributable to the specific design problems, is sufficiently segregable from delays for which the contractor is responsible to avert the normal preclusion of recovery when delays are entirely concurrent." The Board apportioned the award on the basis of a jury verdict.

In *Conner Brothers Construction Company, Inc. v Secretary of Veterans Affairs,* the Court of Appeals for the Federal Circuit confirmed that concurrent critical path delays by both the owner and the contractor deny each the right to the recovery of delay damages.

This was an appeal from two decisions of the Department of Veterans Affairs Board of Contract Appeals. The contract was between Conner, the contractor, and the VA Medical Center (VA). It was for the replacement of a hospital building and tunnels and bridges connecting it to an already existing building.

In attempting to construct one of the bridge connectors, Conner discovered a steam trench and steam lines which interfered with the construction of the bridge connector and were both unanticipated. VA issued field change orders modifying the connector work to reflect the differing site condition. Conner later submitted a claim seeking

48 ABCA Nos. 3236, *et al.*, 1996 WL 368755 (1996).

compensation for direct costs and costs associated with the delay caused by the differing conditions. The claim for direct costs was awarded by the Board but they denied Conner's claim associated with the delays.

In making its determination, the Board found an October 30, 1996 Critical Path Methodology (CPM) to be the most convincing evidence. The Board found that any delay arising as a result of the steam trench and line work was concurrent with delay for which VA was not responsible.

The decision by the Board was affirmed by the court, which denied Conner their requests for delay damages. The court held that the evidence supported the Board's decision that by the fall of 1986 most of the work on the project was time critical and that Conner's delays in performing courtyard work were concurrent with any delay arising from the steam trench and steam line work. Furthermore, in following previous case law, the court found concurrent delay prevents either party from receiving damages.

In *DANAC, Inc.*,[49] DANAC was awarded a fixed price contract for 49,670,000 to renovate and upgrade officer and airmen housing units and the construction of new garages. Numerous changes and problems arose during the course of the project.

The board found that various problems caused by the Government forced DANAC to perform their work out of sequence which caused delays in the project. The Government claimed that concurrent delay bars DANAC from recovery for the delays, citing *William F. Klingensmith, Inc. v. United States*,[50] where the delay was concurrent neither party can recover without clear apportionment of which party caused the delay.

The board, however, rejected the Government's argument, because DANAC did not seek additional time beyond that already given to it in the bilateral modifications. The board held:

> In this context, [the Government's] granting of the time extensions "amounted to a recognition by it that the overall project was delayed to that extent and an administrative determination that the delay in question was not due to the fault or negligence of appellant [the

---

49 ASBCA No. 33,394, 97-2 BCA ¶ 29,184 (1997).

50 731 F.2d 805, 809 (Fed. Cir. 1984).

> contractor]. It also raised a presumption, subject to rebuttal, that respondent [the Government] was responsible for the delay." *Robert McMullan & Son, Inc.*, ASBCA No. 19023, 76-1 BCA ¶ 11,728 (1976), at 55,728. . . . We conclude that here [the Government] has not overcome that presumption. Given our inability to link instances of [DANAC's] deficiencies with periods of project delay . . . the pervasiveness of Government-caused out of sequence work, as well as the failure to assess liquidated damages . . . we cannot say that this regard established concurrent delay baring [DANAC's] recovery.[51]

In *Bechtel Environmental, Inc.*,[52] Bechtel claimed that the Government delayed completion of the project an additional 269 days for which it sought compensation. The Government claimed that the Bechtel schedule presented at trial was simplistic and ignored key facts. The board found that the Government argued:

> generally without any other, connective, probative evidentiary support, that since [certain] activities were delayed, Bechtel's other scheduling assumptions were perforce unreasonable. These arguments [by the government were] not sufficiently developed, comprehensive nor persuasive. . . . It is not enough [for the Government] to point to various job site problems and allege, without adequate evidentiary support, that associated delays impacted the critical path. [Bechtel] has presented a comprehensive analysis.[53]

In general, that board found that Bechtel proved that its initial "as-bid" schedule was reasonable and the board found that in general Bechtel had sustained its burden of proof.

The board then reiterated its decision in *John Driggs Co., Inc.*[54] In *Driggs* the Government was defending against a claim of delay by alleging a number of concurrent delays. There the board held that once an excusable delay had been encountered by a contractor, the contractor

---

51 *DANAC*, ASBCA No. 33,394.

52 ENGBCA No, 6137 *et. al,* 97-1 BCA ¶ 28,640, *recon. denied*, 97-1 BCA ¶ 28,851 (1997).

53 *Id.*

54 ENGBCA No. 4926, 87-2 BCA ¶ 19,833 (1987).

may reasonably reschedule its work without fear that it will be held responsible for a concurrent delay. Furthermore, the board held that:

> A common thread running through all of the alleged "delays" is that Driggs did not complete these particular tasks on the originally-planned and scheduled date. From this, [the Government] concludes that the represent concurrent, contractor-caused delay insulating [the Government] from liability. . . . We disagree. More proof is required to establish the] defense of concurrent delay. When a significant owner-cause delay . . . occurs, the contractor is not necessarily require to conduct *all* of his other construction activities exactly according to his pre-delay schedule, and without regard to the changed circumstance resulting from the delay.
>
> The occurrence of a significant delay generally will affect related work, as the contractor's attention turns to overcoming the delay rather than slavishly following its now meaningless schedule. [The Government] is required to demonstrate that, [despite] the delay caused by [the Government], the contractor *could not* have performed the project in less time, and would necessarily have been delayed to the same extent in any case. . . . Merely speculative or theoretical contractor-caused delays are not adequate to establish a concurrent delay defense. . . .[55]

After quoting this text from *Driggs* the board held that:

> Most of the "concurrent delays" alleged by the Government in this case, either are actually the responsibility of the Government, or are speculative, theoretical, contractor-caused delays of the type described by the Board in Driggs.[56]

In *Cogefar-Impresit U.S.A., Inc.*,[57] Cogefar was awarded a contract for construction of Phase 2 of the Metropolitan Detention Center in Miami, Florida by the Federal Bureau of Prisons of the Department of Justice (FBOP). The board stated the following about concurrent delay claims:

---

55 *Id.*

56 *Id.*

57 DOTBCA No. 2721, 97-2 BCA ¶ 29188 (1997).

As stated above, we have found that delays were caused by the FBOP to the masonry work during the period from December 1992 through early 1993. The masonry also was delayed during the period of the structural repairs . . . . While a contractor cannot recover damages for increased costs resulting from delays during periods in which both it and the Government caused concurrent project delay, the Government is precluded from imposing contract penalties during such a period.[58] Thus, even if Cogefar caused delay to the security electronics system it still would be entitled to a time extension due to the concurrent delay to the masonry caused by the FBOP. In reaching this conclusion we do not condone Cogefar's tardiness under the contract.

Moreover, we find that to the extent that security electronics submissions were late beginning in December 1992, the time that Cogefar began experiencing significant delays to the masonry that was on the critical path, it was reasonable for Cogefar to reschedule its work. When significant Government-caused delay occurs, a contractor is not necessarily required to conduct all of its other activities in accordance with the pre-delay schedule, without regard to the changed circumstances resulting from the delay.[59] The evidence presented demonstrates that the delay in masonry in turn delayed plaster and conduit installation, and that performance of these tasks was necessary prior to installation of the security electronics system. In addition, the FBOP granted several time extensions for causes of delay which it admits were not Cogefar's responsibility. We conclude that the FBOP has not proven that delays in the security electronics submissions caused any project delays. We can only speculate as to whether Cogefar may have made more timely submittals had it not been otherwise delayed.[60]

7. "But For" Test As A Requirement For Extended Duration Claims

Canon Construction Corporation,[61] provides a clear and logical exposition of the methods used to prove an extended duration claim. The Board in *Canon* required a determination of the actual date of work

---

58 *JRR Constr. Co., Inc.*, 88-3 BCA ¶ 20,905 (1988).

59 *John Driggs Co., Inc.*, 87-2 BCA ¶ 19,833 (1987).

60 *Cogefar*, DOTBCA No. 2721.

61 ASBCA No. 16142,72-1 BCA ¶ 9404.

completion. The Board then compared this date with the date the contractor would have completed the contract work, *but for* the delays that were the responsibility of the owner. The difference, of course, represented the compensable period.

Since the *Canon* decision, other authority indicates that a concurrent delay will not foreclose the contractor from recovering delay damages where the concurrent contractor delay does not affect the critical path. In *Fischbach & Moore International Corp.*,[62] the Board stated:

> We take no issue with the application of the *Commerce* rule to the facts of this case insofar as the concurrent delays for which appellant is responsible affected work in the critical path to timely completion of the contract. If the concurrent delays affected only work that was not in the critical path, however, they are not delays within the meaning of the rule since timely completion of the contract was not thereby prevented.
>
> With regard to the alleged intertwining of Government-caused and concurrent delays in this case, we have found, in the critical path analysis offered by appellant, a ready and reasonable basis for segregating the delays. If the delays can be segregated, responsibility therefor may be allocated to the parties. . . . And if there is no basis in the record on which to make a precise allocation of the responsibility, an estimated allocation may be made in the nature of a jury verdict . . . . The seemingly contrary result in *Commerce* is explained by the fact that the Court was unable, on the record in that case, to separate delays for which the Government was not responsible from those for which it was. . . . As will be seen in the discussion that follows, we have no such difficulty in the present case.

Courts and boards have consistently held that only delays to the critical path should be considered. This raises a question as to whether a contractor must give credit back for contractor delays which would have delayed the project but for owner actions. The Board in *Fischbach & Moore* applied the same reasoning used in *Canon* and reiterated the proposition that the contractor cannot recover unless it proves that such costs would not have been incurred *but for* the government action:

---

62 ASBCA No. 18145, 77-1 BCA ¶ 12,300.

> It is axiomatic that a contractor asserting a claim against the Government must prove not only that it incurred the additional costs making up its claim but also that such costs would not have been incurred but for Government action.

Recent authority supports the contractor's responsibility to return credit for any delays, such as strikes or weather, that would have otherwise delayed the project.[63] The "but for" principle recognized by *Canon* and *Fischbach & Moore* was again reaffirmed in *John Murphy Construction Company*,[64] where the Board stated:

> Conversely, it does not appear from the record that but for the Government caused delays Appellant could have completed the work by December 13. It is concluded, therefore, that despite the delays caused by the Government the record establishes that from the time scheduled for commencement of the project, Appellant was at least concurrently responsible for the delay in the process of the work. Appellant must bear the responsibility for the consequences of his search for a less costly source of water and the manner in which he chose to sequence and perform the work.
>
> In this case, the concurrent delays for which Appellant was responsible affected the work in the orderly sequence of events necessary to timely completion of the contract. Although no critical path method of scheduling was called out in the contract, decisions involving such critical path scheduling would appear to be analogous and therefore applicable.

8. *Acceptance Of Concept Of "Buy Back" Time And Sequence Changes*

Courts and Boards of Contract Appeals recognize that a contractor experiencing delay to the critical path may overcome that delay or totally obviate it by additional manloading of equipment, overtime, or performing sequence changes. In other words, the contractor can "buy

---

63 *See, e.g., Haney v. United States,* 676 F.2d 584, 595-96 (1982).

64 AGBCA No. 1418, 79-1 BCA ¶ 13,836.

back" the lost time through acceleration of work or by achieving better efficiency than anticipated in completing subsequent activities on the critical path in shorter durations than those originally planned.[65]

In addition, many of the specifications in use today, such as the Veterans Administration's Master Network Analysis Specification and the Corps of Engineers Specifications,[66] require contractors to calculate on the monthly updates time lost or gained on the critical path. Accordingly, contractors will receive the benefit of time they gain on the critical path to overcome the effect of prior delays to the critical path or to allow completion of the project earlier than detailed on the original critical path for the project. The right of the contractor to complete early (in accordance with CPM analyses) and receive compensation when such early completion is foreclosed is detailed below in the section entitled "The Use of CPM to Establish Early Completion Claims." The monthly evaluation of gains and losses as events occur to the critical path for the project and responsibility for such delays was also acknowledged by a 1984 article.[67]

Among Board authority which has recognized the concept of buy-back time is *Dawson Construction Company, Inc.*[68] There the Board implicitly recognized the contractor's ability to buy back time and receive time extensions for delay even where the contractor modifies the sequence to avoid impact on the critical path. The Board explained:

> Since Appellant combined its differing site condition claims with its weather claims, construction management analyst Warner did likewise. Warner prepared an as-built CPM in bar-chart form to reflect just how the excavation and pouring of footings, grade beams, piers and slabs were affected. In so doing, Warner utilized the actual start and completion dates furnished by Appellant. The analysis shows that post office work through the point of curing the concrete slabs was actually finished about December 22, 1973, as opposed to a planned date of December 14. Carrying the analysis further, Warner found that the

---

65 *See, e.g., Utley-James, Inc.*, GSBCA No. 5570, 85-1 BCA ¶ 17,816.

66 *See, e.g., Corps of Engineers "Contractor Prepared Network Analysis System"* clause, ER 1-1-11 (1985), Appendix A, ¶ (e).

67 *See Nielsen & Galloway, Proof Development for Construction Litigation*, 7 AM. JOUR. TRIAL ADV. 433, 444-5 (1984).

68 GSBCA No. 3998, 75-2 BCA ¶ 11,563.

> actual completion date occurred only one day later than originally planned. The reason for this was that Appellant had actually worked around the particular delays with which it was confronted. Activities which were originally planned in series were in fact done in parallel, and in effect Appellant's work forces were in more places at one time than was originally planned. For that reason Warner felt that it was unfair to award Appellant a time extension of only one day. Many of the tasks involved were considerably delayed. Instead of being finished in the 14 days planned, some 59 were required to accomplish them. Accordingly, Warner determined that in fairness to the fact that Appellant's own efforts had worked around the major delays, it should be granted a time extension of 45 days by reason of claim Items 1, 2 and 6. Warner therefore added 45 days to the start and finish dates of all ensuing critical activities.

The Board confirmed its implicit acceptance of the government delay calculations approach which allowed the contractor the benefit of the time gained through innovative sequence changes:

> We have seen that Warner took a dual approach in its time extension analysis. Following accepted procedures it prepared an as-built CPM and from it analyzed the various non-contractor caused delays. The actual project completion was found to have been delayed only one day as a result of unusually severe weather and differing site conditions. Not liking the result, which appeared unfair in view of Appellant's obviously having made up the delays by working around them, Warner picked a 45-day figure as being a fair time extension for these three claims. To that figure Warner added eight days of critical delay occasioned to the roofing and sheet metal installation by the changes to the roof support angles and the clip angles. This made a total of 53 days delay. The rest of Appellant's claims were found not to warrant time extensions since the delays involved did not affect the critical path of the project. Since substantial completion of the project on September 25, 1973, was only 46 days beyond the August 10th scheduled completion date, the time extension granted was 46 rather than 53 days.

9. Judicial Denial Of Automatic "Per Se" Time Extensions For Changes Issued After Current Contract Completion Date

The objective of CPM analysis is to segregate delays that delay the critical path and, therefore, delay the entire project. However, as of 1974, cases concerning time extensions for change orders issued subsequent to the contract completion date appeared to conflict with this reasoning.[69] Those decisions held that change orders issued after the original completion date automatically entitle the contractor to a time extension for the entire contract to that date when the additional work was completed or reasonably should have been completed. The extensions on these cases were granted without any regard to the cause of the delay in the original contract work. Current authority applying CPM analyses to discriminate between delays which affect the critical path and those that do not, indicate that those decisions are no longer valid.[70]

The argument typically asserted by the contractor in addressing the issue of delays to chains of activities extending after the completion date is that any activities showing negative float are critical, since by definition they will delay the project past the contract completion date. This view of negative float activities, however, fails to acknowledge that there is still a critical path represented by the negative slack activities with the highest numerical designation (e.g., -180 days versus -50 days). The activity chain representing the highest negative slack (e.g., -180 days) represents the longest chain of activities through the project in terms of time. This view is consistent with some of the earliest network analysis materials. For example, in the NASA PERT *and Companion Cost Guide*, at A-1 (October 30, 1962, the critical path was defined as:

> The particular sequence of activities in a network that comprise the most rigorous time constraint in the accomplishment of the end event. The path with the smallest amount of positive slack or largest amount of negative slack.

---

69 *Wickwire & Smith, supra* note 12, at 42-43 (citing FAACAP 66-21, 65-2 BCA ¶ 5306 and cases cited therein).

70 *See Santa Fe, Inc.*, VABCA Nos. 1943, *et al.*, 84-2 BCA ¶ 17,34.

The issue of delays to activities with negative float was considered in the 1984 *Santa Fe* decision, where the contractor asserted that any work sequence or CPM path of activities that ran past the contractually required completion date is critical and delays on these work sequences due to changes are on the critical path.

The contractor argued that the impact of changes or unchanged work cannot be demonstrated by regular CPM rules because "all uncompleted work becomes negative and therefore critical once the scheduled completion date has been reached.[71] In rejecting the contractor's assertion that changes issued after the scheduled completion date automatically entitled it to a time extension the Board noted that delays that do not affect the extended and predicted contract completion dates shown by the critical path in the network should not be the basis for a change to the contract completion date. The Board stated:

> a close examination of . . . cases cited by Appellant reveals that the important issue is not *when* the change order was issued, but the *impact* that change had on the completion of the project.

Santa Fe may not apply to all cases because of the peculiar language in the *Santa Fe* contract requiring that the delay analyses show that the "predicted" completion date was delayed. This naturally would represent the chain of activities with the highest negative slack. However, *Santa Fe* appears to represent a logical extension of the use of network analyses systems to isolate delays affecting the critical path for the project. Because a project extends beyond the contract completion date, this does not mean that the longest chain of activities through the network in terms of time and the delays which affect that chain cannot be determined.

10. Use Of CPM To Establish Early Completion Claims

Where an owner prevents early completion, a contractor can use a CPM presentation to seek and recover additional compensation. A number of recent cases illustrate the trend of authority in this area.

---

71 *Santa Fe*, 84-2 BCA at 86,408.

In *Montgomery-Ross-Fisher, Inc.*,[72] for example, the fact that a contractor completed the project within the contract period did not invalidate its delay claim. The contractor argued that it could have completed the project early, but for government delays. The Postal Service Board held in favor of the contractor, finding that changes had impacted on the contractor's right to complete early. The Board noted that the contractor's CPM expert was more credible than the government's and that the government expert had relied upon erroneous data.

Similarly, in *Green Builders, Inc.*,[73] the contract required completion within one year. The contractor, in developing the CPM schedule, anticipated completion within seven months and based its bid on the early completion schedule. Redesign delayed completion of the project until three months after the contractor's projected completion date -- two months before the contractual completion date. The contracting officer denied the claim because the project was completed before the contract completion date. The Board found the contractor had a right to recover delay costs based on the scheduled early completion date since the contractor showed its performance plan was reasonable.

Three cases decided in 1990 provide further guidance in the area of the use of CPM to establish early completion claims. One of these decisions, *Williams Enterprises v. Strait Mfg. & Welding*,[74] clearly breaks new ground since it holds that the prime contractor can recover from a subcontractor where the subcontractor denies the prime contractor the ability to complete at a date earlier than the date specified in the contract with the owner. This holding was made in circumstances, however, where the prime contractor had specifically consulted the subcontractors to verify the timing and duration of the sub-elements of their work and where the subcontractor had agreed to be specifically bound to the CPM scheduling requirements. Pertinent sections of the court's decision stated:

> (3) Section 6 of the subcontract specifically provides that the subcontractor will prosecute the work efficiently and promptly; further,

---

72 PSBCA Nos. 1033, 1096, 84-2 BCA ¶ 17,492.

73 ASBCA No. 35,518, 88-2 BCA ¶ 17,492.

74 728 F. Supp. 12 (D.D.C. 1990).

that subcontractor agrees to pay to the contractor such other or additional damages as the contractor may sustain by reason of delay by the Subcontractor. This provision is an express statement of the generally recognized common law obligation of parties not to hinder or interfere with one another's performance. *Fuller Co. v. United States,* 69 F. Supp. 409, 411, 108 Ct. Cl. 70 (1947); *see also, United States ex rel. Heller Electric Co. v. Klingensmith, Inc.*, 216 U.S. App. D.C. 408, 670 F.2d 1227 (D.C. Cir. 1982); *Luria Brothers v. United States*, 369 F.2d 701, 608, 177 Ct. Cl. 676 (1966).

(4) In an action between private parties, it suffices to show that a subcontractor caused a substantial delay in performance, that the contract terms forbade such a delay, and that the plaintiff was injured as a result. *United States ex rel. Gray-Bar Electric Co. v. J.H. Copeland & Sons*, 568 F.2d 1159, 1160 (5th Cir. 1978); *District Concrete Co. v. Bernstein Concrete*, 418 A.2d 1030, 1038 (D.C. 1980). Smoot has proved each of these elements.

(5) The steel erection activity was an activity on the critical path of the Project. By definition, a delay to a critical path activity will result in a delay to the entire project resulting in compensable costs. Stephen M. Siegfried, *Introduction to Construction Law*, Chapter 12, page 243 (ALI-ABA, 1987); Bramble and Callahan, *Construction Delay Claims* (John Wiley & Sons 1987), pp. 9-10, p. 145.

(6) *Smoot may properly commit its resources and those of its subcontractors to its projected CPM completion date—even if that date is earlier than the final date required by the contract with the owner—and may recover damages from a subcontractor which causes delay. District Concrete Co. v. Bernstein,* 418 A.2d 1030, 1038 (D.C. 1980); *Grow Construction Co. v. State*, 56 A.D.2d 95, 391 N.Y.S.2d 726, 728 (1977); *Canon Construction Co.*, ASBCA 16142, 72-1 BCA § 9404 (1972).[75]

Another decision, *Sierra Blanca, Inc.*,[76] concerned delay claims associated with the construction of a range operations/instrumentation laboratory for the Naval Facilities Engineering Command. In

75 *Id.* at 22-23 (emphasis added).

76 ASBCA No. 32,161, 90-2 BCA ¶ 22,846.

considering the contractor's claim for compensation, with respect to its right to finish early, the Board stated:

> Appellant's plan to finish early was reasonable and to the extent that the Government's failure to timely approve the masonry block contributed to delay in that early schedule, appellant, as we have concluded is entitled to be compensated. *Owen L. Schwamm Construction Co.*, ASBCA 22407, 79-2 BCA ¶ 13,919 (1979). We have found that the delay in the installation of the masonry block was on the critical path from the time it could have commenced on 12 July 1984, and as such, had an effect upon the overall completion of the contract work.

The final 1990 decision, also a decision of the Board of Contract Appeals, *VEC Inc.*,[77] reminds us that the contractor's claims for damages arising from denial of its ability to complete work early on a project (while a viable cause of action) must have strong *bona fides*. In fact, absent the contractor establishing clear factual bases of its actual intent to finish early, the contractor's claims will be denied. The Board stated in *VEC*:

> We must first address the issue of whether or not appellant is entitled to compensable delays for the performance period exceeding appellant's alleged estimated or desired performance period of 120 days. We have held that for time extension and liquidated damages purposes and for recovery under a SUSPENSION OF WORK clause, a contractor is entitled to the benefit of "cushion" of extra time that would result from planned early completion of the work. *G & S Construction, Inc.*, ASBCA No. 28677, 86-1 BCA ¶ 18,740; *Sydney Construction Co., Inc.*, ASBCA No. 21377, 77-2 BCA ¶ 12,719. However, the factual bases underlying these holdings were that the contractors had established their intent to perform the contract on an accelerated schedule in advance of the contractually mandated completion date and that this intent was supported by the course of the contractors' actions and performance activities during the contract performance that would have led to such an early completion absent Government caused delays.

---

77 ASBCA No. 35,988 (1990).

11. Presumptions Arising From CPM Approvals And Time Modifications

A number of presumptions arise as a result of the government's adoption of CPM techniques, approval of CPMs, and agreement on time modifications to the contract schedule and the CPM schedule. If the parties agree the CPM is a reasonable plan for performing the work, the CPM is presumed correct. A contractor or the government must overcome this presumption if it argues the original plan did not reflect necessary modifications or was overly pessimistic regarding the time scheduled for performance.

For example, in *Santa Fe, Inc.*,[78] the Board noted: "There is a rebuttable presumption of correctness attached to CPMs upon which the parties have previously mutually agreed." Thus, an owner may be responsible for the time allotted in a CPM schedule for owner approvals when the owner approves the schedule.[79]

Most recently in the 1990 decision of the Board of Contract Appeals, *G. Bliudzius Contractors, Inc.*,[80] the Board considered the question of the contractor's entitlement to compensation for delays alleged to have been encountered in the construction of temporary lodging facilities at a naval training center arising from the timing of the approval of the contractor's CQC (contractor Quality Control Plan). In finding that the government was late in providing the necessary approval, the Board looked to the exact time for approvals detailed in the contractor's CPM plan. The Board stated:

> We have found that a period of ten working days for approval of the CQC plan was allowed in appellant's revised CPM network diagram. We conclude that conditional approval of the CQC plan should have been given within that period so that–as we have found–the construction would have started on 20 August.

Contractual modifications to the performance schedule may also create presumptions. An agreement by the government to extend the

---

78 VABCA No. 2168, 87-3 BCA ¶ 20,104.

79 *Fullerton Constr. Co.,* ASBCA No. 12,275, 69-2 BCA ¶ 7876.

80 ASBCA No. 37,707, 90-2 BCA ¶ 22,835.

time of performance for changes to the work creates a rebuttal presumption that the government was responsible for the delay and that the time granted is compensable.[81] In *Elrich Construction Co.*,[82] the Board considered whether appellant was entitled to an equitable adjustment in the contract price to cover indirect costs arising out of government ordered changes. The government argued that appellant could not rely on the 201 days in time extensions granted by modifications for changes as proper proof of compensable delay to the project. The Board rejected the government's argument:

> Government's joinder in contract amendments numbered P00013, P00014, P00035, P00039, and P00045 amounts to a recognition that the overall project performance was delayed in excess of 201 days above and beyond the original contract completion date. Agreeing to extend the time for performance of the contract gives rise to a presumption, subject to rebuttal, that Government was responsible for the delay. *Robert McMullan & Son, Inc.*, ASBCA No. 19023, 76-1 BCA ¶ 11,728.[83]

12. Limitation Of Recoverable Time In CPM Claims Presentations

If multiple owner delays affect independent sequences of work during the same period, the contractor may only recover for the actual delay to the critical path. Accordingly, a contractor may not "double-dip" by aggregating delays to independent activity chains such that the cumulative delay exceed the period of time in question (absent evidence of compression or acceleration to show extra time gained during the

---

81 *See Shuster Engineering, Inc.*, ASBCA Nos. 28,760, 29,306, 30,683, 87-3 BCA ¶ 20,105; *Elrich Construction Co.*, ASBCA No. 29,547, 87-1 BCA ¶ 19,600; *Robert McMullan & Son, Inc.*, ASBCA No. 19,023, 76-1 BCA ¶ 11,728; *Singleton Sheet Metal Works, Inc.*, ASBCA No. 12,402, 69-1 BCA ¶ 7444.

82 ASBCA No. 29,547, 87-1 BCA ¶ 19,600.

83 *Id.*; *see also DANAC, Inc.*, ASBCA No. 33,394, 97-2 BCA ¶ 29,184 (1997); *Bechtel Environmental, Inc.*, ENG BCA No. 6137, et al., 97-1 BCA ¶ 28,640, *recon. denied*, 97-1 BCA ¶ 28,851 (1997), for decisions confirming the proposition of modification time extensions establishing rebuttable presumptions.

period). Thus, compensable delay is limited to the ultimate delay as reflected by the critical path. It is also limited to the total amount of time in the period under consideration.

These principles are fully recognized in two recent decisions, *Ealahan Electric Company, Inc.*,[84] and *Fletcher & Sons, Inc.*[85] In *Ealahan*, the board was faced with the issue of certain aspects of the contractor's claim which it found to have merit. With respect to these items, the board was confronted with the issue of overlapping delays to the critical path. In finding that the contractor's recovery for critical path delays was limited to the time period in question, absent evidence of acceleration, the board stated, "where two or more changes are performed during the same time, a contractor is only entitled to a contract extension for the calendar time taken in performing the changes, absent evidence of acceleration."[86]

A similar result was reached in *Fletcher & Sons, Inc.*. There the Board considered a claim of multiple delays to independent sequence of work during the same time frames. Time extensions had been granted previously for certain of the delays during the concerned period. In rejecting a claim for any further delays to critical path works for the period in question, the Board concluded that the contractor's recovery for delay was limited to the total number of days in the period under consideration: "The fact that multiple delays may have been occurring during that time period does not justify compensation for a greater period than that time span."[87]

13. Ability Of Contractor Or Owner To Negate The Effects Of Delays To The Critical Path Through Risk Taking

The dynamic nature of the CPM process described previously provides significant opportunities for the parties in addressing project delays as they occur during performance. The contractor and the owner are able to react to changing events to modify logic to attempt to bring the project in on time and on budget. In fact, the parties can remove

---

84 Docket No. 1959, 1990 DOT BCA Lexis 31 (1990).

85 VABCA No. 2502, 88-2 BCA ¶ 2677.

86 *Id.*

87 *Id.*

restraints by waiving various approvals to allow construction to proceed in the absence of contractually specified approvals.

Most significantly, the opportunity is present for either the contractor or the owner under the right circumstances to run certain risks which can negate the effect of any delay which might otherwise be the parties responsibility. Such a circumstance was recently confronted by the Board of Contract Appeals in *H & S Corp.*[88] In *H & S*, the contractor and the owner negotiated direct costs for a change order directed by the owner to certain earth work but were unable to agree on the overhead costs of 118 day delay associated with the earth work change. The owner defended the case on the basis that the delay caused by the change was nullified by the time the contractor took to resolve a contractor building load error which related to the footings and foundation.

However, the Board found that the contractor proceeded with the footing construction regardless of the problem with the building load error and that no work was actually delayed on the project. The Board stated:

> Turning first to the concurrent delay issue, we agree with the Government that a contractor-caused delay affecting the construction of the footings would have a significant effect on this contract's critical path. The overwhelming flaw in the Government's concurrent delay argument, however, resides in the premise that appellant's failure to gain approval of the building system caused the delay.
>
> The Government initially chose to return submittal 62 rather than review it (finding 20). When the Government later returned the submittal package, including submittal 80, the Government directed appellant to redesign the foundation to accommodate the apparently heavier building loads (finding 23). However, appellant's earth work subcontractor performed earth work throughout the duration of the alleged concurrent delay and completed the earth work on 30 August (findings 11-13), appellant did not wait for the 7 October approval before proceeding with the footings but started footing work on 9 September (findings 12,17), and the footings were finally poured in the manner and size appellant proposed in the submittal (finding 25). Further, the building load mistake resulted only in confusion--no work

---

88 ASBCA No. 29,688, 89-3 BCA ¶ 22,209.

was delayed or changed (finding 27). Finally, nothing in the contract required Government approval of the building system before pouring the footings.

The Government cannot, from these facts, construct a scenario in which the very real delay caused by the unsuitable soil and additional earth work is nullified by the time it to resolve the building load error. This failure proves fatal to the Government's defense. 'The defense of concurrent delay is valid only when applied to an actually established delay, not merely an alleged delay.' *Essential Construction Company, Inc. and Himount Constructors, Ltd., A Joint Venture*, ASBCA No. 18706, 83-2 BCA ¶ 16,906 at 84,114.

The significance of the *H & S* decision is that it reminds us that the contractor or the owner can frequently negate the effect of delays to the project or delays that would otherwise be charged to that party's responsibility. This can be done by being willing to take the risk of performing certain work or taking certain actions prior to formal approval so that no delay is actually incurred on the project. Of course, in taking such steps, the contractor or owner risks further damage in that if the unapproved installation proves wanting and has to be torn out (or the risk taking step withdrawn), more delay will be charged to that party and charged to the project than would have otherwise been encountered.

14. Necessity To Clearly Delineate All Features Of Plan For Performance In Initial CPM

Four recent decisions confirm the proposition that the contractor must make every effort to delineate all significant features of its plan for performance with respect to resources, durations and logic at the inception of the project. Further, the contractor's intent should be communicated clearly to the owner and its agents.[89]

In *Volpe-Head*, the Board of Contract Appeals confronted a claim by a contractor who asserted that it had planned to construct a station

---

89 *Volpe-Head*, ENG BCA No. 4726 89-3 BCA ¶ 22,105, *VEC, Inc.*, ASBCA No. 35,988, 1990 ASBCA Lexis 323, *LaDuke Constr. & Krumdieck, Inc.*, AGBCA No. 83-177-1, 90-1 BCA ¶ 22,302; *Sierra Blanca, Inc.*, ASBCA Nos. 30,943, *et al.*, 91-2 BCA ¶ 23,990.

entrance, passageway, and center service area for a Washington Metropolitan Area Transportation Authority subway station concurrently with the construction of the main station itself. The owner defended on the basis that the contractor failed to establish that its original plan had provided for concurrent construction and that the delays encountered on the project did not in any way impact the contractor since the contractor had always planned to perform the construction in consecutive stages. In holding for the owner, the Board decision recognized both the burden of proof on the part of the contractor to clearly establish its plan as well as the failure of the CPM plan itself to provide for the concurrent construction staging. The decision stated in pertinent part:

> Respondent maintains that Appellant has failed in its burden of proof. Respondent also put on a case that shows the planning and contemporaneous documentation by Appellant was in part consistent with either concurrent or consecutive excavation and, in other respects, was more consistent with a plan to do the project in consecutive stages. Respondent also offered evidence that Appellant caused a concurrent delay in the CSA&P.
>
> * * *
>
> Appellant supports its argument for concurrent construction planning by contentions that the CPM schedule, support of excavation shop drawings, evidence respecting material deliveries and the minutes of progress meetings all support its case. Analysis of Appellant's contentions, the supporting citations to the record and the record as a whole does not convince us that a plan for concurrent construction has been proved. Appellant has not carried its burden. Rather, the preponderance of the evidence is more consistent with a plan for consecutive execution of the project. (Findings 37, 45-46).

The *Vec, Inc.* Decision concerned the construction of a covered storage addition to an existing storage facility. In this case, the contractor asserted that it had planned to finish well in advance of the contract specified performance time and supported its testimony with a CPM chart prepared the day before the hearing. The Board in rejecting the contractor's position clearly noted its rejection of the hypothetical after-the-fact CPM which was not even consistent with the schedule of progress payments submitted to the owner during construction. The Board decision stated:

> We have held that for time extension and liquidated damages purposes and for recovery under a SUSPENSION OF WORK clause, a contractor is entitled to the benefit or "cushion" of extra time that would result from planned early completion of the work. *G & S Construction, Inc.*, ASBCA No. 28677, 86-1 BCA ¶ 18,740; *C. N. Flagg & Co., Inc.*, ASBCA Nos. 2644, 26655, 84-1 BCA ¶ 17,120; *Sydney Construction Co., Inc.*, ASBCA No. 21377, 77-2 BCA ¶ 12,719. However, the factual bases underlying these holdings were that the contractors had established their intent to perform the contract on an accelerated schedule in advance of the contractually mandated completion date and that this intent was supported by the course of the contractors' actions and performance activities during the contract performance that would have led to such an early completion absent Government caused delays.
>
> We are not persuaded that appellant did establish such an intent here and that its actual course of actions and performance supported such an intent. Except for the critical path chart appellant prepared the day before the hearing and appellant's president's testimony, there was no persuasive evidence that appellant, at the time of award, had a firm commitment or intent to complete the project within 120 days or that it communicated that intent to the Government. There were no charts or workpapers which may have been prepared prior to or immediately after award that would indicate the planning of an early completion but for Government caused or adverse weather delays. Rather, the testimony of appellant's president and the critical path chart he prepared the day before the hearing, in our opinion, represent after-the-fact projections and hypotheticals only. Moreover, its proposed schedule submitted to the Government for progress payment purposes, its schedule of material submittals, and its course of performance are all consistent with a performance schedule of 180 days.

The *LaDuke* decision not only confirmed the necessity that the contractor must clearly display its plan for performance, but also that the logic sequences shown in the CPM, and particularly the critical path, represent a good plan for performance. Specifically, the Board found that the contractor had to show that the plan represented a logical prosecution of the work. The Board decision stated:

> It is not enough for Appellant's principals or their equipment operator to say that the "critical path" ran through the drain field and comfort station. Appellant must additionally show that a logical prosecution of the work required that the drain field and comfort station be completed before any other part of the project could advance. *See Franklin L. Haney v. United States*, 230 Ct. Cl. 148, 167-168; 676 F.2d 584 (1982).

A contractor relying on as-planned schedule should also show it was reasonable and valid. In *Sierra Blanca, Inc.*,[90] the Board found that appellant failed to show that the as-planned schedule was a valid schedule for performance. "In order to use a schedule showing early completion as a basis for proving delays, appellant was obliged to show that it was 'feasible and attainable.'"[91] The Board also determined that there was no credible evidence regarding the extent of delays to the contract resulting from subsurface rock conditions. Also, it is well established that agreement to a new completion schedule precludes subsequent claims for time extension and delay costs based on causes that existed prior to the date of that agreement.

15. Fragnets/Windows/Time Impact Analyses

Fragnets, windows, and time impact analyses are techniques for looking at individual segments of CPM networks. Fragnets are subnetworks, used to break one or more activity shown on a CPM diagram into a finer level of detail to develop the individual subactivities necessary for completion of the activity shown on the critical path. For example, the description of an activity on a critical path network for mechanical rough-in or structural steel erection may be made up of a number of different activities.

Fragnets also show added activities (necessary as a result of additional work) or logic revisions (required by delays or a determination that the original plan for performance is not workable). Since the 1960s, fragnets have been recognized as worthwhile analytical devices to compare portions of CPM schedules as planned with the

---

90 ASBCA Nos. 30,943, *et al.*, 91-2 BCA ¶ 23,990.

91 *Id.*

actual events occurring in the field as they are impacted by changes, delays, contractor inefficiencies, or other causes.[92] Fragnets prove useful for claim presentations and time impact analyses because they present a visual picture of the effect of individual delays on isolated activities. These presentations first show the isolated impact ,then use other visual symbols to show places on the as-planned and as-built schedules where the entire network has been impacted by the delays reflected in the fragnet. The 1979 *Corps of Engineers Modification Impact Evaluation Guide*, EP 415-1-3, at 3-8 through 3-19, provides excellent examples of time impact analyses utilizing fragnet presentations.

Window analyses operate in the same manner as the schedule update procedures required by the Corps of Engineers and Veterans Administration (VA) scheduling specifications. Like the VA and Corps procedures, window presentations focus on the effects of delays on specific periods by looking at the gains or losses to the critical path as they occur within each update period.[93]

In *Cogefar-Impresit U.S.A., Inc.*,[94] Cogefar, the contractor, was awarded a contract for construction of the second phase of a detention center in Miami, Florida by the Federal Bureau of Prisons of the Department of Justice (FBOP). The contract required Cogefar to submit Time Impact Analysis when delays were experienced, change orders were initiated, or if Cogefar wanted to revise the project schedule. The contract further stated that if a Time Impact Analysis was not submitted, an extension of time would not be required.

The FBOP terminated Cogefar's right to continue performance for default on the basis of the failure to meet scheduling requirements. The board, however, disagreed with the termination, stating that "[n]o credible evidence was presented showing any impact to project performance due to any failure by Cogefar to meet the contract's scheduling requirements." Furthermore, the FBOP did not terminate the contract until nine months after Cogefar failed to submit a Time Impact

---

92 *See, e.g., NASA PERT and Companion Cost Guide* (October 30, 1962).

93 *See Nielsen and Galloway, Proof Development for Construction Litigation*, 7 AM. J. TRIAL ADV. 433, 444-48 (1984).

94 DOTBCA No. 2721, 97-2 BCA ¶ 29,188 (1997).

Analysis as required by the contract. Accordingly, the FBOP "waived any right to terminate the contract . . . for this failure."

On these matters the board held that "Cogefar attempted to comply with the scheduling provisions" and "the FBOP was largely responsible for the delays during performance." Although Cogefar's Time Impact Analysis did not satisfy the FBOP or meet "the letter of the contract," Cogefar's Time Impact Analysis would not "provide a basis for termiantion" six months later, when FBOP actually terminated Cogefar. Accordingly, the Board found "no basis to sustain the termination for default due to non-compliance by Cogefar . . ."

Cogefar did submit two Time Impact Analyses in March of 1993. The FBOP found these to be insufficient and rejected them. The board found evidence that FBOP failed, however, to respond to Cogefar's request to meet with them to clarify the basis of FBOP's rejection. On these matters the board held that: "Cogefar attempted to comply with the scheduling provisions" and "the FBOP was largely responsible for the delays during performance." Although Cogefar's Time Impact Analyses did not satisfy the FBOP or meet the letter of the contract," Cogefar's Time Impact Analyses would not "provide a basis for termination" six months later, when the FBOP actually terminated Cogefar.

Accordingly, the Board found "no basis to sustain the termination for default due to non-compliance by Cogefar with the contract's scheduling requirements."

The board did comment on FBOP's attempt to use a window type analysis to evaluate delays on the project. This was rejected, however, due to the failure of the analysis to reflect actual conditions on the project in the update windows:

> Having set forth these several areas that caused delay to Cogefar on the project, we need to determine the amount of and responsibility for the delay caused to the overall project.
>
> Dr. Ma, respondent's expert, used a contemporaneous time frame analysis to evaluate delay at the time of the event based on two successive schedule updates. Dr. Ma used Cogefar's schedule to determine the critical path. The contract required that the critical path be determined from the most recent approved schedule update. (Tr. 2229) Specification 012211.E) However, Dr. Ma did not use the logic

changes made by Cogefar subsequent to the September 11, 1992 schedule because of the FBOP's failure to approve the changes. A use of a different logic can result in a different critical path. Dr. Ma determined that Cogefar caused 181 of the 301 days of contract delay. The 120 days of time extension granted by the FBOP were correct according to Dr. Ma's analysis. Dr. Ma's method showed the critical path in the concrete structure for the first 49 days of performance, into the precast panel for 63 days, and then to the electronic security items for 87 days. Dr. Ma's analysis did not place the structural failure on the critical path at any time. However, Dr. Ma found that this event delayed Cogefar 56 days, since the FBOP had granted Cogefar a time extension for that length of time.

To evaluate a project A CPM schedule must be current. *Fortec Constructors v. United States* [32 CCF § 73,702], 8 Cl. Ct. 490 (1985), *aff'd*, 804 F.2d 141 (Fed. Cir. 1986). The CPM schedule must reflect actual performance to be a reliable basis for evaluating delay. *J.A. Jones Construction Co.*, 72-1 BCA § 9261 (ENG BCA 1972).

While we find both of the expert analyses heavily slanted toward the position of the party that employed that expert, we find the analysis of Mr. Hutchinson for Cogefar to be a more reliable indication of the delay caused to the project as a result of the various events that occurred. Mr. Hutchinson's more thorough review looked at the actual events to plot the critical path. Dr. Ma, relying on Cogefar's schedule updates, which had not been changed to reflect the actual logic on the job, would seem to yield a less precise result. This is borne out by the fact that Dr. Ma's analysis did not place the structural failure on the critical path at any time. Both parties admit that this event's impact to the critical path was significant. Dr. Ma failed to use a current CPM schedule to evaluate the delay on the project. This is mandatory to achieve an accurate assessment of whether an item of delay affects the overall completion of the project. In addition, we find that the evidence is overwhelming that masonry was on the critical path at the time of the significant delays and disruptions to masonry.

16. An Industry Crisis: Abuses In Microcomputer Programs

The explosion in the use of microcomputers on project sites along with powerful microcomputer programming, such as Primavera or Aldergraph have presented great benefits for the construction community. However, such programs present the opportunity for significant abuses because of the opportunity: to not properly update the project; to record the actual completion dates; the opportunity to override the project logic; and to remove delays to the critical paths on monthly updates without modifying the actual logic of the diagram.

The Primavera model details a simple project for the excavation of certain footings then forming the footings along with the installation of the rebar and placement of the concrete in the footings and finally setting steel on top of the footings. As detailed on the plan for performance on the activity description and activity diagram, the project was to be performed in a sequential manner with the excavation of footings as the first activity; the second activity was for form, rebar, and pour footings; and finally the last activity was to set the steel. The duration of the first activity was five days. The duration of the second activity was ten days and the final activity five days again. The project was to take from January 1 until January 28 with consideration for weekends.

However, with our first update, we can see that the project, as is frequently the case, has not proceeded per the plan. Our update which is on January 4, just three days after the commencement of the activity, reflects a minor sequence change with an early start for forming of certain of the footings, one of which was discovered after two days of excavation that not all of the excavation would have to be completed for all of the footings prior to starting certain of the smaller footings on the western third of the project site. The update, therefore, reflects a change in the logic with an early start to the forming of the footing. This, in turn, would have normally reflected an early completion in construction prior to the January 28 date originally anticipated on the original plan. However, on the third day of the project, on the date of our update, a delay has been encountered on the project caused by footing redesign for the larger footings on the eastern two thirds of the project site. The updated chart, therefore, shows an extended duration of an additional four work days to February 1 resulting from the seven work day delay occasioned by the footing redesign.

Most interesting, however, is our third progress chart which is an update for the project also on 4 January at the inception of the delay caused by the footing redesign. This update, however, does not reflect a delay to the project completion. The reason for the failure to disclose the interdiction of the forming of the footings over the eastern two thirds of the site is the fact that the logic of the diagram has been overridden and that the owner's expert in this case has chosen to assert without modification to the logic in the network that excavation was not longer a restraint to the remaining footings and that the forming of the footings could take place at the same time as the excavation. This, of course, represents a physical impossibility since the excavation was a condition precedent.

The purpose of this illustration is to detail exactly how careful parties must be in the use of the extremely valuable CPM tool. If we look at the last activity report and diagram, the material appears to be very professional and appropriate. However, unless the parties and the professionals assisting them are vigilant, no one may pick up the fact that a progress override has been utilized on updates to avoid the effect of delays. For this reason, the parties must always check the basic precepts upon which CPM plans, updates, and analyses are prepared.[95]

17. Requirement That Scheduling Reflect Resource Leveling

The fact that contractors are dealing with the real world with real limitations as to personnel and resources is a key component of CPM scheduling. In fact reasonable resource leveling is a necessary component for a CPM to be used as a project planning tool or to evaluate delays.

Two decisions illustrate this point. This first decision, *Utley-James, Inc.*,[96] concerned a dispute arising out of the McNamara Federal Office building, and provided an essay on CPM scheduling that included consideration of the important concept of resource leveling:

> Another concept of importance to CPM scheduling is resource leveling, also called resource loading. Consider, for example, the interior finish work in an office tower like the McNamara Building. The placing of each floor slab is logically restrained by the placing of the slab beneath it, but that is not necessarily true of interior finish work. There is no

---

95 *See also Bramble & Callahan, Construction Delay Claims*, § 11.12 (2d ed.).

96 GSBCA No. 5370, 85-1 BCA ¶ 17,816 (1984).

> reason in theory why a carpeting contractor could not come in and carpet every floor of the building on the same day. But there are a variety of reasons why that is not done, and one of the principal reasons is that it makes no sense to supply that much material and that many workers to do all the floors at once when the work can be sequenced with all sorts of other work and handled over a convenient period of time with a suitably sized work force. To give an accurate indication of the actual planned job schedule, a CPM schedule must take resource leveling into account. Otherwise it could, for example, depict the same early finish date for all carpet work and the same early start date for every activity directly restrained by the carpet work, even though the carpet work itself is actually going to be performed in a phased sequence, floor by floor.

The second decision, *Neal & Co., Inc. v. United States*,[97] by the Court of Federal Claims demanded that CPMs include resource leveling constraints as an element of the scheduling process. In addition, the court found that a previously approved CPM should not be used as a basis for evaluating delay where necessary crew constraints and resource allocation were not included in the approved CPM.

This case concerned a United States Coast Guard project begun in 1987. The project, the the Lake Louise Housing Project, called for the construction of a subdivision consisting of thirty individual units on a hill overlooking Lake Louise in Kodiak Island, Alaska. NCI was the lowest bidder on the contract which was scheduled for initial completion of August 3, 1989. The contract required the contractor to submit for approval a Network Analysis Schedule (NAS) utilizing the critical path method (CPM).

In order to illustrate proof of delay, NCI referred to the critical path in the NAS it originally submitted. However, the court found that use of that NAS is improper because it did not properly project crew constraints or resource allocations needed for a realistic project scheduling tool. The court found that this failure to build in crew restraints and resource allocation created an unrealistic amount of "float," the amount of time an activity can be delayed without affecting completion of the project. In the words of the court:

---

97 36 Fed. Cl. 600 (1996).

The NAS was corrected by NCI's scheduling consultant in order to add crew constraints, but even the use of a six day week reflected a time overrun, so, as Adkins explained, this discrepancy was avoided, he felt illegitimately, by showing some of the tasks as requiring less time. Adkins viewed these reductions as improper, and the court agrees. In any event circumstances did not bear out the contractor's optimism. The NAS schedule thus never became a useful planning tool for the contractor. Because it lacked crew constrains, it lacked realistic late start and end dates. Because of the unrealistic late start and end dates, an unrealistic amount of float appeared for a too-limited number of critical-path items. Framing work, for example, that the contractor in fact planned to do immediately after foundation work, was not required, under the approved plan, to be finished until very late in the project. When the contractor later furnished crew constraints at the CG's request, they were not tied to the original NAS. In short, NCI's complaint that the CG did not act in a timely way to respond to requests for clarification in order to update the NAS falls flat. NCI never had a useable schedule to update.

18. Owner Issued Schedule: Performance Or Design Specification

One recent state court decision came down hard on the side of owners in rejecting owner responsibility for owner furnished schedules. This occurred even though the owner provided specific sequencing and scheduling information to the contractor in the bid documents. In *Willamette Crushing Co. (d/b/a Wildish Southwest Construction. Co.) v. State of Arizona*,[98] the Arizona Department of Transportation invited bids on a construction project which included a Traffic Control Pattern (TCP). This TCP specified the stages of construction. The bid invite specified completion of the project within 360 working days.

Wildish submitted a bid of $22.1 million which was $2.6 to $3.1 million lower than the State's cost estimate and the bids of two of its competitors. After being awarded the contract, Wildish performed a critical path method (CPM) analysis using the State's TCP. Because of the time-consuming and expensive nature of CPMs they are not normally completed before a contract is awarded. Based on the CPM analysis, it was calculated by Wildish that 542 working days was needed

98 932 P.2d 1350 (Ariz. Ct. App. 1997).

to complete the project using the State's TCP. Wildish asked for an extension which was refused by the State. However, the State did approve a TCP created by Wildish.

Using its own TCP, Wildish completed the project in 357 days, but at a cost which exceeded its bid by $2.9 million. Wildish blamed the cost overrun on inefficiencies inherent in the TCP it had to devise to meet the State's due date and filed a claim with the State for the cost overrun. The claim was rejected and Wildish filed this lawsuit in which the State was granted summary judgment. Wildish appealed.

The contract specified that Wildish would be entitled to price adjustments if the State expressly or impliedly ordered modifications which "significantly changed the character of the work." Wildish claimed that the State's TCP had a hidden defect because only after being awarded the contract and performing the CPM did Wildish discover that it could not meet the due date using the State's TCP. The court concluded that by assuming that the State's TCP sequences were appropriate for the job and if followed would allow the work to be completed within the time required, Wildish bore the risk of any deficiencies in the State's TCP.

Wildish's argument relied on a line of cases beginning with *United States v. Spearin*,[99] which held that when a government specifies how the work is to be done, it impliedly warrants that adherence to the specifications will result in 'satisfactory completion of the work.' The court distinguished the Spearin line of cases by pointing out that they apply only to design and not performance specifications. The court also stating that design specifications precisely detail the manner in which the work is to be done, while performance specification set forth an objective and allow the contractor to determine how to achieve it.

The court held that Wildish could not possibly prove that the State's TCP was a breach of the warranty implied in design specifications for three reasons: (1) even though using the State's TCP would have taken more time, Wildish could have constructed the project using the State's TCP; (2) it is undisputed that Wildish was permitted to deviate from the State's TCP and use its own; and (3) the Spearin rule requires something akin to a hidden physical defect in the design specifications which the State's TCP did not have.

---

99 248 U.S. 132 (1918).

The court stated that Wildish was confined by the contract due date, but that date is not a design specification but a performance specification. Furthermore, the court followed the reasoning in *American Ship Building v. United States*,[100] that as a matter of law the due date in a public contract is not a warranty by the government, but a warranty by the bidding contractor that if they are awarded the contract, they can do the work in a specified time. But that as a matter of fact, Wildish could and did perform on time, but without profit at the price it bid. For these reasons, summary judgment was affirmed.

The court also denied Wildish's claim that the contract was commercially "impossible or impracticable."

III.
Scheduling Specifications

Few people realize the contribution well drafted contract provisions play in achieving the end goal of a successful construction project. Even fewer people realize the significance of scheduling specifications to a successful project.

Well drafted scheduling specifications are critical to the ability of the parties to achieve their joint goals of timely and economical construction. Further, well drafted scheduling specifications can meaningfully address necessary elements of the scheduling process. These include forcing the parties to: (1) require the general contractor (including its construction project manager or superintendent) to commit the necessary skills to the scheduling process; (2) develop a plan for executing the construction and a network schedule that implements the basic scheme or plan for construction, with predecessor and successor relationships (resource loaded with equipment and manpower); and (3) force a commitment of the parties to an approved schedule for executing the work, measuring the progress of performance, and grappling with necessary issues during construction.

---

100 654 F.2d 75 (Ct. Cl. 1981).

A. Necessary Characteristics Of Scheduling Specifications

CPM scheduling specifications should be used to address two basic types of projects. The first is smaller projects (approximately $10 million or less) where a ten thousand activity network is neither desired nor practical. On these projects, network scheduling will be of help in coordinating the activities of the various trades, and a network of several hundred to a thousand activities will more than suffice. The second type is larger projects of up to several hundred million dollars. Here, detailed network coordination of the general contractor and the various trades is an absolute necessity. In this type of detailed scheduling effort, a larger network is typically required with more activities, along with a summary network of fewer than a hundred activities, typically time scaled, for executive level review.

In the case of either the larger or smaller project, the CPM scheduling specification should address certain key attributes of the scheduling process. These should include the feasibility of the schedule specified. If the schedule specified does not present a feasible and pragmatic tool by which the parties can plan the project and measure performance, the schedule is essentially useless.[101]

Schedule specifications also must clearly detail the specific type of diagram desired and the desired network planning technique (*e.g.*, CPM or precedence-activity on arrow or activity on node). In one case, a specification which stated that the "schedule shall show the order in which the contractor proposes to carry on the work, the dates on which he will start the several salient features (including procurement of materials, plant and equipment), and the contemplated days for completing same" was held to only require a bar chart.[102]

The CPM scheduling specification should also address the number of activities. The size of the project should play a role in the number of activities specified for the diagram. Current techniques to specify the desired level of indenture or detail for activity breakdowns include listing a specific number of activities, limiting the duration of activities to a specific period of time (typically no more than fifteen or thirty days

---

101 *Chaney & James Constr. Co.*, FAACAP No. 67-18, 66-2 BCA ¶ 6066 (1967).

102 *H.I. Homa Co.*, ENGBCA No. PCC-41, 82-1 BCA ¶ 15,651 (1982).

apart from long lead item procurement items), and limiting the activity to specific dollar limits.

Specifications should establish whether formal approval is required and identify the party to provide the approval. Some owners believe that approvals should not be provided. However, absent an approved schedule, there is no agreed upon baseline for evaluating delays. Further, silence by the owner in response to a schedule which has been submitted may well be interpreted as assent to the content of the schedule.[103]

A specification which merely states that it will be updated at appropriate intervals totally fails to meet the needs of the parties and the project. Specifications should clearly call out the periodic frames of time where progress will be recorded, and a plan projected for the remainder of the project. Remember that the purpose of the update is to provide answers to three questions: Where have we been? Where are we now? Where are we going?

Specifications should be clear as to whether the activities in the network are to be cost loaded for payment purposes, and also, whether the contractor will be paid from the cost loaded CPM. Care by owners should be taken to provide for an approval of the cost loaded CPM to avoid any problems of unbalanced loading of progress payments.

Another significant area that the scheduling specification should address relates to the reporting requirements for float and the provision for the use of float. In this area, the specification should address the desire of the parties for periodic reporting of the different types of float, including total (or project float), free float, and activity specific float. Further, the scheduling specification should make it clear that float is an expiring resource available to the parties on a non-discriminatory basis (so long as the parties act in good faith towards each other). This concept which inherently recognizes the dynamic nature of the CPM process (*e.g.*, that things that were not formerly on the critical path may later become critical as events evolve during the project due to the expiration of float).[104]

---

103 *Fullerton Constr. Co.*, ASBCA No. 12,275, 69-2 BCA ¶ 7876 (1969).

104 *Weaver-Bailey Contractors, Inc. v. United States*, 19 Cl. Ct. 474 (1990); *see also Jon Wickwire, et al., Construction Scheduling: Preparation, Liability and Claims*, § 9.16 (1991).

Special problems may arise in the context of float provisions in scheduling specifications where the specification indicates that the contractor is not entitled to a time extension until all float has expired for the activity (or chain of activities). This may give rise to issues of contractor entitlement to time extensions, where the project is already in a negative slack or float condition, and the activity in question does not represent the highest negative float condition on the project for the time period in question (*e.g.*, on a particular update, an activity is in a -50 condition while activity B is in a -150 condition). Other special problems which may arise relate to special language included in the current Veterans Administration scheduling specification, which may effectively deny the contractor the right to damages when the owner delays it from finishing early, and also deny the contractor the benefit of "buy back" or additional float time it has created in the schedule. This issue relates to the portion of the Veterans Administration specification that requires the contractor to show that the delay in question has impacted both "the extended and predicted completion dates shown by the critical path in the network."

Schedule specifications must address the circumstance when major revisions are required (by major project delays) or desired in the CPM network. Further, the specification should identify the party responsible for making the revisions, and whether approval is required for certain types of revisions.

In addition, the CPM specifications must be closely coordinated with time extension provisions, and address the baseline and methodology for evaluating time extensions. For example: (1) will hindsight only be used for evaluating time extensions?; (2) will time be quantified by forward pricing, just like the forward pricing of an equitable adjustment for future change order work?; or (3) or will time be priced based upon an evaluation of its impact, based upon a contemporaneous evaluation of its impact, utilizing the current update (or updates) for the time period in question?

Remember that the reference point for measuring the merit and quantum of time extensions will often have an effect upon answering the questions of "whether" and "how much" time will be granted.[105]

---

105 *Blackhawk Heating & Plumbing Co.,* GSBCA No. 2432, 75-1, BCA ¶ 11,261; *Wickwire, et al., Use of Critical Path Method Techniques in*

B. Typical Areas of Conflict In Scheduling Specifications

A number of areas of conflict typically arise within the context of scheduling specifications. Some of these conflicts arise due to poorly drafted (or not well thought out) specifications. Other conflicts may be due to the inevitable clash of events. Still other conflicts may be due to bad faith conduct by a party.

1. The Reasonableness Of The Schedule

One typical conflict is the reasonableness of the schedule, *i.e.*, will the schedule actually be used to run the project, or merely represent the software exercise of a computer technician? An extremely serious problem exists when contractors lack commitment to utilize the scheduling process to plan and execute the work.

This problem might also be described as the problem of giving the job to the "low bidder." Specifically, since the project goes to the low bidder, some generals (and particularly the generals who may not perform significant work themselves) may attempt to just facially satisfy the scheduling requirements of the contract.

Such facial compliance may translate into a failure on the part of the general contractor to actually commit the financial and management resources required to acquire (or employ) the services of the experienced (or "master") scheduler. Rather, the general contractor may decide to engage the cheapest CPM computer processing concern for purposes of complying with the specification. The cost of experienced scheduling resources are much greater than that of mere facial or lip service compliance with scheduling specifications. However, the importance of the experienced scheduler working with the project team in developing a well planned project schedule cannot be overemphasized.

Another unfortunate model for dealing with the preparation of the network diagram is for the junior project engineer to be told to prepare a CPM schedule to comply with the specifications. The junior engineer may prepare the schedule with little input from the project manager or construction superintendent. During actual performance, the project is

---

Contract Claims: Issues and Developments, 1974 to 1988, 18 PUB. CONT. L.J. 338, 367-75 (1989).

run off of weekly or monthly "look ahead" schedules, with no real reference or relation to the overall game plan developed by the junior engineer. The problem with using "weekly or monthly look ahead schedules" that bear no relation to the overall plan for performance is that such look aheads may well represent constant recovery schedules that ultimately give rise to acceleration claims.

As noted by Ralph Johnson of Turner Construction, the commitment of the construction superintendent to the preparation of the network schedule is vital to the success of the project.

In viewing the various federal scheduling specifications, or other specifications, it is necessary for the bidder to see what kind of scheduling effort is requested. Has the owner specified a real commitment to the scheduling process? Will a bidder be at a competitive disadvantage if it includes in its price the kind of scheduling resources to properly plan and execute the work? Does the project or good management require a real commitment to the scheduling process?

2. *Approval Versus Nonapproval*

Owners who fail to provide for approval of the project schedule put both the owner and the contractor at a disadvantage. The owner may fear that it may somehow warrant the schedule by its approval, however, there is no real authority that exists to indicate that the schedule approval means that the owner has taken over liability for the success of construction means or methods, much less the viability of the schedule for the project.[106]

By contrast, demands by the owner to abandon the CPM schedule developed by the contractor for the project in favor of unlimited project acceleration have led to at least one holding of direct liability by the owner, even to subcontractors, as a concomitant of having assumed control over the construction means and methods.[107]

Most significantly, the failure of the owner to exercise (through the scheduling specification) the right to approve the schedule deprives the owner of the right to reject unreasonable plans for performance. It also

---

106 *See, e.g., Utley James v. United States*, 14 Cl. Ct. 804 (1988); *Titan Pacific v. United States*, 17 Cl. Ct. 630 (1989).

107 *Natkin & Co. v. George A. Fuller Co.*, 347 F. Supp. (W.D. Mo. 1972).

exposes the owner to potential claims for early completion (which may be based on defective or unrealistic logic) and denies both to the owner and the contractor a baseline from which to evaluate project problems and delays.

3. The Resolution Of The Approval Standoff

There is simply no excuse for the major construction project that is months or years into performance to lack an approved schedule by which the parties can evaluate performance.

The approved schedule provides a game plan for the project for all to see, and then provides the benchmark for the parties to measure the progress during performance.

From the time of being small children, we all know that things go "bump in the night" and that things do not turn out as planned - sometimes better and sometimes worse. However, with the approved CPM and the update, the contractor, in consultation with the owner, can best assess steps to mitigate damage from unanticipated problems.

Obviously, owners can include remedies in the specifications to force the contractor, by the power of the purse, to make desired revisions in the schedule (subject to contractor objections or reservations). However, the contractor may also frequently have a problem with a lack of action on the part of the owner. It is important, therefore, for the specification to set times for both submittal (and resubmittal) of the project schedule, as well as definite time frames by which the owner is required to act on the schedule submitted.

4. Failure To Require Initial And Continuing Involvement With Major Trade Contractors

One area that scheduling specifications frequently fail to address is the initial and continuing involvement of the major trade contractors in the scheduling process. Regardless of whether the specification requires such involvement, the general contractor significantly increases its ability to coordinate the subcontractors, as well as bind the subcontractors to the schedule, by involving the major trade contractors in the schedule preparation and updating process.

Where the specifications or the general contractor failed to provide for involvement by the major trade contractors in the scheduling process, the general contractor's ability to force the subcontractor to comply with the schedule may be limited. In one case, a court denied the general contractor's claim for delays on the project because it characterized completion dates, set without any consultation with the subcontractors, as "arbitrary."[108] However, where the general contractor, by contract, firmly binds the subcontractor to the CPM schedule, and consults with the subcontractors "to verify the timing and duration of sub-elements of the work," the general contractor has been allowed to recover from the subcontractor, even for damages prior to the contract completion date.[109]

Obviously, the involvement of the major trade contractors in the scheduling process is extremely important to the success of the project. Today, most of the labor risk on projects is associated with the performance of the major trade contractors. The days when the general contractor could sign up subcontractors and utilize scheduling language such as "follow the progress of the work" are long gone.

5. Defective Updating Procedures

The monthly update is one of the most important aspects of the CPM scheduling process. In fact, the update has three main purposes. First, the update will tell you where the project has been or, to put it another way, how you got to the point where you are currently at in the life of the project. Second, the monthly update will tell you where you are currently located in your journey from the beginning to the end of the life of the project. This is a snapshot of time reflecting the progress that has been accomplished. Third, and finally, the update should reflect the game plan as to how you intend to accomplish the remainder of the work.

The updating process, in practice has suffered from a number of maladies. These maladies, however, are entirely preventable. They include the failure of the owner to verify information contained in

---

108 *United States v. F. D. Rich Co.*, 439 F.2d 895, 900 (8th Cir. 1971).

109 *Williams Enter., Inc. v. Strait Mfg. & Welding, Inc.*, 728 F. Supp. 12 (D.D.C. 1990).

submitted updates, the failure of the general contractor to properly record actual completion dates, and the use by the general contractor of logic override (options in current software programs) without making appropriate logic changes in the network.

The problem created by the owner's failure to verify dates shown on the submitted update is frequently associated with lack of knowledge, or lack of commitment, on the part of the owner to the CPM schedule, but is easily correctable. Specifically, the best device for the parties to address this problem is the joint updating meeting, where both parties are required to review the information contained on the update that is proposed for the current period.

The problem created by the general contractor's failure to record actual completion dates sometimes occurs when using programs such as Prima Vera. This may be addressed by having the scheduling specification and owner representatives require the general contractor to actually record the completion dates.

Finally, the problem with respect to computer programs that allow the general contractor to override the logic in the planned CPM (*e.g.*, changing sequential activities to concurrent activities) without making appropriate logic changes, requires continuing attention by both the contractor and the owner to the CPM update, as well as, perhaps, the use of appropriate contract provisions, and logic tracking programs (which are currently available in the industry) to make sure when the general contractor only makes authorized revisions to the network.

6. *Failure To Specify Clearly A Definite Baseline And Methodology For Approval And Incorporation Of Logic Revisions*

Logic revisions may concern addition or deletion of activities, changes in sequencing of work, or durations of work. In this area, the failure of owners to specify clearly the requirement that logic revisions be submitted for approval can cause major conflicts or misunderstandings between the parties. As noted previously, the contractor, using some of the techniques and certain of the many computer programs which allow the overriding of the logic of the network without logic changes, may avoid showing a delay to the project by modifying otherwise valid sequential logic (of a line-up of predecessor and successor activities) to one of concurrent logic (with concurrent activities). This is frequently done without any notification to or knowledge on the part of the owner. For this reason, owners should require, for their own and the general contractor's benefit, that any significant logic revisions must either be submitted for approval, or brought to the attention of the owner, prior to incorporation of the revised logic into the network.

7. *Failure To Specify Clearly A Definite Baseline For Approval And Incorporation Of Request For Additional Time*

One of the most intense and unproductive areas of conflict in the area of scheduling specifications relates to the issue of project time and the failure to define the baseline and methodology for evaluating time extension requests on projects.

Specifically, the failure to provide a definite methodology for evaluating time extensions leads to unnecessary and completely unproductive gamesmanship on the part of the participants to the project. What is most tragic is that millions of dollars and a contractor's well being may rest in the balance based upon the decision as to the type of baseline and the methodology to be utilized in evaluating the time impact. In point of fact, almost all of the cases since the mid-1970's have espoused a dynamic view of the CPM project schedule, with an inherent recognition that time impact of delays should be evaluated concurrent with the events in question, viewing the project from beginning to end as it chronologically unfolds, and further utilizing the updating process as a point of reference. We can describe the approach as both chronological and cumulative.[110] This trend has continued into 1992, with decisions recognizing that the impact of the delay must be determined at the time of the notice to proceed on the modification.[111] However, at least one decision involving the Washington Metropolitan Area Transit Authority, issued by a divided panel, suggested that the owner may have the right to pick and choose whether it wished to use a

---

110 *See, e.g., Gulf Contracting, Inc.*, ASBCA Nos. 30,195, 32,839, 33,867, 89-2 BCA ¶ 21,812, *on recon.*, 90-2 BCA ¶ 22,393 (1989); *Hull-Hazard, Inc.*, ASBCA No. 34,645, 90-3 BCA ¶ 12,173 (1990).

111 *Batteast Constr. Co., Inc.*, ASBCA Nos. 35,818, 36,609, 92-1 BCA ¶ 24,697 (1991).

hindsight or contemporaneous view of evaluating delays.[112] Obviously, the owner chose the type of approach or methodology in evaluating delays that was most favorable to the owner's position with predictable harm to the contractor.

In any event, it is extremely important for the contract to specify clearly the precise methodology by which delays will be evaluated on the project. Projects where the parties argue over methodology of evaluating time extensions are one of the worst tragedies in the construction industry.

B. *The Issue Of The Early Completion Network: The Original Submitted Schedule, Or The Update*

The contractor's right to recovery, when it is denied.[113] In fact, federal authority has even recognized the contractor's right to recovery where the contractor finished early, but not as early as it had originally anticipated.[114]

In recent years, owners have become increasingly concerned about contractors' submission for approval of project schedules showing early completion. owners are not totally without the ability to deal with these circumstances effectively. First, the owner obviously has the right to reject a schedule that is submitted showing an early completion if the schedule is unreasonable or unrealistic.[115] Further, as will be discussed in more detail below, the Veterans Administration use of special language in its time adjustment provisions—that the actual delay must affect the "extended and predicted contract completion dates shown by the Critical Path and the network"—may well eliminate the contractor's right to finish early and also remove any incentive to develop buy-back time (time regained to ameliorate prior delays) to the project.[116]

---

112 *Harrison Western & Franke-Denys, Inc. (JV)*, ENG BCA Nos. 5556, 5576, 93-1 BCA ¶ 25,382 (1992).

113 *Maurice L. Bein, Inc. v. Housing Auth.*, 321 P.2d 753, 764 (Cal. 1958); *Metropolitan Paving v. United States*, 420 F.2d 241 (Ct. Cl. 1963).

114 *Green Builders, Inc.*, ASBCA No. 35,518, 88-2 BCA ¶ 20,734 (1988).

115 *Sierra Blanca, Inc.*, ASBCA No. 32,161, *et al.*, 90-2 BCA ¶ 22,846 (1990).

116 *Robglo, Inc.*, VABCA Nos. 2879, 2884, 91-1 BCA ¶ 23,357 (1990).

IV.
Significant Trends In CPM Analysis

Since 1974, a body of decisions illustrate trends in CPM analysis. Some of the more significant trends include the following concepts. First, CPMs which do not include reasonable resource leveling and appropriate logic constraints are unrealistic and do not constitute a basis for evaluating project delays. In addition, float is an expiring resource available to all parties on the project on a non-discriminatory basis who act in good faith. These decisions also indicate that CPMs are dynamic. Things never go exactly as planned and the critical path may change from month to month. When project delays occur which create additional float in the schedule, the contractor is not required to "hurry up and wait" and slavishly follow the original plan for performance. Delays are best evaluated on a chronological and cumulative basis taking into account the status (and critical path) of the project at the time of the delay in question. With this methodology and protocol, all parties on the project live with the events, actions and "sins" of the past. True concurrent critical path delays by an owner and contractor entitle both to additional time but no compensation for delay. Two concurrent delays which involve one delay by an owner to the critical path and the second by the contractor to an activity with float, entitle the contractor (or owner in a reverse scenario) to recovery for both delay damages as well as time extensions.

V.
Choices For Proof Of Delay Claims

There are a variety of potential choices for the use of schedules to analyze project delays.

A. Bar Chart

Bar chart analyses of delays are typically helpful on projects where we have a few activities with linear type relationship between those activities. This type analysis is of little help (without extensive supporting testimony) when we are confronted with modern construction projects with a variety of activities, with complex logic relationships (predecessor, successor, contemporaneous) and with varying resource requirements and availability.

B. "But For" Analysis / Collapsed As-Built CPM

This analysis calls for a protocol where we: (1) compare the reasonable planned schedule to the as-built schedule with all delays encountered on the project; (2) remove the delays of the other party from the as-built schedule; (3) review the collapsed as-built for anomalies representing "why hurry up and wait" type of activities (rescheduled due to prior delays) and make appropriate adjustments to remove the anomalies.

The advantage of this approach is that it is based on the actual events on the project; it allows us to the compare a reasonable plan for performance with the as-built. We can then identify variances between the plan and actual, look for causes of the variances, and finally determine what, if any, effects of the variances may be. We can also compare our "But For" collapsed as-built with the plan and actual to get an idea not only of delay effects to the critical path, but also to make judgments relative to issues of lost efficiency.

There are two primary disadvantages of the collapsed as built cpm. First, the after the fact approach fails to address the need to address the issue of time extensions on a real time basis as required to address events on the project. Second, adjustments for anomalies in the adjusted schedule require experienced judgement and may be subject to dispute by experienced experts for others on the project.

C. "Total Time" Analysis

Total Cost pricing which asks for the other party to pay for all losses on a project (actual costs plus markups less contract price) is subject to serious objections due to underlying assumptions. For example, the approach assumes that the contractor's original bid was reasonable, that the contractor performed as efficiently on the project as anticipated in its bid, and that all the causes of the increased costs on the project were the sole responsibility of the owner.

Likewise, a "Total Time" Analysis which compares the contractor's plan with the actual as built schedule and blames the owner for all the variances is subject to similar attack for the underlying assumptions. These include the assumption that the original plan was reasonable, that the contractor performed at least as efficiently as planned, and that all the variances between the plan and actual are the responsibility of the owner (not due to excusable or contractor delays).

D. "Impacted As Planned" CPM

This approach which purports to present a fair picture of responsibility for owner delays on the project by impacting the original cpm on the project solely with owner delays encountered during performance suffers from one fatal flaw. It ignores what actually happened on the project. This approach totally ignores excusable delays and delays by the contractor. Actual performance by all parties must be considered.

E. Chronological And Cumulative Approach/Time Impact Analysis

This approach was developed to address the needs of the construction process to recognize time adjustments due the contractor in a timely manner and to provide the ability to resolve disputes prior to some exhaustive after-the-fact analysis reconstruction after the completion of the project.

In this approach, we start with a reasonable as-planned CPM; next, we status the project prior to the onset of a delay which we wish to evaluate to identify the location of the critical path; the original as-planned schedule is then modified from that point forward, incorporating the time impact which may or may not cause a project extension based on the location of the critical path and its relation to the delay in question; when the next delay is evaluated, the project network (as revised by the prior analysis and taking into account actual events on the project) is again statused immediately prior to the advent of the delay to locate the critical path; the second delay is then introduced into the network to see what result it may have.

This approach has been widely accepted and has significant merit. However, care must be taken to make sure that during the period of the analysis of individual delays, other delays which may be the responsibility of the contractor or owner do not overtake the delay in question on the critical path. One way to avoid this problem is through the use of measuring points such as monthly updates.

F. "Window Analysis"

This type of analysis is a variant of the chronological and cumulative approach detailed above. In this approach, we look at the status of the project immediately prior to the advent of the delay, such as the monthly update of the CPM for September. We then introduce the actual events and delays into a window of time, typically an update as they occurred. In our example, we would introduce events and delays as they actually occurred in October and see which delays impacted the critical path as well as which delays may have represented concurrent delays to the critical path. Like the chronological and cumulative approaches, this is an extremely valuable tool.

VI.
How To Prepare For Direct And Cross-Examination Of The CPM Expert At Deposition And Trial

A. Ground Rules For All Experts

The ground rules for CPM or scheduling experts in preparing or defending scheduling claims include: the necessity for a fair and complete review of the relevant project data and the preparation of analyses that take into account major and controlling delays of all the parties on the project. The expert should not eaccept any engagement that requires him or her to ignore the aims of the expert's client. In addition, the ground rules for the CPM expert in preparing or defending scheduling claims include the necessity for clearly establishing a clear role and charter the expert as well as the requirement that the expert pursue four broad phases: (1) Familiarization Phase; (2) Investigation and Data Gathering Phase; (3) Fact Finding and Evaluation; and (4) Presentation of Report and Conclusions.

The Familiarization Phase includes a number of major tasks:

(1) Review of project requirements.

(2) Review of overall submission with legal counsel. In this task attention should be directed to:
   (a) Identification of the issues offered to support the claimed theories of recovery;
   (b) Factual presentations offered for each claim issue;
   (c) Elements of proof required to support each claim issue;
   (d) The methodology used in presenting any comparisons of planned and as-built or as-adjusted schedule analyses, and the specific procedures and techniques used by the claimant to isolate and quantify specific delays and delay to the overall project;
   (e) Any delays accepted by the claimant as her responsibility and how her delay is treated with regard to the alleged compensable delays;
   (f) The position taken on the availability of float;
   (g) Identification and treatment of concurrent and offsetting delay situations;
   (h) Time extensions requested, granted, pending, and denied (in total or partial);
   (i) Compliance with notice and claim documentation requirements of the contract;
   (j) Identification of individuals who are most knowledgeable about the claimant's performance and the alleged problems that were encountered;
   (k) Timeliness of change orders and contract modifications (issuance and responses) and, consequently, the potential delay to and impact on job progress;
   (l) Methods to prove or refute alleged inefficiency and related costs;
   (m) The type of delay damages being sought and the proofs offered;
   (n) The use of reservation of rights to impact and/or acceleration costs; and
   (o) Identification of potential contractual and performance defenses that could be relied upon.

(3) Meetings with key project personnel for further background information.

(4) Visits to the actual construction site (if practicable).

(5) Review of project records systems.
(6) Methodology for assessing overall claim.

The Investigation and Data Gathering Phase includes:

(1) Review of project records.
(2) Chronological listing of claim issues.
(3) Prioritizing fact-finding and analysis requirements.
(4) Development of delay/schedule issue files.
(5) Review of contract provisions relevant to the scheduling analysis such as:
    (a) Responsibility for schedule preparation and approval (noting any conflicts among responsible parties);
    (b) The quality of response expected;
    (c) The involvement of subcontractors in preparation of the schedule;
    (d) The approval status of the schedule (Does an approved schedule exist?);
    (e) The existence of owner protective clauses in the contract (Does a no damage for delay clause exist? Does a notice clause exist?);
    (f) Quality of schedule maintenance (Does the schedule reflect how the project was constructed? Do actual start and finish dates exist for each activity? Are there any broken logic sequences?);
    (g) Frequency of schedule updates and how they were conducted (Were they conducted jointly? Did subcontractors participate?);
    (h) The contract procedure for incorporating network changes (What does the contract say about changes and float?);
    (i) Whether requests for time extensions were made (If any were granted, on what basis? Were any time extension requests denied? Are any pending?); and
    (j) Reasons for abandoning the schedule (if an issue).
(6) Assistance in discovery process. Here assistance in the area of admissions, interrogatories and depositions may be directed to such issues as:

(a) Establishing the facts relating to preproject planning, preparation, and approval of the schedule; periodic updatings; actual performance; documentation; specific delays and corresponding costs; and time extension analysis;
(b) Identification of specific records the opposition will use to support or defend the facts and their position for both time and cost impact;
(c) Determining the thrust of expert testimony to be used in the areas of schedule impact and delay-related costs; and
(d) Identification of evidentiary tools (such as scheduling and delay analysis exhibits) to be used by the opposition during the dispute resolution process (including those already prepared and those that are anticipated for the trial or hearing).

The Fact Finding and Evaluation Phase includes such major activities as:

(1) Establishment, confirmation, and evaluation of as-planned schedule.
(2) Detailed analysis of project records.
(3) Identification of data comparing actual and planned performance.
(4) Establishment/confirmation/evaluation of as-built schedule. This process may include gathering of information such as:
    (a) Activity actual start;
    (b) Activity actual finish;
    (c) Delays encountered and caused by other parties and records of start and finish dates of each such delay;
    (d) Delays encountered and caused by the party responsible for documentation;
    (e) Specific actions or directions given to resolve problems or delays;
    (f) Correction or noncorrection of improper or defective work;
    (g) Non-productive time;
    (h) Lack of materials or lack of labor;
    (i) Lack of a manufacturer's representative;

(j) Changes in management or reorganization of phases;
(k) Inefficient work periods;
(l) Efficient work periods;
(m) Testing (finish) or rework;
(n) Modifications (work performed and resources required);
(o) Crew size (classification and number of workers);
(p) Equipment required, actually used, or idled;
(q) Weather conditions and acts of God;
(r) Strikes or other job actions;
(s) Unforeseeable site conditions;
(t) Suspensions of work;
(u) Periods of waiting for instructions for continuation of work efforts to mitigate delays encountered;
(v) Activities upon which a delayed activity is dependent;
(w) Activities that are dependent on a delayed activity's completion; and
(x) Changes in logic sequencing.

(5) Researching Claim Issues and development of Time Impact Analyses.
(6) Summary Analysis of time delays.
(7) Computation of final delay damages.
(8) Optional fact finding and evaluation phase activities may include evaluation of labor inefficiency and quantification of extended equipment costs.

The Presentation Phase for the expert normally includes:

(1) Preparation of overall time and cost impact report
(2) Preparation/confirmation/evaluation of summary as planned, as built and adjusted schedules
(3) Consideration of use of special graphics. These may include some of the following:
    (a) A series of charts presenting the details of major and controlling delays on an individual basis. These charts would be coordinated with the detailed and summary schedule presentations.

(b) A chart plotting and analyzing change orders. A similar presentation of requests for information (RFI's) and responses can also be helpful.
(c) A manpower chart showing a comparison of planned versus actual resources expended, including overtime and multiple shift hours. The expert also should attempt to show causation between planned and actual manhours.
(d) A chart showing a comparison of the promised and actual delivery dates for equipment and materials.
(e) A chart comparing concurrent delay existing on a series of paths and noting excusable and nonexcusable delay and compensable and noncompensable delay periods.
(f) A chart showing a correlation between problems encountered and operating results.
(g) A chart showing the effect of work performed out of sequence and its delay, impact and disruption.
(h) A model which can prove to be a highly persuasive means for demonstrating the facts of a case.

B. Ten Commandments the Expert Cannot Violate

(1) Thou Shall Not Rely Upon A Schedule That Was Not Followed During The Project.[117]
(2) Thou Shall Consider Actual Performance.[118]
(3) Thou Shall Avoid "As-Planned Plus Impacts" Analyses.[119]
(4) Thou Shall Establish A Correlation Between The Plan, Changes, Actual Performance And Contemporaneous

---

117 *See, e.g., Pathman Constr. Co.,* ASBCA No. 23,392, 85-2 BCA ¶ 18,096 (1985).

118 *See, e.g., Ealahan Elec. Co.,* DOTBCA No. 1959, 90-3 BCA ¶ 23,177 (1990).

119 *Titan Pac. Constr. Corp.,* ASBCA Nos. 24,148, 24,616, 26,692, 87-1 BCA ¶ 19,626 (1987), *summary judgment granted,* 17 Cl. Ct. 630 (1989), *aff'd,* 899 F.2d 1227 (Fed. Cir. 1990); *see also Freeman-Darling, Inc.,* GSBCA No. 7112, 89-2 BCA ¶ 21,882 (1989); *Ealahan Elec. Co.,* DOTBCA No. 1959, 90-3 BCA ¶ 23,177 (1990).

Records.[120]

(5) Thou Shall Consider Your Own Delays In A Delay Analysis.[121]
(6) Thou Shall Keep Schedules Current And Reflect Delays As They Occur.[122]
(7) Thou Shall Involve The Right People In The Delay Analysis.[123]
(8) Thou Shall Be Objective And Avoid Adversarial Interests That Damage Credibility.[124]
(9) Thou Shall Recognize The Right To Finish Early.[125]
(10) Thou Shall Recognize Reasonable Resource Leveling.[126]

C. Your Scheduling Expert's Experience

Any individual who takes a day or two day course on construction scheduling may hold him or herself out as a CPM expert. The use of such an individual as your expert, however, frequently leads to disaster. For this reason it is suggested that you consider some of the following guidelines in evaluating the experience and capabilities of an expert:

---

120 *See, e.g., Titan Mountain States Constr. Corp.,* ASBCA No. 23,095, 85-1, BCA ¶ 17,931 (1985).

121 *See, e.g., Gulf Contracting, Inc.,* ASBCA Nos. 30,195, 32,839, 33,867, 89-2 BCA ¶ 21,812, *aff'd on recon.*, 90-1 BCA ¶ 22,393 (1989).

122 *See Fortec Constructors v. United States*, 8 Cl. Ct. 490 (1985); *Ballenger Corp.,* DOTBCA Nos. 74-32, 74-32A, 74-32H, 84-1 BCA ¶ 16,973 (1983); *Preston-Brady, Co.*, VABCA Nos. 1892, 1991, 2555, 87-1 BCA ¶ 19,649 (1987); *see also Continental Consolidated Corp.,* ENGBCA Nos. 2743, 2766, 67-2 BCA ¶ 6624 (1967); 68-1 BCA ¶ 7003 (1968); *J.A. Jones Constr. Co.,* ENGBCA Nos. 3035, 3222, 72-1 BCA ¶ 9261 (1972).

123 *See, e.g., Turner Constr. Co.,* ASBCA Nos. 25,447, 29,472, 29,591, 29,592, 29,830, 29,851, 29,852, 90-2 BCA ¶ 22,649 (1990).

124 *See, e.g., Nello L. Teer Co.,* ENGBCA No. 4376, 86-3 BCA ¶ 19,326 (1986).

125 *See, e.g., Green Builders, Inc.,* ASBCA No. 35,518, 88-2 BCA 20,734 (1988).

126 *See, e.g., Utley-James, Inc.,* GSBCA No. 5370, 85-1 BCA ¶ 17,816 (1984); *Neal & Co., Inc. v. United States*, 36 Fed. Cl. 600 (1996).

(1) A comprehensive understanding of the basic scheduling techniques utilized by the construction industry is necessary, as is practical experience with the development and successful application of such techniques to specific contracts, projects, and large-scale programs.
(2) A working knowledge of the various project delivery systems, the relationships created among the project participants, the scheduling options that are used, and the strengths and weaknesses for making schedules work successfully.
(3) A working knowledge of the contract documents and the contemporaneous project record keeping systems that should be maintained and how they can be used to substantiate the plans and performance of the project participants.
(4) An understanding of the budgeting, cost estimating and cost control techniques and procedures that are used to plan, budget, control, track and report costs.
(5) Extensive experience with the development, implementation, control and periodic reporting of status versus established plans and schedules.
(6) The ability and experience to conduct a proper delay analysis; determine cause, effect and responsibility for delay, determine the relationship of a particular delay and prior and/or parallel delays that are occurring, and determine whether the delay is excusable and compensable.
(7) An understanding of the various labor trades and craft jurisdictions. In addition, knowledge of and familiarity with industry sources for labor and equipment productivity factors.
(8) A working knowledge of scheduling specifications and what they should include to be effective. In addition, knowledge of the common pitfalls that contribute to the failure or abuse of the schedule.
(9) A thorough understanding of key issues such as float, concurrent delay, offsetting delay, constructive acceleration, impact on unchanged work, and the cumulative effect of changes.
(10) Construction field experience and familiarization with construction methods and sequencing.

(11) A working knowledge of the legal aspects of the schedule and familiarity with recent case law regarding scheduling issues, delay, impact, and disruption claims.
(12) Familiarity with the various legal forums (*i.e.* jury trial, arbitration, A.D.R., Armed Services Board of Contract Appeals, etc.) and their implications and effects on the expert's testimony.
(13) Past experience with claims, dispute forums, similar projects, similar delay theories, written reports, presentations, and expert testimony.
(14) Experience teaching and educating others in the principles, fundamentals, and techniques for planning, scheduling, delay analysis, and controlling and reporting for construction projects.
(15) Good communication skills and the ability to articulate complex issues with simplicity.
(16) Experience advising trial counsel prior to and during trial by serving as a technical advisor, and developing cross-examination questions for adversarial experts during the conduct of a trial or arbitration.
(17) Experience with the development and use of visual evidence that is demonstrative and effective to a judge or jury.
(18) Effectiveness in presenting opinions on direct examination and defending them on cross-examination.
(19) The expert must be courteous, but firm and confident, and able to avoid the appearance of seeming to be pompous or arrogant.

D. Direct Examination: Establishing The Experts

The main areas of examination of a CPM expert on direct examination typically include:

(1) Qualifications
(2) Charter for Expert Opinion or Analysis
(3) Data Base Utilized By Expert in Developing Analysis
(4) Methodology of Experts Analysis
(5) Legal Basis or Other Justification for a Particular Methodology Used on Instant Matter

(6) Actual Conclusions as well as Reasoning and Data Relied Upon by Expert in Reaching Conclusions

E. Cross Examination: How To Use The Opposing Expert To Validate Your Position

1. The "Wind In The Willows" Test

One of the most frequently used and fruitful areas of examination of opposing experts is the presentation to that expert of inconsistent methodology utilized by experts in prior cases or presentations. One example of such recent foolishness was reflected by the testimony of an expert in *Harrison Western Corp. and Franki-Denys, Inc. (JV)*.[127] The Board commented on the inconsistent approaches taken by the contractor's expert in an arbitration with a supplier and in the separate action against the owner.

> Appellant has pursued a claim against a supplier on this project for alleged breach of warranty in an arbitration proceeding separate from this appeal. Evidence, including exhibits, from the arbitration was admitted in this hearing. It appears that Appellant's cause in the arbitration proceeding included claims for delay.
>
> * * *
>
> Finally, we note that Appellant's scheduling witness referred, in the arbitration, to the results of the scheduling analysis he did under Appellant's theory of the case in these appeals as 'meaningless.' (Finding 33). (He used as-built data in his analysis in the arbitration). An attempt was made to rehabilitate and explain this statement by Appellant's witness, and the result was testimony that the result is meaningless 'if the schedule . . . doesn't represent in fact what has happened to that point in time or near that point in time . . .' (Finding 34). A difficulty with this answer is that it appears to exclude evidence of later events that could change the critical path. He made this clear in his response to a succeeding question:
>
> Q: And does that problem exist in this case? The 5556 and 5576?
>
> A: No, absolutely not. The schedule was contemporaneous.
>
> * * *

---

127 ENGBCA Nos. 5556, 5576, 93-1 BCA ¶ 25,382 (1992).

> He added that if there had been a considerable period of time beyond when the plan differed from actual progress, 'that's when I say . . . an analysis . . . required by the . . . specifications yields a meaningless answer.' (Finding 34).

2. The Defective Database Test

This technique may relate to either using incorrect records, not looking at relevant and necessary records, or even incorrectly interpreting records.

An example of this type of problem was contained in *Gulf Contracting*,[128] where the Board stated:

> In support of its claim, Gulf relied on the analysis of its expert, Vinson. Vinson claimed to have performed an in-depth review of all correspondence and data pertinent to the progress of the project (finding 81). If this was indeed the case, we do not believe that Vinson could have missed the well documented performance problems that existed between Gulf and Hughes. Vinson said that he was instructed to "exclude other disruptions from the claim" (finding 92). We have found that Vinson's analysis systematically excluded all delays and disruptions except those allegedly caused by the Government (finding 92). We conclude that his analysis was inherently biased, and could lead to but one predictable outcome. For our purposes therefore, we deem Vinson's analysis to be totally unreliable. To be credible, a contractor's CPM analysis ought to take into account, and give appropriate credit for all of the delays which were alleged to have occurred. *Hanley v. United States* [30 CCF ¶ 70,189], 676 F.2d 584 (Ct. Cl. 1982); *Pathman Construction Co.*, ASBCA No. 23392, 85-2 BCA ¶ 18,096 (CPM analysis with "built-in-bias" rejected).

3. The Methodology Truth Test / Decisions / Manuals And Industry Practices

A variety of reported decisions provide fruitful material for use in testing the methodology utilized by the expert, as well as a basis for asking the expert why the principle detailed in the decision should not

---

128 *See, e.g., Gulf Contracting,* ASBCA Nos. 30,195, 89-2 BCA ¶ 21,812, *aff'd on recon.*, 90-1 BCA ¶ 22,393 (1989).

apply in the instant matter.[129] In addition, the Corps of Engineers Modification Impact Guide and the Veterans Administration VACPM Handbook should be considered, along with other secondary source materials.[130]

Perhaps one of the most successful tools in detailing some of the key principles of CPM scheduling, the dynamic nature of the CPM process, the necessity for resource leveling, and "Buy Back" time, is the *Utley James* decision.[131] The *Utley-James* "CPM Essay" has been used in the deposition examination of the critical owner representative who had been responsible for abandoning the contractually required CPM schedule early in the project in favor of instituting "one week look ahead" schedules that were a "so-called better tool," another device for a total crash acceleration of the project. The ideas contained in the CPM essay were used in the deposition questioning to establish key CPM principles.

---

129 *See, e.g., Fortec Constructors v. United States,* 8 Cl. Ct. 490 (1985); *Wilner Constr. Co.*, ASBCA No. 26,621, 84-2 BCA ¶ 17,411 (1984); *Weaver Bailey Contractors, Inc. v. United States*, 19 Cl. Ct. 474, *recon. den.*; 20 Cl. Ct. 158 (1990); *Continental Consol. Corp.*, ENGBCA Nos. 2743, 2766, 67-2 BCA ¶ 6624 (1967), 68-1 BCA ¶ 7003; *Utley-James, Inc.*, GSBCA No. 5370, 85-1 BCA ¶ 17,816 (1984), aff'd, 14 Cl. Ct. 804 (1988).

130 Construction Scheduling: Preparation, Liability and Claims; Bramble & Callahan, *Construction Delay Claims*, John Wiley & Sons (2d ed. 1992); Wickwire, Driscoll & Hurlbut, and the 1974 and 1989 Public Contract Law Journal Articles on the Use of CPM on Contract Claims: Wickwire & Smith, *The Use Critical Path Method Techniques in Contract Claims,* 1 PUB. CONT. L.J., (1974); Wickwire, Hurlbut & Lerman, *Use of Critical Path Method Techniques in Contract Claims: Issues and Developments, 1974 to 1988*, 18 PUB. CONT. L.J. 338 (1989).

131 GSBCA No. 5370, 85-1 BCA ¶ 17,816.

CHAPTER 11

PROVING (AND DEFENDING) CONSTRUCTION DAMAGES

Julian F. Hoffar III
Robert G. Watt

I.
Acceptable And Unacceptable Methods Of Damage Calculation

In an action for damages, the contractor bears the burden of proving both the existence and the amount of the damages incurred.[1] Where the existence of damages is clearly established, a contractor's inability to prove the precise amount of those damages does not preclude recovery.[2] This concept is particularly applicable to major construction disputes involving large labor inefficiency and delay claims.[3] The general rule is that the injured party must establish the extent of its damages with "reasonable certainty."[4] As the Supreme Court of Washington aptly observed:

---

1 *See United States ex rel. Gray-Bar Elec. Co., Inc. v. J.H. Copeland & Sons Constr., Inc.*, 568 F.2d 1159, 1161-62 (5th Cir. 1978).

2 *Mobil Chem. Co. v. Blount Bros. Corp.*, 809 F.2d 1175 (5th Cir. 1987).

3 *See John W. Johnson, Inc. v. J. A. Jones Constr. Co.*, 369 F. Supp. 484 (E.D. Va. 1973); *Aetna Cas. & Sur. Co. v. Doleac Elec. Co., Inc.*, 471 So. 2d 325 (Miss. 1985).

4 *See Wunderlich Contracting Co. v. United States*, 351 F.2d 956 (Ct. Cl. 1965); *Appeal of Tutor-Saliba-Perini*, PSBCA No. 1201, 87-2 BCA ¶ 19,775 (1987); *Lincor Contractors, Ltd. v. Hyskell*, 692 P.2d 903 (Wash. Ct. App. 1984).

> The difficulty of calculating damages should not be confused with proof of damage as a necessary element of the plaintiff's case. Once the fact of damage has been established by a preponderance, the plaintiff is obligated to produce the best evidence available which will afford the jury a reasonable basis for estimating the dollar amount of his loss.[5]

The contractor must calculate the amount of damages incurred with the highest degree of certainty possible. This chapter will examine the appropriateness and requirements of several methods of calculating construction contract damages which have been accepted by courts. Generally, the method of calculation which a contractor will utilize is determined by the quality and detail of the contractor's project records and/or the nature of the contractor's cause of action.

A. General Rule Of Damages

The basic aim of the measure of damages for breach of a construction contract is to place the aggrieved party in the same financial position which it would have occupied had the contract been fully performed.[6] Generally, recovery of both the losses incurred by the contractor and the gains prevented as a result of the breach may be considered,[7] however, only to the extent those damages were a foreseeable result of the breach at the time of contracting.[8] Section 351 of the *Restatement (Second) of Contract* classifies foreseeability of damages as follows:

> Loss may be foreseeable as a probable result of a breach because it follows from the breach
>
> (a) in the ordinary course of events, or
>
> (b) as a result of special circumstances, beyond the ordinary course of events, that the party in breach had reason to know.[9]

---

5 *Seattle W. Indus., Inc. v. David A. Mowat Co.*, 750 P.2d 245, 249 (Wash. 1988).

6 RESTATEMENT (SECOND) OF CONTRACTS § 344 (1981).

7 *Perfecting Serv. Co. v. Product Dev. & Sales Co.*, 131 S.E.2d 9, 21 (N.C. 1963).

8 *Hadley v. Baxendale*, 156 Eng. Rep. 145 (1954).

9 RESTATEMENT (SECOND) OF CONTRACTS § 351(2) (1981). Unlike breach

Recoverable contract damages are separated into two distinct categories: (1) "general" damages (those which naturally arise from the breach); and (2) "consequential" damages (those which may not naturally flow from the breach but which may reasonably have been in the contemplation of the parties at the time of contracting). This distinction, which has been applied by the courts in virtually all jurisdictions,[10] is discussed more specifically as follows:

> There are two broad categories of damages *ex contract*: direct (or general) damages and consequential (or special) damages. Direct damages are those which arise "naturally" or "ordinarily" from a breach of contract; they are damages which, in the ordinary course of human experience, can be expected to result from a breach. Consequential damages are those which arise from the intervention of "special circumstances" not ordinarily predictable. If damages are determined to be direct, they are compensable. If damages are determined to be consequential, they are compensable only if it is determined that the special circumstances were within the "contemplation" of both contracting parties. Whether damages are direct or consequential is a question of law. Whether special circumstances were within the contemplation of the parties is a question of fact.[11]

As a result of the foreseeability of damages rule, the contractor indeed runs the risk of less than full compensation for its damages. The general damages which naturally arise from a breach and which any contractor would expect to incur as a result of breaches of similar contracts may not fully compensate the contractor for its damages. Furthermore, consequential damages are recoverable only if

---

of contract damages, damages for negligence include compensation for all injuries "proximately caused" by the negligent acts without regard to whether the amount or extent of damages were contemplated or foreseen. *See generally* Philip L. Bruner, *Construction Claims Recovery Measures: Untying the Gordian Knot*, 20 THE FORUM 278 (ABA 1985).

10 *See, e.g., Arctic Contractors, Inc. v. State*, 564 P.2d 30, 44-45 (Alaska 1977); *Burnett & Doty Dev. Co. v. C. S. Phillips*, 148 Cal. Rptr. 569, 571-72 (Cal. Ct. App. 1978); *State Highway Dep't v. Knox-Rivers Constr. Co.*, 160 S.E.2d 641, 646 (Ga. Ct. App. 1968); *Pipkin v. Thomas & Hill, Inc.*, 236 S.E.2d 725, 730-33 (N.C. Ct. App. 1977), *aff'd in part, rev'd in part*, 258 S.E.2d 778 (N.C. 1979).

11 214 S.E.2d 155, 160 (Va. 1975) (citations and footnote omitted).

contemplated by the parties at the time of contracting and if reasonably proven.[12] Examples of consequential damages which may not be recoverable are certain lost profits, certain interest costs, diminution in the value of business, and lost future work.[13]

To enhance the possibility of recovery, the contractor may expressly inform the owner at the time of contracting of any peculiar circumstances which may exist *and* of the damages which would result from those circumstances in the event of a breach. Otherwise, the contractor may not be made whole even if it should prove that it incurred the damages, because the damages would not be foreseeable. In complex construction litigation, both negligence and breach of contract theories are usually asserted to avoid the result of the foreseeability rule. Courts generally look to the "dominant" or "essential" cause of action in determining which measure of recovery to

---

12 *See Topco, Inc. v. Montana Dep't of Highways*, 912 P.2d 805 (Mont. 1996) (proof of lost profits was "speculative" and "uncertain").

13 *See, e.g., Rocky Mountain Constr. Co. v. United States*, 218 Ct. Cl 665 (1978). In *Rocky Mountain*, the contractor sought lost profits as a result of its inability to bid on and possibly obtain other contracts. The court denied the consequential damages, stating:

> The lost profits relate not to the particular contract involved, but to other contracts [the contractor] had not yet entered into but which he anticipated receiving. Such consequential damages are not recoverable because they 'would not have been reasonably foreseeable by the defendant at the time when the breaches of contract . . . were committed.' *Id.* at 666 (citations omitted).

Similarly, in *Roanoke Hospital Ass'n*, the owner sought to recover from the contractor for its damages for inexcusable delay. 214 S.E.2d at 155. The owner's damages included a claim for the costs of additional interest paid by the owner on the construction loan as a result of the extended performance period, additional interest paid as a result of an increase in interest rates, and the loss of interest which the owner could have made through the investment of its funds which were tied up in the project for the extended period of time. *Id.* at 160. The court allowed recovery of the extended financing costs, but denied recovery of damages due to the increase in interest rates. *Id.* at 161-62.

apply. A negligent breach of contract theory is generally not recognized.[14]

B. Methods Of Calculating Construction Damages

1. *Actual Cost Method*

The most accurate and reliable method of quantifying damages is by the use of definite calculations based on actual historical cost records which have been regularly maintained as the costs are incurred. Courts have clearly expressed their preference for actual costs as the basis for damage awards.[15] This preference extends to supply contracts controlled by the Uniform Commercial Code (UCC).[16]

The cost records which should be utilized in determining actual costs include payroll records, time logs, equipment logs, progress schedules, invoices, and work accounts. Informal records which are systematically kept to record the costs of a particular activity impacted by a modification, delay, or breach also may be used to establish actual costs.[17] Actual costs calculated from the above regularly maintained records are presumed to be reasonable.[18] The burden thereafter shifts to the owner to establish that the actual costs incurred were unreasonable.

To insure the accurate calculation of actual costs incurred, the contractor should institute at the outset of the project a detailed and effective system of cost accounting. The cost accounting system should record costs as they occur and compile those costs on a regular basis.

---

14 *See, e.g., Zontelli & Sons, Inc. v. City of Nashwauk*, 353 N.W.2d 600 (Minn. Ct. App. 1984), *rev'd on other grounds*, 373 N.W.2d 744 (Minn. 1985).

15 *See, e.g., Plumbers & Fitters, Local 761 v. Matt J. Zaich Constr. Co.*, 418 F.2d 1054 (9th Cir. 1969), *Appeal of Select Contractors, Inc.*, ENGBCA No. 3919, 82-2 BCA ¶ 15,869 (1982); *Appeal of Bregman Constr. Corp.*, ASBCA No. 15,020, 72-1 BCA ¶ 9411 (1972).

16 *See District Concrete Co., Inc. v. Bernstein Concrete Corp.*, 418 A.2d 1030 (D.C. 1980).

17 *Appeal of Gary Constr. Co., Inc.*, ASBCA No. 19,306, 77-1 BCA ¶ 12,461 (1977).

18 *Bruce Constr. Corp. v. United States*, 324 F.2d 516 (Ct. Cl. 1963); *Appeal of Cal Constructors*, ASBCA Nos. 21,179, 21,180, 21,532, 78-1 BCA ¶ 12,992 (1977).

These costs should be accumulated *at least* on a monthly basis. Where a particular area of the project has been impacted by a change in the contract requirements, a delay, or directed acceleration, the contractor should isolate the costs of that area and compile those costs on a bi-weekly or weekly basis.

The cost accounting system also should establish as many separate cost accounts as the nature of the project will practically permit. The cost accounting system should assign equipment costs and labor costs into each task or function for which they are expended. The supporting documentation for those accounts should also specify the location of the equipment use and the length of time the item was used at each particular location.

The sophistication and complexity of the cost accounting system will necessarily depend upon the magnitude of the project and the resources of the contractor. The widespread availability of the computer places at the disposal of every contractor the capacity to segregate additional construction costs almost at will.[19] By endeavoring to systematically record the costs of the project in distinct and specialized accounts, the contractor will enhance its ability to accurately calculate damages.

2. *Total Cost Method*

A contractor's inability to precisely calculate its damages based on actual historical costs will not necessarily preclude recovery. Where actions by the owner clearly cause damages to the contractor, but the amount of those damages is impossible to precisely define, courts have reluctantly awarded damages without segregation of costs under the total cost method. This method allows damages to be calculated by subtracting the original estimated cost for performing the entire project (the bid) from the total actual cost of the performance without segregation of costs.

The difference between actual cost and estimated cost for the entire project simply is assumed to have been caused by a combination of delay, changes, or disruptions attributed to the owner. Precisely because

---

19 *See Youngdale & Sons Constr. Co. v. United States*, 27 Fed. Cl. 516 (1993) ("Cost accounting has sufficiently advanced to the point where the contractor's additional and ananticipated costs . . . could have been separated out from the original cost estimate with reasonable accuracy.").

of this assumption, the total cost approach generally is not favored by courts and boards of contract appeals, and is applied only as a method of last resort.[20] This disdain for the total cost method of calculating damages was succinctly stated by the United States Court of Claims in *WRB Corp. v. United States*.[21] According to the court, the total cost method "has never been favored by the court and has been tolerated only when no other mode was available and when the reliability of the supporting evidence was fully substantiated."[22] Courts therefore limit use of the total cost method to cases when these elements are present: (1) the owner's liability is clearly established; (2) no alternative method of calculating damages exists and determination of loss on a segregated basis under the particular circumstances is impossible or highly impracticable; (3) the original bid was reasonable; (3) the actual costs incurred were reasonable; (4) the original bid was accurately computed, with no mistakes in its preparation; and (5) the contractor was not responsible for the extra costs incurred.[23] A court will utilize any other method to calculate damages before it will accept a total cost approach.[24]

---

20 *Servidone Constr. Corp. v. United States*, 19 Cl. Ct. 346, 384 (1990), *aff'd*, 931 F.2d 860 (Fed. Cir. 1991); *Youngdale & Sons Constr. Co. v. United States*, 27 Fed. Cl. 516 (1993); *Palazzi Corp. v. Stickney*, 619 A.2d 1001 (N.H. 1992); *John F. Harris Co., Inc. v. School Dist. of Philadelphia*, 460 A.2d 260 (Pa. 1983).

21 183 Ct. Cl. 409 (1968).

22 *Id.* at 426.

23 *J. D. Hieden Constr. Co. v. United States*, 347 F.2d 235 (5th Cir. 1965); *Appeal of Wilbur Smith & Assoc.*, ASBCA No. 35,301, 89-3 BCA ¶ 22,025 (1989).

24 *See Skip Kirchdorfer, Inc. v. United States,* 14 Cl. Ct. 594 (1988), in which the court declined to use the total cost method, unusually advocated by the defendant, where the plaintiff was able to show a direct correlation between its bid price and the number of service calls it relied upon in making its bid. Damages were thus calculated on the basis of the increased number of calls the plaintiff was required to make over the erroneous amount it was told to bid on, as opposed to recovering only for its excess out-of-pocket cost over bid price. *Id.* at 607. The court noted that, although the plaintiff's claim was based on expert analysis, the analysis was produced from records of its actual contract performance which constituted reasonable proof of cost. The total cost method was thus rejected since it is "tolerated only when no other mode of recovery

For instance, in *Miller Elevator Co., Inc. v. United States,*[25] the court declined to use the total cost method because damages could be accurately approximated by the use of actual cost data. *Miller Elevator* involved an elevator maintenance services contract at a federal office building. Subsequent renovation of the building by another contractor resulted in a substantial increase in the maintenance efforts required by the plaintiff. The contractor sought application of the total cost method.[26] The court recognized, however, that when the "computation of damages relies on estimates involving actual costs, actual costs and not total costs control."[27] The court therefore rejected the use of the total cost method, finding that there was sufficient documentation of actual costs upon which the contractor could have based its damages calculation.[28]

In contrast, the United States Court of Federal Claims did allow use of total cost method in *American Line Builders, Inc. v. United States,*[29] which involved a contract to replace 180 miles of electrical transmission lines. At issue was whether the contractor was entitled to additional compensation for work which was performed pursuant to amendments and delays in contract work schedules issued by the government.[30] The case resulted in a lengthy trial with complex issues which created volumes of documentary evidence and trial testimony. Because the presentation of facts and issues during the trial was less than clear under the circumstances, and due to the difficulty in applying a different method, the court allowed the total cost method for all claims in which the court found liability or the government conceded liability.

When using the total cost method, the contractor must demonstrate that its bid was prepared in a reasonable manner and that the bid was reasonable in relation to other bids on the project. Several methods may be utilized to demonstrate reasonableness: (1) testimony of competent witnesses;[31] (2) use of accepted industry standards;[32] (3) comparison

---

[is] available." *Id.* at 605 (internal quotations omitted).

25 30 Fed. Cl. 662 (1994).

26 *Id.* at 702.

27 *Id.* at 706.

28 *Id.*

29 26 Cl. Ct. 1155, 1182 (1992).

30 *Id.* at 1176.

31 *See McDevitt & St. Co. v. Department of General Services.,* 377 So. 2d 191 (Fla. Dist. Ct. App. 1979), in which the court held that the total cost

with costs on related work;[33] and (4) averaging of bids and estimates.[34] Moreover, courts will not compensate a contractor for its own inefficiencies or improper performance simply because the owner may have impacted some other, possibly unrelated, aspect of the project.[35]

---

method may be proper where the contractor presents "testimony of individuals who prepared appellant's bid concerning their experience and their methods." *Id.* at 193. The *McDevitt* decision is in accord with the United States Court of Claims' decision in *J.D. Hedin Constr. Co. v. United States.*, 347 F.2d 235 (Ct. Cl. 1965), where the court looked to the qualifications of the person that prepared the estimate and the closeness of that estimate to the other bids submitted on the project in determining the reasonableness of the original estimate.

32 *See Pebble Bldg. Co. v. G.J. Hopkins, Inc.*, 288 S.E.2d 437 (Va. 1982), in which the court awarded damages to an electrical contractor on the total cost basis upon a showing that the original estimate was prepared from an industry estimating manual.

33 *See Seattle Western Indus., Inc. v. David A. Mowat Co.*, 750 P.2d 245 (Wash. 1988) (the requirement of reasonableness, both for the original bid and for actual cost, was satisfied by a comparison of profit, manhours, and experience gained on the phase of the project which gave rise to the damages with those of a later phase of the same project which was virtually identical).

34 *See Great Lakes Dredge & Dock Co. v. United States,* 96 F. Supp. 923 (Ct. Cl. 1951) (court allowed recovery of actual costs in excess of the average of other bids and the government's estimate).

35 *See McMillin Brothers Constructors, Inc.*, EBCA No. 328-10-84, 91-1 BCA ¶ 23,351 (1990), *aff'd*, 949 F.2d 403 (Fed. Cir. 1991), in which the Board of Contract Appeals rejected a contractor's use of the total cost method in its cumulative impact claim because it held that to do so might have relieved the contractor of additional costs for which it was responsible. Similarly, in *Appeal of Fanning, Phillips & Molnar,* VABCA Nos. 3856, 3964, 96-1 BCA ¶ 28,214 (1996), the Board of Contract Appeals declined to accept the total cost method when the contractor presented weekly and daily reports of hours its employees expended on the project, but failed to distinguish between hours spent on the original design work and any rework caused by the change in the scope of work. The Board objected to this manner of proving damages, because it required the Board to accept "on blind faith" that the contractor's work was one hundred percent efficient. *Id.* That, according to the Board, "presumes too much." *Id.*

Some state courts also allow recovery under the total cost method.[36] Like some courts, juries and arbitration panels demonstrate less reluctance to apply total cost recovery.[37] In contrast, federal courts indicate a willingness to award on a simple total cost basis where the bid was authenticated and the cause of action pervaded the contractor's entire performance.[38]

3. *Modified Total Cost Method*

The modified total cost method represents a somewhat more acceptable way of calculating contractor damages than the total cost approach. Under this method, the actual cost of performance for impacted work activities is adjusted by the contractor to eliminate any actual costs which are not the responsibility of the owner. The owner bid also must be adjusted to eliminate any mistakes or inaccuracies. This refinement generally involves the segregation of the work activities which are impacted by the claim from those which are unaffected.[39]

---

36 In *State Highway Comm'n of Wyo. v. Brasel & Sims Constr. Co., Inc.*, 688 P.2d 871 (Wyo. 1984), for example, the Wyoming Supreme Court upheld a two million dollar differing site conditions award based on a pure total cost calculation of damages. *Id.* at 880-81. Similarly, in *Prichard Bros., Inc. v. Grady Co.*, 436 N.W.2d 460 (Minn. Ct. App. 1989), the Minnesota Court of Appeals affirmed a construction manager's total cost award arising out of a negligence action against an architect/engineer. *Id.* at 467-68.

37 *See, e.g., Nebraska Pub. Power Dist. v. Austin Power, Inc.*, 773 F.2d 960 (8th Cir. 1985) (jury); *Scherbenske Excavating, Inc. v. North Dakota State Highway Dept.*, 365 N.W.2d 485 (N.D. 1985) (arbitration).

38 *See e.g., Chicago College of Osteopathic Med. v. George A. Fuller Co.*, 776 F.2d 198 (7th Cir. 1985); *Paul N. Howard Co. v. Puerto Rico Aqueduct Sewer Auth.*, 744 F.2d 880 (1st Cir. 1984).

39 *See E.C. Ernst Inc. v. Koppers Co., Inc.*, 520 F. Supp. 830 (W.D. Pa. 1981); *see also Appeals of J&T Constr. Co., Inc.*, DOTCAB No. 73-4, 75-2 BCA ¶ 11,398 (1975), where the contractor calculated damages based on the difference between the estimated and actual costs of performing only those work activities impacted by a differing site condition. The Board of Contract Appeals noted that the contractor's theory of recovery was "something much different than the pure 'total cost' and 'total time' theories which seek to measure the amount of an equitable adjustment on the basis of the difference between a contractor's

In addition to separating the impacted work activities from the unaffected activities, the contractor may refine the modified total cost calculation further by engaging an independent, expert estimator to re-estimate the cost of performing the impacted work activities. This re-estimate must be based on the original specifications without regard to the actual jobsite events. Ideally, the expert should not have access to the contractor's bid or estimates. The damages may then be calculated by subtracting the independently estimated cost of performing the impacted activities from the actual cost of performing those activities. Use of this independent estimate adds credibility to the claim by eliminating the possibility of a windfall to a contractor who submitted an unreasonably low original estimate. In any event, the failure to prove the reasonableness of the original or revised estimate will be the death knell to a modified total cost approach.[40]

II.
Excusable Delay And Damages Presentation

A. Basis Of Delay Damages

A contractor ordinarily is entitled to recover its increased costs resulting from delays to its contract performance caused by another party that were unforeseeable and outside the contemplation of the parties at the time the contract was entered into. Delays due to causes outside the control of the other party afford no basis for recovery of delay damages.[41] The delay claim may be against the owner, designer, engineer, construction manager, or another contractor. Various actions such as unreasonable delays in issuing change orders, suspension of work orders, and the issuance of faulty specifications and drawings can cause delays to the work for which the resulting expenses may be recovered.

---

total bid price and the actual costs of performing the *entire* contract." *Id.* Instead, the contractor's cost analysis focused the contractor's claim "exclusively upon the excavation and haul items," which were the only activities affected by the differing site conditions. *Id.*

40 *See Appeal of Servidone Constr. Corp.*, ENGBCA No. 4736, 88-1 BCA ¶ 20,390 (1987)

41 *See Triangle Sheet Metal Works v. James H. Merit & Co.*, 588 N.E.2d 69 (N.Y. 1991).

The basis of the right to claim delay damages may be either an express contractual provision or an implied covenant of non-interference with the other party's work. As stated in *Kenworthy v. State*,[42] "In every building contract which contains no express covenant to the contrary, there is an implied contract that 'the contractor shall be permitted to proceed with the construction of the building in accordance with the other terms of the contract without interference by the owner."[43]

When dealing with federal government contracts, costs for delay caused by the owner may be claimed under changes or suspension of work clauses. A constructive suspension may be found on the same elements and with the same effect and consequences as an actual suspension.[44]

---

42 46 Cal. Rptr. 396 (Cal. Ct. App. 1965).

43 *Id.* at 400 (citation omitted).

44 For instance, in *Hilltop Elec. Constr. Co., Inc.*, DOTCAB 76-47, 77-2 BCA ¶ 12,676 (1977), a contractor was not allowed by an air traffic controller to begin construction work on an airfield. The air traffic controller's refusal was the result of the government's failure to make the proper arrangements. In granting the contractor delay damages, the Board of Contract Appeals explained:

> The Suspension of Work Clause gives the contractor the right to a price adjustment, if the Government orders a suspension of the work, total or partial, which is unreasonable in duration, and which causes additional costs to the contractor. This applies not only where a suspension of work is ordered in writing, but also where the Contracting Officer fails to issue the suspension order that should have been issued. The Board will consider the matter as if a proper suspension order had been issued. The constructive suspension of work doctrine has long been recognized by the contract appeals boards. We find here that the failure to issue a work suspension order for 2 days constructively amounted to a suspension of work.

Id. (citations omitted). The Board of Contract Appeals explained this element further in *J.R. Pope, Inc.*:

> A constructive suspension is usually granted only when the Government has caused the delay. However, even in the absence of Government fault, constructive suspensions have been found when a delay lasts so long that the contractor cannot reasonably be expected to bear the risk and costs of the disruption and delay.

B. Proof Of Delay

Regardless of whether the claim is under a contract clause, a breach of contract action, or a tort action, the elements of recovery are the same. To prove that delay damages are owing, a contractor must show (1) that an act or failure to act by the other party or its authorized representative caused unforeseen delay to the contractor; (2) that such delay was unreasonable; and (3) that this delay was the "necessary" cause of the increase in the contractor's costs.[45] The United States Court of Claims outlined the contractor's burden in *Wunderlich Contracting Co. v. United States*:[46]

> A claimant need not prove his damages with absolute certainty or mathematical exactitude. It is sufficient if he furnishes the court with a reasonable basis for computation, even though the result is only approximate. Yet this leniency as to the actual mechanics of computation does not relieve the contractor of his essential burden of establishing the fundamental facts of liability, causation, and resultant inquiry . . . . Broad generalities and inferences to the effect that defendant *must* have caused some delay and damage because the contract took 318 days longer to complete than anticipated are not sufficient.[47]

In *Joseph Penner*,[48] the Board of Contract Appeals, quoting *Wunderlich Contracting*, denied a contractor's claim for delay damages caused by twenty-three alleged defects in government-furnished plans and specifications, and held:

> We find that the appellant has failed to provide the necessary proof which is a prerequisite to entitlement. In essence, appellant has failed to link the alleged cause (defective Government specifications) with any determinable effect, the extent of delay and increased costs attributed to the project. We are unable to find any evidence in the record as to how many days the project was delayed by the various

---

DOTCAB No. 78-55, 80-2 BCA ¶ 14,562 (1980).

45 *See United States v. Mueller*, 113 U.S. 153 (1885).

46 351 F.2d 956 (Ct. Cl. 1965).

47 *Id.* at 968-69 (citations omitted).

48 GSBCA No. 4647, 80-2 BCA ¶ 14,604 (1980).

alleged errors, or any legal basis on which increased costs might be computed.[49]

If a delay had concurrent causes, or the amount of delay cannot be specifically attributed to the various sources, damages will not be allowed.[50] Most courts will, however, attempt to apportion concurrent delays between the responsible parties and award a party damages for that portion of the delay caused by the defendant.[51] The burden is on the party seeking recovery to apportion the delay.[52]

Under the generally accepted method of computing delay time, the first day of delay is omitted but the last is included.[53] Usually, the delay period starts when the date set for completion passes without full performance. Where the contractor can prove that it intended to finish early and told the owner so in contemporaneous schedules, the period of delay may be measured from the contractor's earlier intended completion date rather than the contract completion date.[54]

---

49 *Id.*

50 *S.J. Groves & Sons Co. v. Warner Co.*, 576 F.2d 524, 527 (3rd Cir. 1978).

51 *United States ex rel. Heller Elec. Co., Inc. v. William F. Klingensmith, Inc.*, 670 F.2d 1227, 1231 (D.C. Cir. 1982).

52 *Blinderman Constr. Co., Inc. v. United States*, 695 F.2d 552 (Fed. Cir. 1982); *see also TechDyn Systems Corp. v. Whittaker Corp.*, 427 S.E.2d 334 (Va. 1993), in which a contractor in Virginia filed suit against its subcontractor for breach resulting in project delays. The subcontractor counterclaimed for unreimbursed delay damages and lost profits. With regard to the burden of proof required to establish damages where more than one party may have contributed to project delays, the court found that the contractor met its burden through the credible testimony of a CPM expert who tied the delays complained of solely to the subcontractor's actions or inactions. *Id.* at 338. In addition, the court denied the subcontractor's counterclaim, ruling that the subcontractor's proof of damages depended on uncorroborated, hearsay evidence. *Id.* at 339-40.

53 *Midstate Constructors, Inc.*, 81-1 BCA ¶ 14,898, PSBCA No. 913 (1981).

54 *See Owen L. Schwam Constr. Co., Inc.*, ASBCA No. 22,407, 79-2 BCA ¶ 13,919 (1979) where the Board of Contract Appeals stated that the delay costs could be measured from the time the contractor reasonably expected to complete performance if the schedule was feasible and

C. Escalation Costs

An obvious element of delay damage is cost escalation for labor, materials, and supplies attributable to delay. Escalation is the additional cost incurred because labor is performed and/or materials and supplies are purchased in a later more expensive period than originally contemplated. Normally, without a price escalation clause, the increased costs of labor, materials, and supplies are not recoverable if the contract

---

attainable "but for" the delay. This measure of delay time was appropriate despite the fact that the contract was completed prior to its stated completion date and despite the fact that the contract was completed only a little later than the contractor's initial estimate, if the contractor could have completed earlier in the absence of any delay. As the Board in *Owen L. Schwam* explained:

> In *Sydney Construction Co., Inc.*, ASBCA No. 21377, 77-2 BCA ¶ 12,719, we recognized a contractor's right to recover delay costs measured from a date preceding the originally specified contract delivery date where there existed a realistic schedule for completing early. *See also Barton and Son Co.*, ASBCA Nos. 9477 and 9764, 65-2 BCA ¶ 4874, citing *Metropolitan Paving Co. v. United States*, 163 Ct. Cl. 420 (1963). In both *Sydney* and *Barton, supra*, the contractors finished early, and the Government was never informed that appellant intended to take less than the full period allowed by the contract for completing the contract work. We nevertheless held that to the extent appellant had shown it was unreasonably delayed by the Government and that its costs were increased thereby, appellant was entitled to recover an equitable adjustment under the respective suspension of work clauses of the contracts involved. Here, appellant communicated its intention to deliver early to the Government at the outset of contract performance. The evidence establishes appellant's schedule as feasible and attainable and the Government's actions and inactions we have described above prevented appellant from meeting its planned early completion date. Accordingly, the delays of the Government in resolving the aforementioned deficient drawings and specifications rendering appellant incapable of meeting its planned early delivery date are remediable under, not only the Suspension of Work clause, but also the Changes clause of the contract. *Id.* (citation omitted); *see also Appeal of BECO Corp.*, ASBCA No. 27,090, 82-2 BCA ¶ 16,124 (1982); *see also Appeal of ACS Constr. Co., Inc.*, ASBCA No. 35,872, 89-1 BCA ¶ 21,314 (1988).

is performed on schedule.[55] However, if the contractor is delayed by the owner or others, recovery may be sought.

1. *Labor Escalation*

Wage escalation is recoverable if the contractor is required to pay its labor force higher wages as a result of delays in the contract performance.[56] Three elements are necessary to prove labor escalation: (1) an anticipated manloading schedule; (2) an actual manloading schedule; and (3) the wage rates actually paid over the life of the project and the time periods during which those rates were paid.

The first and second elements, the anticipated and actual manloading schedules, are elementary in concept, but are frequently difficult to plot. Many contractors prepare an anticipated manloading schedule at the time of bid or prior to the start of work. If there is no anticipated manloading schedule, an *ex post facto* schedule must be prepared from the original contract schedule. It should allocate the number of workers needed to complete each activity on the schedule. The total labor force should then be plotted on a time/labor force chart on a monthly or weekly basis. The actual manloading schedule is prepared from daily reports, payroll records, or other comparable documents.[57] The information from the documents is then plotted as actual manloading data on the same time/labor force chart as the anticipated schedule.

A problem which arises in preparing anticipated and actual manloading curves is that the anticipated and actual numbers of work hours are almost never the same. It is important to remember that a labor escalation claim is only for the increase in wage rates on hours actually worked; it does not contemplate the recovery of the cost difference between the total anticipated and the total actual hours worked. In order to correct this problem, the anticipated manloading curve must reflect the actual hours expended in accordance with the anticipated curve.

---

55 *Brazier Lumber Co.*, ASBCA No. 18,601, 76-2 BCA ¶ 12,207 (1976).

56 *United States Steel Corp. v. Missouri Pac. R.R. Co.*, 668 F.2d 435, 441 (8th Cir. 1982) *Luria Bros. & Co. v. United States*, 369 F.2d 701, 712 (Ct. Cl. 1966); *Appeal of Sydney Constr. Co., Inc.*, ASBCA No. 21,377, 77-2 BCA ¶ 12,719 (1977).

57 Remember that these documents should be admissible into evidence.

Another problem with anticipated and actual manloading curves is whether they should be prepared on a trade-by-trade or composite basis. Obviously, if there is only one trade for a contractor, or somewhat more likely, if only one trade is affected by escalation, the method is obvious. On most contracts, however, there are multiple trades. The best method is to prepare a chart for each trade, unless the amount of time and effort is prohibitive. In such a case, a composite chart should suffice.

The third element of labor escalation is the wage rate. Obviously, the contractor can only recover when more manpower is being expended in a later and more expensive time period. Therefore, it is necessary to determine the anticipated wage rates for the contract performance period and the actual wage rates.

On union contracts, the anticipated and actual wage rates are easy to determine. The union agreements provide the amount of wage rates and the time period each such rate is effective. On non-union projects, the contractor's task is more difficult, at least for the anticipated rates which must be based on what the estimator estimated. The actual rates can, of course, be readily ascertained from the payroll records. The difficulty is that a wide range of wage rates apply on non-union projects, and the wages tend to increase in a piecemeal fashion rather than all (or all of one trade) increasing on a particular date. To overcome this difficulty, the contractor should use averages. Averages are computed by totaling all the labor costs on the project for a specific period of time and dividing by the number of labor hours worked during that time frame. The best method would be to calculate the average wage rates for the original contract period (*i.e.*, the period in which work should have been completed) and the average wage rates for the extended period. Alternatively, rates could be calculated for various periods of time corresponding to general wage increases.

Since most delay/escalation claims also include some additional hours worked for which the contractor is entitled to compensation, it is important to avoid duplication. In this regard, all hours for which the contractor has been paid its actual costs incurred must be subtracted from the computation.

Once the labor hour allocation has been completed and the wage rates established, the actual computation is quite simple. It is simply necessary to multiply the number of hours by the appropriate wage rates on both the anticipated and actual schedules and subtract the difference.

2. *Material And Supply Escalation*

In addition to labor costs rising before performance is complete, a delayed contractor may face higher material and supply prices. Here, too, escalation is allowed and is calculated in a similar manner to labor costs.[58]

A preliminary step to recovering material and supply escalation is to establish that each item could not have been purchased earlier. In *Paccon, Inc.*[59] the Board of Contract Appeals stated:

> A mere showing of the increased costs resulting from paying a higher price for material purchased during the stretch-out period would not be enough, as appellant would have to show that the delay in purchase of the materials until the time of the higher price was caused and rendered necessary by the suspensions of work. A suspension in the performance of work at the project site would not necessarily delay the procurement of materials. As an example of how the work suspensions could have delayed the purchase of materials until after a price increase went into effect, the Board suggested that appellant might show that it was prevented from purchasing the materials at an earlier date because of lack of plans from which it determined its requirements for the particular material involved.[60]

Thus, in order to collect material escalation, it must be shown that there were some reasons why the materials could not have been purchased at an earlier date. As set forth in *Paccon*, the reason might have been defective specifications. Limited storage space on the project site may also prevent purchase at an earlier date.

In most cases, contractors enter into purchase orders with suppliers for a fixed period of time. When that period expires, the prices usually increase. Thus, if owner delays prevent delivery during the life of the purchase order, the contractor can usually recover, especially if there is no readily available storage area. A common example of a material that cannot be purchased at an earlier date is ready-mix concrete.

When quantifying the escalation costs, it is best to present item-by-item documentation as to all price increases through purchase orders,

---

58 *Luria Bros.*, 369 F.2d at 701; *Canon Constr. Corp.*, ASBCA No. 16,142, 72-1 BCA ¶ 9404 (1972).

59 ASBCA No. 7890, 65-2 BCA 4996 (1965).

60 *Id.*

invoices, and the contractor's own books.[61] However, if necessary, industry-wide indices which illustrate the generally prevailing prices during the period of delay can be used. A common example is the Engineering News Record *Quarterly Compilation of Construction Costs*.

D. Extended Performance Costs

Another aspect of delay-induced expense is the cost of continuing work on a project over a longer time period than first anticipated. Additional costs related to labor, equipment, and overhead may be incurred simply because performance had to be spread out over time, sometimes leaving workers and equipment idle or forcing personnel to remain on the job when they had been scheduled to work elsewhere. Before these continuing project costs are recoverable, however, they must be reduced by the amount of productive work performed during the delay period.[62]

In the latter *Hardeman-Monier-Hutcherson* case, the Board took a broad view of productive work, stating:

> [R]ecovery for costs of men and material, during the extended periods, must be offset by the cost of necessary work performed during those periods. We do not mean thereby only work which advanced the job along the critical path. Much support work is required, such as maintenance and repair of equipment and other "housekeeping" jobs. This work must be done in any event.[63]

The position of the Board in *Hardeman-Monier-Hutcherson* appears to expand the previously held theory that the Board of Contract Appeals would only discount a claim for work performed along the critical path.[64]

---

61 *Ingalls Shipbldg. Div., Litton Sys., Inc.*, ASBCA No. 17,579, 78-1 BCA ¶ 13,038 (1978) (used actual cost records).

62 *Laburnum Constr. Corp. v. United States*, 325 F.2d 451, 459 (Ct. Cl. 1963); *Hardeman-Monier-Hutcherson, A Joint Venture*, ASBCA No. 11,785, 67-1 BCA ¶ 6210 (1967); *Hardeman-Monier-Hutcherson, A Joint Venture*, ASBCA No. 12,392, 68-2 BCA ¶ 7220 (1968).

63 *Hardeman-Monier-Hutcherson*, 68-2 BCA ¶ 7220.

64 *See Lane-Verdugo*, 73-2 BCA ¶ 10,271, ASBCA Nos. 16,327, 16,328 (1973).

1. *Idle Labor*

The cost of having to maintain a work force, despite not being able to utilize it, is recoverable. This cost is recoverable, however, only if the labor could not be discharged without risk of later unavailability or could not be moved to an interim project without exorbitant expense.[65] The full salaries of supervisory personnel have been allowed since releasing such people for an unknown length of time is not deemed practical.

The recovery amount for idle labor usually is based on the total amount spent on personnel during the period the contract was delayed.[66] However, other methods to compute the recovery amount may be accepted. For instance, one contractor determined the amount for idled labor by putting the amount of the direct labor costs that should have been incurred over the direct labor amount that was actually incurred. Even though this was an uncommon method of calculation, the Board of Contract Appeals accepted the contractor's formula because nothing indicated that the formula distorted the true result, and because the other party (the government) had accepted the formula.[67]

2. *Idle Equipment*

In a delay claim, reimbursement for assembled but unused equipment may also be allowed. Additional equipment hours must first be calculated and then multiplied by the appropriate equipment rate.[68]

Idle equipment hours for a delay claim may be calculated in two basic ways. The first is to assume that each piece of equipment was used for a typical work day of eight hours for each day of delay. The second method is used if there has been a complete stoppage of work. In this case, the number of claimable equipment hours would be the number of equipment working hours per day for each *working* day of delay.

---

65 *Hardeman-Monier-Hutcherson*, 67-1 BCA ¶ 7220.

66 *Luria Bros. & Co., Inc. v. United States*, 369 F.2d 701 (Ct. Cl. 1966).

67 *Fluor Corp., Ltd.*, 69-1 BCA ¶ 7626, ASBCA No. 13,166 (1969).

68 *See* Bruner & Jacobsen, *Pricing Equipment Claims*, CONSTR. BRIEFING No. 92-5 (Fed. Pubs., Inc. Apr. 1992).

It is not enough just to show the number of idle hours, however. A contractor must also establish that it was necessary to have kept the equipment on the jobsite. As stated in *J.D. Shotwell Co*:[69]

> An allowance of standby ownership expense must be supported by a showing that the equipment for which compensation is claimed was reasonably and necessarily set aside and awaiting use in performing the contract. Stated differently, we must be able to conclude that a contractor would prudently have held the equipment for which claim is made in readiness to perform the Government contract rather than making some other disposition of it, in order to find that the Government should pay the charges for that equipment.[70]

Once the number of hours has been determined and justified, this figure must be multiplied by the appropriate equipment rate.

3. Jobsite Overhead

When performance on a project is delayed, a contractor often continues to incur expense at the jobsite even in the absence of productive work. Bills for items such as supervision, office space, storage, utilities, telephone service, and insurance continue to accrue. All of these costs are time related and are recoverable.

To calculate the amount of extended jobsite overhead, a total of all time related or fixed costs must be determined. Each item is usually reflected as a specific line item on a budget, so the calculation is usually not difficult. This total is then divided by the number of days over which the costs were accrued to arrive at a daily overhead rate. The daily overhead rate is then multiplied by the number of days of delay to arrive at the claim amount. For example, a contractor's extended overhead would be calculated as follows:

I. OVERHEAD COSTS	
A. ADMINISTRATIVE & GENERAL	COST TO 02/01/97
Automobiles	$ 14,048.64
Employee Maintenance	41,481.41
Health & Medical Insurance (Office)	44,788.39

69 ASBCA No. 8961, 65-2 BCA ¶ 5243 (1965).
70 *Id.*

	Professional Services	19,131.35
	Office Supplies	31,561.40
	Property Rental	15,576.41
	Telephone	28,967.82
	Postage	2,607.95
	Travel Expense	31,024.22
	Photos	9,206.22
	Office & Apartment Rental	12,393.32
	Office Maintenance	29,393.32
	Miscellaneous Expense	8,627.38
	Dues & Subscriptions	12,690.14
	Consultants	32,633.71
	Public Relations	3,478.60
	SUBTOTAL	$ 337,610.00
B.	HEAT, LIGHT, & POWER	COST TO 02/01/97
	Primary Power	$ 62,751.19
	Miscellaneous Power, Lighting, Heating	67,208.48
	Water and Sewage	6,601.49
	SUBTOTAL	$ 136,561.00
C.	OVERHEAD LABOR	
	1. Non-Union	COST TO 02/01/97
	Project Manager	$ 400,000.55
	Project Engineer	250,141.85
	Field Engineer	35,210.48
	Office Engineer	5,164.63
	Draftsman	3,827.27
	Safety Engineer	23,688.74
	Office Manager	25,875.00
	Purchasing Agent	30,940.00
	Clerks	22,940.00
	EEO Officer	45,961.05
	Typist	27,408.00
	SUBTOTAL	$ 871,158.00
	Labor Burden at 8% (871,157.57 x .08)	$ 69,693.00
	2. Union	COST TO 02/01/97
	General Superintendents	$ 346,085.00
	Superintendents	588,699.00
	Foremen	700,991.00
	SUBTOTAL	$1,635,775.00

Labor Burden at 20% (1,635,775 x 20.00%)		$ 327,155.00
TOTAL JOBSITE OVERHEAD		$3,377,952.00
II. DAYS OF CONTRACT PERFORMANCE 01/01/94 - 02/01/97 = 1126 days		
III. DAILY OVERHEAD $3,378,000/1126	=	$ 3,000/day
IV. CLAIM AMOUNT $3,000/day x 245 days	=	$ 735,000.00

This formula has been adopted in many cases.[71] However, care should be taken when calculating jobsite overhead that double recovery does not result by including some of the claim's expenses in the overall home office overhead claim amount. In *Two State Construction Co.,*[72] markups for overhead and profit had been included in the jobsite overhead amount. Although such costs are normally administrative and treated as indirect expenses (put into home office overhead), this procedure is not mandatory. The important point is not to claim the same amount twice.

4. Home Office Overhead

Home office overhead consists of indirect costs such as officers' salaries, legal and accounting expenses, rent, insurance, and other general administrative costs. *Overhead* includes costs a contractor expends for the benefit of its business as a whole. *Unabsorbed overhead* encompasses those overhead costs that are needlessly absorbed by a partially or totally idle contractor. In other words, unabsorbed overhead is the "amount of indirect expense actually incurred which would not have been allocated to the contract had the delay not occurred, and is not recoverable in the revenue of other work."[73] *Extended overhead* refers to an increase in overhead costs due to an extension of time in contract performance. The additional time spent on the project requires the

---

71 *See, e.g., Shirley Contracting Corp.*, ASBCA No. 29,483, 85-1 BCA ¶ 17,858 (1984); *Kemmons-Wilson, Inc.*, ASBCA No. 16,167, 72-2 BCA ¶ 9689 (1972).

72 DOTCAB Nos. 78-31, 1006, 1070, 1081, 81-1 BCA ¶ 15,149 (1981).

73 HOWARD WRIGHT & JAMES P. BEDINGFIELD, GOVERNMENT CONTRACT ACCOUNTING 347 (1979).

contractor to allocate more overhead to the contract than contemplated in its original bid.

When one party suspends or delays contract performance, home office overhead costs continue to accrue, even though the contractor's income from the project is suspended or reduced. This decrease in the contractor's payments for direct costs can create unabsorbed overhead, unless the contractor cuts its home office expenses or obtains additional work during the suspension or delay period.

Courts and boards of contract appeals have long permitted contractors to recover overhead incurred during periods of delay caused by the government. In the past, when dealing with an extended performance situation, the courts and boards of contract appeals have held that "a contractor is entitled to recover as damages the amount of overhead on a daily basis allocable to the period of overrun for which the government is responsible."[74]

The usual computation of home office overhead costs has often been expressed as a percentage rate of the direct costs associated with the particular change. However, in a delay situation, the time of performance is significantly extended without necessarily causing a corresponding increase to the contractor's direct costs on the project. The Board of Contract Appeals acknowledged this difference in *Therm-Air Manufacturing Co., Inc.*, stating:

> "[U]nabsorbed" overhead has been allowed to compensate a contractor for work stoppages, idle facilities, inability to use available manpower, etc., due to Government fault. In such delay situations fixed overhead costs, e.g., depreciation, plant maintenance, cost of heat, light, etc., continue to be incurred at the usual rate, but there is less than the usual direct cost base over which to allocate them.[75]

Thus, the usual percentage markup would not accurately reflect the possible increases to home office overhead allocated to the project caused by delay.

---

74 *J.D. Hedin Constr. Co. v. United States*, 347 F.2d 235, 259 (Ct. Cl. 1965); *see also Contracting & Material Co. v. City of Chicago*, 314 N.E.2d 598 (Ill. App. Ct. 1974), *rev'd on other grounds*, 349 N.E.2d 389 (Ill. 1976).

75 ASBCA Nos. 15,842, 17,143, 74-2 BCA ¶ 10,818 (1974).

In claims for unabsorbed overhead, courts and boards of contract appeals have concentrated on whether the contractor can reasonably establish that, absent the breach by the government, the construction project would have absorbed more overhead. Once a contractor is able to prove entitlement to the recovery of unabsorbed overhead costs, the amount of the overhead recovery must then be calculated. Since unabsorbed overhead is an item of damage that cannot be easily quantified, courts boards have attempted to determine the appropriate measure and quantum of damage. The purpose of developing a method for measuring extended or unabsorbed home office overhead is to merely place the contractor in the financial position it would have been in had the compensable delay not occurred.

In the past decade, culminating with *Satellite Electric Co. v. Dalton*,[76] the recovery of unabsorbed home office overhead on construction projects has come under increasingly sharp criticism. While recognizing that recovery of overhead costs is often necessary to fully compensate a contractor for changes to its work, there has been uncertainty as to the best method to calculate these costs. For years, courts and Boards of Contract Appeals have considered and utilized various methods for calculating the amount of unabsorbed overhead.

The most widely used method of calculating home office overhead is the *Eichleay* formula.[77] This method determines the amount of overhead incurred during the contract period and breaks it down to a daily figure. That figure is then used to calculate overhead costs due to delay:

1)	CONTRACT BILLINGS Total Billings for Contract Period	x	Total Overhead Incurred During Contract Period	=	Overhead Allocable to Contract
2)	OVERHEAD ALLOCABLE TO CONTRACT Actual Days of Performance			=	Daily Contract Overhead
3)	Daily Contract Overhead	x	Number of Days of Delay	=	Total Unabsorbed Overhead

76 105 F.3d 1418 (Fed. Cir. 1997).

77 *Eichleay Corp.*, ASBCA No. 5183, 60-2 BCA ¶ 2688 (1960), *aff'd*, 61-1 BCA ¶ 2894 (1961).

The *Eichleay* formula was applied consistently by federal courts and boards of contract appeals during the late 1960s and 1970s,[78] and more recently by some state courts.[79]

Other state courts have rejected the *Eichleay* formula, however. It is significant that these state cases dealt with situations where there were delays but no evidence of corresponding additional costs.[80] In *Berley Industries, Inc. v. City of New York,*[81] for example, the court disallowed the contractor's use of the *Eichleay* formula because the contractor had "provided no substantiation that even a single penny's increase in overhead beyond what the rest of the plaintiff's business would in any event have required had in fact been incurred."[82] The court noted that the *Eichleay* formula contained no component which represented an actual item of increased costs and that no attempt had been made to prove that the formula was logically calculated to produce a fair estimate of actual damages.

An alternative method of calculating unabsorbed home office overhead discussed by the Board of Contract Appeals in *Capital Electric Co.*[83] is the *Allegheny* method. This method permits recovery of unabsorbed overhead by utilizing an allocation rate differential between anticipated and actual allocation rates for either the delay period alone or during the actual period of contract performance multiplied by an acceptable allocation base, such as contract manufacturing costs, contract billings, or direct labor costs. This "burden fluctuation" method "has the advantage of at least simulating in the accounting period of the

---

78 *See, e.g., Luria Bros. & Co, Inc. v. United States*, 369 F.2d 701 (Ct. Cl. 1966); *Dawson Constr. Co., Inc.*, GSBCA No. 4956, 79-2 BCA ¶ 13,989 (1979); *W. Schroyer, Inc.*, ASBCA No. 21,859, 78-2 BCA ¶ 13,513 (1978); *Sydney Constr. Co., Inc.*, ASBCA No. 21,377, 77-2 BCA ¶ 12,719 (1977).

79 *See Central Florida Plastering & Dev. Co. v. Sovran Constr. Co.*, 679 So. 2d 1226 (Fla. Dist. Ct. App. 1996); *Gold Landscaping, Inc. v. Century Constr. Co.*, 696 P.2d 590 (Wash. Ct. App. 1984).

80 *Guy James Constr. Co. v. Trinity Indus., Inc.*, 644 F.2d 525, 532-33 (5th Cir.), *modified on other grounds*, 650 F.2d 93 (5th Cir. 1981); *W.G. Cornell Co. of Wash., D.C., v. Ceramic Coating Co., Inc.*, 626 F.2d 990, 994 (D.C. Cir. 1980); *Berley Indus., Inc. v. City of N.Y.*, 385 N.E.2d 281, 283 (N.Y. 1978).

81 385 N.E.2d at 281.

82 *Id.* at 282.

83 GSBCA Nos. 5316, 5317, 83 BCA ¶ 16,548 (1983).

delay the effect of the performance delay, suspension, or extension on allocable overhead."[84] Be aware, however, that the *Allegheny* method does not account for balancing of costs deferred into subsequent accounting periods.

In 1992, the *Capital Electric Co.* decision was modified somewhat by the United States Court of Appeals for the Federal Circuit in *C.B.C. Enterprises, Inc. v. United States*.[85] The Federal Circuit affirmed the decision of the Court of Claims, holding that the contractor's use of the *Eichleay* formula was improper when the extension of the project was due to additional work under a contract modification, rather than delay, disruption, or suspension caused by the government. In the former situation caused by the government, the court held that the contractor would be compensated for its additional home office overhead expenses by a percentage markup of the direct costs, which would adequately reimburse the contractor for its home office overhead costs during the period of time which the modification extended the contract. Unlike a disruption or suspension of work caused by the government, in the case of a contract modification, there was no "element of uncertainty" arising from the modification which would make it impractical for the contractor to take on other work during the delay. The court held that the use of the *Eichleay* formula to calculate extended home office overhead was not permissible where the contractor negotiated a change order with the government which extended contract performance for a brief, known period of time, during which the contractor experienced no suspension of work, no idle time, and no uncertain periods of delay.[86] The court also clarified that the use of the *Eichleay* formula was the exception rather than the rule and that the formula was "an extraordinary remedy" limited to suspensions or disruptions of work caused by the government.[87]

---

84 *Id.; see also Allied Materials & Equip. Co., Inc.*, 75-1 BCA ¶ 11,150, ASBCA No. 17,318 (1975).

85 24 Cl. Ct. 187 (1991), *aff'd*, 978 F.2d 669 (Fed. Cir. 1992).

86 *Id.* at 193.

87 *C.B.C. Enters., Inc.*, 978 F.2d at 675; *see also Community Heating & Plumbing Co., Inc. v. Kelso*, 987 F.2d 1575, 1582 (Fed. Cir. 1993) (the court, applying the standard in *C.B.C. Enters., Inc.*, did not allow the *Eichleay* formula where the claim did not involve a suspension or "hiatus in performance").

In 1994, in the landmark decision of *Wickham Contracting Co., Inc. v. Fischer,*[88] the United States Court of Appeals for the Federal Circuit disregarded all other formulae and held *Eichleay* to be the *only* acceptable methodology for calculating unabsorbed home office overhead on federal construction projects.[89] The court acknowledged that it is "impossible" to calculate the precise amount of unabsorbed overhead, but that the *Eichleay* formula is "feasible, equitable and predictable."[90] The court further stated:

> By allocating a pro-rata share of the overhead expense to the delayed contract, the *Eichleay* formula fairly distributes the overhead. By reimbursing the contractor based on the length of the delay, the *Eichleay* formula compensates the contractor, but only for unabsorbed overhead, recognizing that a portion of the overhead allocable to the delayed contract will be absorbed by the direct costs of that contract.[91]

Thus, the Federal Circuit held that the *Eichleay* formula was the "exclusive" method to be used when computing unabsorbed home office overhead.[92]

The *Eichleay* formula may not be used if the contractor cannot prove that it was unable to take on (or even bid on) additional work during the period of delay, that it was impossible or impractical to shift its work forces to a new job or that its bonding capacity or capital prevented reassignment of resources.[93]

In *Mech-Con Corp. v. West,*[94] the Federal Circuit reaffirmed the three components necessary to establish damages under the *Eichleay* formula, but shifted the burden of the third element of mitigation to the government. Despite this shift in burden, however, the onus remains on

---

88 12 F.3d 1574 (Fed. Cir. 1994).

89 *Id.* at 1580-81.

90 *Id.* at 1580.

91 *Id.*

92 *Id.* at 1580-81.

93 *Capital Elec. Co. v. United States*, 729 F.2d 743, 745 (Fed. Cir. 1984). As stated by the Board of Contract Appeals in *Appeal of R.G. Beer Corp.*, "We agree in principle that the contractor has an obligation to mitigate damages by absorbing indirect costs through performance of other work where it is possible to do so effectively, productively and practicably." ENGBCA No. 4885, 86-3 BCA ¶ 19,012 (1986).

94 61 F.3d 883, 886 (Fed. Cir. 1995).

the contractor to prove all elements, including that it was unable to obtain other work.[95]

Finally, in *Satellite Electric Co.*,[96] the Federal Circuit recently held that a contractor was not entitled to overhead damages under the *Eichleay* formula because the contractor had actively bid on other projects during the delay period, thereby defeating the third *Eichleay* requisite: that the contractor was unable to take on other work. Although there were two suspensions lasting a total of 228 days for which the contractor was obligated to remain on "standby," the court observed that if the company was aggressively seeking new contracts, it was reasonable to infer that it had the capacity to perform whatever work it could obtain.

III.
Subcontractor Damages In The Composite Claim

A. Subcontractor's Claim As Part Of The Prime Contractor's Claim

Typically, where an owner's acts or omissions increase the cost of performance of a prime contractor's subcontractors, those costs are legitimate bases for requests for price adjustments.[97] Unless restricted or totally precluded contractually, a prime contractor is entitled to apply its normal G&A and profit markups to subcontractor claims and changes.[98] If the prime contract specifies a percentage of profit or overhead to be applied to subcontract costs, such clauses are binding.[99] If the contract is silent, courts will allow a reasonable percentage of costs as profit on subcontract claims.[100]

Usually, a subcontractor is barred from submitting its claim directly to the owner because no legal relationship exists between the two parties. Thus, the subcontractor must rely upon the prime contractor to

---

95 *Altmayer v. Johnson*, 79 F.3d 1129, 1133 (Fed. Cir. 1996).

96 105 F.3d 1418 (Fed. Cir. 1997).

97 *United States ex rel. Mariana v. Piracci Constr. Co., Inc.*, 405 F. Supp. 904, 905 (D.D.C. 1975).

98 *Nager Elec. Co. v. United States*, 442 F.2d 936 (Ct. Cl. 1971).

99 *Ferelli v. Weaver*, 26 Cal. Rptr. 439, 443 (Cal. Ct. App. 1962).

100 *Heaton v. Imus*, 608 P.2d 631, 634 (Wash. 1980).

present its claim to the owner. Under certain limited circumstances, however, a subcontractor can proceed directly against the party responsible for the subcontractor's increased costs, such as an owner's architect/engineer or construction manager, notwithstanding the lack of a direct contractual relationship with that party. For example, an architect/engineer can be directly liable to a subcontractor where the source of a subcontractor's increased costs is the architect/engineer's negligence. Generally, such negligence entails (1) defective plans and specifications; (2) failure to properly inspect work; or (3) improper coordination of trades giving rise to delays.[101]

B. Consistency Of Subcontractor/Prime Contractor's Claims

Although often overlooked, a critical consideration in the preparation and submission of large, multifaceted claims is the method by which the subcontractor has structured its claim. Care should be taken to ensure that a subcontractor's theories of relief and pricing procedures are consistent with the prime contractor's claim. It often appears to a subcontractor that performance problems are attributable to a combination of acts by the owner, architect/engineer, construction manager, or prime contractor. In fact, it is not unusual to find that all the above parties bear some responsibility for a subcontractor's cost increases. Thus, when a subcontractor embarks on a consolidated claim effort with the prime contractor seeking compensation from the owner, it is important that the prime contractor retain the right to review and revise the subcontractor's claim to ensure that it is consistent with the consolidated claim and that it contains nothing to jeopardize overall entitlement.

C. Pass Through Agreements

With respect to claims or disputes with owners that involve subcontractors, prime contractors are likely to require that the subcontractor enter into a claims handling or pass through agreement. Such agreements usually provide that the subcontractor equally assume all risks of non-recovery from the owner, waive any and all rights

---

101 *See, e.g., Lane v. Geiger-Berger Assocs., P.C.*, 608 F.2d 1148, 1150-51 (8th Cir. 1979).

against the prime contractor, and accept as a full accord for its claim whatever amount is received from the owner on its behalf.

Where the prime contractor has incorporated the prime contract's disputes and changes clauses (or other claim related clauses) into the subcontract, the prime contractor has contractually assumed the obligation of acting as a conduit for all reasonable and legitimate claims of a subcontractor arising out of the owner's acts or omissions. If the above clauses are absent in the subcontract, the pass through agreement is usually necessary to effectively administer subcontractor claims. Since the pass through agreement can materially affect the rights of the parties to seek redress from each other, a prime contractor and subcontractor must carefully assess each other's performance prior to engaging in a consolidated claims approach.

When a prime contractor passes through its subcontractor's claim, it is generally referred to as a "sponsorship appeal" because the claim is being *sponsored* by the prime contractor. As previously mentioned, in circumstances where the subcontractor has a claim against the owner, a sponsorship appeal is generally the only remedy available to the subcontractor. Section 44.203(c) of Title 48 of the Code of Federal Regulations implicitly recognizes sponsorship appeals:

> Contracting officers should not refuse consent to a subcontractor merely because it contains a clause giving the subcontractor the right of indirect appeal to an agency board of contract appeals if the subcontractor is affected by a dispute between the Government and the prime contractor. Indirect appeal means assertion by the subcontractor of the prime contractor's right to appeal or the prosecution of an appeal by the prime contractor on the subcontractor's behalf. The clause may also provide that the prime contractor and subcontractor shall be equally bound by the contracting officer's or board's decision. The clause may not attempt to obligate the contracting officer or the appeals board to decide questions that do not arise between the Government and the prime contractor or that are not cognizable under the clause at 52.233-1, Disputes.[102]

In sponsorship appeal situations, the prime contractor will want to protect itself from risk by including some type of exculpatory clause into the subcontract. Under an exculpatory clause, the prime contractor agrees to sponsor the subcontractor's claim with the stipulation that the

102 48 C.F.R. § 44.203(c).

prime contractor be released from liability for damages resulting from government actions. This release may be either a complete release or limited to the extent of any recovery from the government.

D. Certification Of Subcontractor Claims On Federal Projects

An area affecting the prime/subcontractor relationship is the requirement for certification of claims on federal government projects. The Contract Disputes Act[103] requires that all claims in excess of $50,000 be certified.[104] Under the Contract Disputes Act, a prime contractor who submits a subcontractor's claim either separately or as a part of its claim to the government must certify it.[105] The language of the Contract Disputes Act requires that: 1) the claim be submitted in good faith; 2) the claim be supported by accurate and complete data; and 3) the prime contractor certify that the "amount requested accurately reflects the contract adjustment for which the contractor believes the government is liable."[106]

The dilemma in which a prime contractor may find itself is when the prime contractor does not believe in the merits of the subcontractor's claim. The prime contractor is faced with the option of certifying a claim in which it may not believe, or forcing a breach of contract action by the subcontractor for failure to prove its claim. To alleviate this predicament, the Federal Circuit has held that, with regard to a subcontractor's claim, the prime contractor need only certify that there are "good grounds" for the claim, not that the claim is certain.[107] Thus, the prime contractor may certify a claim even if it does not believe the subcontractor is entitled to recover on the claim. It is only necessary that the prime contractor believes there are good grounds for recovery.

Similarly, the Armed Services Board of Contract Appeals has held that a prime contractor, when certifying the claims made by its subcontractors, must provide the "necessary assurance of its own belief

---

103 41 U.S.C. § 605(c).

104 41 U.S.C. § 601, *et seq.*

105 *United States v. Turner Constr. Co.*, 827 F.2d 1554 (Fed. Cir. 1987).

106 41 U.S.C. § 605(c)(1).

107 *Dillingham Constr., N.A., Inc. v. United States*, 33 Fed. Cl. 495, 500 (1995), *aff'd*, 91 F.3d 167 (Fed. Cir. 1996); *Transamerica Ins. Corp., Inc. ex rel. Stroup Sheet Metal Works v. United States*, 973 F.2d 1572, 1580 (Fed. Cir. 1992); *Turner Constr. Co.*, 827 F.2d at 1561.

in the Subcontractor's claims."[108] In *Tutor-Saliba-Perini,*[109] the prime contractor was entitled to rely upon the subcontractor's attorney's evaluation of a claim and was not required to conduct a separate investigation. It would seem, however, that if the prime contractor had knowledge that the subcontractor's claim was improper he would be violating the Contract Disputes Act by certifying it.

E. The *Severin* Doctrine Defense

The goal of the prime contractor should always be to pass through the claims of its subcontractors to the owner. Initially, this was not possible on federal government projects unless the prime contractor could establish that it had actual liability to the subcontractor.[110] In *Severin v. United States,*[111] the prime contractor brought claims on behalf of its subcontractor when, in the subcontract, the prime contractor and subcontractor had agreed to hold each other harmless for any damages caused by the government. The court ruled that the prime contractor was not allowed to pursue the claim on behalf of its subcontractor. The theory underlying *Severin* is that the prime contractor can assert a claim only if it suffered actual damages. Over the years, however, the *Severin* defense has been almost completely eroded on federal projects.

Courts have consistently held that a contractor need not allege that it is liable to its subcontractor in order to pass through a claim for damages against the owner on behalf of the subcontractor. The courts have uniformly held that, under the *Severin* doctrine, "*non*-liability is an affirmative defense that the owner must plead and prove."[112]

In *Frank Briscoe Co., Inc. v. County of Clark,*[113] the owner of a wastewater treatment plant argued under the *Severin* doctrine that a liquidating agreement made between a prime contractor and its subcontractor precluded the prime contractor from bringing a suit on

---

108 *Robert E. McKee, Inc.*, ASBCA No. 42,917, 92-2 BCA ¶ 24,976 (1992) (internal quotation omitted).

109 PSBCA No. 1201, 87-2 BCA ¶ 19,775 (1987).

110 *Severin v. United States*, 99 Ct. Cl. 435 (1943).

111 *Id.*

112 *Gilbert Pac. Corp. v. State ex rel. Dep't of Transp., Highway Div.*, 822 P.2d 729, 731 (Or. Ct. App. 1991).

113 772 F. Supp. 513 (D. Nev. 1991).

behalf of the subcontractor since the agreement served to release the prime contractor from liability to the subcontractor. The court held that if the prime contractor has agreed to reimburse the subcontractor for its damages at the hands of the government, even if this reimbursement is contingent on the prime contractor's recovery, then the contract clause or release does not completely exonerate the prime contractor from liability to its subcontractor, and the owner's *Severin* doctrine defense will fail.[114] Likewise, a subcontract provision which provides that the prime contractor will not be liable to the subcontractor unless the prime can recover from the government has been held sufficient to bar application of the defense.[115]

Furthermore, the *Severin* doctrine has been held not to apply to claims based on the changes clause,[116] the differing site conditions clause,[117] and the suspension of work clause.[118] In *Atlantic States Construction, Inc. v. Hand, Arendall, Bedsole, Greaves & Johnson,*[119] the combined effect of a subcontract clause which incorporated the prime contract's equitable adjustment provisions and a subcontract's exculpatory provision which limited the prime contractor's liability to the subcontractor to the extent the government was found liable to the prime contractor was sufficient to avoid the *Severin* doctrine. Similarly, in *Oconto Electric, Inc.,*[120] the Board of Contract Appeals found that an agreement which provided for the subcontractor's recovery only if the prime contractor recovered from the government was sufficient to allow the prime contractor to proceed on the subcontractor's behalf.

Despite rejection of the *Severin* doctrine by several courts, the United States Court of Appeals for the Federal Circuit upheld the

---

114 *Id.* at 516-17.

115 *See, e.g., Morrison-Knudsen Co. v. United States*, 397 F.2d 826, 841-42 (Ct. Cl. 1968).

116 *Caddell Constr. Co., Inc.,* ASBCA Nos. 46,231, 48,363, 95-2 BCA ¶ 27,772 (1995); *Aydin Corp*, EBCA No. 355-5-86, 89-3 BCA ¶ 22,044 (1989).

117 *Turner Constr. Co. ex rel. Bryan-Durham Elec. Co., Inc.*, 87-ASBCA No. 25,602, 3 BCA ¶ 20,192 (1987).

118 *Atlantic States Constr., Inc. v. Hand, Arendall, Bedsole, Greaves & Johnston,* 892 F.2d 1530 (11th Cir. 1990); *Jordan-DeLaurenti, Inc.*, ASBCA Nos. 45,467, 46,589, 94-3 BCA ¶ 27,031 (1994).

119 892 F.2d at 1539.

120 ASBCA No. 45,856, 94-3 BCA ¶ 26,958 (1994).

doctrine in *George Hyman Construction Co. v. United States*.[121] In this case, the court found that a general release appearing as an endorsement on the prime contractor's final payment check to a subcontractor sufficiently freed the prime contractor of liability to the subcontractor for purposes of the *Severin* doctrine. The court determined that the subcontractor's act of depositing the check without reservation prevented the prime contractor from later pursuing a claim on its behalf.

Although the *Severin* defense is no longer an issue on federal projects, it has been applied, although not always by name, in other jurisdictions. For example, in *Mars Associates., Inc. v. New York City Educational Construction Fund,*[122] the court held that a prime contractor could not prosecute a claim against the owner on behalf of its subcontractors unless the prime contractor admitted its liability to the subcontractors to the extent it recovered from the owner or, alternatively, had already been found liable to the subcontractors and was seeking indemnification. The Virginia courts have gone even further. In *Apac-Virginia, Inc. v. Virginia Department of Highways & Transportation,*[123] the court held that subcontractor claims could not be prosecuted against a public owner because there was no privity of contract between the subcontractor and the owner.

F. Protecting Yourself

To the extent a subcontractor or prime contractor suspects the other party was the actual cause of its increased costs, the risk of non-recovery from the owner should not be assumed. This can interfere with recovery because the unreserved cooperation in a consolidated claims effort is often essential to a successful claims effort. However, to the extent a subcontractor feels the prime contractor's performance may have increased its costs, the subcontractor must clearly reserve its right to look to the prime contractor for compensation to the extent any portion of its claims is not recoverable against the owner.

---

121 30 Fed. Cl. 170 (1993), *aff'd*, 39 F.3d 1197 (Fed. Cir. 1994).

122 513 N.Y.S.2d 125, 135 (N.Y. App. Div. 1987).

123 388 S.E.2d 841, 842-43 (Va. Ct. App. 1990).

G. Lump Sum Settlements

Consolidated claims are often settled on a lump sum basis, *i.e.*, the constituent elements of the original claim lose their identity in the final negotiated "lump sum" amount. After settlement with the owner is obtained, serious and often more intense negotiations remain between the prime contractor and subcontractors regarding the division of proceeds.

Claims handling or pass through agreements often provide for a pro rata disbursement of lump sum claim proceeds. The pro rata disbursement is often based on the original or audited claim amount. This can result in inequities as parties employ different pricing techniques which tend to inflate claim amounts. Also, such formulae do not take into account factors not caused by the owner which may have contributed significantly to each party's increased costs.

While there is some value to arriving at an equitable disbursement formula prior to commencing the consolidated claims effort, a formula satisfactory to all parties is extremely difficult to develop. If an owner will not negotiate on a segregated claim basis (as is often the case), usually it is in everyone's best interest to leave the question of proceeds disbursement open until after negotiations are concluded.

H. A Word For The Subcontractor

In submitting its portion of the consolidated claim to the prime contractor, a subcontractor must clearly establish that: (1) it expects to attend any and all settlement negotiations with the owner that in any way relate to its claim; and (2) it will not be bound by any settlement with the owner that is reached without its prior concurrence. While it is the prime contractor's contractual right to settle any claims with the owner without subcontractor approval, it is important to establish on the record that the subcontractor will not be bound by a unilateral settlement. While this will not foreclose such a settlement, it will tend to influence a prime contractor in its "closed door" negotiating sessions with the owner, and keep the subcontractor actively involved in the claims process.

IV.
The Claim Submittal As A Trial Exhibit

A. Requirements For Claim Submission

Unlike other types of legal actions, most construction contract claims do not start with the filing of a lawsuit; instead, they start with the submission of a claim for additional costs. Indeed, most standard form construction contracts require the submission of a written claim and a ruling on that submission by the appropriate party, such as the architect or the contracting officer, as a condition precedent to the initiation of legal proceedings.

There are several benefits to a detailed claim presentation. The preparation of the claim package requires the preparer to organize its claim and evaluate the support for its position. It also allows the engineer and the other party to effectively evaluate the claimant's presentation.

B. Organization Of Claim Package

The primary purpose of a claim package is to convince the reviewing party of the merit of the claim. In order to achieve that purpose, the claim package must be realistic, it must be accurate, and it must be organized.

A realistic claim is one where the method of calculation and the total amount of the claim are such that the claim is not immediately rejected by the owner, prime contractor, or other reviewing party. For example, a simple extra work claim that includes an amount for lost profits that is four or five times the base claim amount is likely to be totally rejected. This does not mean that there may not be cases where the damages flowing from an action are not large in comparison to the immediate direct costs or contract amount, but they should be the exception.

In addition to being realistic, a claim must also be accurate. In other words, the base information and the calculation should be accurate. Nothing affects the validity of a claim as much as an obvious mathematical error. Such carelessness does far more harm to the overall claim than the amount of the actual error.

Most construction claims include numerous detailed calculations. For example, many different categories of labor hours are multiplied by different labor rates, many items of equipment usage hours are multiplied by varying equipment rates, and material units are multiplied by various unit prices. There is substantial room for error in both the calculation of quantities and rates or prices and in the actual computations. Next, the direct costs are totaled and multiplied by various percentage markups, fees, labor burden, overhead, profit, and bond. Not only do these involved calculations offer substantial room for error, but when any error is discovered, it frequently requires recomputation of the entire claim. For instance, a single change to the unit price of an item of material may require retotaling of all the direct costs and recomputation of the various markups. Thus, a party is often faced with the decision of submitting its claim with a known minor error or delaying the presentation while the error, which may result in only a few dollars difference in the total claim, is corrected and retyped or printed.

An organized claim is a fairly obvious matter. The "acid test" of organization is whether the reviewing party can read the claim and locate all the backup data without extraneous explanation. Different types of claims, of course, require different organization and presentation. A simple extra work claim, for example, may be nothing more than a total of signed extra work tickets. A logical format would be a cover sheet showing the date of each extra work ticket, the number of labor hours, and the wage rate. Each ticket would be assigned an exhibit number and included under a separate tab. For example:

Exh. No.	Date	Hours		Rate	Extension
1	09/01/97	40	x	$24/hour	$ 960
2	09/02/97	36	x	$24/hour	864
3	09/03/97	48	x	$24/hour	1,152
					$ 2,976
		Markups @ 10%			298
					3,274
		Overhead and Profit @ 20%			655
		TOTAL			$ 3,929

Most claims, however, are not simple extra work claims. Instead, they require the compilation of many different cost elements, most of which are discussed in the previous sections of this chapter. A claim document should be a pyramid. The peak of the pyramid is a summary sheet of all the additional costs claimed. Each portion of the summary sheet should then refer back to additional calculations. Those calculations should then refer back to sub-calculations, the sub-calculations should refer back to compilations of basic data, which should in turn refer back to the basic entry level data. A sample summary is set forth below.

SUMMARY OF INCREASED COSTS

A.	DIRECT LABOR	
	Excavation Labor (Exhibit 1)	$ 959,175
	Concrete Labor (Exhibit 2)	798,700
	SUBTOTAL	1,757,875
	Small Tools @ 5% (Exhibit 3)	87,894
	TOTAL DIRECT LABOR	$ 1,845,769
B.	EQUIPMENT	
	Excavation Equipment (Exhibit 4)	$ 835,363
	Concrete Equipment (Exhibit 5)	296,060
	TOTAL EQUIPMENT	$ 1,131,423
C.	MATERIALS (Exhibit 6)	
	Ace Supply & Equipment	$ 8,925
	Jones Sand & Gravel	2,400
	Construction Form Rental Co.	28,000
	Smith Lumber Yard	32,000
	TOTAL MATERIALS	$ 71,325
D.	SUBCONTRACTORS AND CONSULTANTS (Exhibit 7)	
	Rebar Rebending Company	$ 68,575
	Ground Test Co., Inc.	8,500
	TOTAL SUBCONTRACTORS AND CONSULTANTS	$ 77,075
E.	ESCALATION	
	Labor Escalation (Exhibit 8)	$ 380,800
	Material Escalation (Exhibit 9)	5,100
	TOTAL ESCALATION	$ 385,900
F.	SUBCONTRACTOR CLAIMS	
	Haul-Away, Inc. (Hauling Subcontractor) (See separate submittal)	$ 25,000
	TOTAL DIRECT COSTS	$ 3,536,492
G.	JOBSITE OVERHEAD	
	10% of Direct Costs (Exhibit 10)	$ 353,649
H.	EXTENDED JOBSITE OVERHEAD	
	$1,300.47/day x 182 days (Exhibit 11)	$ 236,686

	TOTAL DIRECT COSTS AND JOBSITE OVERHEAD	$ 4,126,827
I.	HOME OFFICE OVERHEAD 7.5% of Direct Costs and Jobsite Overhead (Exhibit 12)	$ 309,512
J.	EXTENDED HOME OFFICE OVERHEAD $794.30/day x 182 days (Exhibit 13)	$ 144,563
	SUBTOTAL TOTAL DIRECT COSTS AND OVERHEAD	$ 4,580,902
K.	PROFIT 10% (Exhibit 14)	$ 458,090
	SUBTOTAL	$ 5,038,992
L.	BOND .6%	$ 30,234
	SUBTOTAL	$ 5,069,226
M.	INTEREST 04/01/97 to 07/01/97 @ 10% per annum	$ 126,731*
	TOTAL CLAIM	$ 5,195,957

* This portion of the claim will continue to accrue at the rate of $42,244.00 per month until paid.

As can be seen from the above example, each direct cost item on the summary sheet refers to an exhibit. Exhibit 3, for example, refers to a percentage computation for small tools. The actual Exhibit 3 would show the computation for the small tools percentage *(i.e.*, total small tools cost divided by total direct labor cost). The exhibit which shows this calculation should then reference as a sub-exhibit the small tools cost and the labor cost.

All references to exhibit numbers should be readily referable to an exhibit tab which contains the precise information referenced. If intermediate steps between the information set forth in the summary and the exhibited material are needed, the calculation should be shown. On the small tools example, it would be inadequate to simply list the cost of small tools and the cost of direct labor. The division equation which results in the percentage must be shown.

C. How Much Information To Include

Like the general format of the package, the amount of information to include depends a great deal on the party with whom one is dealing.

Some owners are very specific as to the type of information they want. For example, they may request that purchase orders, daily reports, equipment logs, or even canceled checks be submitted. If such information is wanted either by contract or by custom, the answer is clear: provide everything that is requested.

In most claim presentations, however, the provision of detailed information down to purchase orders and daily reports, is "too much of a good thing," especially if a claim is large. The presentation of too much information can make the claim totally unwieldy. It might also invoke a minute and detailed review which could delay the claim processing time and reveal minor discrepancies which impugn the validity of the claim without necessarily being significant. Nevertheless, the contractor should have all the backup information readily available for review. It would be extremely detrimental to the negotiation process if detailed support information is requested and is not readily available. This makes it appear to the reviewer that the claim preparation was less than conscientious.

1. Legal Authority

A question which frequently arises during claim preparation is whether to cite legal authorities in support of the claim's various elements. As a general rule, it is preferable not to include citation of cases or other legal authorities unless the claim will be reviewed in the first instance by the other party's legal representative. The party who might well be prepared to negotiate a claim solely on its technical or cost elements will be prompted to refer the entire matter to its attorney or corporate legal department if the claim package includes legal citations. This will tend to prompt a response citing opposing legal authority. The result is that project or even management level negotiations are impeded in favor of a legal resolution.

In certain instances, it might be appropriate to cite legal authority. For example, if a contractor or other party is claiming an item of damage which is unique or based on a recent court ruling, it might be wise to cite that authority. Legal authority is also appropriate is when a particular claim element is allowed by statute. For example, if interest is recoverable on a claim under a specific state statute at a prescribed rate, it would be appropriate to refer to the statute. In most cases, however, citations to legal authority should be avoided.

The avoidance of legal authority in the actual claim presentation does not mean that the claim need not be legally correct. In some jurisdictions, there are substantial penalties for filing improper claims. For example, filing a mechanic's lien for amounts not properly due may result in the party filing the lien being held for damages if the lien prevents a profitable sale of property. Likewise, the Contract Disputes Act [124] prescribes very severe penalties for submitting an improper claim. It is therefore extremely prudent to have an attorney review a claim package prior to submittal to be sure it is legally correct.

2. *Audit Report*

The inclusion of a formal audit report with the claim is a similar situation to the citation of legal authorities. Again, as a general rule, a formal audit report probably should not be submitted with the initial claim package. The reasons for this are twofold: (1) an audit report only covers very limited items of the claim; and (2) it invites a detailed audit response (similar to the legal response discussed above) instead of management level negotiations.

In most claim situations, a formal audit can only establish that the actual costs were incurred. It can establish the amount of direct labor, the amount of direct material, the overhead rates, the interest rate paid, and similar costs. An auditor cannot, however, establish that the costs incurred were the result of the claimed actions. Therefore, unless there is reason to believe that the reviewer will doubt the fact that the costs were incurred, the inclusion of a formal audit may not advance the claim negotiation process. It should also be noted that under the Contract Disputes Act, a contractor is required to certify its claimed costs.[125] Thus, if costs are improperly claimed, the contractor may be subject to severe financial and criminal penalties. In this situation an audit before certification may be a prudent step.

When, however, a party is certain that the other party will conduct an audit of its claim, an auditor should be retained to assure that the claim will withstand a critical audit and to coordinate with the other party's auditor. It is essential that the claimant's auditor conduct a thorough entrance/exit interview with the reviewing party's auditor.

---

124 41 U.S.C. § 601, *et seq.*

125 41 U.S.C. § 605(c)(1).

This ensures that the audit ground rules are clear to all concerned parties prior to the examination of the records, and also that there are no misunderstandings with respect to the record keeping or cost accounting procedures. Once the reviewer's audit is completed, the claimant's auditor should review it to ensure that acceptable audit procedures were used and to prepare a response, if appropriate. It is not unusual on construction claims for the auditors for the two parties to agree at least on the amount of costs incurred and the allocation of those costs to specific cost categories.

3. Expert Report

In certain types of claims, it may be appropriate to retain experts at the claim preparation stage to assure that the computation of damages may later be verified by the expert. If, for example, a claim is based on the total cost method, an outside expert estimator should certainly verify the reasonableness of the bid by comparing it to the owner's estimate or the bid submitted by the other contractors on the project. If the bid did contain errors or omissions, the expert can correct them prior to the submission of the claim. Again, it may not be appropriate to have the expert include a report with the claim, but he or she should have performed a review.

The reason why the expert's analysis should not be included with the claim differs somewhat from the reasons for not including legal citations or an audit report. Experience has dictated that an expert's evaluation of a particular project will undergo a considerable maturation process as the expert reviews and studies the project. Since the initial claim package is only the first step in the claim resolution process, an expert's final opinion may differ from the original conclusions. Few things are more damaging to an effective claim presentation than an expert who has changed his or her opinion. If, for any reason, an expert's report is included with the initial claim package, it should contain sufficient qualifications so that it can be changed without impugning the entire claim.

Another type of expert which might be consulted at the claim preparation stage is a productivity or construction techniques expert. If a portion of the claim is based on labor loss of efficiency due to delays, disruptions, acceleration, multiplicity of changes, stacking of trades, undue overtime, etc., an expert may be needed to arrive at the loss of efficiency factor. Such experts can also be utilized to verify the reasonableness of the original performance plan and the cost associated with that plan. In addition, an expert can be very helpful in simplifying a contractor's data and presenting it in a manner that highlights the causes for the extra cost. Again, however, it would be highly unusual to have an expert opinion included in the original claim document. As set forth above, an expert's opinion should fully ripen before it is presented to the opposing party.

D. Admissible Evidence

All the damages information must be grounded on admissible evidence. Another advantage of the pyramid system is that it leads to the original documents of entry which are normally admissible into evidence as business records. The documents can then be traced up to the peak of the pyramid, establishing the chain of admissibility all the way. It is important to remember that the base of the pyramid must always be supported by admissible information or, hopefully, by stipulations at some level above the basic entry documents.

E. Charts And Other Graphic Material

Many items of damage are readily explained with charts and graphs. For example, labor escalation is very easily explained by using anticipated and actual labor hour curves as set forth above. Charts and graphs of these types should be included in the original claim package. For trial purposes, it is often advisable to expand these charts and graphs so they can be explained by the witnesses. Whenever large-scale charts or graphs are used, it is important that individual copies be provided to each of the triers of fact. This is especially important in arbitration panels where the three different arbitrators frequently live in different cities and do not have ready access to the main oversized exhibit. In addition, it allows them to place explanatory notes on their copies without defacing the primary trial exhibit.

V.
Trial Presentation Of Damages

If the negotiation process fails and the claim is not settled during the pre-trial stages, the final step is litigation. Regardless of whether the litigation is before a state or federal court with a judge or jury, a board of contract appeals, or an arbitration panel, an effective damages presentation will have the same elements. As with the initial claim package, the key is to make the damages presentation readily understandable.

A. Damages Overview

Since a damages presentation is composed of many individual numbers and calculations, the first step to a successful case is to present a summary of the damages incurred. This summary should serve as an outline of the proof that will be submitted. The summary must list only items that can later be proven. It is very damaging to the overall presentation if an item of damage must be deleted from the summary because of lack of proof.

A summary of all the increased costs, similar to that set forth above,[126] should be marked for identification at the beginning of the damages portion of the case. This document will typically not be admissible into evidence until the conclusion of the damages presentation because it contains summaries of materials that have not yet been introduced into evidence. It will, however, set forth for the court, board, or arbitration panel the damages elements to be proven.

It is sometimes a good idea to have the summary enlarged and set on an easel. This way the entire court can see the elements. It is also a good idea to check off each element of damage as it is established or as evidence in support of each element is admitted. This process gives a sense of finality to the element; it is particularly effective before juries. It also helps the attorney and the court keep track of which elements have been supported by admissible evidence.

Another preliminary step in the damages presentation is to introduce into evidence and explain the company's basic cost records. Since the majority of the damages computations are made from the accounting

---

126 *See supra* Section IV.B.

records, it is important that they be admitted into evidence before there is testimony on the individual computations.

If at all possible, the parties should stipulate to the actual costs incurred on the project. Indeed, most courts, especially federal courts and the various boards of contract appeals, actively urge the parties to stipulate as to actual costs. If such a stipulation is reached, it is not necessary to introduce all the basic cost records.

B. Overview Witnesses

The overview witness is a party who testifies as to the summary. Although the person who will fill the role of the overview witness will vary depending on the contractor's organization, it should be someone who is both familiar with the preparation of the claim and the workings of the cost accounting system.

The overview witness should review the summary sheet and briefly explain how each of the computations were made. For this reason, the party who will serve as the overview witness should usually be intimately familiar with the claim preparation. The overview witness does not actually have to perform all the calculations, but he or she should be the in-house supervisor of claim preparation. In most cases, it is advisable not to have an outside expert as the overview witness. An expert at this stage tends to separate the claim from the actual loss on the project. It is much better to have a witness who has "lived" the damages.

The overview witness should briefly explain each of the calculations. He or she should state who prepared each of the specific claim elements. If the witness personally prepared any portion, he or she should explain how it was calculated. Preferably, the same witness who explains the summary can testify as to the accounting system, if there has not been a stipulation as to actual costs. Otherwise, it may be necessary to have a secondary overview witness. This is typically the company treasurer or comptroller who can explain how the accounting system functions. He or she should testify as to the preparation of the basic entry documents in the field such as the foremen's daily reports and how those reports are eventually accumulated into the system.

C. Fact Witnesses

After the overview witness, the next step is to prove each of the individual elements of the claim. The witnesses who participated in the preparation of each of the elements of increased costs should testify as to their individual sections. The following witnesses might be needed:

A. Direct Labor	The foreman who prepared the daily labor reports to testify as to how the reports were prepared; a project engineer to explain how the actual costs were computed for the claim areas; an estimator to explain the anticipated costs.
Labor Burden @ 10%	A payroll clerk.
Small Tools @ 10%	Anyone who is familiar with the cost reports and the system for purchasing and dispensing tools.
B. Equipment	The company equipment manager to testify how the equipment was allocated to the particular job and how usage records were kept; an accountant or an expert consultant to explain how the hourly rates were devised. If the equipment rates were based on outside rentals, a purchasing agent should be able to explain the computations.
C. Materials	A purchasing agent.
D. Subcontractors and Consultants	A subcontractor purchasing agent, administrator, or project manager.
E. Escalation	The project estimator.
F. Subcontractor Claims	Each subcontractor should be allowed to present its claim as a separate element.
G. Extended Jobsite Overhead	The project cost engineer.
H. Home Office Overhead	The company treasurer or comptroller.
I. Extended Home Office Overhead	The company treasurer or comptroller.
J. Bond	The company treasurer or comptroller and the company's bonding agent.
K. Profit	The project estimator or a company executive who can testify as to the profit history of the firm.
L. Interest	The company treasurer or comptroller, and perhaps a representative of a banking institution.

When each of these persons has testified, the overall summary sheet can be introduced into evidence.

It is not unusual that the person preparing the calculations is not familiar with the underlying formula. For example, the company comptroller may not be familiar with the *Eichleay* formula.[127] In such instances, there should be no hesitancy to testify that he or she performed the calculations in accordance with a formula provided by a consultant or counsel.

D. Expert Witnesses

Once the basic calculations have been explained, it may be necessary to verify portions of them with expert witnesses. Typical types of experts for damages presentations are: (1) auditors; (2) efficiency experts; (3) estimators; (4) technical experts; and (5) scheduling consultants. As set forth above, the expert witnesses should be contacted at the claim preparation stage although their reports will typically not appear in the initial claim package. If this is done, preparing the experts for trial is an easy process.

Auditors may be required to testify as to the validity of the actual costs incurred. Efficiency experts may be called upon to testify as to loss of efficiency computations. Expert estimators may be called upon to testify as to the contractor's bid on all or various portions of the contract when the total cost or modified total cost approach is being used. Technical experts may be called upon to testify as to various types of equipment usage or the need for certain extraordinary supplies or materials. Scheduling consultants may testify as to the amount of delay incurred. Before an expert of any nature testifies, all the records upon which his opinion is based should already be admitted into evidence. Therefore, the damages presentation usually closes with an expert witness.

---

127 For a more detailed discussion about the *Eichleay* formula, see *supra* Section II.D.4.

VI.
Defense Of Damages: Assumption Of Risk

A. Labor

A contractor is generally not excused for delays attributable to labor difficulties or availability. Courts and boards of contract appeals have strictly adhered to the rule that a contractor assumes the risk of hiring and retaining a competent work force.[128]

One noted exception to this rule is labor difficulties resulting from strikes. Delays resulting from strikes are generally considered excusable delays unless the contract specifically provides otherwise.[129] The controlling test as to the excusable nature of delays resulting from strikes is whether the strike was an unforeseeable cause beyond the control of the contractor. It is possible that when a strike was caused by the contractor's bad faith negotiations with its workers, the resulting delay will not be excused.

A similar rule is applied to subcontractor delays. The contractor is responsible for the acts and delays of its subcontractors and cannot excuse its delay in performing the contract on the grounds of its subcontractor's defective or delayed performance.[130] In order for such delays to be excusable to prime contractors, they must be beyond the control and without the fault or negligence of both the prime contractor and the subcontractor. Thus, when a prime contractor asserts the right to a time extension for delays, it has the burden of establishing the excusability of both its and the subcontractor's cause of delay.[131] However, where the owner specified a sole source subcontractor or supplier, the subcontractor's delay was held excusable.[132] Generally, the excusability in this instance depends upon whether the contractor made every reasonable effort to insure timely performance of the sole source subcontractor.

---

128 *Yarling*, AGBCA No. 382, 75-2 BCA ¶ 11,540 (1975).

129 *Contracting & Material Co. v. City of Chicago*, 314 N.E.2d 598, 602-603 (Ill. App. Ct. 1974), *rev'd,* 349 N.E.2d 389 (Ill. 1976).

130 *J.J. Brown Co. v. J.L. Simmons Co.*, 118 N.E.2d 781, 785 (Ill. App. Ct. 1954); *Fritz v. Woldenberg*, 225 N.W. 700, 702 (Wis. 1929).

131 *BECO, Inc.*, ASBCA Nos. 9702, 9734, 1964 BCA ¶ 4493 (1964).

132 *D & E Constr. Co.*, VCBCA No. 561, 67-2 BCA ¶ 6558 (1967).

B. Economic And Financial Difficulty

A contractor is expected to have the financial ability to perform its contract. Lack of working capital does not excuse a failure of performance.[133] A contractor's delay, or failure or inability to perform, because of a lack of sufficient funds is inexcusable regardless of its lack of control over the cause, *i.e.*, economic conditions or the financial failure of a third party on whom the contractor relied for support.[134]

As an exception to the general rule denying relief for delays because of lack of financing, the contractor will be granted relief when undercapitalization is caused by owner actions such as the failure to make progress payments. In *Stamell Construction Co., Inc.*,[135] the Board of Contract Appeals noted this exception and extended it to situations where there was only a threatened withholding:

> A wrongful refusal or failure to make progress payments can amount to a breach of contract justifying abandonment of the work by the contractor or can amount to excusable delay constituting a valid defense against a subsequent assessment of liquidated damages or against a default termination action.
>
> * * *
>
> A mere threat to withhold progress payments, with nothing more, would not necessarily justify a work stoppage by the contractor. But here there was more. Not only did the Contracting Officer write a letter threatening to reject the fabricated, partially-tensioned T-beams, and to withhold progress payments for that completed work, but at the same time the Government's inspector at the job site issued a field memorandum in which he strongly "suggested" that appellant suspend further post-tensioning, which meant in effect suspending any further precast operations.[136]

---

133 *Tucker v. Bitler Bros., Inc.*, 197 N.Y.S.2d 899, 903 (N.Y. App. Div. 1960).

134 *Marionneaux v. Smith,* 163 So. 206 (La. Ct. App. 1935); 17A AM. JUR. 2D *Contracts* § 680 (1991).

135 DOTCAB No. 68-27J, 75-1 BCA ¶ 11,334 (1975).

136 *Id.*

Similarly, in *William Green Construction Co., Inc. v. United States,*[137] the Court of Claims reasoned: "Construction contractors cannot be expected to continue working if the Government improperly holds back on substantial interim payments which enable the project to proceed."[138]

C. Unusually Severe Weather And Acts Of God

Delays occasioned by adverse weather conditions are not excusable to the contractor. Contractors are deemed to be aware of the normal weather patterns in a particular area and should consider their adverse affect on performance when estimating costs. Thus, in *Gross v. Exeter Machine Works,*[139] the defendant contractor was not excused when winter snow storms caused a delay in transportation of material to a project site. It was established that the weather was not unusually severe when measured by the usual winter conditions in northern Pennsylvania. The court noted that winters in that region are expected to be severe and the contractor should have considered that when preparing its bid.[140]

A situation similar to unusually severe weather is a delay caused by an "act of God" or natural disasters. "Acts of God" are often specifically mentioned in construction contracts as justifying time extensions. Absent such provision, "acts of God" are, nevertheless, ordinarily excusable since "acts of God" are: (1) per se beyond the control and without the fault or negligence of the contractor; and (2) not events which a contractor can foresee or anticipate. Natural disasters such as earthquakes, tornadoes, floods, and fires typify "acts of God."[141]

D. Bargained For Risks

1. Delay Damages

The most common and significant bargained for risk in the modern construction contract is represented by the no damages clause. The no

---

137 477 F.2d 930 (Ct. Cl. 1973).

138 *Id.* at 938.

139 121 A. 195 (Pa. 1923).

140 *Id.* at 197.

141 *See, e.g., Tennessee Elec. Power Co. v. White County,* 52 F.2d 1065 (6th Cir. 1931); *Ellis Gray Milling Co. v. Sheppard,* 222 S.W.2d 742 (Mo. 1949).

damages clause has taken several forms, but in essence, it purports to deny a contractor the right to recover delay costs against the owner or its agents regardless of the owner's responsibility for the delay.

The clause is often quite broad with regard to the nature or causes of the delays. An extreme example of a no damages clause provides:

> No payment, compensation or adjustment of any kind (other than the extensions of time provided for . . .) shall be made to the Contractor for damages because of hindrances or delays from any cause in the progress of the work, whether such hindrances or delays be avoidable or unavoidable and the Contractor agrees that he will make no claim for compensation, damages or mitigation of liquidated damages for any such delays, and will accept in full satisfaction for such delays said extension of time.[142]

Although an exculpatory clause which grants immunity from the harmful consequences of one's own negligence will be strictly interpreted by any court, the no damages for delay clause has generally been found to be valid and enforceable.[143]

Most states enforce these clauses literally. Washington, however, is one state which has declared the no damages clause to be invalid and unenforceable. In 1979, the Washington legislature, in response to severe contractor delays incurred during construction of several nuclear power projects administered by a political subdivision of that state, enacted a statute which provides, in part:

> Any clause in a construction contract . . . which purports to waive, release, or extinguish the rights of a contractor, subcontractor, or supplier to damages or an equitable adjustment arising out of unreasonable delay . . . which delay is caused by the acts or omissions of the contractee or persons acting for the contractee is against public policy and is void and unenforceable.[144]

---

142 *A. Kaplen & Son Ltd. v. Housing Auth. of City of Passaic*, 126 A.2d 13, 14 (N.J. Super. Ct. App. Div. 1956).

143 *See, e.g., John E. Green Plumbing & Heating Co., Inc. v. Turner Constr. Co.*, 742 F.2d 965, 966 (6th Cir. 1984); *John Burns Constr. Co. v. City of Chicago,* 601 N.E.2d 1024, 1029-30 (Ill. App. Ct. 1992).

144 WASH. REV. CODE § 4.24.360 (1988).

Courts which have upheld the no damages clause even in the face of unforeseeable or unreasonable delays have circumvented the public policy argument by reasoning that the contractor cannot render meaningless an express condition of the contract which it knowingly and freely accepted. It has also been argued that an owner has likely paid a premium for the clause in the form of added contingencies for delay costs in contractors' bids. However, as previously discussed, these clauses are not enforced without exception and will be construed strictly against the party seeking exemption from liability.

2. *Site Conditions*

Construction contracts often allocate the risk of adverse site conditions to the contractor. The typical vehicle for this risk allocation is the site investigation clause. An example of such a clause may be found in Article 1.2.2 of the American Institute of Architects Document A201. Article 1.2.2 states:

> 1.2.2 Execution of the Contract by the Contractor is a representation that the Contractor has visited the site, become familiar with local conditions under which the Work is to be performed and correlated personal observations with requirements of the Contract Documents.[145]

In spite of such contract language, contractors often fail to investigate the site and instead accept the owner's representations set forth in the contract documents as accurate. Courts have upheld owners' defenses that this clause protects the owner from liability when there are inaccuracies in the specifications as to site conditions.[146]

However, these clauses are not enforced without exception. For example, in *Hollerbach v. United States,*[147] the contractor entered into a contract with the government to repair a dam. In addition to a requirement that bidders investigate the work site, the *Hollerbach* contract included the following clause:

---

145 American Institute of Architects Document A201.

146 *See, e.g., H.B. Mac, Inc. v. United States*, 36 Fed. Cl. 793 (1996); *Neal & Co., Inc. v. United States*, 36 Fed. Cl. 600 (1996), *aff'd*, 121 F.3d 683 (Fed. Cir. 1997); *Weeks Dredging & Contracting, Inc. v. United States*, 13 Cl. Ct. 193 (1987), *aff'd*, 861 F.2d 728 (Fed. Cir. 1988).

147 233 U.S. 165 (1914).

> The dam is now backed for about 50 feet with broken stone, sawdust, and sediment to a height of within 2 or 3 feet of the crest, and it is expected that a cofferdam can be constructed with this stone, after which it can be backed with sawdust or other material.[148]

Once work had begun, the contractor discovered that the material was not of the character that had been represented. The court concluded that the contractor was justified in relying upon the positive representations in the specifications. The court noted that if the government had wished to place upon the claimants the burden of discovering the actual conditions, it should have omitted the unequivocal representation as to the character of the filling material.

If the specifications contain disclaimers as to the accuracy of site data contained in the pre-bid documents, the site investigation clause will be granted full force and effect. In *Archie & Allan Spiers, Inc. v. United States,*[149] the contractor made no pre-bid examination of the worksite. The relevant clause provided the following:

> 1-21. *Examination of Premises.* Before submitting proposals, bidders are expected to inspect carefully the work in place and satisfy themselves as to the character and amount of work to be removed, renewed or replaced, and of new work.[150]

An additional clause supported the owner's assertion that the contractor was to have assured itself of the nature and conditions of the work, regardless of the information provided in the bid package:

> (b) *Changed Conditions.* Information respecting the site of the work given in drawings or specifications has been obtained by Government representatives and is believed to be reasonably correct but the Government does not warrant either the completeness or accuracy of such information and it is the responsibility of the Contractor to verify all such information; *Provided,* That in case of subsurface, latent, or unknown conditions, this contract may be modified . . . .[151]

---

148 *Id.* at 168.
149 296 F.2d 757 (Ct. Cl. 1961).
150 *Id.* at 759.
151 *Id.*

Once work had proceeded, the contractor discovered that the drawings supplied by the government were inaccurate, thereby necessitating significant additional work. The court was unsympathetic to the contractor's plight and referred to the above two warning clauses:

> These warning clauses were not idly placed in the contract. They were meaningful and necessary in the light of the nature of the contract and the condition of the old piers. Therefore, if the contractor miscalculated or failed to examine the site and verify the measurements, which were called to his attention by the contract itself, no claim could arise even though the contractor sustained a loss.[152]

The court distinguished *Hollerbach* and its progeny[153] as follows:

> What the cases generally hold is that where the Government has made a positive representation of a fact of which it should have knowledge, and where in the circumstances the contractor could reasonably rely on the Government's representation without an investigation of its own, the Government will not be relieved of responsibility merely because of the presence of a clause in the specifications admonishing the contractor to inspect the premises. Such is not the case here . . . .[154]

The *Archie & Allen Spiers* court concluded that the contractor, by its clear warnings, did not misrepresent the conditions. In short, the court concluded that the contractor had not been misled.[155]

The central weakness of the contractor's claim in *Archie & Allan Spiers* was the contractor's failure to inspect the premises. In claims where the site has been investigated, yet the alleged changed condition remains undetected, an additional inquiry is required into the thoroughness and reasonableness of the contractor's actual inspection. Such an issue arose in *James Julian, Inc. v. President & Commissioners of Town of Elkton*.[156] In *James Julian*, the contractor entered into a contract with the defendant to lay two sewer lines. Prior to bid

---

152 *Id.* at 763.

153 *See United States v. Spearin*, 248 U.S. 132 (1918); *Christie v. United States*, 237 U.S. 234 (1915).

154 *Archie & Allan Spiers,* 296 F.2d at 763.

155 *Id.; see also Lang-Miller Dev. Co.*, AGBCA No. 81-129-3, 81-2 BCA ¶ 15,433 (1981).

156 341 F.2d 205 (4th Cir. 1965).

submission, the contractor inspected the job site pursuant to its customary practice by walking the site and taking test borings at various designated manhole sites. During this investigation, the contractor failed to discover an old subsurface wharf, which ultimately caused increased performance costs during construction. The contract contained a changed conditions clause. The primary defense of the owner was that the contractor negligently conducted the pre-bid site inspection. The appellate court rejected the trial court's conclusion that the contractor should have discovered the wharf. Instead, the court concluded that "[the] record . . . as a whole . . . [lacked] substantial evidence to support a finding that the pre-bid examination of the plaintiff was careless or negligent or that it was below the standard prevalent in the industry."[157]

Often, the physical placement of the exculpatory clause in the contract documents will determine whether the contractor was adequately warned. In *E.H. Morrill Co. v. State,*[158] the special conditions of the specifications indicated the existence of boulders of a specific size. It was later discovered that the boulders were significantly larger than had been represented. The contract also included the following general conditions clause:

> The bidder shall examine carefully the site of the work and the plans and specifications therefor, and shall satisfy himself as to the character, quality, and quantity of surface and subsurface materials or obstacles to be encountered. He shall receive no additional compensation for any obstacles or difficulties due to surface or subsurface conditions actually encountered.[159]

Although the court determined that the owner had placed responsibility for incorrect representations on the contractor, the court, nonetheless, found for the contractor. The court relied on the fact that the owner had buried this exculpatory clause in the fine print of the general conditions while making the misrepresentation in the special conditions. In doing so, the court recognized the realities of a contractor's standard bidding

---

157 *Id.* at 209; *see also McCormick Constr. Co., Inc. v. United States*, 18 Cl. Ct. 259 (1989), *aff'd,* 907 F.2d 159 (Fed. Cir. 1990).

158 423 P.2d 551 (Cal. 1967).

159 *Id.* at 552-53.

procedure which is to carefully scrutinize special and technical provisions and generally review standard form provisions.

3. *Misrepresentation As To Soil Data*

Owners, by contract, can successfully allocate the risk of actual subsurface conditions to the contractor.[160] In one case however, the time period available to the bidder to confirm the owner's representations affected the force of the exculpatory clause. In *Public Constructors, Inc. v. New York,*[161] a case involving a highway construction contract, the owner had indicated to the bidders that the subsurface soil consisted primarily of coarse grained material which had the capacity to shed moisture. In reality, the soil consisted primarily of fine grained materials and hence, lacked the capacity to permit compaction in moist conditions. The true nature of the subsurface soil necessitated a timely and costly change in procedure for which the contractor sought additional compensation. The owner defended by asserting the existence of a site investigation clause as well as a provision which advised bidders that they were not entitled to rely upon the accuracy of descriptions of subsurface conditions contained in bidding documents. The court rejected the owner's position and concluded that the documents furnished to the bidders contained misrepresentations. Furthermore, the court held that the bidders could not reasonably have been expected to discover in a bid period of three and one-half weeks the kind of information which the owner possessed as a result of testing for several years. In other words, time may not allow for the verification of extensive or complex data.

Thus, when the owner represents that it has conducted such extensive research and proffers misleading results therefrom, a general exculpatory clause will not insulate the owner from liability for an inaccurate evaluation of the site. Additionally, the contractor may allege that the owner made so positive a misrepresentation as to constitute a warranty and that the contractor was, in fact, misled by the assertion. The relative strength of the contractor's position in asserting such a

---

160 *See S & M Constructors, Inc. v. City of Columbus*, 434 N.E.2d 1349, 1353-54 (Ohio 1982); *James McHugh Constr. Co.*, ENGBCA No. 4600, 82-1 BCA ¶ 15,682 (1982).

161 390 N.Y.S.2d 481 (N.Y. App. Div. 1977).

claim, however, will depend heavily upon the contractor's personal knowledge concerning the area.

CHAPTER 12

PREPARING AND PRESENTING DEMONSTRATIVE EVIDENCE AT TRIAL AND DEMONSTRATING THE IMPACT OF TECHNOLOGY ON THE TRIAL PRESENTATION OF THE COMPLEX SURETY CASE

John H. Hinderaker

I.
Introduction

The trial of a substantial construction case poses daunting challenges to a trial lawyer. In addition to the many issues of fact, inference, and credibility that are common to any civil case, the finder of fact in a construction case is often called upon to sort out difficult questions involving engineering, accounting, labor productivity, and causation—often in situations that are mind-numbing complex. Few fact-finders, whether jurors or judges, bring any particular qualifications to this table and even the most conscientious listener has only a finite attention span and ability to absorb new information.

Under the rules of evidence now prevailing in virtually all courts, it is relatively easy to admit into evidence the raw data on which a construction case is based—time cards, job diaries, drawings, specifications, job cost reports, weather data, and the welter of correspondence that a troubled project inevitably generates. To access this raw data and place it before a court and jury is generally not difficult. What is difficult, however, is to organize the raw material into a presentation that is simple enough to be understood by the average person; brief enough to convey in a reasonable time; and persuasive enough to carry the day.

This chapter is addressed to the most common legal issues that arise when one party objects that its adversary has somehow gone too far in organizing, summarizing, or creating evidence. Some of the governing authorities include Rule 1006 of the Federal Rules of Evidence, pertaining to summaries; Rule 702 of the Federal Rules of Evidence, which sets the limits on expert testimony; Rule 403 of the Federal Rules of Evidence, which requires the court to balance the positive value of evidence against any potential for prejudice; and Rule 30(b)(2) of the Federal Rules of Civil Procedure governing the video presentation of deposition evidence. Consideration of these authorities follows.

II.
Summaries Under Rule 1006

Rule 1006 of the Federal Rules of Evidence provides:

> The contents of voluminous writings, recordings, or photographs which cannot conveniently be examined in court may be presented in the form of a chart, summary, or calculation. The originals, or duplicates, shall be made available for examination or copying, or both, by other parties at reasonable time and place. The court may order that they be produced in court.[1]

A. Requirements For Admitting Summaries As Evidence

Rule 1006 and the cases interpreting it have imposed several requirements that must be met before a summary can be admitted as evidence.[2] These requirements include: 1) the underlying writings,

---

1 FED. R. CIV. P. 1006.

2 Note that a trial court's admission or exclusion of a summary under Rule 1006, like other evidentiary rulings, is reviewed for an abuse of discretion. *See, e.g., Harris Market Research v. Marshall Marketing & Communications, Inc.*, 948 F.2d 1518, 1525 (10th Cir. 1991); *United States v. Duncan*, 919 F.2d 981, 988 (5th Cir. 1990); *Gomez v. Great Lakes Steel Div., Nat'l Steel Corp.*, 803 F.2d 250, 257 (6th Cir. 1986). Furthermore, the trial court's ruling will not be reversed unless the party objecting to the court's ruling suffered prejudice as a result. *See, e.g.*,

recordings, or photographs on which the summary is based must be too voluminous to conveniently examine in court; (2) the underlying writings, recordings, or photographs must be made available to the other parties; (3) the underlying writings, recordings, or photographs must otherwise be admissible; and (4) the authenticity and accuracy of the summary must be established.

1. Voluminous

Rule 1006 requires that the writings, recordings, or photographs upon which a summary is based be too voluminous to conveniently examine in court. This does not mean that a summary is admissible only when it is *impossible* to examine the writings, recordings, or photographs underlying the summary.[3] Instead, Rule 1006 requires that the writings, recordings, or photographs cannot be examined *conveniently* because it would be too time-consuming or burdensome.[4]

Many types of evidence have been deemed by courts to be too voluminous to examine conveniently in court and therefore appropriate as summaries. Written summaries of insurance company records,[5] composite tapes prepared from over two hundred hours of taped broadcasts,[6] composite tapes prepared from twenty hours of recorded conversations,[7] a written summary of numerous manufacturers' asbestos

---

United States v. Winn, 948 F.2d 145, 157-58 (5th Cir. 1991); *United States v. Miller*, 771 F.2d 1219, 1238 (9th Cir. 1985).

3 *United States v. Scales*, 594 F.2d 558, 562 (6th Cir. 1979).

4 *See United States v. Stephens*, 779 F.2d 232, 239 (5th Cir. 1985). According to the Fifth Circuit in *Stephens*, "The fact that the underlying documents are already in evidence does not mean that they can be 'conveniently examined in court.'" *Id.* (quoting *United States v. Lemire*, 720 F.2d 1327, 1347 (D.C. Cir. 1983)); *see also* FED. R. EVID. 1006 advisory committee's note ("The admission of summaries of voluminous books, records, or documents offers the only practicable means of making their contents available to judge and jury.").

5 *Duncan*, 919 F.2d at 988.

6 *United States v. Bakker*, 925 F.2d 728, 736-37 (4th Cir. 1991).

7 *United States v. Gorel*, 622 F.2d 100, 106 (5th Cir. 1979).

products sold to a shipyard over a period of years,[8] a chart summarizing damages suffered in a construction case due to bidding errors and time delays,[9] and written summaries of licensing and processing fees due under numerous license agreements[10] have all been deemed appropriate because the originals were too voluminous to examine conveniently.

In addition to looking at the total number of writings, recordings, or photographs to determine whether they can be examined conveniently, courts also look to whether the jury would be able to understand the original writings, recordings, or photographs without some type of summary to aid them. If the "average jury" is likely to be confused by the evidence, the court will be more likely to allow the summary on the basis that the original writings, recordings, or photographs cannot conveniently be examined in court.[11] For example, evidence of a complicated insurance fraud scheme[12] and evidence of a complicated conspiracy to kidnap[13] have been deemed appropriate as summaries because the underlying documents would be confusing to the average jury.

If the underlying writings, recordings, or photographs can be conveniently examined in court because they are few in number and/or easy to understand, the proponent of a summary will not be allowed to introduce the summary into evidence. For example, in *United States v.*

---

8 *Fagiola v. National Gypsum Co. AC & S, Inc.*, 906 F.2d 53, 56 (2d Cir. 1990).

9 *C.L. Maddox, Inc. v. Benham Group, Inc.*, 88 F.3d 592, 597-98 (8th Cir. 1996).

10 *Harris Market Research v. Marshall Marketing & Communications, Inc.*, 948 F.2d 1518, 1523 (10th Cir. 1991).

11 *United States v. Winn*, 948 F.2d 145, 158 (5th Cir. 1991) ("the average jury cannot be rationally expected to compile on its own such charts and summaries which would piece together evidence previously admitted and revealing a pattern suggestive of criminal conduct"); *United States v. Duncan*, 919 F.2d 981, 988 (5th Cir. 1990) ("We cannot rationally expect an average jury to compile summaries and to create sophisticated flow charts to reveal patterns that provide important inferences about the defendants' guilt.").

12 *Duncan*, 919 F.2d at 988.

13 *Winn*, 948 F.2d at 158.

Carr,[14] the United States Court of Appeals for the Seventh Circuit upheld the district court's refusal to admit expert testimony summarizing five recorded conversations.[15] Because the conversations were "uncomplicated conversation[s] in English," and because the transcripts of the conversations totaled less than eighty pages, the transcripts and recordings could conveniently be examined in court and the expert summaries would be inappropriate.[16] Similarly, in *Daniel v. Ben E. Keith Co.*,[17] the United States Court of Appeals for the Tenth Circuit upheld the district court's refusal to allow the plaintiffs to introduce summary evidence based on eighteen one-page pulmonary test results. The court's ruling was based in part on the fact that the original test results had already been used by the defense counsel "without undue difficulty."[18]

2. *Making The Originals Available*

The writings, recordings, or photographs on which a summary is based, as well as the summary itself, must be made available to the opposing party prior to introduction of the summary into evidence. Furthermore, the originals and summaries must be make available at a reasonable time and place.[19] For example, in *Daniel v. Ben E. Keith Co.*,[20] the United States Court of Appeals for the Tenth Circuit upheld the district court's refusal to allow summary evidence where the opposing counsel had only had an opportunity to review the proffered summary during the lunch hour prior to the introduction of the summary into evidence.[21] In *Square Liner 360 Degrees, Inc. v. Chisum*,[22] the Eighth Circuit held that the trial court erred in admitting summary

---

14 965 F.2d 408 (7th Cir. 1992).
15 *Id.* at 412.
16 *Id.*
17 97 F.3d 1329 (10th Cir. 1996).
18 *Id.* at 1335.
19 *Square Liner 360 Degrees, Inc. v. Chisum*, 691 F.2d 362, 376 (8th Cir. 1982).
20 97 F.3d 1329 (10th Cir. 1996).
21 *Id.* at 1335.
22 691 F.2d 362 (8th Cir. 1982).

evidence when the original documents on which the summaries were based were not provided to opposing counsel in advance of trial.[23] In contrast, in *United States v. Bakker*,[24] the Fourth Circuit upheld the admission of composite tapes where the opposing party had at least six months prior to trial to review the original tapes.[25]

The opposing party's right to examine the summaries and underlying writings, recordings, or photographs is not limited by its failure to request the underlying documentation during discovery.[26] Even if the party did not request the underlying documentation during discovery, the party offering the Rule 1006 summary has an affirmative duty to make the underlying documentation available before it introduces a summary based on that documentation at trial.[27] The United States Court of Appeals for the First Circuit has expanded the affirmative duty under Rule 1006 even further, holding that a party planning to invoke Rule 1006 for admission of a summary exhibit must "identify its exhibit as such, provide a list or description of the documents supporting the exhibit, and state when and where they may be reviewed."[28]

3. *Underlying Evidence Otherwise Admissible*

In order to admit a summary, the writings, recordings, or photographs on which the summary is based must be admissible into evidence. The rule does not require that the underlying evidence actually be *admitted*, however, just that the underlying evidence is admissible. A few decisions do indicate that the writings, recordings, or photographs underlying a Rule 1006 summary *cannot* be admitted into

---

23 *Id.* at 376.

24 925 F.2d 728 (4th Cir. 1991).

25 *Id.* at 736.

26 *Air Safety, Inc. v. Roman Catholic Archbishop of Boston*, 94 F.3d 1, 8 (1st Cir. 1996); *Square Liner 360 Degrees, Inc. v. Chisum*, 691 F.2d 362, 376 (8th Cir. 1982).

27 *Air Safety, Inc.*, 94 F.3d at 8; *Square Liner 360 Degrees, Inc.*, 691 F.2d at 376.

28 *Air Safety, Inc.*, 94 F.3d at 8.

evidence if the summary itself is to be admitted.[29] However, these decisions are the exception; most cases do not mandate that the underlying documentation *cannot* be admitted into evidence, nor do they require that the underlying documentation *must* be admitted.[30] Instead, the underlying writings, recordings, or photographs must only be *admissible*.

The most common problem with the admissibility of the underlying writings, recordings, or photographs is that those underlying documents are hearsay.[31] The most common way to combat this objection to a summary is to argue that the underlying writings, recordings, or photographs fit within the business records exception to the hearsay rule,[32] found in Rule 803(6) of the Federal Rules of Civil Procedure.[33]

---

29 *See, e.g., United States v. Grajales-Montoya*, 117 F.3d 356 (8th Cir. 1997) ("[R]ule [1006] appears to contemplate, however, that a summary will be admitted instead of, not in addition to, the documents that it summarizes").

30 *Compare C.L. Maddox, Inc. v. Benham Group, Inc.*, 88 F.3d 592, 601 (8th Cir. 1996) (summary prepared from evidence already admitted into evidence was admissible); *Harris Market Research v. Marshall Marketing & Communications, Inc.*, 948 F.2d 1518, 1525 (10th Cir. 1991) (summary prepared from evidence already admitted into evidence was admissible); *United States v. Duncan*, 919 F.2d 981, 988 (5th Cir. 1990) (summary prepared from evidence already admitted into evidence was admissible) *with Air Safety, Inc. v. Roman Catholic Archbishop of Boston*, 94 F.3d 1, 7 n.14 (1st Cir. 1996) (evidence underlying summary need not be admitted into evidence); *United States v. Bakker*, 925 F.2d 728, 736-37 (4th Cir. 1991) (summary prepared from evidence not admitted into evidence was admissible).

31 *See Soden v. Freightliner Corp.*, 714 F.2d 498, 506 (5th Cir. 1983) (excluding summary evidence based on hearsay).

32 *See, e.g., Martin v. Funtime, Inc.*, 963 F.2d 110, 116 (6th Cir. 1992); *Paddack v. Dave Christensen, Inc.*, 745 F.2d 1254, 1259 (9th Cir. 1984).

33 Rule 803(6) provides:

A memorandum, report, record, or data compilation, in any form, of acts, events, conditions, opinions, or diagnoses, made at or near the time by, or from information transmitted by, a person with knowledge, if kept in the course of a regularly conducted business activity, and if it was the regular practice of that business activity to make the memorandum, report,

Sometimes summaries are based on writings, recordings, or photographs that are partially admissible and partially inadmissible. In such a situation, the portions of the summary based on admissible writings, recordings, or photographs is admissible, and the portion based on inadmissible information is not. However, if the proponent of the summary cannot establish which information in the summary came from the admissible writings, recordings, or photographs, the entire summary will be excluded. For example, in *Paddack v. Dave Christensen, Inc.*,[34] the plaintiffs sought to introduce a summary into evidence based on three underlying sources of information.[35] Although each of the three sources constituted hearsay, two of the sources fit within the business records exception to the hearsay rule. The third source did not fit within any hearsay exception, however.[36] Because the witness who prepared the summary was unable to differentiate which information in the summary came from the admissible business records sources and which came from the inadmissible source, the court excluded the entire summary.[37]

4. Authenticity And Accuracy

Before a summary can be admitted into evidence, the proponent of the summary must lay the proper foundation to establish the accuracy and authenticity of the summary and of the underlying writings, recordings, or photographs.

---

record, or data compilation, all as shown by the testimony of the custodian or other qualified witness, unless the source of information or the method or circumstances of the preparation indicate lack of trustworthiness. The term "business" as used in this paragraph includes business, institution, association, profession, occupation, and calling of every kind, whether or not conducted for profit.

FED. R. EVID. 803(6).

34 745 F.2d 1254 (9th Cir. 1984).

35 *Id.* at 1259.

36 *Id.* at 1259.

37 *Id.* at 1260.

The writings, recordings, or photographs underlying the summary must be authenticated pursuant to Rule 901 of the Federal Rules of Evidence.[38] For example, in *United States v. Gorel*,[39] where a two hour summary of over twenty hours of recorded conversations were presented as evidence, the FBI agents who wired the recording devices testified to authenticate the recordings.[40] The agents testified as to their training and experience in the use of recording devices, the type of equipment and procedures used to record the conversations, and the chain of custody of the tapes.[41]

The summary itself must also be authenticated. Two common methods have been deemed appropriate to lay the foundation for the authenticity of the summary. One method to authenticate the summary is to introduce the summary through the witness that prepared it.[42] Another method to authenticate the summary is to admit the underlying writings, recordings, or photographs into evidence.[43]

The summary must also be accurate. For example, in *Vasey v. Martin Marietta Corp.*,[44] the court refused to admit a summary chart offered by the plaintiff when the court was unable to verify the accuracy of the chart because the plaintiff's counsel was unable to explain to the court the meaning of its contents.[45] Similarly, the court in *United States v. Bakker*[46] refused to admit the defendant's summary charts where the

---

38 *See Fagiola v. National Gypsum Co. AC&S, Inc.*, 906 F.2d 53, 58 (2d Cir. 1990).

39 622 F.2d 100 (5th Cir. 1979).

40 *Id.* at 106.

41 *Id.*; *see also United States v. Segines*, 17 F.3d 847, 854 (6th Cir. 1994) (providing similar authentication of summary tape recordings).

42 *State Office Sys., Inc. v. Olivetti Corp. of Am.*, 762 F.2d 843, 845 (10th Cir. 1985). Some courts have expressly noted that a summary cannot be prepared by a lawyer trying the case, but must be prepared by a witness available for cross-examination. *See, e.g.*, *United States v. Grajales-Montoya*, 117 F.3d 356, 361 (8th Cir. 1997).

43 *See, e.g.*, *United States v. Duncan*, 919 F.2d 981, 988 (5th Cir. 1990); *United States v. Stephens*, 779 F.2d 232, 238 (5th Cir. 1985); *United States v. Howard*, 774 F.2d 838, 844 (7th Cir. 1985).

44 29 F.3d 1460 (10th Cir. 1994).

45 *Id.* at 1468-69.

46 925 F.2d 728 (4th Cir. 1991).

government demonstrated two ways in which the charts were inaccurate.[47]

As long as the summary and the underlying documentation are accurate and authentic, the summary will be admitted even if it may not be completely representative of the underlying materials. For example, the court in *Bakker* admitted summary tapes despite the defendant's objection that the tapes were unrepresentative composites of the original tapes; the court held that such objection went to the *weight* to be accorded the summary tapes, not to the *admissibility* of the tapes.[48] The defendant had ample opportunity to show how and why the tapes were unrepresentative, and therefore to be given little weight, by playing the original tapes for the jury.[49]

A jury instruction can eliminate the concern, such as the defendant's in *Bakker*, that undue influence will be placed on a summary that is not completely representative of the underlying documentation. For example, in *Fagiola v. National Gypsum Co. AC & S, Inc.*,[50] where a summary presented did not contain all of the available evidence and was also not prepared from a random sampling of the available evidence, the judge cautioned the jury:

> [I]t's important to recognize that . . . the record[s] summarize[d] are not all of the records . . . or even a random selection of the records . . . You should not assume that these are all, or even a random selection from these records . . . [W]ith those caveats, you may consider these summaries and give them such weight as they deserve.[51]

Such a cautionary instruction ensures that the jury is aware of the limitations of the summary and will not give it undue weight.[52]

---

47 *Id.* at 737.

48 *Id.*

49 *Id.*

50 906 F.2d 53 (2d Cir. 1990).

51 *Id.* at 56.

52 *Id.* at 57-58.

B. Use Of Summaries As Jury Aids

The use of summaries as jury aids differs from the use of them as actual evidence.[53] Rule 1006 of the Federal Rules of Evidence deals only with summaries presented as evidence. The use of summaries as jury aids or "pedagogical charts" is not governed by Rule 1006. Unlike summaries presented under Rule 1006, the evidence underlying jury aids *must* be admitted into evidence.[54] The chart merely summarizes or organizes the testimony or documents that have already been admitted.[55]

Because most courts no longer mandate that the documentation underlying Rule 1006 summaries cannot be admitted into evidence,[56] it is often difficult to differentiate between Rule 1006 summaries for which the underlying documentation has been admitted into evidence and summaries used as jury aids. However, most courts still require that summaries used as jury aids instead of as Rule 1006 summaries be accompanied by a limiting instruction to the jury stating that the summary itself is not evidence.[57] For example, the court may instruct the jury:

> Summary testimony by a witness and the charts or summaries prepared by him and admitted into evidence are received for the purpose of explaining facts disclosed by testimony or exhibits which are evidence

---

53 *See, e.g.*, *United States v. Winn*, 948 F.2d 145, 159 (5th Cir. 1991) ("it is appropriate to distinguish between charts and summaries presented pursuant to Rule 1006 . . . and mere pedagogical charts"); *Gomez v. Great Lakes Steel Div., Nat'l Steel Corp.*, 803 F.2d 250, 257-58 (6th Cir. 1986); *Pierce v. Ramsey Winch Co.*, 753 F.2d 416, 431 (5th Cir. 1985).

54 *See Winn*, 948 F.2d at 159; *United States v. Bakker*, 925 F.2d 728, 736 (4th Cir. 1991).

55 *Pierce*, 753 F.2d at 431.

56 *See supra* Section II.A.3 (discussing the requirement that the evidence underlying Rule 1006 summaries be *admissible*, but not necessarily admitted or not admitted).

57 *Daniel v. Ben E. Keith Co.*, 97 F.3d 1329, 1335 (10th Cir. 1996); *United States v. Paulino*, 935 F.2d 739, 753 (6th Cir. 1991); *Gomez*, 803 F.2d at 257-58; *United States v. Howard*, 774 F.2d 838, 844 n.4 (7th Cir. 1985).

> in the case. Such charts or summaries are not evidence in the case. They are not in and of themselves evidence or proof of any facts.[58]

Accordingly, when a summary is clearly a jury aid instead of a summary under Rule 1006, a cautionary instruction will be required.

C. Prejudicial Effect Of Summaries

Summaries, whether admitted as evidence under Rule 1006 or as jury aids, can make a strong impression on juries. Presentation of evidence in summary form can vest the summaries "with 'an air of credibility' independent of the evidence purported to be summarized."[59]

Even if a summary has potential prejudicial effect on the jury, courts often deem such prejudice "neutralized" by an instruction by the court. The court may instruct the jury that, even though the summary has technically been admitted into evidence, the jury should not treat the summary as evidence.[60] The court may further instruct the jury to disregard any summary that is not supported by the other evidence admitted in the case.[61]

III.
Using Computer Animations And Simulations As Demonstrative Evidence

A. Using Computer Animations And Simulations

Computer animations and simulations have been used as demonstrative evidence in many types of cases.[62] Courts have admitted

---

58 *Winn*, 948 F.2d at 157 n.30.

59 *United States v. Means*, 695 F.2d 811, 817 (5th Cir. 1983).

60 *United States v. Duncan*, 919 F.2d 981, 988 (5th Cir. 1990).

61 *United States v. Massey*, 89 F.3d 1433, 1441 (11th Cir. 1996); *Duncan*, 919 F.2d at 988.

62 Demonstrative evidence is that evidence which is used to "illustrate some verbal testimony." BLACK'S LAW DICTIONARY 432 (6th ed. 1990). Demonstrative evidence illustrates expert or other testimony as opposed to providing the basis for such testimony.

computer animations and simulations depicting the differences between printing presses,[63] a fatal automobile accident,[64] a fatal accident involving an automobile and a train,[65] a fire inside the engine of a plane that crashed,[66] and the events leading up to the shooting of an individual by a police officer.[67]

A trial court likely will view the proposed animation or simulation outside the presence of the jury before it will admit the animation or simulation into evidence.[68] The court can thereby assess admissibility issues before the animation or simulation is presented to the jury. In this way, the court acts as the "gatekeeper" with respect to evidentiary issues.[69]

B. Admissibility Issues

Several admissibility issues arise with respect to the use of computer animations and simulations as demonstrative evidence. A computer animation or simulation used as demonstrative evidence must be authenticated under Rule 901 of the Federal Rules of Evidence, it must be relevant under Rules 401 and 402, and the probative value of the animation or simulation must not be substantially outweighed by the risk of unfair prejudice to the opposing party under Rule 403.[70] As with

---

63 *Rockwell Graphic Sys., Inc. v. DEV Industr., Inc.*, No. 84-C-6746, 1992 WL 330356, at *1 (N.D. Ill. Nov. 4, 1992).

64 *People v. McHugh*, 476 N.Y.S.2d 721, 722 (N.Y. Sup. Ct. 1984).

65 *Robinson v. Missouri Pacific R. Co.*, 16 F.3d 1083, 1086 (10th Cir. 1994).

66 *Datskow v. Teledyne Continental Motors Aircraft Prods., A Div. of Teledyne Indust., Inc.*, 826 F. Supp. 677, 685 (W.D.N.Y. 1993).

67 *Hinkle v. City of Clarksburg, W.V.*, 81 F.3d 416, 424 (4th Cir. 1996).

68 *See id.* at 425 (encouraging trial courts to examine animations or simulations outside the jury's presence); *Robinson,* 16 F.3d at 1088 (same).

69 *See Robinson*, 16 F.3d at 1089.

70 Note the when the computer animation or simulation is being used as scientific evidence and/or the basis for expert testimony, as opposed to demonstrative evidence, it must also meet the expert testimony requirements of the Federal Rules of Evidence or the applicable state rules. *See, e.g., Commercial Union Ins. Co. v. Boston Edison Co.,* 591

other evidentiary rulings, a district court's rulings with respect to the admissibility of computer animations and simulations will be upheld absent an abuse of discretion that resulted in prejudice to the party opposing admission or exclusion of the particular evidence.[71]

1. *Authentication*

The computer animation or simulation must be authenticated pursuant to Rule 901 of the Federal Rules of Evidence. The authentication requirement of Rule 901 is satisfied by "evidence sufficient to support a finding that the matter in question is what its proponent claims."[72]

If an animation or simulation is not authenticated, it will be excluded. For example, in *Bledsoe v. Salt River Valley Water Users' Ass'n*,[73] the court held that the trial court erred in admitting a computer simulation of a bicycle accident.[74] The proponent of the computer simulation did not lay an adequate foundation as to the authenticity of the simulation.[75] According to the court, the proponent of a computer simulation must, at a minimum, "show that the computer simulation fairly and accurately depicts what it represents."[76]

N.E.2d 165, 168-70 (Mass. 1992); *Schaeffer v. General Motors Corp.*, 360 N.E.2d 1062, 1066-67 (Mass. 1977).

71 *See Hinkle*, 81 F.3d at 425; *Strock v. Southern Farm Bureau Cas. Ins. Co.*, No. 92-2357, 1993 WL 279069, at *1 (4th Cir. July 12, 1993).

72 FED. R. EVID. 901(a).

73 880 P.2d 689 (Ariz. Ct. App. 1994). Although *Bledsoe* is a state case decided under state rules of evidence, it illustrates the need to provide an adequate foundation for computer animations and simulations regardless of the evidentiary rules being used.

74 *Id.* at 691.

75 *Id.* at 692.

76 *Id.*; *see also* FEDERAL JUDICIAL CENTER, MANUAL FOR COMPLEX LITIGATION, § 34.32 (3d. ed. 1995) (proponent of computer animation or simulation "may be required to present both the factual information relied on and the process by which this information was used to produce the animation.").

2. *Relevance*

Relevant evidence is any evidence which tends to make any fact of consequence to the case more probable or less probable.[77] A computer animation or simulation, like other evidence, must, under Rule 402 of the Federal Rules of Evidence, be relevant to be admitted.

3. *Seeing Is Believing: The Potential Prejudicial Effect Of Computer Animations And Simulations*

Under Rule 403 of the Federal Rules of Evidence, relevant evidence may be excluded if its "probative value is substantially outweighed by the danger of unfair prejudice, confusion of the issues, or misleading the jury, or by considerations of undue delay, waste of time, or needless presentation of cumulative evidence."[78]

a. The Prejudicial Effect

Because of the significant potential prejudicial effect of computer animations and simulations, such evidence can be, and often is, excluded under Rule 403. Four characteristics of computer animations and simulations can lead to a ruling that that an animation or simulation must be excluded under Rule 403. First, the impact of visual evidence on the jury may cause the animation or simulation to be too prejudicial. Second, the cost of computer animations or simulations could lead a court to rule that their use by one financially able party is prejudicial to the opposing party. Third, the animation or simulation could actually be an attempted recreation of the event depicted. Finally, the belief that

---

77 Rule 401 of the Federal Rules of Evidence provides:

> "Relevant evidence" means evidence having any tendency to make the existence of any fact that is of consequence to the determination of the action more probable or less probable than it would be without the evidence.

FED. R. EVID. 401.

78 FED. R. EVID. 403.

computers are infallible machines could lead jurors to place too much impact on the credibility of a computer animation or simulation.

Visual computer animations and simulations are a "powerful evidentiary tool"[79] that have a very strong impact on juries and on the outcomes of the cases in which they are presented. Studies indicate that jurors remember only ten percent of oral testimony after a three-day period.[80] By contrast, after a three-day period, jurors remember a full sixty-five percent of evidence presented both orally and visually.[81] According to the Federal Judicial Center, "[t]he powerful impression left by computer animation may be difficult for an opposing party to overcome." In fact, "[o]nce jurors view the animation, it can become difficult to persuade them to accept a different version of events."[82] The strong impact this evidence may have may lead a court to exclude it as too prejudicial under Rule 403.

Computer animations and simulations can be very expensive. Often a party will lack the monetary resources to produce this very compelling evidence. The Federal Judicial Center has cautioned that where one party has presented a computer animation or simulation, but the opposing party lacks the resources to present its own computer animation or simulation to the jury, "particular attention needs to be given to the fairness of the presentation."[83] In other words, the lack of financial ability to respond adequately to a computer animation or recreation may lead a court to exclude the animation or recreation as too prejudicial.

Computer animations and simulations are appropriately used as demonstrative evidence to visually illustrate testimonial evidence of how a particular event occurred. However, if the animation or

---

79 *Robinson v. Missouri Pacific R. Co.*, 16 F.3d 1083, 1088 (10th Cir. 1994).

80 Rebecca White Berch, *A Proposal to Amend Rule 30(b) of the Federal Rules of Civil Procedure: Cross-Disciplinary and Empirical Evidence Supporting Presumptive Use of Video to Record Depositions*, 56 FORDHAM L. REV. 347, 372 (1990).

81 *Id.*

82 FEDERAL JUDICIAL CENTER, MANUAL FOR COMPLEX LITIGATION, § 34.32 (3d. ed. 1995).

83 *Id.*

simulation purports to actually recreate the event, such that the jury is lead to believe that "they are seeing a repeat of the actual event," the court may deem the animation or simulation too prejudicial.[84]

The concern that jurors will believe that computers are infallible machines may also lead a court to exclude computer animations and simulations as too prejudicial. According to the New York Supreme Court in *People v. McHugh*,[85] "[c]omputers are simply mechanical tools—receiving information and acting on instructions at lightening speed."[86] This view, which may be shared by jurors as well as judges, ignores the fact that the computer programmer controls the manner in which the computer manipulates data. In fact, "computer-animated images are ultimately human-created and thus susceptible to human biases."[87]

b. Eliminating The Prejudicial Effect

Several methods exist to eliminate a ruling excluding a computer animation or simulation. First, it must be remembered that, in order to exclude evidence under Rule 403, the danger of prejudice must *substantially outweigh* the probative value of the evidence.[88] In other words, the court cannot exclude evidence under Rule 403 unless the danger of prejudice is much greater than the probative value of the evidence.

Second, the prejudicial effect of computer animations and simulations may be eliminated by a cautionary instruction to the jury.

---

84 *See, e.g.*, *Datskow v. Teledyne Continental Motors Aircraft Prods., A Div. of Teledyne Industr., Inc.*, 826 F. Supp. 677, 686 (W.D.N.Y. 1993); *see also Racz v. R.T. Merryman Trucking, Inc.*, Civ. A. No. 92-3404, 1994 WL 124857, at *5 (E.D. Pa. Apr. 4, 1994) (excluding computer simulation because "the viewing of the computer simulation might more readily lead the jury to accept the data and premises underlying the defendant's expert's opinion").

85 476 N.Y.S.2d 721 (N.Y. Sup. Ct. 1984).

86 *Id.* at 722-23.

87 FEDERAL JUDICIAL CENTER, MANUAL FOR COMPLEX LITIGATION, § 34.32 (3d. ed. 1995).

88 FED. R. EVID. 403.

For example, in *Datskow v. Teledyne Continental Motors Aircraft Products*,[89] in response to the concern that a computer animation was being offered as an exact recreation of the events leading up to a plane crash, as opposed to simply an animation of an expert's opinion of the events, the court ruled that a cautionary instruction eliminated any potential prejudice such a concern caused.[90] The court instructed the jury that the animation was "not meant to be a re-creation of the accident" but instead consisted of "computer pictures to help [them] understand" the expert's opinion of the events leading up to the plane crash.[91]

Similarly, the court in *Hinkle v. City of Clarksburg, West Virginia*[92] upheld the district court's admission of a computer animation illustrating an expert's version of the events leading up to the shooting of an individual by a police officer.[93] The court instructed the jury that "[t]his animation is not meant to be a recreation of the events, but rather it consists of a computer picture to help you understand [the expert's] opinion."[94] The court went on to instruct the jury, "to reinforce the point, the video is not meant to be an exact recreation of what happened during the shooting, but rather it represents [the expert's] evaluation of the evidence presented."[95] The Fourth Circuit held that the district court's instruction eliminated any possible prejudicial effect the animation had.[96]

The court in *Robinson v. Missouri Pacific Railroad Co.*[97] also upheld the admission of a simulation of a fatal accident involving a car and a train where the district court had given a cautionary instruction to the jury:

---

89 826 F. Supp. 677 (W.D.N.Y. 1993).
90 *Id.* at 685.
91 *Id.*
92 81 F.3d 416 (4th Cir. 1996).
93 *Id.* at 424.
94 *Id.* at 425.
95 *Id.*
96 *Id.*
97 16 F.3d 1083 (10th Cir. 1994).

> Again, I want to emphasize to you that you're not to take this as a recreation of the accident because there's no way that a year or two years after the fact that an accident could be recreated. The only reason this is shown is that the [expert] witness will testify about certain principles that he feels that this video would show to the jury and perhaps it would be helpful to you. So just bear in mind that you cannot recreate an accident.[98]

The court's instruction, combined with the ability of the opposing party to cross-examine the expert witness upon whose testimony the simulation was based, eliminated any prejudice caused by the admission of the simulation.[99]

Early disclosure of the computer animation or simulation to the opposing party also will make the court more likely to admit the animation or simulation. If the proponent of the animation or simulation attempts to introduce it at trial with little or no prior disclosure to the opposing party, the court will likely find that the animation or simulation is too prejudicial to admit at trial.[100]

IV.
Videotaped Depositions

Rule 30(b)(2) of the Federal Rules of Civil Procedure provides:

> The party taking the deposition shall state in the notice the method by which the testimony shall be recorded. Unless the court orders otherwise, it may be recorded by sound, sound-and-visual, or stenographic means, and the party taking the deposition shall bear the cost of the recording. Any party may arrange for a transcription to be

---

98 *Id* at 1086.

99 *Id.* at 1088.

100 *See, e.g.*, *Van Houten-Maynard v. ANR Pipeline Co.*, No. 89-C-0377, 1995 WL 317056, at *12 (N.D. Ill. May 23, 1995); *Baugh v. Gulf Air Transport, Inc.*, 526 So. 2d 1239, 1240-41 (La. Ct. App. 1988) (noting that the opposing party would not be able to effectively cross-examine based on the late disclosure of the exhibit).

> made from the recording of a deposition taken by non-stenographic means.[101]

This rule, enacted as part of the 1993 amendments to the Federal Rules of Civil Procedure, is a substantial change from the previous rule regarding depositions. The previous rule, enacted as part of the 1980 amendments to the Federal Rules, provided that deposition testimony could be taken by means other than stenographic means only if the parties stipulated as such or if the court ordered it upon motion.[102] Based on the 1993 amendment to the rule, which allow for "non-stenographic" depositions by right, videotaping depositions has become "routine practice."[103]

If a party intends to offer a videotaped deposition as evidence at trial, a transcript of the deposition will be required pursuant to Rule 26 and Rule 32.[104] Rule 26(a)(3) provides:

> [A] party shall provide to other parties the following information regarding the evidence that it may present at trial other than solely for impeachment purposes:
>
> * * *
>
> (B) the designation of those witnesses whose testimony is expected to be presented by means of a deposition and, if not taken stenographically, a transcript of the pertinent portions of the deposition testimony . . . .[105]

Rule 32(c) provides that a party offering deposition testimony as evidence at trial "may offer it in stenographic or nonstenographic form,

---

101 FED. R. CIV. P. 30(b)(2).

102 FED. R. CIV. P. 30(b)(4) (1980 version); *see also Gillen v. Nissan Motor Corp.*, 156 F.R.D. 120, 122 (E.D. Pa. 1994). Note that some states still follow the older version of the rule on depositions, allowing videotaped depositions only if the parties have agreed or by leave of the court. *See, e.g.*, ARIZ. R. CIV. P. 30(b)(4); WASH. R. CIV. P. 30(b)(4).

103 *Gillen*, 156 F.R.D. at 122.

104 FED. R. CIV. P. 30 advisory committee's note.

105 FED. R. CIV. P. 26(a)(3).

but, if in nonstenographic form, the party shall also provide the court with a transcript of the portions so offered."[106]

Court recognize several benefits of videotaped depositions. First, videotaped depositions, when used at trial, allow the fact finder to better assess the credibility of the deponent. The fact finder has the benefit of assessing a deponent's non-verbal communication, including facial expressions, voice inflection and intonation, gestures, and body language, in addition to the deponent's words.[107] Moreover, a videotaped deposition better reflects an evasive deponent, since a videotape captures the deponent's pauses and any coaching or notes from counsel, while a simple transcript captures none of these things.[108] Finally, a videotaped deposition can reduce the boredom of listening to a recitation of a stenographic deposition transcript, thus ensuring that the fact finder is truly listening to the evidence being presented.[109]

---

106 FED. R. CIV. P. 32(c).

107 *Riley v. Murdock*, 156 F.R.D. 130, 131 (E.D.N.C. 1994); *Mishkin v. Peat, Marwick, Mitchell & Co.*, No. 86 CIV. 4301, 1988 WL 117448, at *1 (S.D.N.Y. Oct. 27, 1988) (decided under the 1980 version of Rule 30); *Rice's Toyota World, Inc. v. Southeast Toyota Distributors, Inc.*, 114 F.R.D. 647, 649 (M.D.N.C. 1987) (decided under the 1980 version of Rule 30); *In re American Broadcasting Cos., Inc.*, 537 F. Supp. 1168, 1172 n.10 (D.D.C. 1982) (decided under the 1980 version of Rule 30).

108 *Riley*, 156 F.R.D. at 131.

109 *Rice's Toyota World, Inc.*, 114 F.R.D. at 649.

CHAPTER 13

SHAPING THE ISSUES, THEMES, AND PSYCHOLOGY FOR TRIAL

Joyce Tsongas

It is always a daunting task to consider the challenges of presenting any complex or technical case to a jury. Complex surety cases are no exception. Even the mediators of those matters, with biases obtained through their own personal industry experience, or judges with no particular surety experience, may present their own sets of challenges when counsel attempts to explain the fine points of an indemnity agreement.

Complex surety cases present many of the generic types of problems faced in other legal areas such as patent litigation, environmental coverage litigation, or products liability litigation. However, some of the difficulties in presenting complex surety cases are somewhat unique. Jurors must often sort through multiple parties, multiple claimants and claims, technical engineering and construction issues, and substantial financial issues.

The process of jury deliberations is essentially the process by which jurors try to reach a consensus as to what happened and what, if anything, they should do about it. Jurors must try to construct a group story, which will have characters, themes, motives, a plot, sub plots, and an ending in the form of a verdict.

In a complex case the key is to clarify, to simplify, to amplify, and ultimately to persuade. The discussion which follows will consider adding clarity and persuasiveness to complex surety cases by focusing on three elements of the case presentation: (1) the themes; (2) the story; and (3) the witnesses.

I.
The Themes

The key to developing effective case themes lies in the ability to get distance, perspective, and balance in looking at all you know about the matter. The more you know and the longer you have been involved in discovery and research, the more difficult this task may become. At some point all case elements, case facts, and arguments may seem equally important and unimportant.

There are many different paths to take and stories to tell. For example, a complex surety case can include the indemnity agreement story, the story from each of several project locations, the bad faith story, the engineering story, the construction timeline story, the quality of work story, the financial story, and the story for each claim. The list of potential viable stories for any one case appears unending. The problem is that these various stories must be blended into one overarching story that is as clear and persuasive to the jurors as possible.

One effective way to get distance and see the big picture is to step back from the case and force yourself to state your entire case in a few words or phrases. Three seems to work well. "Three" is an ideal number. For example, three major overarching themes are used in political campaigns and in many advertising campaigns. Some examples of possible case themes in a complex surety case include saying "This case is about…."

- A clear agreement
- A clear breach of an agreement
- Substantial resulting damages

- Trust
- Responsibility
- Unjustified blame

- An important timeline
- Costly delays
- Excuses

- A high quality design
- Low quality work
- Expensive consequences

When brainstorming about case themes, each lawyer involved in the case should write down several sets of themes. Each set should be shared so the members of the trial team can reach a consensus on the optimal word themes by combining their various lists to produce one that resonates. Many of the remaining words and phrases may become themes for a particular witness, demonstrative exhibit, or specific claim or counterclaim.

II.
The Story

After you have a vision for general case themes you have an overall guideline for framing the case story. It may be helpful to begin by writing a simple story sentence such as: "This is a case about a contractor who should have finished the job he started" or "This is a case about the owners of a development who paid for quality but did not get it" or "This is a case which proves the saying that timing is everything." When giving your opening statement, utilize the story sentence after you state the themes. For example, your opening statement may begin "This is a case about three things," stating the themes, followed by the story sentence. Structuring your opening statement this way gives jurors a general framework for your story and sets up a schematic for them to interpret the evidence in a manner congruent with your story. If you begin by discussing the nature of a time impact claim, the definition of an indemnity agreement, or the importance of a technical engineering issue, you will not give the jurors a general framework for the story.

Several important decisions must be made in trying to create one central, simple story from the set of all the possible stories. These decisions are concerned with such issues as characters, plots, motives, action, and all of the other basic elements of any good story. When does your best story start and end? Who should be the central character or characters? How can the action and facts tell the basic story? Should the surety be a central character?

Once you have decided on a plan for the story, it can be helpful to list the main elements of the story from one to ten for all members of the trial team to see. Only allow yourself ten items in the plot or story. This process of elimination helps you focus on the tall trees and incorporate them into the story. Point number ten ends the story the way you want it to end with a favorable verdict. An example of a final story statement would be, "The plaintiff should be repaid lost business revenues and profits due to the delays." When you finish listing the ten story points it is useful to try to write a paragraph or two which tells the ten-point story, and to tell that brief, focused story to the jury before you launch into the complexities of the surety matter.

III.
The Witnesses

A. Generally

As jurors begin to hear about a complex surety case some basic things are clear: 1) there is a disagreement involved in this case; 2) several people and parties are involved; and 3) these multiple parties are blaming one another for problems which occurred. Jurors need to start sorting out, based on what the lawyers say and how the witnesses testify, who is wearing a white hat, a black hat, or a gray hat. If the surety and its interests in the litigation are obscured from jurors, the surety "becomes" one of the other parties, such as its principal. If witness testimony from the surety itself will be minimal, the surety's counsel or expert witness then "becomes" the surety. Jurors will attribute the characteristics of credibility or lack of it to the surety based on how those individuals do in court. The surety and its credibility essentially become defined by others over whom the surety's counsel may have little or no access or control in preparing for trial.

In general, there are many ways in which witness communication patterns combine to create favorable or unfavorable impressions on others. Consider, for example, some of the specific combinations of visual, vocal, and verbal traits that would create the following negative or positive impressions:

- Nervous vs. Confident
- Defensive vs. Proactive
- Evasive vs. Direct
- Not believable vs. Credible
- Hesitant vs. Fluent
- Arrogant vs. Approachable
- Uncertain vs. Confident
- Deceptive vs. Honest
- Argumentative vs. Expository
- Boring vs. Interesting
- Easily led vs. Assertive
- Monotone vs. Animated
- Overly technical vs. Clear
- Stiff vs. Conversational
- Emotional vs. In control
- Unresponsive vs. Responsive
- Too talkative vs. Succinct
- Uncooperative vs. Cooperative

Testimony given by a nervous witness who fidgets in the chair, avoids eye contact, and speaks in a tremulous voice may be perceived as noncredible or even evasive. The same testimony delivered by a witness who sits straight in the chair, makes good eye contact, and speaks in a clear and confident tone will be perceived much more positively. The adage, "It's not just what you say, but how you say it," is certainly appropriate when applied to witness testimony.

Communication research demonstrates that when two people are having face-to-face communication, such as between witness and jury or between witness and opposing counsel, messages take place in three channels. The first is the visual message the witness sends through facial expressions, body language, posture, gestures, movement, clothing, and other visible signals. Everything the jury or opposing counsel sees about the witness communicates a message. The second way in which a witness sends messages is through the *speaking voice itself.* Volume, rate of speech, pitch, frequency of pauses, and vocal patterns transmit a wide variety of messages to the listener. Finally, the content of what a person is saying sends more specific messages.

About fifty percent of a witness' message is communicated visually, about thirty percent is transmitted through paralinguistic cues or through the speaking voice, and only about twenty percent comes through the actual words spoken by the witness. Thus, it is possible for a witness to say the word "yes" and sound like it means "no." It also is possible for a witness to say "I don't know" and sound like he or she really does know the truthful answer.

There is a dynamic relationship among what a witness says, how the witness sounds, and how the witness looks when he or she says it. Witness preparation sessions can produce significant improvements in witness performance in all three channels.

B. Witnesses For The Surety In Bad Faith Claims

One type of claim that provides an example of the problems that arise with surety cases is a bad faith claim. Bad faith claims, which give rise to significant testimony from the surety, raise witness issues that also apply to other claims in complex surety cases.

While surety claims witnesses face some of the same generic challenges mentioned in the prior discussions above, they are also subject to unique circumstances which affect the credibility of their testimony. First, claims witnesses are acutely aware of the importance of effective deposition testimony in creating a favorable record during discovery, and the extent to which "blowing the deposition" can have a serious negative impact on the outcome of a case before or during a trial. Claims witnesses are also well aware of the need to testify effectively in trial testimony and of the ways in which bad testimony can easily translate into a substantial verdict against a defendant. Because of that knowledge and awareness, fear of not doing well in deposition or at trial is particularly stressful for claims witnesses, a fear that is exacerbated by having to testify for one's employer.

A second stressor can be a heightened feeling of responsibility to give perfect or textbook testimony because of having had extensive litigation, trial, and case-related job experience. It is common under these circumstances to set unrealistically high self-expectations for giving outstanding testimony. Those higher expectations can create a fear of not meeting those high standards. Serious symptoms of stage fright may appear before, during, and after testimony. It is easy to be your own worst critic when the self-imposed standard may be something close to perfection.

Third, claims witnesses are also aware of the extent to which they are likely addressing a highly unsympathetic audience in the jury box. It doesn't take much imagination to picture a group of jurors with little sympathy for a surety company that jurors will perceive as being an insurance company defendant. Fear of a hostile or negative audience can be a significant factor in creating fear of rejection, and may result in symptoms of stage fright exhibited in the witness' demeanor and answers.

Another stress factor is the nature and language of the litigation and issues inherent in bad faith claims. Claims witnesses must testify about their past behavior and job performance in front of a group of strangers who will pass judgment on how well they did. Anxiety about performance censure can magnify an already tense situation.

Some of the language used to characterize claims witnesses and their behavior may include the following terms and phrases:

- Guilty of negligence
- Rejection
- Intentional deceit
- Unfair
- Disregard
- Bad faith
- Arbitrary
- Failure to disclose
- Refusal
- Misleading
- Unreasonable
- Conflicted
- Improper
- Capricious

- Consciously risking loss
- Failure to settle

These words and phrases, when used by opposing counsel in questioning in deposition or at trial, may lead directly to defensive or argumentative responses from claims witnesses.

As a result of common fears and the pressures unique to claims witnesses, claims witnesses are often tempted to communicate in ways that are ineffective, inappropriate, or actually harmful to the case. While their intent is to appear competent, helpful, and responsive, the effect of such inclinations may be just the opposite. Claims witnesses are prone to fall prey to several common temptations when giving deposition and trial testimony.

During deposition testimony, a common tendency among claims witnesses is to chat with the opposing counsel as a peer, to talk too much, and to explain too much. A second inclination is to use defensive denial language rather than positive proactive descriptions of their conduct and what he or she *did* do well. Attempting to convince the opposing counsel that the claims witness is a very fair-minded, reasonable person can also lure a witness into a defensive-sounding posture. This is often accompanied by the temptation to access and review information beyond the scope of what counsel asked the witness to review for the deposition.

During trial testimony, the temptation to address the opposing counsel rather than the judge or the jury is a frequent error among claims witnesses who, obeying social convention, direct their answers to the inquirer. Claims witnesses may also use complex legal, technical, or suretyship terms without explaining them. When an uncertain witness becomes flustered, there is a tendency to gain control by using technical language. Going beyond the scope of the questions asked in direct, cross-examination, and adverse questioning is a common temptation among claims witnesses who desire to be helpful or thorough. The tendency to second-guess the surety's trial attorney can be equally harmful. Finally, the temptation to argue the case to the jury or to argue with opposing counsel can create a costly negative impression.

A number of effective techniques that can be used by counsel and claims witnesses to help control and improve testimony in deposition and at trial, including: (1) framing a positive story; (2) controlling the content of answers to questions; (3) familiarity with documents and exhibits; and (4) testimony practice sessions.

1. Frame A Positive Story

Think about presenting a positive proactive description of what did happen rather than to think in terms of denying the opposing counsel's very negative description and characterization of what happened. Consider, for example, the positive impact of using the following words and phrases to describe actions that are in question:

- Due consideration
- Care
- Diligence
- Honesty
- Good faith
- Protection
- Negotiation
- Trust
- Accepting
- Reasonable compromise
- Skill
- Intelligence
- Conscientiousness
- Candid
- Common interest
- Responsibility
- Respect
- Responsible

2. Control Content Of Answers

In deposition testimony, witnesses can maintain control of their responses by answering questions in one or two sentences; avoiding one-word answers which can sound defensive, closed, and evasive; and setting a slow, measured pace for answers. Witnesses should make certain they understand the questions and never guess about their answers.

In trial testimony, witnesses should work at testifying in ordinary clear language. Witnesses can orient their posture toward the jury and make eye contact with each juror while testifying. The witness should answer questions during cross-examination or adverse testimony in complete sentences to avoid the trap of allowing plaintiff's counsel to testify for the witness by asking a series of leading questions. Each of these tactics will allow witnesses to control the content of their answers.

3. Be Familiar With Documents And Exhibits

Be certain witnesses are very familiar with the documents relevant to the file, with the underlying case, and with the bad faith claim. When witnesses know names, dates, facts, and figures without having to look at them, the witness creates the impression of being attentive to the concerns of the claimants. When a witness has to fumble for important facts, it gives the impression that he or she did not give adequate consideration to the claim in the underlying case.

4. Testimony Practice Sessions

It is critical for key witnesses to prepare for deposition or trial by actually practicing deposition and trial testimony. It is not enough to discuss substantive issues, what the answers will be, and how to deal with certain issues. It is important to actually go through questioning practice sessions. Practice sessions are most productive after a witness has been substantively prepared. When a witness is subjected to vigorous adverse questioning without adequate preparation, it is easy to shake his or her confidence beyond repair.

One of the most effective ways to practice testimony is to videotape the session. This allows counsel and the witness to view the testimony from the perspective of the audience, and the process provides immediate feedback when addressing communication problems. Suggestions for conducting a videotaped practice session include:

- Ensure the witness is substantively prepared regarding case facts, the purpose and themes for the witness's testimony, and general areas of inquiry before a video practice session begins.
- Place the camera so the picture to be replayed later is seen from the perspective of the most important audience, whether that is the opposing counsel, the jury, or the court. The monitor should be turned away from the witness so it is not distracting during the videotaping process.
- Videotape in three to five minute segments of questions and answers. Discuss methods for improving the content or demeanor aspects of the testimony when the camera is off between videotaped segments.
- When three to five segments have been videotaped, play back parts of each segment to note improvements and to identify areas for further work.
- Focus on the most critical or difficult substantive topic areas to be covered in the testimony.

- Use a wide variety of questioning types, styles, and techniques in the session so the witness is able to practice responding appropriately in each format.

The configuration of the room used for the practice sessions is also important. Set up the room to reflect the seating arrangement you expect at the actual deposition or trial. For deposition, the witness should be seated on one side of the table directly across from opposing counsel. The defense attorney should be seated next to the witness. The video camera should be directed at the witness only.

For jury trial testimony practice, the witness should be seated as he or she would be in the witness chair. The witness should be instructed to direct answers to the right or left of the room, depending upon the anticipated seating of the jury. For bench trials, the witness should be advised to direct his or her practice testimony to the witness' immediate right or left, depending upon the anticipated location of the judge's bench. For arbitration, set the room up as it will be at the arbitration and let witnesses practice testifying to the arbitrators.

Remember that an attorney telling a witness what to do is simply not enough. As is true when learning to drive a car or play a musical instrument, it is necessary to actually practice the skill in order to achieve a reasonable level of confidence to perform effectively at deposition or in court.

IV.
Summary

Although surety litigation presents multiple challenges for case synthesis, juror comprehension, and effective case presentation, application of specific trial preparation methods adds clarity and persuasiveness to the most complex surety case. Careful consideration of the themes which best represent the issues simplifies and amplifies the essential elements of the case. Creation of a concise story establishes a framework for jurors to interpret the evidence and helps promote a "conclusion" in the form of a favorable verdict. Finally, attention given to the preparation of witnesses and consideration for the placement of witness testimony within the story synthesizes the evidence. Combined with powerful themes and a compelling story, this creates a persuasive presentation out of a complex surety case.

Bibliography

S.L. Bordsky et al., *Jury selection in malpractice suits: An investigation of community attitudes toward malpractice and physicians*, 14 INT'L J.L. & PSYCHIATRY 215-22 (1991).

J. Boyll, *Psychological, cognitive, personality and interpersonal factors in jury verdicts*, 15 LAW & PSYCHOL. REV. 163-84 (1991).

J.D. Caspar et al., *Juror decision making, attitudes, and the hindsight bias*, 13 LAW & HUM. BEHAV. 291-310 (1989).

C. Cather et al., *Plaintiff injury and defendant reprehensibility: Implications for compensatory and punitive damage awards*, 20 LAW & HUM. BEHAV. 189-205 (1996).

R.J. CRAWFORD, THE PERSUASION EDGE (1989).

S.M. Fulero & S.D. Penrod, *Attorney jury selection folklore: What do they think and how can psychologists help?*, 3 FORENSIC REP. 233-59 (1990).

J. Goodman et al., *Matters of money: Voir dire in civil cases*, 3 FORENSIC REP. 303-29 (1990).

E. Greene, *On juries and damage awards: The process of decisionmaking*, 52 LAW & CONTEMP. PROBS. 225-46 (1989).

V.P. Hans, *The jury's response to business and corporate wrongdoing*, 52 LAW & CONTEMP. PROBS. 177-224 (1989).

R. HASTIE, INSIDE THE JUROR: THE PSYCHOLOGY OF JUROR DECISION MAKING (1993).

R. HASTIE ET AL., INSIDE THE JURY (1983).

J. Holstein, *Jurors' interpretations and jury decision making*, 9 LAW & HUM. BEHAV. 83-100 (1985).

N.L. Kerr & J.Y. Huang, *Jury verdicts: How much difference does one juror make?*, 12 PERSONALITY & SOC. PSYCHO. BULL. 325-43 (1986).

E. KRAUS & B. BONORA, JURYWORK: SYSTEMATIC TECHNIQUES.

R.J. MATLON, COMMUNICATION IN THE LEGAL PROCESS (1988).

M.T. Nietzel & R.C. Dillehay, *Psychologists as consultants for changes of venue: The use of public opinion surveys*, 7 LAW & HUM. BEHAV. 309-35 (1983).

P.V. Olczak et al., *Attorneys' lay psychology and its effectiveness in selecting jurors: Three empirical studies*, 6 J. SOC. BEHAV. & PERS. 431-52 (1991).

J.W. Pendell, *Enhancing juror effectiveness: An insurer's perspective*, 52 LAW & CONTEMP. PROBS. 311-22 (1989).

S.D. Penrod, *Predictors of jury decision making in criminal and civil cases: A field experiment*, 3 FORENSIC REP. 261-77 (1990).

A.B. Pettus, *The verdict is in: A study of jury decision making factors, moment of personal decision, and jury deliberations—from the jurors' point of view*, 38 COMM. Q. 83-97 (1990).

A. Reifman et al., *Real jurors' understanding of the law in real cases*, 16 LAW & HUM. BEHAV. 539-54 (1992).

R.L. Wissler et al., *Explaining "pain and suffering" awards: The role of injury characteristics and fault attributions*, 21 LAW & HUM. BEHAV. 181-208 (1997).

CHAPTER 14

PREPARING AND PRESENTING THE COMPLEX SURETY CASE TO A JURY, PRESERVING ISSUES FOR APPEAL AND PRESENTING THE APPEAL

Hugh E. Reynolds, Jr.
John J. Petro

I. Introduction

The purpose of this chapter is to discuss generic topics relating to the trial of a complex surety dispute. For purposes of selection, we will address disputes involving a surety who has issued a construction contract bond since most, but not all, complex surety cases arise in that context. To avoid duplication, discussions concerning the causes of action, the organization and control of documents, quality of work issues, time impact claims, construction damages, and demonstrative evidence will be limited to those instances where such comments are necessary for context.[1]

---

1 Since the subject of this chapter is so broad, this chapter assumes for the sake of space and clarity that a design professional is not a party to the jury trial. Instead, this chapter concentrates on the surety and contractor versus owner scenario with occasional comments on major difference in approach to the other scenarios.

A. The Major Types Of Surety Disputes Resulting In Complex Litigation

There are generally three types of disputes that may result in complex litigation. Sometimes the litigation involves more than one of these circumstances.

1. The Surety Takes Over And Completes A Project Although The Defaulted Contractor Objects

In the first type of case, the surety takes over and completes a project although the defaulted contractor objects. In this scenario, both the surety and its principal are likely to have claims against each other. Who is the plaintiff usually depends on who sued first. Issues that almost always arise in such cases are whether the owner had the right to terminate and whether the surety was guilty of bad faith in depriving its principal of available defenses, issues, or claims. Such cases almost always involve the reasonableness of completion costs. If there were problems with work in place, issues over these problems will be litigated. More often than not, responsibility for delay will be an issue. From the surety's point of view, a principal problem here is that the surety must take the owner's position in many of these disputes without any guarantee of any sort of enthusiastic assistance from the owner or its representatives.

2. The Owner Defaults The Contractor And The Surety Refuses To Complete

When the owner defaults the contractor and the surety refuses to complete, the central issue is almost always whether the owner had the right to terminate. There are usually claims for damages asserted by the owner against the contractor and by the contractor against the owner, and, of course, by the owner against the surety. These types of disputes often involve serious problems about the scope of the contractual obligation of the contractor, responsibility for delay, and reasonableness of the costs incurred by the owner to complete the project. Finally, there are often claims of alleged design defects or for alleged non-conforming work.

3. *Multi-Party Disputes Involving Four Or More Parties Who Are Active In The Prosecution Or Defense Of The Litigation*

Multi-party disputes usually involve the owner, the surety, the general contractor, the design professionals, and some major subcontractors and major suppliers. In multi-party disputes, all of these claims may arise directly or indirectly. In addition, where major subcontractors or suppliers are involved, the dispute broadens to warranty issues, issues of whether products or workmanship are conforming, issues of coordination, and interference. Almost always, such a complex case involves scheduling issues concerning access, ability to perform work, interference, and disputes over what is the critical path.

II.
Final Planning For The Case Presentation

A. Consider The Personality And Qualifications Of The Witnesses Whom You Will Present To The Jury And Prepare Them With Care

Throughout the months or even years that you have been working with the individuals in your case, you should always be evaluating their potential assets and liabilities as witnesses. This is particularly true of the key witnesses whom you must present to the jury. To some degree, you may have a choice among individuals. To the extent you do not have a choice, you need to prepare an individual who is a necessary witness from the very first point of contact until they have left the witness stand to make the most favorable possible impression. It is essential that the witness be clear, understandable, and persuasive in the presentation they offer on your behalf.

In accomplishing this mission, one must always have in mind the likely composition of the jury that will hear the case. In the state court in Indiana, for example, one would expect a very different panel depending on whether the case were tried in Marion County (Indianapolis); Dubois County (where the Circuit Court judge's name was Hugo Songer, the best restaurant is the Schnitzelbank, and the local bank is the German American Bank); St. Joseph County (South Bend in northern part of the state), or Franklin County (where one of the authors once argued successfully that it was not defamatory to call someone a bootlegger).

Some witnesses describe the facts. Some are or claim to be victims of the outrageous misconduct of the owner or have suffered severe financial loss (or will suffer severe financial loss if you lose the case). In these cases, the more the jury can identify with the witnesses, the more likely the witnesses will receive a favorable reception. This is not necessarily true of experts or people who supposedly bring specialized skill or knowledge to their testimony, however. There the relationship and jury acceptance is a more complex process. The ability to communicate clearly and in a way that suggests competence, fairness, and truthfulness is much more important than whether the jury can identify a kindred spirit.

You must try to get a sense of the overall impression the witness is likely to make. Whether you personally like or believe the witness is often a good guide to assess how a jury will be likely to feel. Of course, your feelings are not always dispositive. Furthermore, sometimes you do not have a choice as to your witnesses. In preparing witnesses, listen to how well each witness understands and responds to questions. Ultimately, that capacity will be reflected on the witness stand. If you have a witness who listens carefully to questions and responds appropriately, this is a very good sign. If you have a necessary witness who does not listen carefully to the questions or whose responses are often inapplicable, you need to start from day one to work toward changing the witness's habits when responding to questions. In the case of key witnesses, it is often a very good idea to conduct a cross-examination of the witness as though you were the opposing lawyer. This prepares the witness for the experience as well as how to respond to the difficult questions he or she may be asked.

Beware of witnesses who try to explain away a fact that harms your side when it cannot be explained away, witnesses who have a tendency to evade the responses to difficult questions, and witnesses who just do not answer questions truthfully. Sometimes you cannot avoid putting such a witness on the stand. However, you need to work as hard as you can to persuade each witness (whose responses are an attempt to persuade the jury to the witness's view of things) that responses that are evasive, argumentative, or just plain wrong will not persuade a jury. Witnesses like this have to be told that when there is a bitter pill to swallow, the quicker you swallow the pill and, if possible, if you swallow the pill with a smile on your face, the less likely it is to hurt going down.

The most important thing to remember in preparing witnesses is that in a complex construction case there may be a few jurors who understand as much as fifty percent of what is being presented to them. The average juror, however, understands thirty percent of what you present, and there are others who understand even less. The key to a complex case is to make sure, to the extent you can, that the jurors understand the issues in which they believe your side is telling the story accurately. If you are lucky, the jury may believe the other side is not telling the story accurately. In such cases, the jury will usually resolve that vast portion of the case they do not understand in your favor because they believe you have been candid with them on the portion of the case they do understand.

In this regard, in the process of selecting witnesses to testify, you will, to some degree, select witnesses based on the issues you will emphasize. The issues you will seek to minimize are also based, at least in part, on the strengths and weaknesses of the witnesses you will present on those issues.

A problem can arise when some witnesses who, theoretically, personify your side are really not under your control and may be hostile. This is particularly true when a surety needs to rely upon former employees of its contractor principal or subcontractors or suppliers of its contractor principal who have had some distasteful experience in dealing with the contractor principal or, in some cases, with the surety.

It is especially important that the surety present the best possible witness where the surety's own acts or omissions are issues in the case and an employee of the surety must testify. The lawyer will probably find several surety employees involved in the case. The lawyer should immediately begin to assess which of these individuals the lawyer wants to be the principal witness on behalf of the surety in front of the jury. With this evaluation, the lawyer can often shape the access to knowledge to bring about such a result.

Every complex surety case is document intensive. This may tend to make testimony tedious and the case difficult to manage. It does have some advantages, however. Your witnesses (and many of the witnesses for the other side) have probably expressed their views contemporaneous with the events in a variety of documents, or they may have attended meetings at which views were expressed. Documents should be prepared for use either chronologically within issues or chronologically including all issues in the case. This organization provides an ideal base for examining your witnesses and determining how best to evaluate, prepare, and present them. Be particularly concerned about a witness who consistently attempts to deny, minimize, explain away statements the witness has made previously in the documents. That is not to say that witnesses cannot supply a context not readily apparent from the document which makes the statements said therein more meaningful and more accurate as to what was being said or thought at the time. This requires careful analysis and thought, however.

In a complex surety case, it is likely that all of the major witnesses have been deposed prior to trial. The depositions of your witnesses involve the downside that the witnesses may make mistakes that can rise to haunt you later or that the deposition will reveal information helpful to the other side. But depositions also provide an advantage. Depositions allow you to see how opposing parties intend to approach the case and, in particular, how they intend to approach the case with this witness. This enables you to prepare the witness to make a better presentation in front of the jury than her or she makes in the deposition.

In many complex surety cases, witnesses who are deposed are not from the jurisdiction where the case is to be tried. A so-called "discovery deposition" in many states can be used as a deposition to be presented as evidence at trial as well where the witness is from another state. This is something to bear in mind when you are taking depositions of hostile witnesses and when other people are taking depositions of your witnesses. It gives you a selection process. You may elect to use the deposition of a witness who makes a particularly bad appearance, lives in another state, and has been deposed, giving you some opportunity to avoid the ill effects of that witness's live appearance and still have the required testimony. You usually will then be able to use the depositions and documents the witness has authored to speak for the witness better than the witness would speak for him or herself.

Many of the witnesses you will use, particularly experts, will have testified in court before. It is almost always a good idea to ask people who have seen this witness testify about the witness's appearance and skill. The same expert can testify in several different cases and, depending on how the expert was prepared, can make a very good impression and appearance in one case and a very bad impression and appearance in another case.

Two final important notes. First, witnesses get in trouble more often because they do not listen to or understand the question than for any other reason. Make sure you do not put a witness on the witness stand who has not had that maxim beaten into them. It may not insure that they will pay attention to it, but one can only do the best that one can. Second, if the jury believes a witness is lying or unduly evasive, the witness' testimony may be of little or no value unless it is clearly corroborated by independent testimony of others or irrebuttable documentary evidence. There are often a variety of ways to present unpalatable truths. Some are better than others. Obviously, you prefer to select the one that is the best for you.

B. In Planning Trial Strategy, Who Goes First Is An Important Consideration: You Must Consider Your Witness And Evidence Presentation And Your Theory Of Persuasion With This Reality In Mind

At least ninety percent of the time, the surety in a complex surety case presents its evidence last, or at least not first. This means many of your witnesses may already have testified before you present your case. In the situation in which the principal is fighting alongside of you, many of the surety's witnesses will also be witnesses for the principal who will be telling your side of the story.

Often, however, the principal files the initial lawsuit and then the owner brings both a counterclaim against the principal and an action against the surety. Under this scenario, the surety's evidence will be presented after the principal and the owner have presented their cases. There are advantages and disadvantages to the surety in this scenario. One of the principal disadvantages is that witnesses who are important to you may be examined by somebody else who is not trying to tell your story. You then have a problem cross-examining those witnesses. You may want to put things in a logical, chronological, or emphasis of issue order that is most persuasive for your theory of the case. This often involves repeating testimony that has already occurred. Most judges will allow a cross-examiner to cover territory already examined by asking essentially the same questions (at least once). But it is very important to understand how a particular trial judge is likely to rule. Some judges will closely control cross-examination so that you can cover the *issues* that have been covered on direct, but you cannot ask identical questions for context over objections from your opponents. Many times opponents will allow you to ask identical questions for context without objection. But you have no guarantee that this will occur.

As a generality, try to structure the evidence the jury hears to most nearly approximate a chronological presentation, while presenting strong witnesses when the jury is most likely to be impressed by their testimony and weak witnesses when the jury is least likely to pay attention. You do not have this kind of control when you do not present your case first. However, you should still keep this in mind when structuring the timing of your cross-examination, the timing of making witnesses available in an opposing party's case, etc.

Although likely to change as the case goes forward, you should map out the order of witnesses you would call in your case. In a complex surety case, you will have many problems about timing and preparation. Most courts will require disclosure of witnesses one day in advance and, in a complex case, as many as two days in advance. This is most valuable to the party who goes last, which is usually the surety. Both direct and cross-examination should be organized generally by subject matter. Often it is best to have a separate file on each subject matter and to organize the files in the order in which you wish to raise each issue. In this way, you are less likely to be confused and more likely to be persuasive. If the volume of documents permits, have the documents you intend to use with the witness in the particular file. If you have to look in many different places for the necessary documents during your examination, you will be more likely to miss something or ask the questions ineffectively. On the other hand, courts in complex cases often require counsel to make copies of each document counsel intends to offer for each juror and for the alternates. For instance, in Indiana, this means eight copies plus copies for opposing counsel, a copy for the record, and a copy for the judge. This volume of paper makes it very difficult not to have your file on the materials you intend to cover and a separate file or series of files on the documents you intend to cover. The better organized you are, particularly during your opponent's case, the more you will gain credibility with the jury. In a complex case, you will often find that your opponent is not well organized. This is particularly true of the party that presents its case first. You can establish a contrast to improve your position in the client's and jury's eyes. A client with confidence in your demonstrated trial skills is very important to a successful outcome.

Finally, be sure to work with your principal so you have a good understanding of how the principal intends to present its case. You have an opportunity in your own cross-examination of the principal's witnesses to develop favorable points or to take the sting out of points raised in the owner's cross-examination. The normal order of examination in such instances when the principal puts a witness on the stand is likely (1) principal direct; (2) owner cross; and (3) surety cross. This is certainly the preferable position from the surety's point of view.

C. Adverse Witnesses That A Party May Call

Your witnesses, having been called by the owner as adverse witnesses, should be well prepared to properly present your view of the case where the questions reasonably call for it. Because context is so important and often the owner's counsel will not cover all of the areas in which your witness is expected to testify, you may prefer to hit a couple of really good points for you in cross-examination of your witness (called by the owner) and then present a more logical, well formed line of testimony that the jury can understand and follow by putting the witness on in your own case. This is particularly true if you are dealing with a strong witness.

By the same token, many times the owner will have a witness who the owner does not choose to call to testify. You will call the witness. You will expect the testimony to be favorable to you, but the witness will certainly want to slit your throat at every available opportunity. First, any witness who is this important should have been deposed in detail before the trial so you are reasonably sure of the answers you will get. Second, to the extent possible where there are documents supporting your view and involving the witness, structure your examination of this witness by taking him or her through the document trail with the least opportunity to stray.

There are two things you need to remember. First, if you have a client representative who has knowledge of the facts and who is present in the courtroom, that representative should be prepared for the possibility that he or she will be called to the stand by either the owner or perhaps even your principal. Although this may not require detailed preparation, it certainly requires adequate preparation to make sure that some kind of untoward disaster does not occur because the witness has not been prepared for the basic issues which may arise and how to address such examination. Often, courts will allow you an opportunity to prepare a witness if it is clear that an ambush has occurred. However, you cannot rely upon that protection. Second, when examining adverse witnesses is it is not always bad for those witnesses to tell their side of the story when they are pinned down by documents or by earlier depositions. Usually, though not always, such attempts cause a jury to question the credibility and objectivity of the witness.

D. Plan To Emphasize Those Legal And Factual Issues That Most Help Persuade The Trier Of Fact That You Should Win

From the time you acquire a case, you should develop working hypotheses of your views of the issues. What are the important facts? What is available to prove your case or disprove your opponent's case? This does not necessarily mean that you are preparing lists and outlines on a daily, weekly, or perhaps even monthly basis. It does require some form of organization other than just keeping it in your head, however. By the time you are starting the last-minute trial preparation, if you have performed this procedure properly in an organized fashion and reduced at least the summary of your thoughts to writing, you should be in a position to identify those issues that work best for you. You also need to consider those issues that are difficult for you. Finally, you need to consider how to handle issues on which you must present evidence, even though they are not dispositive issues in the case, but are simply a necessary predicate to other evidence.

In this process, do not ignore bad facts or issues that work against you. Rather, try to cover them in the best possible way and the shortest possible time whether in an opening statement, during presentation of a witness who has information on the subject matter, or in the handling of documents. If you have an opportunity to put your best foot forward even on a bad issue, that will often be very helpful.

However, in all events, concentrate on those issues that work the best for you. Spread them throughout the case. Emphasize them where you can, particularly if they can be emphasized without causing the jury to believe that you are gilding the lily. Plan your order of witnesses to maximize your advantages in the key areas.

A key area often may be an attack on the credibility of one of the other side's principal actors. This is ideal because such evidence will normally be spotted throughout the trial. It will appear in many documents where you can show that the witness' testimony is wrong, controversies involving the witness, or that the witness has impeached him or herself. It will also appear in the testimony of a number of witnesses. Normally the best way to handle this evidence is bring it out, limiting your personal comments about the nature of the evidence to a limited number of instances where the court and the jury will feel your comments are appropriate under the circumstances. Remember, the jury's sympathy is always with the witness and against the lawyer unless and until the jury feels the witness is not entitled to that sympathy.

In construction cases, those factual issues that work for you may be very complex. It is important to simplify these complex issues with graphs, summaries, and witnesses who can best identify and elaborate on a complex issue in a simple way. Examples are particularly helpful. Often, having a good witness draw or write elements of his or her testimony can be very effective. If you plan to do that, however, be sure to practice before the jury sees the witness.

E. Why Plan At All

There are several reasons to plan. First, the planning process if carefully followed will often alert you to areas in which you need to get additional witnesses, consider changing witnesses, work hard on a witness, examine the documents more carefully, or look for additional documents to use from among the thousands of documents you have. In other words the planning process helps you find and fill the holes in your case before the jury sees them. Second, planning also allows you to consider the issues and examine them to determine which issues work best for you and which work against you. You can then plan to maximize those issues that are to your advantage and minimize those that are not. Third, planning enables you to organize the documents, the presentations, and the questions, and to examine depositions you may use to refresh or impeach a witness. Juries do not expect lawyers who are supposedly specialists in a complex area to be stumbling around not finding documents, citations to depositions, and so forth. Only planning will place these items in an organized fashion where you can act promptly as you go through the testimony of witnesses and the presentation of documentary evidence.

Part of your planning should consider what your opponent is likely to do and being prepared for it. You may have issues you want to bring out which are best brought out first from a particular opposing witness. Think that through. Obviously you want to go after the other side's weakest witness on such an issue.

Planning also enables you to try to tie the issues and the evidence you expect to present to the jury instructions that you intend to tender to the court. Further you must, to some degree, put in key pieces of evidence with the thought that the evidence will be a vital part of your final argument. Only planning can maximize the advantages to be gained from this technique.

III.
Last Minute Procedural And Preparation Tasks

A. The Motion *In Limine*

A motion *in limine* is addressed to the trial judge, often prior to the commencement of trial, but on some occasions during the course of the presentation of evidence. The purpose of the motion *in limine* is to avoid the injection into the trial of anticipated evidence that is potentially prejudicial and which is or may be irrelevant or otherwise inadmissible. Its successful use avoids the age-old problem of harmful matter reaching the attention of the trier of fact, subject only to the court's subsequent admonition that it was not admissible evidence and should be ignored.

In most jurisdictions, the motion *in limine* is neither a creature of statute nor of procedural rule.[2] Rather, the motion and the time for filing

---

2 In *Luce v. United States*, 469 U.S. 38 (1984), the court pointed out that a motion *in limine* is not a creation of the Federal Rules of Civil Procedure. *Id.* at 41 n.4. It arises from the court's inherent power to manage and conduct trials. One Ohio court has held that the authority for the court to grant or deny a motion *in limine* is grounded in the inherent power and discretion of the court to control proceedings before it, citing Rules 103(a) and 611(a) of the Ohio Rules of Evidence. Note that the Ohio Rules of Evidence are virtually identical to the Federal Rules of Evidence. *City of Cincinnati v. Contemporary Arts Ctr.* 566 N.E.2d 214, 216 (Ohio Mun. 1990).

it is likely founded on local rule, custom, or practice, and may be subject to the eccentricities of the particular trial judge's procedural edicts. It is therefore of critical importance to be aware of the pretrial requirements that a judge may enforce in matters pending before him or her, particularly as those requirements deal with the time for filing motions and, more importantly, the time after which motions may not be presented to the court.

An order *in limine* is merely a tentative evidentiary ruling which states the court's anticipated treatment of an evidentiary issue. Of course, if special circumstances occur, the court may apply a different ruling to the same evidentiary question when it arises in the actual trial of the case. The ruling on the motion *in limine* cannot be definitive until the presentation of evidence in the trial has commenced so that the context and foundation for the proposed evidence has been fully developed. Consequently, the granting of a motion *in limine* is prospective in nature and does not necessarily determine the ultimate question of the admissibility of the evidence in question.[3]

If a motion *in limine* is sustained, the party who is prevented by the court's order *in limine* should request, during the trial of the case and outside of the hearing of the jury, the right to put the material into evidence. If the court again bars the introduction of the proposed evidence, then an adequate offer of proof should be made outside of the hearing of the jury so that an appellate court can assess the propriety of the trial court's exclusionary ruling.[4] Obviously, if the excluded evidence is not available to the reviewing court, the court cannot judge the effect on the proponent's case of the exclusion of the evidence.

On the other hand, if the motion *in limine* is overruled, the moving party should object anew at trial when the questionable evidence is offered in order to preserve the issue for appeal. There does seem to be a division of authority on this subject. In some jurisdictions, it is clearly the rule that a pretrial denial of a motion *in limine* does not preserve the issue of the improper admission of the evidence for a subsequent appeal. In those jurisdictions, an objection must be asserted at trial.[5] There are

---

3 *See Brunett v. Albrecht*, 810 P.2d 276, 282 (Kan. 1991); *Feely v. Davis*, 784 P.2d 1066, 1067 (Okla. 1989).

4 *Braden v. Hendricks*, 695 P.2d 1343, 1349 (Okla. 1985).

5 *Thweatt v. Ontko*, 814 F.2d 1466, 1470 (10th Cir. 1987); *Hendrix v. Raybestos-Manhattan, Inc.*, 776 F.2d 1492, 1503 (11th Cir. 1985);

other jurisdictions in which it has been held that, if the trial court makes a definitive ruling before trial and there is no indication from the court that it will reconsider the matter, an objection at trial is repetitious and without purpose and the prior denial of the motion *in limine* preserves the issue for subsequent review by an appellate court.[6] Obviously, the safe practice is to renew the objection at trial when the offending evidence is offered.

The motion *in limine* is clearly an effective tool for the trial attorney. The primary advantage is to prevent the trier of fact from hearing evidence that is likely to be prejudicial. Making the motion and obtaining an order *in limine* before a jury is seated has the added advantage that the evidence is excluded outside the presence of the jury, thus avoiding the appearance that something is being kept from the jury. Finally, a favorable ruling on the motion gives the movant a better opportunity to concentrate the preparation of his or her case on the evidence that he or she knows is likely to be admitted.

There are, of course, concomitant disadvantages that must be taken into account. It is possible that the opponent is unaware of or has overlooked the significance of the evidence that is sought to be barred. The motion thus serves to educate the opponent on evidence or issues that he or she previously thought unimportant. It is also possible that the opponent, spurred by the court's order *in limine*, will concentrate his or her efforts and obtain additional relevant evidence so that, at trial, he or she will be better prepared to demonstrate the relevance and admissibility of the barred evidence. Because the order *in limine* is tentative or preliminary, the successful movant cannot afford to totally ignore the barred evidence, but must be prepared to counter it on the chance that, at trial, the court will conclude that the evidence is relevant or non-prejudicial and permit it to be offered.

B. Witness Preparation

Early in the handling of the matter, the trial attorney, after a considered analysis, should have formulated a theme to the case to be

---

Northwestern Flyers, Inc. v. Olson Bros. Mfg. Co., Inc., 679 F.2d 1264, 1275 n.27 (8th Cir. 1982).

6 *Palmerin v. City of Riverside*, 794 F.2d 1409, 1413 (9th Cir. 1986); *American Home Assurance Co. v. Sunshine Supermarket, Inc.*, 753 F.2d 321, 324-25 (3rd Cir. 1985).

presented to the trier of the fact on behalf of his or her client. That theme should have been part of the basis for the discovery conducted throughout the litigation, and now that theme, having been adjusted to reflect any new found facts, is in its final form and becomes the foundation for the preparation for trial.

One might best prepare for trial of the case by first conceiving the closing argument that will be persuasively delivered to the trier of fact at the end of the evidence. Working backward from the closing argument, one can then plan the presentation of the evidence that will lead to the factual conclusions that, in closing argument, the trier of fact will be urged to consider to be the only logical conclusions that may be reached from the evidence.

The evidence presented to the trier of fact, for the most part, will come from the testimony of witnesses. In a well prepared case, the presentation of the testimony of the witnesses should occur with few, if any, surprises. To accomplish that end, those witnesses must be prepared, not simply called, to testify. Preparation of witnesses for testimony begins by the trial attorney determining what facts must be proven to carry out the theme of the case and to support the final argument that has been planned. Knowing those facts necessary to be proven, the trial attorney is in a position to determine what witnesses are available to testify to prove a fact, and can then turn his or her attention to preparing each witness to testify.

There are literally hundreds of trial practice manuals that describe the "best" method of preparing a witness for trial. All of those manuals agree that there is no adequate substitute for having the trial attorney meet with and prepare those witnesses that he or she is going to examine at trial. The witnesses' testimony should be organized, perhaps by question and proposed answer or by a narrative outline of the witnesses' testimony. The question and answer method has the advantage that the questions are prepared in advance and therefore likely to be in a proper form. The disadvantage obviously is that the presentation of the testimony using this method has no spontaneity and may appear rehearsed. The narrative outline method has the advantage that the questions are formulated as the testimony is elicited, and therefore the questions seem fresh and spontaneous. The disadvantage to this method is that the questions, not having been prepared in advance, may prompt an objection as to form or content and may divert the trial attorney's attention. An additional disadvantage is that the witness may respond to a question with a totally unexpected answer.

In preparing a witness for the delivery of his or her testimony, review any exhibits that the witness is to identify. Explain that you must authenticate the document in order to lay a foundation for its admission so that the witness understands why certain questions must be asked. Review with the witness his or her previous testimony by deposition or by written interrogatory, and review any prior statements or declarations that the witness may have made in order to determine that the testimony to be presented is consistent with prior statements or testimony. If inconsistent, develop an explanation for the apparent inconsistency. The witness should understand that, on direct examination, the trial attorney cannot ask leading questions, particularly where the trier of fact is a jury. Consequently, the witness must understand that he or she has to deliver the testimony by his or her own words and not by yes or no answers to the trial attorney's questions. The witness should likewise understand that, unless he or she is an expert witness who can testify to conclusions and opinions based on his or her expertise, the witness can testify only about those things that he or she saw, heard, and did.

Satisfied that the witness is properly prepared for direct examination, the next step is preparation for cross-examination. The trial attorney should explain to the witness that, as opposed to direct examination, the examiner may ask leading questions on cross-examination. Emphasize to the witness that he or she must listen to the question that has been asked, and to answer that question and only that question. The witness must understand that the cross-examiner's role is to either attempt to turn the witness' testimony to benefit the cross-examiner's side of the case or discredit the testimony so that the trier of fact will ignore it. It is critical that the witness understand that he or she must, at least outwardly, appear calm and respectful of the cross-examiner. It is the rare witness that can "match wits" with a skillful cross-examiner. Those who attempt to do so generally court disaster.

The trial attorney must personally meet with and prepare each witness that he or she is going to examine. Depending upon the nature, importance, and extent of the witness' testimony, a series of preparation meetings may be necessary to review the witnesses' direct testimony. By the same token, with respect to witnesses who are delivering significant testimony, the preparation meetings should include another member of the firm to conduct cross-examination.

Note that it is possible to over-prepare a witness. Some signs of the over-prepared witness are: (1) the witness delivers testimony in an almost flat, lifeless fashion; (2) the witness begins to use legalese or phrases that one would not anticipate; (3) the witness shows signs of inflexibility so that, if the order of questions or the manner in which the questions are phrased is changed, the witness appears lost; (4) the witness becomes contentious, arrogant, or acts like a know-it-all. If trial preparation has resulted in the creation of a witness who reflects one or more of these signs, the trial attorney must make a fresh start with the witness. The trier of fact must "like" the witness, and the delivery of his testimony must be convincing. The over-prepared witness is likely to score low on both counts.

C. Preparation for Cross-Examination

Just as effective direct examination is the product of planning and witness preparation, so too is effective cross-examination. The use of discovery in the pretrial stages of the case should make it possible for the trial attorney to reasonably anticipate the direct testimony of his or her opponent's witnesses. In fact, it is probably a worthwhile exercise in preparing cross-examination to contemplate how you would use the witness if he or she was to be your witness on direct examination. What documents can the witness identify and authenticate? What facts would you attempt to establish in his or her direct testimony?

The cross-examination of significant witnesses will probably involve three areas: (1) admissions which will favor the cross-examiner's case; (2) favorable subjects for cross-examination which will raise doubt as to the witnesses' direct testimony; and (3) impeachment. The preparation for cross-examination should include a review of: (1) documents authored by the witness or in which the witness is mentioned; (2) answers to interrogatories given by the witness or in which the answer given by others mentions the witness; (3) the deposition testimony given by the witness; (4) testimony that may have previously been given in a related case; and (5) declarations or statements made by the witness or attributed to the witness by others concerning one or more subjects about which the witness will likely testify on direct examination.

With technical or expert witnesses, a review should be made of any testimony given in other cases involving a similar or related subject, articles published by the witness, and standard treatises or texts discussing the subject of the witness' probable opinion testimony.

Armed with this information, the preparation of likely cross-examination can begin. The cross-examiner must consider whether the witness can give testimony that is favorable to the cross-examiner's position in the case, including whether the witness is able to identify or authenticate a document or lend credence to the existence or content of a document that the cross-examiner has or will offer into evidence. In truth, the opportunities to establish a substantial element of the cross-examiner's case with an opponent's witness are few and far between. After all, the opponent produced the witness after preparing the witness for direct examination. It is unlikely, at best, that the opponent is going to produce a witness who can offer substantial assistance on critical facts to the other side of the case. That rare opportunity presents itself when the opponent must use a witness to establish a fact that can be established only by the testimony of that witness, so that the opponent must take the risk that the good that the witness will serve to his or her case far outweighs the harm that will come from skillful cross-examination.

In preparing the cross-examination, the trial attorney should consider whether there are areas of likely direct examination that the trier of fact may think improbable, and whether there are areas of likely direct examination that conflict with documents to be offered into evidence or conflict with the testimony given or to be given by witnesses to be presented by the cross-examiner. These are subjects for potential favorable cross-examination which, handled properly, may cause the trier of fact to discredit some or all of the witnesses' direct testimony.

The final exercise is for the preparer of cross-examination to take into account areas of likely direct examination that conflict with prior statements, either in discovery or elsewhere, made by or attributed to the witness. These are areas of cross-examination that may lead to the impeachment of the witness or may put the witness in an uncomfortable position of having to explain away contradictory or conflicting statements.

D. Use Of Contentions And Answers To Interrogatories In Planning During The Trial

Interrogatories are a pretrial discovery tool that can assist the trial attorney prior to, during preparation for, and during the trial of the case. Prior to trial, interrogatories allow the questioning party to identify witnesses, lay and expert; define the contentions of the opposing party; and discover relevant facts. Carefully prepared interrogatories, when answered, can provide a plan from which the case may be developed by identifying witnesses and depositions that should be taken, matters that are admitted or at least are not in controversy, and areas in which pretrial motions may be helpful to limit issues for trial. Investigative or "source" interrogatories are often helpful in particularizing vague portions of pleadings, which is an important function since pleadings are now "notice" pleadings and no longer contain allegations of fact. In addition, answers to interrogatories provide information that may be used in planning the direct examination of your witnesses and the cross-examination of the opponent's witnesses in order to put their testimony in question or to impeach them. Interrogatories are particularly useful for corporate defendants where the information sought may be within the knowledge of more than one individual.

Contention interrogatories are a similarly useful tool in preparation of the case for trial. These interrogatories seek the opponent's contentions relating to the facts of the case or the application of the law to the facts. Contention interrogatories narrow the issues for trial and avoid wasteful preparation time. A contention interrogatory may not be used to obtain legal conclusions.

At trial, answers to interrogatories are generally admissible in evidence as the admissions of a party. Since the interrogatory answer is in writing, it can be used as a trial exhibit, thus becoming tangible evidence to be referred to in the closing statement and taken by the jury into the jury room when beginning its deliberations. Be advised, however, that interrogatory answers are not part of a trial record unless they are offered and received into evidence.

The fact that answers to interrogatories are admissible against a party does not make those answers binding admissions. An amendment to the answer may be filed at any time prior to trial. Even if a party does not amend its answers, many courts will allow the presentation of inconsistent evidence and inconsistent contentions, unless the propounding party can show substantial and justifiable reliance of the answer given to the interrogatory. If the responding party is permitted to offer evidence or contentions which are inconsistent with prior interrogatory answers, the propounding party is then in a position to use the answers to interrogatories to put the witness' testimony in question or impeach the witness. It would then be up to the jury to determine the credibility of the witness and the truth of his or her testimony.

E. Jury Instructions

Jury instructions provide a guide to the jury regarding the law applicable to the facts of the case. It is not often that a court insists that a party submit proposed jury instructions; although in a case which is out of the ordinary, *i.e.,* one to which the court's typical "canned" instructions will not apply, the parties may be specifically directed to submit proposed jury instructions.

The court will have a set of canned instructions relative to the duty of the jury, the function of the court, the credibility of witness, the difference between circumstantial evidence and direct evidence, the weight to be attributed to the evidence, and the burden of proof. The standard instructions available to most judges will likely provide substantive instructions in every kind of criminal case and in many civil actions which are repeated on the court's docket. Few judges have access to jury instructions in cases involving sureties, however. Most judges do not have a standard instruction that defines the duty of a surety, the obligation of an indemnitor, or an instruction that applies, even in the most general sense, to construction contract disputes.

The trial attorney should take advantage of the opportunity to submit proposed jury instructions. Submitting jury instructions provides the trial attorney with a chance to persuade the judge regarding the applicable law and, in some measure, facts which the attorney contends have been proven. Moreover, to the extent that the submitted jury instructions are actually given to the jury, the trial attorney has an opportunity to persuade the jury through the authority of the judge.

Care should be taken that the trial attorney is familiar with the procedural statutes and local rules of the jurisdiction in which the case is being tried. Many courts, through their local rules, require that jury instructions be filed at the time of final pretrial or at some other time prior to the actual commencement of the presentation of evidence. Rule 51 of the Federal Rules of Civil Procedure differs from the rules in many state courts in that the submission of proposed jury instructions is required "[a]t the close of the evidence or at such earlier time *during the trial* as the court reasonably directs, any party may file written requests that the court instruct the jury on the law as set forth in the requests."[7] The rule further provides that the court "shall inform counsel of its proposed action upon the requests prior to their arguments to the jury."[8]

At the inception of the matter, the trial attorney developed a theme to the case. The theme played a role in planning discovery and should have been "fine tuned" following discovery to be consistent with the information developed. As the trial attorney began the final trial preparation, both the direct and cross-examination were prepared in order to develop the case theme. If the jury instructions which have

---

7 FED. R. CIV. P. 51 (emphasis added).

8 *Id.*

been proposed are actually given by the judge, the jury should take the theme of the case into the jury room to use during their deliberations.

The primary goal in preparing proposed jury instructions is to develop the most persuasive instructions that are likely to be given by the court. Guidelines to the preparation of jury instructions are generally a reflection of common sense. For example, one should omit unnecessary words and legal terms where possible. Short sentences should be used, and the instructions should be short so that they are easier to understand. Prepare the instructions using the facts that you believe have been proven, rather than using abstract concepts.

IV.
The Trial

A. Opening Statement

The purpose of the opening statement is to give an opportunity to the attorneys for the parties to inform the jury in a "general way" what the lawsuit concerns, the nature of the claims being made, and the nature of the defenses being asserted so that the jury is better prepared, as it the litigation proceeds, to understand the evidence.[9] The office of the opening statement is not to permit the attorneys to argue either the law or the facts. Rather, the opening statement is to be used to tell the jury what the case is about by telling the jury what the evidence will show. In making the opening, the attorney is, as a general rule, permitted to refer to evidence which, in good faith, the attorney believes to be admissible in evidence.[10] In addition to providing an opportunity for each attorney to tell the jury what the case is about, opening statement gives the attorney a chance to develop rapport with the jury and establish credibility.

It has been said, by those who purportedly know, that the statements that have the greatest impact on the listener, and that the listener remembers longest, are those statements that the listener hears and last concerning a subject. These are sometimes referred to as the rule of primacy and the rule of recency. True or not, studies have shown that

---

9 *Best v. District of Columbia*, 291 U.S. 411, 415 (1934).

10 *Hamann v. State*, 324 N.W.2d 906, 911 (Iowa 1982). Even if the evidence which has been referred to is subsequently held to be inadmissible, reference to it in opening statement was not improper so long as the attorney was of the good faith belief that the evidence was admissible. *Sawicki v. Kim*, 445 N.E.2d 63, 66 (Ill. App. Ct. 1983); *Timsah v. General Motors Corp.*, 591 P.2d 154, 163 (Kan. 1979).

far more than fifty percent of jurors form an impression favoring one side or the other at the conclusion of opening statements. Clearly, one must therefore deliver an opening statement that gets the attention of the jury, holds their interest, and persuades the jury that the party for whom the statement is being made is the party that is in the right in the litigation. By the same token, the statement must be delivered in such a way that the jury trusts the attorney making the statement.

Basically, there are two ways to fashion an opening statement. Whichever method is employed, opening statement permits the attorney to convey to the jury the theme of the case that has been the foundation of discovery and subsequent trial preparation. When the attorney employs an "outline of the evidence" method, he or she generally begins most substantive statements to the jury with a phrase something on the order of "We will prove . . ." or "The evidence will show . . . ." This method of making an opening statement is often boring and difficult to follow. There are some judges who believe that this is the only form of a proper opening statement since the office of the opening statement is to inform the jury of what each attorney believes the evidence will show. The story telling method, particularly when done by someone who can tell a story, makes the opening statement (and the case) far more interesting to the jury. Of course, in some cases, telling a story is easy, but, in the typical construction contract dispute, it is much more difficult to do so, because telling the story may require words or phrases unfamiliar to the jury and probably involves an event that is beyond the typical life experience of a juror.

In preparing an opening statement, the attorney for the plaintiff may well want to consider anticipating the opponent's statement in order to refute it in advance. This can be very effective because the jury is told, before the defendant's attorney rises to address them, what they may expect to hear, why the defendant's evidence proves nothing, and why the defendant's story of the case is implausible. Of course, this approach is not without risks. Even in this age of excessive discovery, it is possible that one may totally misapprehend the position of the opponent. Thus the opening statement ends up concentrating in part on a position that the opponent either does not assert or does not emphasize, and the jury is left to wonder whether the plaintiff's attorney has sense or credibility.

Some people suggest that consideration should be given, perhaps not in all cases, to waiving opening statement or, as a defendant, to deferring it (where the local procedure permits deferral) until the plaintiff's evidence has been elicited. The thought is that, by waiving opening statement, the attorney does not take the risk of promising the jury that he or she will produce certain evidence that, in fact, the testimony or documents fail to produce. The argument for deferral is that, by deferring opening statement until the plaintiff's case has been produced, the defendant's case retains a level of fluidity such that a change in the approach to the case can easily be taken mid-stream. There are others who embrace the theory that a short opening statement is to be favored so that there is some surprise left for the jury as the evidence is produced, something for the jury to "discover." Those who want to waive, defer, or shorten their opening statement should give serious consideration to the following. During the trial of a case, the attorneys for the parties have three opportunities to legitimately talk directly to the jury: *voir dire*, opening statement, and final argument or closing. Why would you give up or limit one of these opportunities? Logic tells us that you would not and should not do so.

B. Direct Examinations

Prior to the commencement of trial, the trial attorney spent a significant time preparing the witnesses that the attorney intended to call to testify. In the course of the witness preparation, either all of the likely questions and answers or a narrative outline of the witness' likely testimony was drafted and reviewed with the witness.

Representing the plaintiff, the attorney should have prepared a witness to testify to each fact material to the case and required to be proven in order to meet the burden of proof. At trial, plaintiff's attorney probably called each of those witnesses to testify, unless stipulations made at final pretrial or at the commencement of the trial eliminated the necessity of proving certain facts. The defendant, on the other hand, will probably not call each witness that he or she prepared to testify. There is little sense in calling a witness or examining a witness to establish a fact already established by the testimony of one of the plaintiff's witnesses. Doing so may confuse the jury or may aggravate the jury with the appearance that the defendant's attorney is doing nothing more than going over plowed ground.

The purpose of direct examination is to elicit testimony from a witness, in a persuasive and credible manner, regarding the observations and actions of the witness to establish a necessary fact in a convincing and memorable manner. While many attorneys regard themselves to be at the strongest when they are in the midst of cross-examining the opponent's witnesses ("Boy, I really crushed that guy"), most cases are actually won on the strength of the testimony and exhibits adduced on direct examination.

The selection of witnesses to produce at trial is controlled by a number of factors:

(1) Can the witness testify to a fact that is necessary to be proven in order to establish the theme of the case contended by the party calling that witness;

(2) Can the witness, on cross-examination, testify to facts which tend to establish or lend credence to the opponent's theme of the case;

(3) Is the witness' appearance and demeanor likely to be appealing to the jury?

When there are a number of witnesses who can testify to a fact, the attorney has the luxury of balancing these factors and selecting the witness who best satisfies each of them. Of course, more times than not the testimony of one witness is the only available method of proving a fact, and the attorney must take the risk that the witness is not likable or may testify to something valuable to the opposition.

The direct examination of witnesses should be direct and simple. Questions should be posed in plain language, avoiding technical jargon as much as possible. The questions should be framed to give the witness an opportunity to respond with a narration. The questions on direct examination should be asked in a logical manner, often chronologically, emphasizing what the witness saw and what the witness did in response to it, assuming that the witness' action is relevant.

The interrogator must listen to the responses to the questions. Often, the attorney gets so caught up in the outline of testimony or in the questions that have been prepared that he or she does not listen to the response given by a witness to the question. Failing to listen to the response of a witness can result in the failure to establish the very fact for which the witness was called to testify, or may result in an illogical follow-up question in an attempt to elicit testimony that was already given. That follow-up question makes the witness wonder what he or she failed to say the first time, and now the witness may search for ways to embellish his or her testimony. The result may, of course, be a disaster.

Leading questions must be avoided. A witness who has been "led by the nose" through his or her testimony is of little value to the party who called the witness. Leading questions on direct examination may convince the jury that the testimony is not the observation or action of the witness, but instead is little more than what the witness has been told to say by the attorney examining him or her. This makes the witness non-persuasive, perhaps incredible, and the witness' testimony has little or no value.

C. Hostile Witnesses

Under the traditional rule, a party vouched for the credibility of the witnesses that he or she called and was bound by the testimony that those witnesses delivered. Under the Federal Rules of Evidence, a party no longer vouches for its witnesses' testimony. [11] When a witness is called to testify and his or her testimony, by surprise, is against the party calling the witness, the witness may be declared to be a hostile witness and may be asked leading questions on direct examination as if the witness was on cross-examination.[12]

Care should be taken to determine the applicable rules relative to hostile witnesses in the jurisdiction in which the case is being tried. In some jurisdictions, one must show surprise in trial. Thus, if the attorney has learned prior to commencement of trial that the witness is not going to testify as originally anticipated, surprise has not occurred at trial, and the witness will not be held to be hostile. In order to prove surprise, the attorney may have to demonstrate to the court, out of the presence of the jury, what the attorney anticipated the testimony of the witness to be such that the court can determine that the testimony is materially different. This may be done by the use of a deposition that was taken during discovery, from previous written statements of the witness, or perhaps from the attorney's preparation notes for trial.

If the court declares the witness to be hostile, thus permitting examination as if on cross-examination, the attorney should then proceed to establish before the jury that the witness had previously made statements or given testimony that is inconsistent with the witness' testimony at trial. With that fact established, the attorney should then proceed, in as short an examination as possible, to obtain what testimony of value may be obtained from the witness.

D. Expert Witnesses

The basic approach to direct examination of expert witnesses is not greatly different than the approach to direct examination of lay witnesses. The principal factors to good expert testimony are having a competent, believable expert witness and good preparation between the examiner and the witness. It is probably more important with an expert witness than with a lay witness that leading be avoided. The jury must believe that the witness is stating his or her opinion and not simply parroting the attorney. Moreover, the questions used in the direct

---

11 FED. R. EVID. 607.

12 FED. R. EVID. 611(c).

examination of an expert witness must give the witness plenty of opportunity to explain his or her opinions and conclusions so that the jury can understand that there is a legitimate foundation for the witness' testimony.

The rules for the qualification of an expert witness vary from jurisdiction to jurisdiction. The rule stated in the Federal Rules of Evidence is: "If scientific, technical, or other specialized knowledge will assist the trier of fact to understand the evidence or to determine a fact in issue, a witness qualified as an expert by *knowledge, skill, experience, training, or education,* may testify thereto in the form of an opinion or otherwise."[13]

Rule 702 recognizes that education is not the sole criteria for the qualification of a witness as an expert, thus permitting one who is skilled or experienced, but not necessarily specialized by education, to render an opinion.[14] In direct examination, the attorney should spend sufficient time to qualify the witness as an expert. This effort is made not simply to demonstrate the actual qualification of the witness to express an opinion, but also to make the jury comfortable with the witness and to give the jury confidence in the witness.

In some jurisdictions, the opinion of an expert witness must be stated by the witness in response to a hypothetical question. Where that is the case, extreme care must be given to the preparation of the hypothetical question. The question may well be objectionable, thus preventing the expert from testifying to his or her opinion, if the question does not take into account every fact in evidence relevant to the opinion. Even if an objection to the hypothetical question is not interposed, or if interposed is not sustained, the omission of a relevant fact may be fodder for a skilled cross-examiner. Under the Federal Rules of Evidence, an expert witness may state his or her opinion and state the reasons for the opinion without a prior disclosure of the underlying data on which the expert bases the opinion, it being the burden of the cross-examiner to obtain disclosure of the basis for the opinion.[15] In that connection, Rule 703 of the Federal Rules of Evidence provide:

> The facts of data in the particular case upon which an expert bases an opinion or inference may be those perceived by or made known to the

---

13 FED. R. EVID. 702 (emphasis added).

14 *Rogers v. Raymark Indus, Inc.*, 922 F.2d 1426, 1429 (9th Cir. 1991); *Hammond v. International Harvester Co.*, 691 F.2d 646, 653 (3rd Cir. 1982).

15 FED. R. EVID. 705.

expert at or before the hearing. If of a type reasonably relied upon by experts in the particular field in forming opinions or inferences upon the subject, the facts or data need not be admissible in evidence.[16]

---

16 FED. R. EVID. 703. This rule permits an expert to rely on information that is not in evidence and is hearsay provided it is information on which experts in the field reasonably rely. *Baumholser v. Amax Coal Co.*, 630 F.2d 550, 553 (7th Cir. 1980).

CHAPTER 15

PREPARING AND PRESENTING THE COMPLEX SURETY CASE TO A PANEL OF ARBITRATORS

Patricia H. Thompson
Alicia E. Kirshenbaum

A surety lawyer (who shall remain nameless for obvious reasons) once remarked to those in attendance at an American Bar Association seminar that her surety clients would rather swallow a live frog than participate in arbitration. She then tried to explain why litigation was a safer forum for sureties. Whether or not she was correct, it is true that arbitration can be different from litigation in many ways.

One of the most obvious differences arises because the arbitrator's expertise "pertains primarily to the law of the shop, not the law of the land."[1] This means that the arbitrator may not appreciate the substantive rights and defenses of suretyship law. Similarly, arbitrators often evidence a marked antipathy to the type of adversarial "lawyering" litigators may consider to be good trial strategy. Additional differences result from the rules and procedures that apply to arbitration, the interplay between state and federal law, the involvement of the courts in the arbitral process, and the dissimilar techniques of effective case presentation.
Incidentally, notwithstanding its introduction, this chapter will not discuss the risks and benefits of arbitration as a means of dispute resolution in general or for the surety in particular, despite the passion which can be

1 *McDonald v. City of West Branch, Mich.,* 466 U.S. 284, 290 (1984) (quoting *Alexander v. Gardner-Denver Co.*, 415 U.S. 36, 57 (1974).

invoked on either side of the argument. Both the courts[2] and arbitration[3] have more than adequate advocates and critics. Instead, this chapter will discuss the legal, procedural, and practical issues facing a surety and its counsel when a surety decides or is required to participate in the arbitration of a complex construction claim.

I. What Law Applies

Although arbitration is intended to be a private, prompt, and economical[4] means of resolving disputes "in a final and binding manner outside the traditional court system,"[5] its procedures exist within, and are subject to, an ever changing statutory and judicial framework. Many of the rules and procedures governing arbitration proceedings are dictated by state and federal legislation, which often differ, and like most statutes, are subject to frequent judicial interpretation and enforcement. Because parties to a complex construction arbitration may also find themselves in court at some point before, during, or after arbitration, it is important to

---

2 *See, e.g.*, QUOTABLE LAW. 65 (David Shrager & Elizabeth Frost eds., 1986) (quoting George C. Wallace, PLAINVIEW (TEXAS) DAILY HERALD, Sept. 24, 1971) ("I have nothing but utter contempt for the courts of this land."); Thomas E. Carbonneau, *Arbitral Justice: The Demise of Due Process in American Law*, 70 TUL. L. REV. 1945, 1967 (1996). ("Commercial expediency and privatized justice through arbitration may produce the desired efficiency in the short run, but what manner of political society will the United States be in the twenty-first century if, in desperation and despair, we rid it of the normative function of law and basic procedural justice?").

3 *United States ex rel. Pensacola Constr. Co. v. Saint Paul Fire & Marine Ins. Co.,* 705 F. Supp. 306, 314 (W.D. La. 1988) ("some rights" are "too important to entrust to arbitration."); Warren E. Burger, Address to the American Bar Ass'n, L.A. TIMES, Aug. 27, 1978 ("We should get away from the idea that a court is the only place in which to settle disputes. People with claims are likely people with pains. They want relief and results and they don't care whether it's in a courtroom with lawyers and judges or somewhere else.").

4 AMERICAN ARBITRATION ASS'N, COMMERCIAL ARBITRATION RULES (May 1, 1992).

5 Carbonneau, *supra* note 2, at 1945.

know which statutes govern the proceedings and which courts have jurisdiction over disputes concerning the arbitration.

A. The Federal Arbitration Act

It may surprise some to learn that arbitration is not the invention of modern legal reformers. Arbitration got its original start with the ancient Greeks.[6] In medieval times, arbitration was one of the major, if not exclusive, mechanisms for resolving commercial disputes.[7] At some point, however, arbitration fell into disfavor. It is presumed that English judges, who were compensated based on the number of cases they decided, were the force behind the hostility towards arbitration.[8] Similarly, until the early twentieth century, most courts in the United States were opposed to arbitration.[9] Eventually, however, United States federal courts began to see the need for change, and in 1924 the United States Supreme Court in *Red Cross Line v. Atlantic Fruit Co.*[10] gave its blessing to arbitration as an appropriate alternative non-judicial remedy by upholding New York's arbitration laws as constitutional.[11]

In 1925, Congress declared itself in favor of arbitration as an acceptable alternative to judicial dispute resolution by enacting the Federal Arbitration Act (FAA). The FAA proclaimed that arbitration agreements were valid contracts that constituted a legal exercise of the contracting rights of the parties.[12]

The FAA contains several provisions for judicial enforcement of arbitral rights. For example, Section 3 of the FAA provides for the stay of federal court litigation, upon application of one of the parties, until the corresponding arbitration has been completed.[13] Section 4 permits parties

---

6 *See generally* John C. Norling, *The Scope of the Federal Arbitration Act's Preemption Power: An Examination of the Import of Saturn Distributing Corp. v. Williams*, 7 OHIO ST. J. ON DISP. RESOL. 139, 139-40 (1991).

7 *See* Julius H. Cohen & Kenneth Dayton, *The New Federal Arbitration Law*, 12 VA. L. REV. 265 (1926).

8 *See id.*

9 *See id.*

10 264 U.S. 109 (1924).

11 *See id.*

12 *See* 9 U.S.C. § 2.

13 *See* 9 U.S.C. § 3 ("If any suit or proceeding be brought in any of the courts of the United States upon any issue referable to arbitration under an

to compel arbitration, while Sections 9, 10, and 11 provide provisions for confirming, vacating, modifying, and correcting arbitration awards.[14] Further, Section 5 of the FAA allows either party to petition the district court to appoint arbitrators where the underlying agreements do not provide a method for their selection.[15] Section 7 provides arbitrators with the power to conduct discovery as they deem necessary and, in addition, upon proper petition, the district court may compel attendance of obdurate witnesses pursuant to subpoenas issued by an arbitrator.[16] Moreover, the FAA significantly limits judicial supervision of arbitral awards,[17] thus guaranteeing some measure of finality to an arbitration proceeding and its result. Section 16 allows appellate review in the limited circumstances provided therein.[18]

After passage of the FAA in 1925, the scope of the Act's application remained a matter of debate for some time. Many states continued to have statutes that prohibited the enforcement of contractual arbitration clauses based on public policy. Notwithstanding the equal bargaining power of the parties and the fact that both parties had agreed at the time of contracting to submit any subsequent contract disputes to arbitration, state courts applied these state statutes to prevent enforcement of contractual arbitration clauses.

From the outset, federal courts applied the FAA in all cases having federal subject matter jurisdiction. However, the United States Supreme Court held that the FAA itself did not provide an independent basis for federal jurisdiction.[19] Accordingly, an interesting federalism issue arose: if a federal court had only diversity of citizenship jurisdiction, was the

---

agreement in writing for such arbitration, the court in which such suit is pending, . . . shall on application of one of the parties stay the trial of the action until such arbitration has been had . . . .").

14 *See* 9 U.S.C. §§ 4, 9-11.

15 *See* 9 U.S.C. § 5 (if the agreement does provide a method for the appointment of the arbitrator(s), the court must defer unless delay in the appointment is established).

16 *See Stanton v. Paine Webber Jackson & Curtis, Inc.*, 685 F. Supp. 1241, 1242 (S.D. Fla. 1988) (holding that under Act, arbitrators may order and conduct discovery as they find necessary); *see also* 9 U.S.C. § 7.

17 *See* 9 U.S.C. § 10.

18 *See* 9 U.S.C. § 16.

19 *See Moses H. Cone Memorial Hosp. v. Mercury Constr. Corp.,* 460 U.S. 1, 25 n.32 (1983).

court bound under the *Erie* doctrine[20] by the applicable state arbitration law which might invalidate contractual arbitration clauses, or could the court apply the FAA and enforce the contract? The Supreme Court answered this question in *Prima Paint Corp. v. Flood & Conklin Manufacturing Co.*,[21] reasoning that Congress had authority to "prescribe how federal courts are to conduct themselves with respect to subject matter over which Congress plainly has power to legislate."[22] According to the Court, the FAA was such a directive and as a result, the federal courts had to apply the Act even where the court's jurisdiction was based solely on diversity of citizenship.[23]

Because the FAA does not create an independent basis of federal subject matter jurisdiction, either diversity or a federal question must exist before a federal court will have the power to adjudicate or compel arbitration under the FAA.[24] The scope of the Act, though, is defined in Section 2, which provides:

> A written provision in any maritime transaction or a contract evidencing a *transaction involving commerce* to settle by arbitration a controversy thereafter arising out of such contract or transaction, or the refusal to perform the whole or any part thereof, or an agreement in writing to submit to arbitration an existing controversy arising out of such a contract, transaction, or refusal, shall be valid, irrevocable, and enforceable, save upon such grounds as exist at law or in equity for the revocation of any contract.[25]

Accordingly, the FAA applies to all arbitration agreements in "transaction[s] involving commerce."[26] The interpretation given to the

---

20 *Erie R.R. Co. v. Tompkins*, 304 U.S. 64 (1938) (holding that state substantive law applies in diversity actions).

21 388 U.S. 395 (1967).

22 *Id.* at 405.

23 *See id.*

24 *See Mercury Constr. Corp.*, 460 U.S. at 25 n.32.

25 9 U.S.C. § 2 (emphasis added).

26 *Id., see also* 9 U.S.C. § 1 (defining commerce as "commerce among the several States or with foreign nations, or in any Territory of the United States or in the District of Columbia, or between any such Territory and another, or between any such Territory and any State or foreign nation, or between the District of Columbia and any State or Territory or foreign nation . . . .").

"commerce" definition within the Act is identical to the one given the Commerce Clause of the United States Constitution.[27] Therefore, the FAA applies to any transaction within the reach of the Commerce Clause,[28] and a federal court sitting in diversity is bound by the FAA if it determines that the underlying transaction involves interstate commerce.[29] On the other hand, a federal court sitting in diversity must apply state arbitration laws if the underlying transaction does not involve interstate commerce.[30]

The same result is required if a state court dispute involves interstate commerce. That is, the FAA must also be applied by state courts when the transaction underlying the arbitration dispute comes within the purview of Section 2 of the Act.[31] In other words, the FAA displaces all state arbitration laws when the underlying transaction or contract comes within its scope, but does not pre-empt state arbitration laws when the underlying transaction or contract is not one "involving commerce."[32]

---

27 *See Allied-Bruce Terminix Cos., Inc. v. Dobson*, 513 U.S. 265, 273 (1995) (citing *United States v. American Bldg. Maintenance Indus.*, 422 U.S. 271, 276 (1975), for the proposition that "commerce" means interstate commerce).

28 *See id.* "That phrase—'affecting commerce'—normally signals a congressional intent to exercise its Commerce Clause powers to the full." *Id.* at 273 (citing *Russell v. United States*, 471 U.S. 858, 859 (1985)). The court in *Allied-Bruce Terminix* concluded that although the Act uses the word "involving" instead of "affecting," the words are functionally equivalent. *See id.* at 277; *see also Perry v. Thomas*, 482 U.S. 483, 489 (1987).

29 *See Snyder v. Smith*, 736 F.2d 409, 417-18 (7th Cir. 1984) (holding that court must examine the contract, the complaint, and the facts when determining whether the underlying transaction is one involving commerce within the meaning of the Federal Arbitration Act).

30 *See Bernhardt v. Polygraphic Co. of Am.*, 350 U.S. 198, 202-203 (1956). For an in depth analysis of the interpretation of the "involving commerce" language of the Act by the U.S. Supreme Court see Preston Douglas Wigner, *The United States Supreme Court's Expansive Approach to the Federal Arbitration Act: A Look at the Past, Present, and Future of Section 2*, 29 U. RICH. L. REV. 1499, 1504-1506 (1995).

31 *See Southland Corp. v. Keating*, 465 U.S. 1, 14-16 (1984).

32 *See id.* at 15-16 (holding that Federal Arbitration Act pre-empts state law). In *Allied-Bruce Terminix*, the Supreme Court held that where the contract involves interstate commerce, the FAA displaces or pre-empts any state law to the contrary. 513 U.S. at 276-77. Accordingly, the Court refused to

State law also will govern disputes between the parties to an arbitration agreement when they have expressly agreed to be bound by a specific state's arbitration laws. This was the case in *Volt Information Sciences, Inc. v. Board of Trustees of Leland Stanford Junior University.*[33] Although the underlying transaction involved interstate commerce, the parties had expressly agreed that any contractual disputes would be governed by the laws of California.[34] The Court concluded that even though the FAA would normally pre-empt state arbitration law, the choice of law clause in the contract demonstrated the parties' intent to have any disputes controlled by California arbitration law.[35] Since California law does not forbid arbitration agreements, the Court held that the intention of the parties controlled.[36] Accordingly, the FAA will allow parties to a contract to be bound by the arbitration laws of the state of their choice.[37]

B. State Arbitration Statutes: The Uniform Arbitration Act

At present, most states have adopted some form of arbitration legislation and the majority have passed some version of the Uniform Arbitration Act (UAA).[38] The UAA was approved by the National Conference of Commissioners on Uniform State Laws (NCCUSL) and the American Bar Association in 1955, at a time when arbitration was largely limited to a few specific trades.[39] The purpose of the UAA was to "validate arbitration agreements, make the arbitration process effective,

---

enforce an Alabama law that disallowed all arbitration agreements. *Id.* at 269-70.

33 489 U.S. 468 (1989).

34 *See id.* at 476.

35 *See id.* at 479.

36 *See id.* "There is no federal policy favoring arbitration under a certain set of procedural rules; the federal policy is simply to ensure the enforceability, according to the their terms, of private agreements to arbitrate." *Id.* at 476.

37 *Cf. Mastrobuono v. Shearson Lehman-Hutton, Inc.*, 514 U.S. 52 (1995) (provision of contract that made New York law controlling does not bar award of punitive damages in arbitration where New York law allows for punitive damages by court but not by arbitration award and contract did not specifically prohibit the award of punitive damages).

38 J. Michael Franks & John W. Heacock, *Arbitration and the Contract Surety: Inclusion and Preclusion*, 32 TORT & INS. L.J. 977, 983 (1997).

39 Michael Yablonski, *Conference Considers Overhaul of Arbitration Act*, LITIG. NEWS (A.B.A.), Sept. 1997, at 6.

provide necessary safeguards, and provide an efficient procedure when judicial assistance is necessary."[40]

A number of statutes still contain limiting provisions which are not contained in the FAA. While not intending to be an exhaustive list, the following examples are representative of such limiting provisions. Three states continue to prohibit or significantly limit pre-dispute arbitration agreements: Alabama by statute;[41] Nebraska by court decision, notwithstanding passage of the UAA;[42] and Mississippi by older case law.[43] Some states will not enforce contracts for arbitration if the clause would require an in-state party to arbitrate outside the state.[44] Others will not enforce the clause if the contract has elected to be governed by the law of some other state. Still other states go as far as to refuse to enforce the arbitration clause in a contract if the contract has elected to have the arbitration take place in another state.[45]

The UAA will be undergoing some changes in the not so distant future. The NCCUSL has appointed a committee charged with the weighty task of revising the UAA.[46] Some of the proposed changes concern arbitrability of certain issues, arbitral immunity, interlocutory review of arbitral pre-award orders, and punitive damage awards.[47] The process of amending the UAA is expected to take two years or more.[48] It is, of course, impossible to predict which states will adopt any or all of the changes proposed by the NCCUSL.

---

40 UNIF. ARBITRATION ACT, Prefatory Note, 7 U.L.A. (1995).

41 ALA. CODE § 8-1-41(3) (1994) (providing that an "agreement to submit a controversy to arbitration" cannot be specifically enforced).

42 *State v. Nebraska Ass'n. Of Pub. Employees*, 477 N.W.2d 577, 580 (Neb. 1991).

43 *McClendon v. Shutt*, 115 So. 2d 740, 741 (Miss. 1959) (providing that "either party may revoke an agreement to submit to arbitration at any time before an award has been made").

44 *E.g.*, LA. REV. STAT. ANN. § 38:2196 (West 1994).

45 VA. CODE ANN. § 8.01-262.1 (Michie 1997).

46 Thomas J. Stipanowich, *What's a NCCUSL? The Uniform Law Commissioners Tackle the Uniform Arbitration Act*, 17 CONSTR. LAW. 25 (Jul. 1997).

47 *Id.* at 25-26.

48 Yablonski, *supra* note 39, at 6.

II.
What Rules Apply

Given the differences between federal and state arbitration statutes, the parties to an arbitration should determine as soon as possible what law applies to their proceedings and whether their interests will be affected by that law. Additionally, they should immediately ascertain whether their proceedings are subject to any non-statutory rules and procedures. The most well known are the American Arbitration Association (AAA) Construction Arbitration Rules and the Center for Public Resource (CPR) Institute for Dispute Resolution's Private Adjudication Committee Rules. Due to the popularity of the AAA rules, contracts frequently incorporate them by reference. The CPR rules are primarily used by parties who prefer non-administered arbitration.

A. AAA's Construction Arbitration Rules

One of the AAA's primary functions is to serve as an administering organization for arbitrations. Consequently, in addition to assisting in or controlling the process by which arbitrators are selected, the AAA provides administrative staff to render services required for case handling and to insulate the arbitrators from the parties, determine arbitrator fees and bill the parties for such fees, schedule hearings and send notices of hearings, provide hearing rooms, and distribute documents.[49]

The AAA rules governing construction disputes have recently undergone extensive changes. A few years ago, the AAA convened a Construction ADR task force of fifty-five construction related professionals to undertake a comprehensive review of the services which AAA offered to the construction industry to determine how AAA could be more responsive to the industry's dispute resolution needs.[50] The result was a report, revised rules, a pared down roster of neutrals, and a training program for those neutrals.[51] Generally, the changes were intended to give the arbitrators more authority while simultaneously speeding up the

---

49 *Dispute Resolution Clauses for Business Agreements*, 10 CENTER FOR PUB. RESOURCES, July 1992, at 107.

50 Donald J. Cummings, *American Arbitration Association Revisions to Rules & Panels*, COMMENTS ON CONSTRUCTION (published by Wagner-Hohns-Inglis, Inc.), Winter 1997, at 1.

51 *Id.*

process of arbitration.[52] Some of the salient features of these new rules are described below.

First, all construction disputes are divided into three different categories: the Fast Track; the Regular Track; and the Large/Complex Track. The Fast Track designation is assigned to any claim of less than $50,000.[53] Some of the highlights of the Fast Track system include: a single, pre-qualified, arbitrator appointed by AAA (with party input);[54] a sixty-day time limitation for case completion, requiring the arbitrator to issue an award within sixty calendar days of his or her appointment, unless the parties extend the time by agreement or "extraordinary" circumstances justify an extension;[55] a single day hearing (in most cases) held within thirty days of the arbitrator's appointment;[56] an award to be rendered within seven days of the close of the hearing;[57] and presumption that cases involving less than $10,000 will be heard on a document only basis.[58]

The Regular Track cases include those disputes involving claims between $50,000 and $1,000,000. The changes to the Regular Track will affect about half of the construction arbitrations filed with the AAA, according to the 1994 case filings.[59] Some of the highlights of the Regular Track system include: a single arbitrator will preside over the proceedings, unless otherwise specified in the contract or the AAA so decides;[60] an increased role for the parties in arbitrator appointment;[61] a written breakdown of the award and optionally a written explanation as well, if requested by the parties;[62] a more comprehensive arbitration demand form and a new answer form;[63] and authority given to the arbitrator to control the discovery process.[64]

---

52 George H. Friedman, *Major Changes Coming to AAA Construction Arbitration*, 15 CONSTR. LAW. 25 (Nov. 1995).

53 *Id.*

54 *Id.*

55 *Id.*

56 Cummings *supra* note 50, at 1.

57 *Id.*

58 *Id.*

59 Friedman, *supra* note 52, at 26.

60 Cummings, *supra* note 50, at 1.

61 *Id.*

62 *Id.*

63 *Id.*

64 *Id.*

The Large/Complex Track includes those cases involving claims in excess of $1,000,000. The program features a purportedly elite panel of trained neutrals, and special supplementary procedures to be used in larger cases.[65] In addition to the Regular Track procedure, some of the highlights of the Large/Complex Track system include: the parties have the choice of having the case heard by one or three arbitrators (if parties cannot agree three arbitrators will be used);[66] a mandatory preliminary hearing with the arbitrator to be conducted either by telephone conference or in person for the purpose of discussing statements of claims, stipulated facts, document handling and exchange, identification/schedulingof witnesses, scheduling of hearings, etc.;[67] hearings to be conducted on consecutive and/or block basis;[68] and arbitrators given broad authority regarding discovery, disposition, and management of the proceedings.[69]

Some of the other changes made by the Construction ADR Task Force dealt with the roster of neutrals. In fact, the first finding of the Task Force regarding the Panel was:

> There are too many arbitrators in many Panels. This is especially true in the large urban areas where most of the cases are heard. These panel members do not serve often enough to gain experience. Many of the members of the current panels have been selected without regard to standards of quality . . . .[70]

In attempting to correct this problem, each of the Regions invested considerable effort in removing from its list of neutrals those individuals no longer deemed to be able, interested, and/or qualified to perform as arbitrators.[71] Consequently, each arbitrator's qualifications were reviewed with regard to his or her character, experience, ability, and willingness to serve.[72] Those arbitrators deemed to be most qualified were invited to reapply for Panel membership with the condition that these applicants complete a newly instituted, nationwide arbitrator training program within

---

65 Friedman, *supra* note 52, at 27.
66 Cummings, *supra* note 50, at 1.
67 *Id.*
68 *Id.*
69 *Id.*
70 *Id.* (internal quotations omitted).
71 *Id.* at 2.
72 *Id.*

a reasonable period of time and continue to undergo training on a regular basis.[73]

B. Center For Public Resources Rules For Non-Administered Arbitration Of Business Disputes

As noted above, one of the major differences between the arbitration rules for the AAA and the CPR is that the CPR does not function as an administrative body.[74] Therefore, the functions performed by an administering organization such as the AAA are not performed by the CPR.[75] Instead, these functions are performed by the arbitrators themselves (increasing the importance that they be experienced in such matters) or by the parties' attorneys. The CPR's responsibilities are limited to serving as an appointing authority, deciding disputes concerning the appointment of arbitrators, and other related issues. Under non-administered proceedings, the parties simply agree to the rules that will apply to their dispute, either through their contract documents or through mutual agreement once the dispute arises, in a manner they believe to be best calculated to lead to resolution of the dispute.

C. No Rules

There is no obligation for parties who choose to arbitrate their disputes to use (and pay for) the administrative services of AAA or adhere to any preset rules at all. They can take advantage of private ADR or consulting companies which are now offering such arbitration services, or they may use a mutually trusted third party, who by agreement will function as arbitrator pursuant to what ever procedures the parties require.

III. Where To Arbitrate

Unless the pre-dispute arbitration agreement contains a forum selection clause, it may not be immediately apparent where the parties to a complex construction dispute should arbitrate that dispute. This issue can

---

73 *Id.*
74 *Dispute Resolution Clauses for Business Agreements*, *supra* note 49, at 107.
75 *See supra* note 49 and accompanying text.

be complicated if the parties have their primary offices in different states or in states other than the construction site or sites.

For those proceedings subject to AAA rules, the rules provide that the parties may mutually agree on the locale where the arbitration is to take place.[76] Alternatively, once a party files for arbitration requesting that the arbitration be held in a given locale, the opposing party has only ten days after the notice of the request has been sent to it by the AAA to raise an objection to the requested situs.[77] If no objection is made within the required time limitation, the arbitration will be held in the locale requested by the filing party.[78] If the objection is made in a timely fashion, the AAA has the authority to determine the locale and its decision is usually considered final and binding.[79]

If the parties have not agreed that their arbitration will be governed by AAA rules, and if the FAA applies to their dispute, Section 4 of the FAA provides that a party may file an independent proceeding[80] to compel arbitration in "any United States district court which, save for such [arbitration] agreement, would have jurisdiction under Title 28, in a civil action or in admiralty of the subject matter of a suit arising out of the controversy between the parties . . . ."[81] The district court where the petition to compel arbitration is brought may only direct arbitration to take place within the district in which it is located.[82]

---

76 COMMERCIAL ARBITRATION RULES, *supra* note 4, at 8.

77 *Id.*

78 *Id.* at 9.

79 *Id.*

80 An independent proceeding is one in which the sole issue before the court is the dispute's arbitrability. *Gammaro v. Thorp Consumer Discount Co.*, 15 F.3d 93, 95 (8th Cir. 1994). By contrast, an "embedded proceeding" is not governed by the venue provisions of the FAA, because the legal action was brought prior to the opposing party's attempt to compel arbitration.

81 9 U.S.C. § 4. This standard also applies to an application for the appointment of the arbitrators pursuant to Section 5. *See* 9 U.S.C. § 5.

82 *See Econo-Car Int'l, Inc. v. Antilles Car Rentals, Inc.*, 499 F.2d 1391, 1394 (3rd Cir. 1974); *Federated Rural Elec. Ins. Co. v. Nationwide Mut. Ins. Co.*, 874 F. Supp. 1204, 1209-10 (D. Kan. 1995). However, if the arbitration agreement or contract contains a forum selection clause, that clause will be controlling. *See Bosworth v. Ehrenreich*, 823 F. Supp. 1175, 1180 (D.N.J. 1993) (holding that the forum selection clause requiring the arbitration to take place in New York controlled where one of the parties moved a New Jersey district court, that would otherwise have jurisdiction over the dispute

Therefore, a party who has no forum selection clause in its arbitration agreement and who wants to try to control the venue of the hearing should determine which district courts would have subject matter jurisdiction of the underlying disputes and file a petition to compel arbitration in the competent district court of its choosing. This option would apply only in a proceeding to which the FAA applies and only to an "independent proceeding" (*i.e.,* one filed only for the purpose of compelling arbitration). In an embedded proceeding (one in which arbitration is ordered during the course of an already existing legal action), the party that filed the action will have already selected the forum district.[83]

In those cases subject to state arbitration statutes, the state statutes may provide, as does Section 18 of the UAA, that an action to compel arbitration shall be filed in the county where the adverse party resides or has a place of business. However, Section 5 of the UAA provides that the actual place of the hearing is up to the arbitrators to determine. Unfortunately, the arbitrators may agree on a location more convenient for them than for the parties.

Generally, if the arbitration agreement contains a forum selection clause, that clause will be controlling. Therefore, parties to a construction contract would be well advised to include forum selection clauses in their arbitration agreements to insure a mutually convenient forum for arbitration. If it is neither practical nor feasible to determine the place of arbitration in advance of any dispute, the clause may provide "the place of arbitration will be determined by agreement of the parties upon commencement of an arbitration, provided that if the parties fail to agree, the arbitrator(s) shall designate the place of arbitration."[84]

---

and parties, to compel arbitration, the New Jersey district court transferred the matter to the New York district court, because the New Jersey court did not have the authority to compel arbitration in New York, nor could it compel arbitration in New Jersey, as this would violate the forum selection clause to which the parties had expressly agreed).

83 *See id.*

84 *Id; see, e.g., Jackson Trak Group ex rel. Jacskon Jordan, Inc. v. Mid States Port Auth.*, 751 P.2d 122, 127 (Kan. 1988).

IV.
Arbitrators: Who And How Many

The selection of the arbitrators is a crucially important aspect of the arbitration process. It has been said that "you only get one bite at the apple in arbitration, and the result is only as good as your arbitrator."[85] While some may consider this to be an overstatement, it is certainly true that the wrong arbitrator(s) can turn the proceedings into a legal and procedural nightmare.

The process for selecting arbitrators differs depending on the tribunal or provisions of the agreement for arbitration. All applicable procedures and statutes must be reviewed to determine what options are available for selection of, objection to, and resolution of disputes concerning arbitrators. Regardless of which process is involved, parties should try to learn as much about the arbitrator nominees as is practically possible. Obvious sources for such information are clients, consultants, other lawyers, and members of the construction and ADR industries. However, it may also be possible to obtain from the administering officer tribunal resumes or even an oral summary of the nominees' experience professionally and as a neutral.

In addition to learning a much as possible about specific arbitrator nominees, consideration should be given as to whether arbitrators from certain professions, or with particular professional experience and training, may be better suited than others to understand the issues of a given dispute. For example, if a surety intends to rely upon legal defenses which would require the arbitrators to rule in the surety's favor despite the existence of "equities" to the contrary, it may be wise to insist upon at least one lawyer as an arbitrator, especially a lawyer with sufficient expertise to appreciate and follow applicable surety law. If there is not a lawyer on the panel, especially a lawyer who will at least keep an open mind on suretyship issues, the arbitrators may be hostile to surety defenses that they perceive as "legal maneuvering" and mere technicalities. Non-lawyers are likely to believe it is their duty as arbitrators to "do equity," to be fair to all concerned, and to be realistic problem solvers, rather than make hard decisions of right and wrong. Indeed, they are often quick to

---

85 Carbonneau, *supra* note 2, at 1958.

note that they are under no obligation to follow the law.[86] If the parties are not able to agree on an arbitrator, then recourse may be made to the applicable court for appointment of the arbitrator(s), and it may be that a judge will be particularly sensitive to a surety's need to have a decision maker with legal experience.

Other than suggesting that a surety require a lawyer on the panel, no other recommendations as to the "ideal" arbitrator are easy to make in the abstract. Usually, the parties try to determine what professions would have some familiarity with the subject matter and might theoretically be inclined to "favor" each side of the dispute. This can be an exercise in amateur psychology akin to some of the older, often cited rules for jury selection. For example, an architect recently recommended that the best arbitrator in a complex construction case would be another architect, because architects are generalists who understand everything that happens on a project and are trained in fair and equitable problem solving. It has also been suggested that those arbitrators possessing type-A personalities are more likely to "appreciate the structure imposed by a case management agreement and will instinctively want to move things along" to reach a resolution. Regardless of the personality type, it is good to look for an arbitrator who has experience in handling case administration. It is also safe to say that it might not be of much help to choose a specialist with a narrow range of experience, unless that specialty is relevant to the issues in the case.

Depending upon the rules the parties have selected to control the proceedings and the nature of the claim, they may have no choice as to the number of arbitrators.[87] Assuming, however, that there is a choice to be made, there are several advantages to having three arbitrators rather than one or two. First, a panel of three arbitrators provides the parties with a better balance of opinion and expertise than a single arbitrator.[88] Imagine

---

86 Consequently, parties to an arbitration agreement may consider adding a provision specifically requiring the arbitrators to follow all the applicable substantive law of a specified jurisdiction including that jurisdiction's statutes of limitations, and requiring the arbitrators to issue a reasoned award with statements of fact and conclusions of law.

87 For example, under the AAA Construction Industry Arbitration Rules (1996), if a claim involves more than $1,000,000, unless the parties agree to use only one arbitrator, the proceeding will be resolved by three arbitrators.

88 Tom Arnold, *Avoiding Arbitration Booby Traps*, 8 PRAC. LITIGATOR 31, 35 (1997).

for example the exchange of ideas and analysis which might occur among the members of a panel consisting of an architect, a contractor, and a trial lawyer or law professor.[89] Therefore, it has been argued that the decision made by a panel of arbitrators is of higher quality due to the interaction and debate that takes place among the arbitrators.[90] The fact that the arbitrators will be deliberating would appear to ensure a more reasoned consideration of the evidence as well. "The judgment of a panel of three may also help to stifle eccentricities that might be present in the judgment of an individual neutral."[91]

A second advantage to having a panel of three arbitrators is that it can speed up the appointment process.[92] If the parties are diligent and cooperative, "the arbitrators can be efficiently designated when each party appoints one arbitrator and the third is appointed by those two or [by] an agency."[93] The parties may reach an agreement in choosing three arbitrators in this fashion more quickly than if they had to agree on a single arbitrator.

The value of having three arbitrators is undermined, however, if all three are not truly "neutrals." Having a panel of arbitrators consisting of two partisan, party-appointed arbitrators and only one neutral is simply counterproductive. The arbitrators should be chosen to honestly and ethically resolve a difficult dispute and not simply vote in favor of the party who hired them. Otherwise, the decision will be left in the hands of the third and only neutral arbitrator. In that instance, the parties might be better off saving time and money by simply utilizing one arbitrator.

It may come as a surprise to many who engage in arbitration that neutrality of the appointed arbitrators is not a given.[94] On the contrary, the AAA Domestic Rules as construed provide for the non-neutrality of party-

---

89 *Id.* at 36.

90 *Id.*

91 *Id.*

92 *Id.*

93 *Id.* Some of the better known agencies include: World Intellectual Property Organization, Geneva, CPR, AAA, LICA, Private Adjudication Center, J.A.M.S./Endispute, International Chamber of Commerce, Paris. *Id.*

94 *Id.* "10 or 20 per cent of all party-appointed arbitrators are not impartial even if 'neutral.'" *Id.* Thus, parties are urged "[n]ever [to] use any party-appointed arbitrators except when jointly appointed by both parties." *Id.*

appointed arbitrators.[95] Thus, the arbitrators in some states are permitted to be advocates and to talk to their appointers throughout the proceedings of the arbitration.[96] Because "neutral" arbitrators are critical to the success of an arbitration, the drafters of arbitration contracts may want to include the following clause:

> All arbitrators shall be neutral, independent, disinterested, unbiased and impartial arbitrators, bound by the Rules of Ethics of the ABA and AAA for neutral arbitrators, and none shall have ex parte communications with parties . . . except when the other party was given due notice of a hearing and failed to participate.[97]

There are several reasons why a single arbitrator may be better than a multi-arbitrator panel. First, there is an obvious cost advantage to having a single arbitrator.[98] As one observer stated, "the cost of using three arbitrators is *much* more than merely treble the fee of one arbitrator."[99]

Another, related advantage to using a single arbitrator is the avoidance of scheduling problems that invariably arise with three arbitrator panels.[100] This is more likely when utilizing respected professionals or experienced arbitrators, who are usually extremely busy, and hence have difficulty scheduling lengthy hearings.[101] Because scheduling problems have a tendency to increase exponentially with the number of people and number of days to be accommodated, engaging a single arbitrator may be preferred to a three arbitrator panel.[102]

Furthermore, a single arbitrator is less burdened by the need to accommodate different learning curves than a three arbitrator panel.[103] Three arbitrator panels are usually more reluctant to limit discovery or witness examinations due to their legitimate concerns that "one of the other arbitrators may still be learning from the presentation, that the award

---

95 *Id.* at 44.

96 *Id.*

97 *Id.* at 44-45.

98 *Id.* at 36.

99 *Id.* The author of the cited article suggests that the total cost of using three arbitrators may be as much as ten times greater monetarily, and twelve to fifteen months longer in time than with a single arbitrator. *Id.*

100 *Id.* at 37.

101 *Id.*

102 *Id.*

103 *Id.*

may be set aside if proffered evidence is not accepted, or that one of the parties may feel that its presentation was not fully heard."[104] By contrast, a single arbitrator, who is likely to become impatient with redundant testimony, may be bolder in expediting the discovery or hearing process once he or she is satisfied that fairness has been accomplished on a particular point.[105]

Finally, a single arbitrator is better suited for managing the case administration process. For this reason, if a three arbitrator panel is used, it would be wise to empower one to assume the role of case manager and overall expediter. Failure to have a case manager may result in:

> [A] case [that] is prolonged by the slow pace of needed discovery, by over-liberalized discovery, [or conversely] by a denial of deposition discovery which results in dozens of witnesses called live at the arbitration hearing, or by a failure to schedule the whole case early in the proceeding and manage it throughout.[106]

For those parties who cannot choose between one arbitrator and three, there may be special advantages to the less often used option of a two arbitrator panel. Two arbitrators might provide a healthy balance of having more than one "brain" contribute to the process without sacrificing as much efficiency as is the case of a three arbitrator panel.[107] If the two disagree, it is likely that they will inevitably come together by compromising their differences.[108] If an impasse should result, the parties can deal with that rare occurrence by appointing a third arbitrator or "umpire" to resolve the disagreement.[109] Although not ideal, such an appointment would serve as a "safety-valve . . . that only rarely is needed and that can be reasonably relied upon for practical degrees of efficacy for all its imperfection."[110]

---

104 *Id.*

105 *Id.*

106 *Id.*

107 *Id.*

108 *Id.* at 37-38.

109 *Id.* at 38.

110 *Id.* The author of the cited article stresses looking at the "big picture" in terms of the number of arbitrators to choose. He also favors having fewer arbitrators, rather than more:

V.
Once Again, What Rules Apply: You Decide

As soon as the arbitrators are selected, the parties should request a preliminary conference, if one is not automatically scheduled, so that they and the arbitrators can fashion the rules which will apply to the balance of the proceeding. At the conference, parties should be well prepared to: (1) define logistics; (2) properly frame the issues to be arbitrated; (3) define the type of relief requested and authorized by the agreement or by the parties themselves; (4) control the timing and specify, define, or limit the number of hearing days allowed with an allocation of number of days per side; (5) define discovery; (6) schedule a final hearing; (7) set the ground rules for how evidence should be submitted; (8) educate the arbitrators; and (9) agree upon the type of award to be given by the arbitrators (reasoned, preliminary with comments by the parties, or unreasoned).

If counsel is well prepared, this conference can be used to gain an advantage by setting schedules and establishing rules and procedures for the balance of the proceedings consistent with counsel's specific, predetermined goals and strategy, while at the same time making a good initial impression on the arbitrator(s). This conference should identify each party's witnesses, what further information the parties are required to obtain, and what issues will be arbitrated. The proposed AAA report form usually provided to arbitrators for their use in this conference should be used by the parties themselves as a guide to determine what should be done and what each party should be prepared to agree to and require from the other parties.

> [H]ow much better justice, if any, do you really get from three arbitrators with one qualified neutral? Perfect justice is an impossible dream. It is not obtainable. In the closer cases the decision can be a raffle at best. Neither the predictability of results nor the odds of improving such raffles are materially improved for either side by an election to carry the often-extreme burdens and many hidden costs of having three arbitrators rather than one. But three are unlikely to be as strong in awarding harsh remedies as one—the group dynamics of three seems very commonly to dampen the more extreme views.

Id.

Additionally, considerable attention should be given to the evidentiary aspects of the final hearing. Although it is not often done, it may benefit the surety to request that the parties be required to submit their direct testimony in advance, in writing, together with bound, previously exchanged exhibits. Oral testimony could possibly be limited to cross-examination of those witnesses for whom written testimony was offered by opposing counsel. Parties should be strictly limited as to the number of days they may use for their case and/or for cross-examination.

The parties should try to require the arbitrators to agree in advance as to whether they will give a reasoned or unreasoned award, and whether the arbitrators will agree to submit preliminary awards to the parties for comment prior to issuing a final award. They should also agree in advance whether the hearing will be transcribed, and whether pre- and post-hearing written submissions will be allowed. Generally arbitrators like to simply issue unreasoned awards, but the parties may find it easier to live with the results, and to subsequently enforce the awards, if the awards are reasoned.

Another important issue to be addressed at the initial hearing is the extent of case management authority to be given to the arbitrator(s). The purpose of case management is to provide structure and control over what otherwise is a very loose process. By not providing for case management authority for the arbitrator, the parties run the risk that one party will abuse the process unchecked, and/or that the arbitrator will behave just "like a judge with an overloaded docket-reactively," which will effectively slow down the entire process.[111]

The problem of lack of case management can be resolved in advance in the initial arbitration agreement. Some of the elements that should be included within the agreement to arbitrate are: (1) time-is-of-the-essence provisions; (2) which arbitrator has primary responsibility to manage the process; (3) choice of law or governing law provisions; (4) discovery provisions; (4) applicability of the rules of procedure; (5) payment of fees and expenses to the prevailing party; (6) award of attorney's fees; (7) type award needed (*i.e.,* "reasoned award"); (8) the need for transcript proceedings; (9) time limitations for decisions by the arbitrators both for pre-hearing issues and post-hearing awards; and (10) expressly provide that each of the arbitrators will be neutral, independent, and impartial.[112]

---

111 *Id.* at 39.
112 *Id.* at 38-43.

VI.
Who Decides Your Fate: (It's Never Too Late To Mediate)

Many authorities on the broad subject of alternative dispute resolution discuss arbitration and mediation separately, as if to imply that they are two mutually exclusive means of avoiding conflict. While the success rate of mediation as a means of resolving disputes short of litigation is widely acknowledged,[113] parties in arbitration frequently fail to go through pre-hearing mediation. AAA studies support this conclusion, indicating that many more cases slated for arbitration proceed to final hearing, rather than settle in advance of the hearing, than do cases slated for trial in the courts.[114] Perhaps this difference in practice is due to the fact that there is no mechanism in arbitration to force the parties to mediate, while many jurisdictions require litigants to participate in mediation prior to trial.

While this difference may be explainable, it may also be a mistake on the part of the parties' attorneys.[115] The fact that parties have chosen to arbitrate their disputes does not preclude the possibility that they would be better off resolving their differences in mediation. Indeed, the unpredictability of result, the unfettered discretion afforded arbitrators, and the absence of meaningful appeal are only some of the reasons parties might prefer to mediate construction disputes prior to arbitration.

Even if mediation does not result in resolution of an entire dispute, it may eliminate parties or claims and narrow the issues for subsequent hearing. Additionally, given the minimal amount of discovery allowed in some arbitrations, mediation may allow everyone, especially the parties' decision makers, their only real opportunity to observe and evaluate their opponent's arguments or evidence. Such disclosure would, at the very least, help the parties prepare more effectively for the arbitration hearing if the dispute is not resolved at mediation.

Mediation may offer another benefit. If the parties carefully choose a knowledgeable and respected mediator, they may obtain from the mediator

---

113 For example, according to AAA's statistics, just over ninety percent of all AAA mediations settle at or shortly after mediation. 1 EDWARD A. DAUER, MANUAL OF DISPUTE RESOLUTION, § 8.03 (1994).

114 *Id.* at § 8.03 n.14.

115 Professor Dauer suggests that counsel for parties to an arbitration dispute may owe their clients an ethical duty to consider and recommend mediation if the case appears appropriate for such a procedure. *See id.* at pt. 6.

an independent and confidential analysis of their respective cases, which may assist them in evaluating and valuing their positions and, if necessary, when preparing for the subsequent hearing. Specifically, surety counsel may want to solicit the subjective non-binding opinion of the mediator as to the merits of the surety's claims and defenses.

As a final point about mediation, just in case the question arises, the mediator should not be one of the arbitrators.

VII.
How To Prepare For Final Hearing

The first stage of hearing preparation involves investigation of the relevant claims and defenses, discovery, and exchange of information among the parties.

During the initial investigation of the claims, counsel must determine what the potential witnesses know and relevant documents say, what additional information is needed, and who is likely to have that information. Counsel should interview the surety's claims and underwriting personnel, the principal, its employees, consultants, and anyone else who will agree to be interviewed.

At the beginning of the investigation process, as counsel is developing the theory of the case, and thereafter, during subsequent investigations and hearing preparation, it may be of value to ask a consultant to review the likely "evidence" and claims or defenses to advise what information appears to be most important, what might be missing, and how the case appears from the consultant's more or less objective perspective. Because the parties usually get only one shot at the arbitrators, it is important that their presentations answer all of the arbitrators' likely concerns and questions, give the arbitrators all the information they need to get to the conclusion, and be as persuasive, and concise as possible. The input of an industry consultant, who is likely to be as impatient with legalese as an arbitrator, might be valuable in preparing such a presentation

Formal discovery by way of deposition and documentary subpoena is usually more curtailed in arbitration proceedings than in litigation. While it is not necessarily true that there is no discovery in arbitration, it is true that there is usually less discovery, especially discovery of uncooperative non-parties. The parties have agreed to arbitration, with all of its legal and procedural drawbacks, because it is supposed to be less expensive and less time consuming. For the most part, arbitration is less expensive and time consuming to the extent discovery can be avoided.[116]

State statutes vary as to the discretion allowed arbitrators to issue subpoenas and authorize depositions. The FAA does not appear to allow third party depositions,[117] although it has been construed to allow arbitrators to require the production of documents by third parties.[118] The CPR and AAA rules both give the arbitrators the authority to permit limited discovery as appropriate in the circumstances, and to the extent the parties have agreed to be bound by these rules, a court may enforce those rules, regardless of the parties' subsequent objections.[119]

As discussed above, the parties should resolve the issue of allowable discovery at the preliminary conference. At the conference, the parties should be ready to provide, and to require of each other, ample justification as to the reasonableness and necessity for taking any depositions. Discovery, to the extent it is allowed by the arbitrators or agreed to by the parties, should be used only to obtain the information the parties cannot otherwise obtain. Even if the arbitrators allow the parties to take depositions, they will probably not allow them to read extensively from those depositions at the hearing, nor will the arbitrators likely read the transcripts if they are submitted at the hearing. Therefore, it may be unwise and an unnecessary expense to take numerous, extensive depositions. Instead of formal discovery, the parties should be reasonable and share information and documents upon request and in compliance with the schedule agreed upon at the preliminary conference. It is important to insist upon cooperation at this stage in the proceedings and to be cooperative in return. Not only will the parties obtain needed information if they cooperate, their conduct as representatives of their clients is being watched by the only persons who matter as far as the ultimate resolution of this case is concerned: the arbitrators. Arbitrators

---

116 Money and time are spent mostly in pre-trail discovery. Accordingly, eliminating discovery is an essential part of arbitration." *Id.* at § 11.06.

117 *See Integrity Ins. Co. v. American Centennial Ins. Co.*, 885 F. Supp. 69, 71-72 (S.D.N.Y 1995).

118 *Meadows Indem. Co. Ltd. v. Nutmeg Ins. Co.*, 157 F.R.D. 42, 44-45 (M.D. Tenn. 1994); *Stanton v. Paine Webber Jackson & Curtis, Inc.*, 685 F. Supp. 1241, 1242 (S.D. Fla. 1988).

119 *Wachtel v. Shoney's, Inc.*, 830 S.W.2d 905, 908-909 (Tenn. Ct. App. 1991); *Breeding v. Kraner*, No. 89AP-1297, 1990 WL 61128, at *2 (Ohio Ct. App. May 10, 1990).

usually dislike "Rambo-litigators" who make it difficult for their opponents to obtain needed information and cause numerous hearings over issues which other attorneys in other cases have resolved amicably. Arbitrators decide early on who in the proceedings they can believe and who they can trust to lead them to the right result at the conclusion of the case. They may have difficulty setting aside initial impressions of obstreperous lawyers. Usually, arbitrators buy into the arbitration philosophy of quick and inexpensive resolution of business disputes. They have had training in it, they spend time serving as arbitrators which they might spend more profitably otherwise, and they may resent lawyers who turn the process into something too much like the worst of litigation: expensive, lengthy, and embittering.

The next stage of preparation for the final hearing involves an analysis of the factual and legal information that must be communicated to the arbitrators in order to "win" (or lose as little as possible) and how best to communicate that information. This analysis is different from trial preparation, as it does not primarily concern the issues of how to get evidence admitted and how to exclude the opponent's evidence. For all practical purposes, the rules of evidence are ignored in arbitration. Arbitrators will usually let everything in, unless the proffered evidence is redundant or excessively unreliable.[120] Therefore, it should be remembered when preparing for the final hearing, that the name of the game is to *persuade* the arbitrator(s), not to build a record.[121]

Too often, counsel are misled by the informality of arbitration so that they do not properly prepare for the hearing. While it is true that it is generally unnecessary in preparing for arbitration to worry about admissibility of evidence, prepare jury instructions, draft detailed legal argument and motions *in limine*, or do much of what lawyers do to prepare for trial, that does not mean that preparation for arbitration is less important, it is just different.

One way surety counsel may prepare for the arbitration hearing is to compile a pre-hearing notebook or similarly organized and easily accessed submission to be given to the arbitrators. The pre-hearing notebook should completely explain the case using graphics, charts (arbitrators who are architects and engineers seem to especially appreciate charts and visual displays of information), summaries, photos, chronologies, and indexed

---

120 DAUER, *supra* note 113, at §11.05.

121 *Id.*

documents, all fashioned to quickly convey what happened on the project and why the surety's position in the case is the right one. The pre-hearing notebook can be prepared with the aid of a consultant and possibly one of the graphics and litigation support companies which prepare visual aids for jury trials.

Surety counsel should also prepare and submit a sample award calculation, each component of which is clearly supported in the submission. A proposed award, referenced throughout the hearing, if it is credible and easy to understand, may prove to be just the assistance the arbitrators need to find in the surety's favor.

The arbitrators should be asked whether the parties should submit pre-hearing memoranda of law. If so, the memoranda should be short, easy to understand, and, most importantly, believable. The surety's memorandum should discuss, rather than omit, the cases likely to be cited by its opponent. A good memorandum should appear to be a tool which the arbitrators can use to understand the law and use it to reach the right result. If a memo fails to show that the law in favor of the surety is fair and equitable, the arbitrators may not feel constrained to follow it.

Considerable thought should be given to the use of live testimony at the hearing, whether it is even needed, and if so, how best to present such testimony. Care should be taken not to waste the arbitrators' attention with unnecessary oral testimony. Information should be provided orally only if that is the only and best way to provide it. Many studies show that most information conveyed orally is lost on the listener. Consequently, it may not be wise to entrust the most important aspect of the surety's case to oral proof alone. While it may be useful to have a client testify to personalize the client and allow the arbitrators to get a sense of the client's credibility and validity of its position, the bulk of a case might be best submitted in a visual or summary format, supported by written proof.

Additionally, a surety's counsel should determine whether it is really worth it to proffer any expert witnesses besides the parties who are expected to testify. An expert should be used as a witness only if he or she will enhance the credibility of the surety's arguments. Such an expert is one who will truly be of assistance to the arbitrators, who can answer their questions, and who can explain why the surety's theory of the case is believable. The arbitrators will not likely be impressed by the mere appearance of someone who may be in the business of giving desired opinions. To the contrary, such a witness may make a case appear weaker than it actually is.

Cross-examination of the other parties' witnesses should be limited. It is appropriate to use cross-examination to obtain necessary information which cannot be obtained otherwise. But when preparing to challenge proffered evidence or impeach an opponent's witnesses, surety's counsel should not "sweat the small stuff." Impeachment should be attempted and objections made only to show the arbitrators that they would otherwise be misled. Counsel should avoid the appearance of showing off, wasting time, or, even worse, trying to keep the arbitrators from hearing something important.

In preparing opening and closing arguments, counsel should plan to use the charts, visuals, summaries, and, if appropriate, a proposed award, in reasoned fashion, even if the arbitrators have refused to give a reasoned award. Counsel should not be argumentative. It is important to admit any points undeniably made by one's opponent and concede weaknesses when necessary to maintain credibility, while at the same time explaining why one is nevertheless entitled to win. Overall, the surety's attorney should stress the fairness and equity of the surety's position. Counsel should anticipate and try to answer every question raised by the arbitrators and ask them for any other questions or concerns they may have. If unexpected questions arise which cannot be answered at the hearing, the parties should ask for permission to submit information and argument after the hearing on the unanswered issue.

Above all, everything surety's counsel does before and during the hearing must be tailored to persuading the arbitrators of one's trustworthiness and competence, and the credibility and reasonableness of the surety's evidence and position.

VIII.
Conclusion

Two points of advice are appropriate for the end of this discussion. Whether or not counsel is satisfied with the results of an arbitration, they should consider requesting a "post-mortem" by the arbitrator(s). One of the least used advantages of arbitration is the opportunity it theoretically provides counsel to meet with the arbitrators to find out why they ruled as they did or how they perceived the surety's case. Although the arbitrators are not obligated to meet with the parties after the proceedings, they may be willing to do so once the time has run for appellate review or re-hearing. The insight the arbitrators may offer after the fact as to their impressions of the surety's arguments may be helpful in preparing for the next arbitration.

Finally, because arbitration can be an unfriendly place for sureties no matter how credible their arguments or how trustworthy and persuasive their counsel, it may be that the first and most important thing counsel for a surety can do is to persuade a court to exempt the surety from the arbitration altogether, or limit the issues that can be determined in the arbitration concerning the surety. This can take the form of a claim for a declaratory relief on specific issues of law relevant to the surety's defenses or liabilities. Then, the surety need only eat frog legs, nicely sautéed in butter, rather than swallow a whole, live, wiggly frog.

Exhibit 2.1

Date

Consultant
Street Address
City, State and Zip Code

Re: Performance Bond Surety Claim[s]

Dear ________________:

This letter shall constitute the consulting agreement ("Agreement") between Law Firm and Consultant.

1. Services to be Rendered

Law Firm, on behalf of Surety, retains Consultant as an independent consultant to provide consultation and assistance as requested by Law Firm in connection with potential or anticipated litigation arising out of the above-referenced Performance Bond Surety Claim[s] including assistance in development of the work product of Law Firm and the evaluations necessary to render independent opinions and advice to Surety. Such services may include review, analysis, verifications, confirmations, investigation, or other actions requested by Law Firm. Services will include consultations with Law Firm. Depending upon facts and circumstances that may emerge, other services may be performed by Consultant as authorized by Law Firm.

2. Place of Work

Consultant shall render services hereunder at such time and at such places as are mutually agreeable.

3. Term of Agreement

This Agreement shall be effective as of ___________________, and shall continue until resolution of the Performance Bond Surety Claim[s] between Surety and Obligee. However, Law Firm may terminate this Agreement at any time upon written notice to Consultant that Consultant should cease work, in which case the confidentiality provisions contained herein shall survive such termination and continue in full force and effect.

4. <u>Fees and Expenses/Payment</u>

For Consultant's services under the Agreement, Consultant shall be paid such compensation as set forth in the attached proposal dated __________________, for all services rendered. Compensation shall generally be limited to no more than 10 hours per day, except when additional hours in a single day are necessary and are authorized by Law Firm. Whenever feasible, Consultant shall provide Law Firm with an estimated budget for each phase of the work to be performed under this Agreement prior to the work being undertaken in sufficient time to permit review by the Surety.

Law Firm shall reimburse Consultant for reasonable expenses incurred for meals, lodging, and travel (air coach rates) [consistent with the attached surety guidelines for expenses], which are incurred as a result of the work performed by Consultant at Law Firm's request.

At least once, but not more frequently than twice, each month, Consultant shall submit an invoice showing dates and a general description of work performed on each date, and an itemization of expenses incurred during such period, including copies of receipts for all individual expenditures.

Law Firm shall pay Consultant upon the receipt of reimbursement from Surety. The maximum cumulative compensation including expenses permitted to Consultant for Consultant's performance hereunder shall not exceed $________________ without the prior written authorization of the representative of Law Firm.

5. Confidentiality

Consultant acknowledges that all conversations with Law Firm or conducted at the direction of Law Firm, and all materials disclosed to Consultant by Law Firm (including Surety) and the work Consultant performs hereunder, are confidential and proprietary and constitute lawyer work product and are subject to a work product and/or attorney-client privilege. Consultant will abide by all reasonable restrictions placed by Law Firm (including Surety) on the dissemination of such materials and work product. In the event Consultant is served with a subpoena or other legal process requiring the disclosure of such conversations, materials or work product, Consultant will immediately advise Law Firm and will cooperate with Law Firm in responding to such service. Consultant agrees to sign any required additional confidentiality agreement and to abide by the terms of any related protective order. In the event that Consultant is called upon to give testimony, by subpoena or otherwise, with respect to any aspect of the project[s], Consultant will be represented during such proceeding by Law Firm or other counsel selected by Surety.

6. Entire Agreement

This instrument contains the entire Agreement between the parties with respect to the services to be provided by Consultant to Law Firm in connection with the subject matter, except as may be amended by subsequent modifications executed by both parties.

Very truly yours,

Exhibit 8.1

DATA ENTRY FORM FOR NAMES DIRECTORY

Last Name:
ID Code:
First Name & Initials:

Address:

Phone Number:

Employer:

Position with Employer:

Employer Address:

Superior's Name:

Employer Phone:

START DATE:
END DATE:

Issues:
1.
2.
3.
4.
5.
6.

Location of files: ________________
Custodian of files: _______________
Counsel: _______________________

Attorney Comments:

Exhibit 8.2

PRIVILEGE LOG

Begdoc#	***Docdate***	***Doctype***	***Issues***	***Author***	***Privilege***
L078232	07/23/80	NOTES	OSHA Medical Record Air Quality	Irwin, S.	Atty./Client
SPC534194	01/02/81	GRAPHICS	Medical Record	Lopp, W.	Dr./Patient
L066753	01/23/81	ESTIMATE	Medical Record Breach Damages		Dr./Patient
L054833	07/17/81	MEMO	Water Flow OSHA Medical Record	Good, B	Atty./Client
L056121	09/26/81	MEMO	OSHA Medical Record Air Quality		Atty./Client
L056558	08/21/83	LETTER	Medical Record Air Quality Timing Modification	Falk, M.	Dr./Patient
L047953	11/28/83	LETTER	Medical Record	Falk, M.	Dr./Patient
L009258	01/24/84	LETTER	Water Flow OSHA Medical Record		Atty./Client

Exhibit 8.3

DEPOSITION TRANSCRIPT CONTROL CHECKLIST

Witness:	Attending attorney:	
Case caption/docket/file:		
Deposition date:	Length:	Prior testimony?
Other attendees:		
Court reporter	Phone:	

Transcript Information

	Requested	Done
Standard Transcript	☐	☐
Electronic Transcript	☐	☐
Compressed Transcript	☐	☐
Reviewed for Errors	☐	☐
Errors Corrected	☐	☐
Transcript Signed by Deponent	☐	☐
Transcript Filed with the Court	☐	☐

Physical File

	Requested	Done
Transcript(s)	☐	☐
Exhibit No. _______ thru _______	☐	☐
Related Notes	☐	☐
Other Indicated Wrap-ups	☐	☐

Wrap-Ups

	Requested	Done
Quick Steps ($50–$100)		
Highlight Notes	☐	☐
Immediate Post-mortem	☐	☐
Transcript Skimmed and Tagged	☐	☐
Brief Table of Contents	☐	☐
Selected Photocopies	☐	☐
Structured Debriefing ($100–$200)		
Formal Exit Memo	☐	☐
Completed Debriefing Form	☐	☐
Traditional Digest ($700–$1000) (not recommended)	☐	☐
Integrated Computer Indexing ($300–$800)	☐	☐

Additional Comments: ____

____

____

____

____

Signed by: ______________________________________

DEPOSITION SPECIFICS

Significant topics discussed: ______________________________

____

____

Significant parties mentioned: ______________________________

____

____

Key dates (events or meetings) discussed: ______________________________

____

____

Important exhibits and why: ______________________________

____

____

____

Interesting quotes or admissions: ______________________________

____

____

____

____

____

____

Noteworthy objections, balks, inconsistencies or retreats: ______________________________

____

____

____

____

DEPOSITION CONTROL

Court reporter: ______________________________ Phone number: ______________

Electronic transcript: ☐ Requested ☐ Obtained Compressed transcript: ☐ Requested ☐ Obtained

☐ Transcript reviewed for errors ☐ Corrected ☐ Reviewed/signed by witness

Location of exhibits: ______________________________

Relative deposition importance: ☐ Low ☐ Medium ☐ High

Further treatment recommended? ______________________________

Information provided by: ______________________________ Date: ______________

Witness:	Attending attorney:	
Case caption/docket/file:		
Deposition date:	Length:	Prior testimony?
Other attendees:		

HIGHLIGHTS

Key testimony and observations: ______

FOLLOW-UPS

Documents to be requested/produced: ______

Future questions for this deponent: ______

Questions for other deponents/witnesses: ______

Other uses of testimony: ______

DEMEANOR

Appearance, tone, recall, control, and credibility: ______

Exhibit 10.1

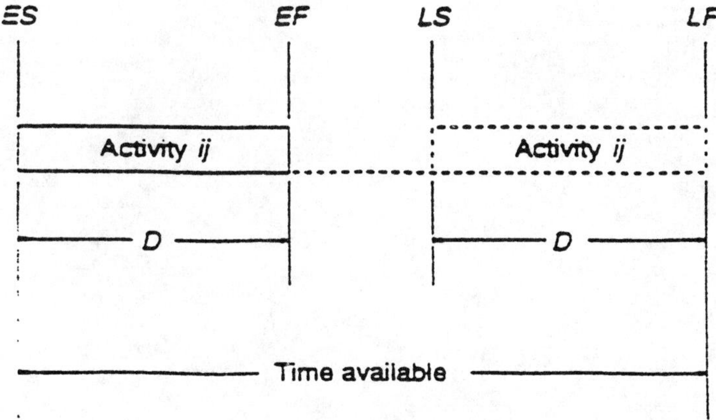